Organic Name Reactions

A Unified Approach

Organic Name Reactions
A Unified Approach

Goutam Brahmachari

Alpha Science International Ltd.
Oxford, U.K.

Goutam Brahmachari
Department of Chemistry
Visva-Bharati University
P. O. Santi Niketan
West Bengal, India

Copyright © 2006

Alpha Science International Ltd.
7200 The Quorum, Oxford Business Park North
Garsington Road, Oxford OX4 2JZ, U.K.

All rights reserved. No part of this publication may be reproduced, stored in a retrieval system, or transmitted in any form or by any means, electronic, mechanical, photocopying, recording or otherwise, without prior written permission of the publisher.

Printed from the camera-ready copy provided by the Author.

ISBN 1-84265-304-0

Printed in India

To
My little *Asanjan*
and
his *Grandfather*

Foreword

Organic chemistry encompasses a very large number of compounds, which in turn get involved in multitude of chemical reactions (often intimidating) that define organic chemistry. A chemical reaction is associated with reactant/s or substrates, reagent/s, reaction conditions and product/s. These are the essential tools of a chemist, and to use these tools effectively, chemists must organize them in a logical manner and look for patterns of reactivity that allow us make reasonable predictions.

Most of the chemical reactions occur at special sites of reactivity known as functional groups, and these constitute an organizational set-up that helps us catalog and remember reactions. Ultimately, the best way to achieve proficiency in organic chemistry is to understand how reactions take place, and to recognize the various factors that influence their course. This is best accomplished by perceiving the reaction pathway or mechanism of a reaction.

One of the main difficulties students have with organic chemistry is organizing the information in their minds. It is critical that students take time to not only organize the information, but also to understand it. There is a general misconception that one has to memorize many things in organic chemistry. There is logic, there is rationale and by truly learning something, one will be able to apply concepts beyond what one is memorizing.

Chemistry, over the years, has made efforts to systematically name chemical compounds, however, it has not been very successful in developing a nomenclature of chemical reactions. As a result, many reactions are known by the name of the chemists who discovered, studied and popularized them. Many of these reactions formed the basis of the organic chemistry of today. Every beginning organic chemistry student as well as those pursuing graduate studies learn and become familiar with the organic chemistry 'name' reactions.

In recent years a few useful books have come out delineating different aspects of Name Reactions. Different from other books on name reactions in organic chemistry, "ORGANIC NAME REACTIONS: A UNIFIED APPROACH" by Dr. Goutam Brahmachari deals with more than 100 classical as well as contemporary significant name reactions in organic chemistry, which are of considerable synthetic significance. These are covered under three sections: Detailed Discussion on Ten Name Reactions; Concise Discussion on Forty Name Reactions and Short Account on Sixty Name Reactions. Mechanistic aspects of each reaction are dealt with carefully and special emphasis has been placed on synthetic applicability of the reactions. Critical analysis and recent significant developments on such reactions have also been taken into consideration. Additionally each reaction is supplemented with the original and the latest references.

This book is not intended to replace an organic chemistry textbook but that is ideally suited for easy reading and learning for undergraduate and postgraduate students of organic chemistry and is also a good reference work on name reactions to researchers.

S. Chandrasekaran
Professor and Chairman
Chemical Science Division
Indian Institute of Science
Bangalore

Preface

Chemical reactions and the concepts of their mechanistic pathways are the basis of modern organic chemistry. This fascinating branch of science has already undergone tremendous growth, and still is being enriched rapidly; the knowledge of organic chemistry continues to move ahead on multidirectional fronts. Organic chemistry embodies numerous reactions; the most significant ones among them are customarily recognized after the names of the chemists who discovered, and studied them in-depth. Organic name reactions are the perfect means for learning the principles of organic chemistry. They are too much fascinating for their far-reaching utilities, particularly in organic syntheses. Detailed knowledge on their mechanistic aspects as well as critical survey on each of them is very much essential not only to the undergraduate and postgraduate students, but to everyone who deals organic chemistry from any perspective. From the platform of a student and also of a teacher, I had felt for a systematic and comprehensive coverage on the subject for which one is compelled to consult various sources that seems to be much troublesome so far as the students are concerned. Here lies the intention to write this book — *Organic Name Reactions: A Unified Approach* — the very first edition of which covers a good number of such reactions, specially enjoying considerable synthetic significance.

The book covers over 100 classical as well as contemporary significant name reactions in organic chemistry. Reactions are classified into three sections — Detailed Discussion, Concise Discussion, and Short Account — on the basis of the extent of discussion. Mechanistic aspects of each reaction are dealt with proper approach. Special emphasis is given to the synthetic applicability of the reactions. Critical views and recent significant developments on such reactions are taken into consideration. Moreover, each reaction is supplemented with the original and the latest references. I hope the undergraduate and postgraduate students studying organic chemistry will be benefitted from this book, and at the same time this one will also serve the purpose of a good reference for researchers, professionals, and all other chemists interested in name reactions.

Ten reactions have been carefully screened under *Section-1*, and they have been discussed in detail. They have been discussed critically in every aspects including mechanistic approach, reaction conditions, limitations, utilities, modern advancement and extensions with ample of examples. Each of them bears a plenty of references for further study. Each discussion has been started from the very beginning stage and gradually developed. This section will be helpful to the undergraduate & postgraduate students, researchers as well as the teachers. The reactions are in accord with the undergraduate & postgraduate syllabus along with those of NET (UGC-CSIR), GATE, SLET and other competitive examinations. The *Section-2* of the book discusses forty name reactions (most of which are enlisted in the undergraduate and postgraduate syllabus of most of the Universities). A concise discussion for each of them is available here. They are treated under the classified subtitles — Introduction, Mechanism, Critical Views and Applications. The subtitle "Critical Views" is looked upon in such a manner that a student can get a clear idea regarding a particular reaction — most of all the possible critical points are summarized. Each reaction is supplemented with the original and the latest references. Further, in *Section-1* and S*ection-2* the references arranged in 'bullets' form are included to afford the sources for more information; subject matters of these references are not covered up in the present text. The *Section-3* particularly deals with sixty reactions having notable synthetic applications. They are discussed in a very nutshell. Each reaction will bear an introduction, mechanistic approach, and one or two examples along with references. The main motto of this section is to make the readers well acquainted with some of the reactions frequently used by the organic chemists.

I would like to express my deep sense of gratitude to Prof. S. Chandrasekaran (Indian Institute of Science, Bangalore) for his keen interest in my manuscript and for writing a foreword to the book. I am also grateful to Prof. U. Maitra (IISc., Bangalore), Prof. M. Chakraborty (Bose Institute, Kolkata), Prof. B. C. Ranu (IACS, Kolkata), Prof. S. Thakur (University of Burdwan, Burdwan), Prof. K. S. Mukherjee (Visva-Bharati, Santiniketan), Prof. B. Basu (North Bengal University, Darjeeling), Dr. S. Mallik (University of Central Florida, USA), Dr. G. S. Singh (University of Botswana, Gaborone, Botswana), and Dr. F. A. Khan (IIT, Kanpur) who looked into a major portion of the manuscript, and conveyed their valuable comments that encouraged me a lot. I am also thankful to my research fellows — most specifically names of Arindam and Sadhan are to be mentioned — for their constant inspiration during the period of writing. Thanks are due to my colleagues and students also.

Special thanks are due to Mr. N. K. Mehra, Managing Director and Publisher, Narosa Publishing House Pvt. Ltd. for his earnest interest in publishing the work; I am also thankful to the Staff-members of the Publishing House.

My effort will be successful only when it is found helpful to the students and teachers. Every step has been taken to make the manuscript errorless; in spite of that, some errors might have crept in. Any remaining error is, of course, of my own. Constructive comments on the content and approach of the book from the readers will be highly appreciated (my e-mail address is brahmg2001@yahoo.co.in).

Last but not the least, I wish to thank my family members, very particularly to my wife, Piyasi, for their well understanding and support throughout the entire period of writing. Without their support, this work would not have been possible.

Santiniketan   **Goutam Brahmachari**

Contents

Foreword	*vii*
Preface	*ix*

Section 1 (Detailed Discussion) — 1-165

1.	Baeyer-Villiger Oxidation	1
2.	Barton Reaction	24
3.	Beckmann Rearrangement	39
4.	Claisen Rearrangement	56
5.	Diels-Alder Reaction	65
6.	Favorskii Rearrangement	93
7.	Fries Rearrangement	106
8.	Norrish Type I and Type II Reactions	122
9.	Paterno-Buchi Reaction	138
10.	Stork Enamine Reaction	152

Section 2 (Concise Discussion) — 166-363

11.	Arndt-Eistert Homologation	166
12.	Benzilic acid Rearrangement	172
13.	Benzoin Condensation	177
14.	Birch Reduction	184
15.	Bischler-Napieralski Reaction	191
16.	Bouveault-Blanc Reaction	198
17.	Cannizzaro Reaction	203
18.	Chichibabin Amination Reaction	209
19.	Claisen Condensation	212
20.	Clemmensen Reduction	219
21.	Cope Rearrangement	225
22.	Dakin Reaction	233
23.	Di-Pi methane Rearrangement	237

24.	Etard Reaction	242
25.	Fischer Indole Synthesis	245
26.	Friedel-Crafts Reaction	251
27.	Gabriel Synthesis	260
28.	Gattermann-Koch Reaction	265
29.	Haller-Bauer Reaction	268
30.	Hell-Volhard-Zelinsky Reaction	272
31.	Hofmann Rearrangement	277
32.	Houben – Hoesch Reaction	283
33.	Hunsdiecker Reaction	287
34.	Knoevenagel Reaction	291
35.	Mannich Reaction	297
36.	Meerwein-Ponndorf-Verley Reduction	304
37.	Michael Reaction	307
38.	Oppenauer Oxidation	312
39.	Perkin Reaction	315
40.	Pinacol Rearrangement	319
41.	Reformatsky Reaction	325
42.	Reimer-Tiemann Reaction	328
43.	Sandmeyer Reaction	332
44.	Sharpless Asymmetric Expoxidation	336
45.	Stevens Rearrangement	340
46.	Stobbe Condensation	345
47.	Williamson Ether Synthesis	348
48.	Wittig Reaction	351
49.	Wolff-Kishner Reduction	357
50.	Yamaguchi Esterification	361

Section 3 (Short Account)   364-452

51.	Allan-Robinson Reaction	364
52.	Amadori Rearrangement	366
53.	Angeli-Rimini Hydroxamic Acid Synthesis	368
54.	Baker-Venkataraman Rearrangement	369
55.	Bamberger Rearrangement	371
56.	Bamford-Stevens Reaction	372
57.	Bardhan-Sengupta Synthesis	374

58.	Bartoli Indole Synthesis	376
59.	Baylis-Hillman Reaction	378
60.	Blanc-Quelet Chloromethylation Reaction	380
61.	Boord Reaction	381
62.	Bouveault Aldehyde Synthesis	382
63.	Brook Rearrangement	383
64.	Bucherer Reaction	385
65.	Carroll Rearrangement	386
66.	Chapman Rearrangement	388
67.	Chugaev Reaction	389
68.	Corey-Kim Oxidation	390
69.	Corey-Winter Reaction	392
70.	de Mayo Reaction	394
71.	Demjanov Rearrangement	395
72.	Dienone-phenol Rearrangement	397
73.	Eglinton Reaction	399
74.	Elbs Persulphate Oxidation	400
75.	Fischer-Hepp Rearrangement	401
76.	Friedlander Synthesis	403
77.	Hoch-Campbell Ethyleneimine (or aziridine) Synthesis	405
78.	Hofmann-Loffler-Freytag Reaction	406
79.	Jones Oxidation	408
80.	Koch Reaction	410
81.	Kolbe Electrolysis Reaction	411
82.	Kolbe-Schmitt Reaction	413
83.	Leuckart-Wallach Reaction	414
84.	Lieben Haloform Reaction	415
85.	Lossen Rearrangement	416
86.	McMurry Reaction	417
87.	Mitsunobu Reaction	419
88.	Neber Rearrangement	420
89.	Nef Reaction	422
90.	Pechmann Reaction	423
91.	Prins Reaction	425
92.	Ritter Reaction	426
93.	Rosenmund Reduction	427
94.	Schotten-Boumann Reaction	429

95.	Simmons-Smith Reaction	430
96.	Simonini Reaction	431
97.	Sommelet-Hauser Rearrangement	432
98.	Stephen Reaction	433
99.	Strecker Synthesis	434
100.	Suzuki Coupling	436
101.	Swern Oxidation	437
102.	Thorpe (Ziegler) Reaction	439
103.	Tiemann Rearrangement	440
104.	Tollens Reaction	442
105.	Ullmann Diaryl Synthesis	443
106.	Vilsmeier-Hack Reaction	445
107.	von Braun Reaction	446
108.	von Richter Rearrangement	447
109.	Wagner-Meerwein Rearrangement	449
110.	Wohl – Ziegler Reaction	451

Subject Index 453

1

Baeyer – Villiger Oxidation

Introduction

The conversion of carbonyl compounds to esters by means of peroxy acids is a novel and useful synthetic reaction. Such type of oxidation reactions was first reported by Adolf von Baeyer and Victor Villiger[1] in the year 1899, and widely known as *Baeyer–Villiger oxidations*. Such type of reaction is an example of rearrangement reaction involving the migration of a group from carbon to electron-deficient oxygen atom.

$$RCOR' + R''COOOH \longrightarrow RCOOR' + R''COOH$$
Ketone   Peroxy acid   Ester   Carboxylic acid

Thus, a ketone can easily be oxidised to an ester by this method; esters so obtained can be hydrolyzed to the corresponding acids and this provides an alternative route to convert a ketone into a carboxylic acid as well. The reaction is applicable to open chain, cyclic as well as aromatic ketones. Straight chain ketones give carboxylic esters while cyclic ketones furnish lactones. Carboxylic esters or lactones are developed due to *"insertion of oxygen"* — an oxygen atom from the peroxy acid (oxidant) is inserted between the carbonyl group and one of the carbons attached to that carbonyl functionality present in the reacting substrate molecule.

cyclopentanone —Baeyer-Villiger oxidation→ δ-valerolactone   [Ref. 2]

α-Diketones take part in *Baeyer-Villiger reaction* to give anhydrides[3]; but enolisable β-diketones do not take part in this reaction.

$$CH_3-CO-CO-CH_3 \xrightarrow{RCO_3H} CH_3-CO-O-CO-CH_3$$

α-Naphthaquinone + RCO$_3$H →

α-Naphthaquinone

Baeyer–Villiger oxidation of aldehydes usually yields carboxylic acids owing to hydrogen migration to oxygen (analogous to a hydride shift in a carbocation); for example —

$$p\text{-}O_2N\text{-}C_6H_4CHO \xrightarrow{\text{peracid}} p\text{-}O_2N\text{-}C_6H_4COOH$$

Reagents

Some of the recommended reagents for *Baeyer–Villiger oxidations* are *Caro's* acid (H_2SO_5); a solution of peracetic acid in acetic acid containing sulphuric acid or *p*-toluenesulphonic acid as a catalyst; perbenzoic acid; *m*-chloroperbenzoic acid (MCPBA); peroxomonophosphoric acid; peroxytrifluoroacetic acid or monopermaleic acid in CH_2Cl_2 solution; 90% H_2O_2 and certain metal complexes; etc.

Peroxytrifluoroacetic acid is generally the reagent of choice because reactions with this reagent are rapid and clean, giving high yields of product. However, use of this peracid may bring some complications due to the occurrence of *transesterification*, which takes place between the ester formed and trifluoroacetic acid —

$$RCOOR' + CF_3COOH \rightleftharpoons RCOOH + CF_3COOR'$$

This *transesterification* is of no concern if the crude ester product is to be hydrolyzed; but if isolation of ester is desired, it is then necessary to add a buffer such as disodium hydrogen phosphate (Na_2HPO_4) to the reaction mixture to get rid of this complication — the buffer reacts with the trifluoroacetic acid to form a salt and hence minimize *transesterification*.

Examples

A few examples of *Baeyer–Villiger rearrangement* reactions are cited below:

Baeyer-Villiger Oxidation

a) acetyl substituted L-tyrosine →(H₂O₂, NaOH; migration of aryl moiety in preference to 1° alkyl (-CH₃))→ aryl acetate intermediate → L-dopa (a drug for the treatment of perkinson's disease)

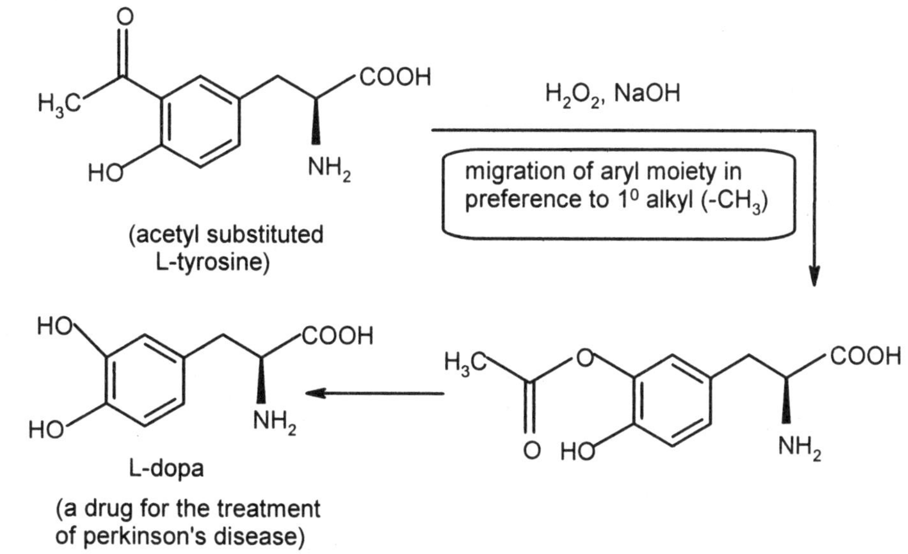

b) bicyclic cyclobutanone with C=C →(H₂O₂, AcOH)→ fused lactone
(the C=C double bond remains particularly unreactive; the reaction is facilitated on relieving strain of the smaller four-membered ring in the fused bicyclic substrate)

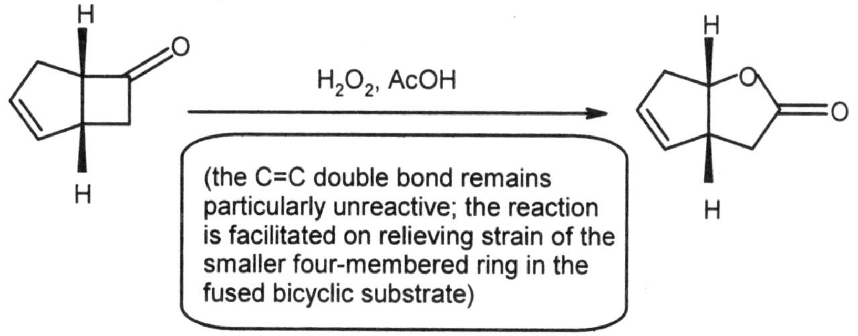

c) $p\text{-}O_2N\text{-}C_6H_4COC_6H_5$ $\xrightarrow[\text{H}_2\text{SO}_4 \atop \text{CH}_3\text{CO}_2\text{H},\ 25°]{\text{CH}_3\text{CO}_3\text{H}}$ $p\text{-}O_2N\text{-}C_6H_4COOC_6H_5$ 95% Ref. 4

d) cyclopropyl-COCH₃ $\xrightarrow[\text{CF}_3\text{CO}_2\text{H},\ \text{Na}_2\text{HPO}_4 \atop \text{CH}_2\text{Cl}_2,\ \text{reflux}]{\text{CF}_3\text{CO}_3\text{H}}$ cyclopropyl-O-COCH₃ 53% Ref. 5

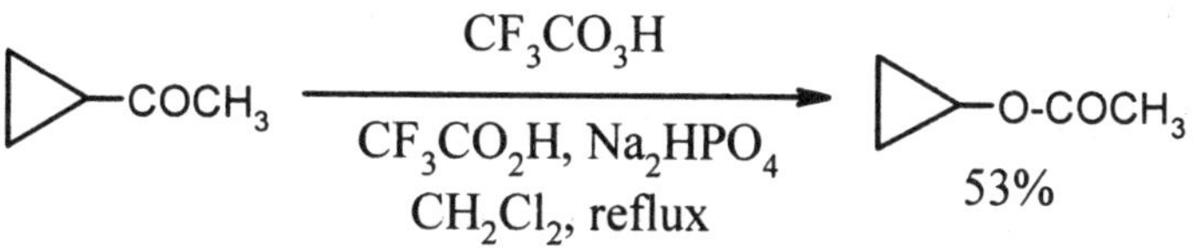

e) 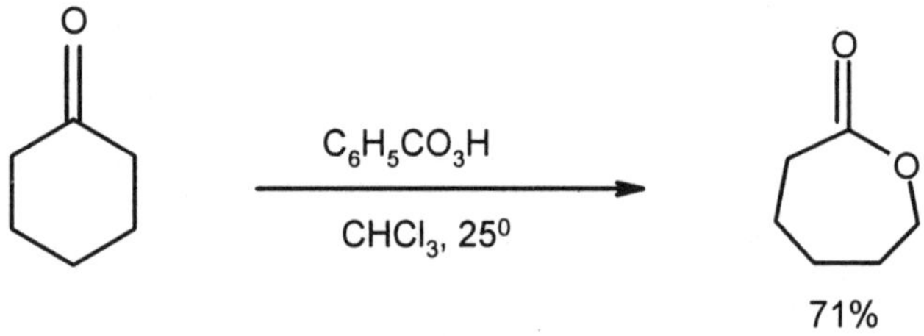 cyclohexanone $\xrightarrow[\text{CHCl}_3,\ 25^\circ]{\text{C}_6\text{H}_5\text{CO}_3\text{H}}$ ε-caprolactone, 71% Ref. 6

f) 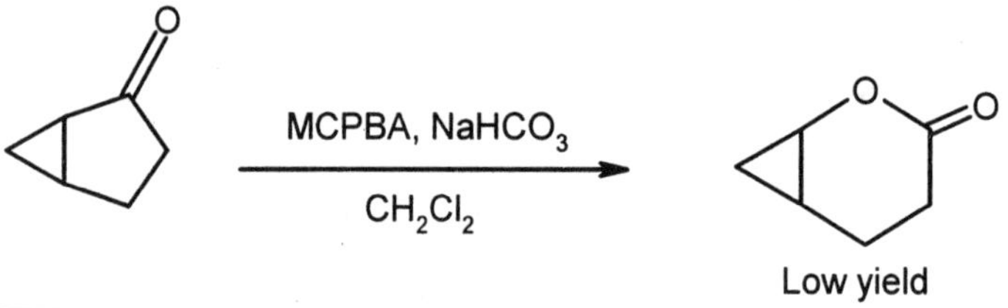 bicyclo[3.1.0]hexan-2-one $\xrightarrow[\text{CH}_2\text{Cl}_2]{\text{MCPBA, NaHCO}_3}$ bicyclic lactone, Low yield Ref. 7

g) 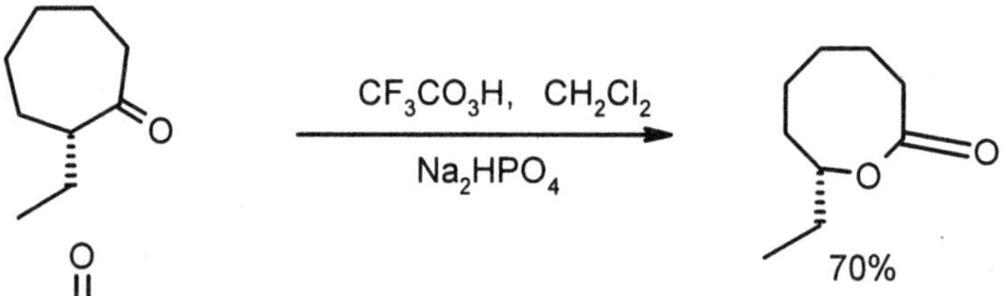 (S)-2-ethylcycloheptanone $\xrightarrow[\text{Na}_2\text{HPO}_4]{\text{CF}_3\text{CO}_3\text{H},\ \text{CH}_2\text{Cl}_2}$ lactone, 70% Ref. 8

h) 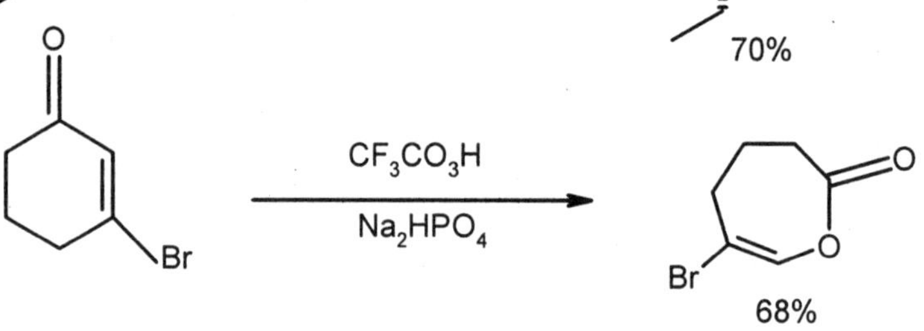 3-bromocyclohex-2-enone $\xrightarrow[\text{Na}_2\text{HPO}_4]{\text{CF}_3\text{CO}_3\text{H}}$ bromo-enol-lactone, 68% Ref. 9

i) 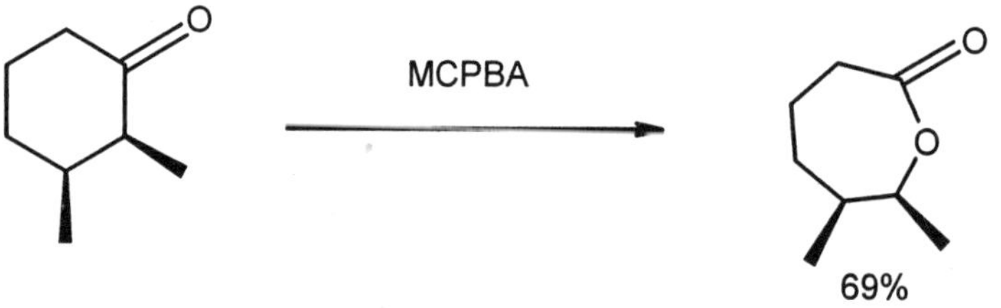 trans-2,3-dimethylcyclohexanone $\xrightarrow{\text{MCPBA}}$ lactone, 69% Ref. 10

j) 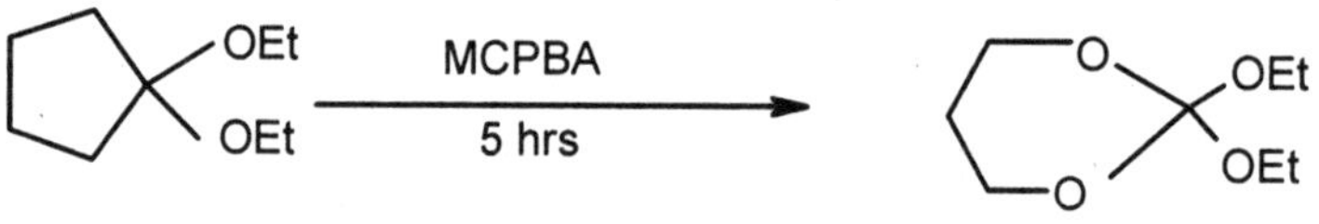 1,1-diethoxycyclopentane $\xrightarrow[\text{5 hrs}]{\text{MCPBA}}$ 2,2-diethoxy-1,3-dioxepane, 25% Ref. 11

(*MCPBA = *meta*-chloroperbenzoic acid)

(A unique example of a ketal undergoing *Bayer- Villiger* oxidation)

k) [cyclopentanone with C_6H_{13} substituent] →(Magnesium monoperphthalate hexahydrate (MMPP)) [δ-lactone with C_6H_{13}] 95%   Ref. 12

l) [cyclobutanone derivative] →(Zr (salen), UHP / CH_2Cl_2, 24 hrs.) [γ-butyrolactone with R substituent] 81 – 87%   Ref. 13

Zr (salen) complex bearing a biphenyl unit has been found to catalyze asymmetric *Baeyer-Villiger oxidation* of prochiral and racemic ketones enantioselectively when UHP (urea-hydrogen peroxide) is used as the oxidant.

Mechanism

For the first fifty years since the first report on *Baeyer–Villiger reaction* there was a controversy regarding its exact mechanistic pathway; more than one mechanisms were proposed in different times by different groups of chemists. However, a variety of studies on *Baeyer–Villiger reaction* from different angles resulted a generally accepted mechanism[14] (**Scheme 1**) for the rearrangement reaction. The reaction has been found to being catalyzed by acid, and the rate of oxidation is accelerated by electron-donating groups in the carbonyl compound and by electron-withdrawing groups in the peroxyacid. It begins with nucleophilic addition of the peroxy acid to the carbonyl group to create a tetrahedral "*Criegee intermediate*"[15], which subsequently rearranges to ester or lactone.

Initial Step: In this step carbonylic oxygen of carbonyl compound is protonated —

R—C(=O)—R + H⁺ ⇌ R—C(=O⁺H)—R ↔ R—C⁺(OH)—R
(resonance stabilized)

Step I: The peroxy acid adds to the electron-deficient carbonyl carbon to form a tetrahedral '*Criegee intermediate*". This step is a nucleophilic addition analogous to *gem*-diol and hemiacetal formation.

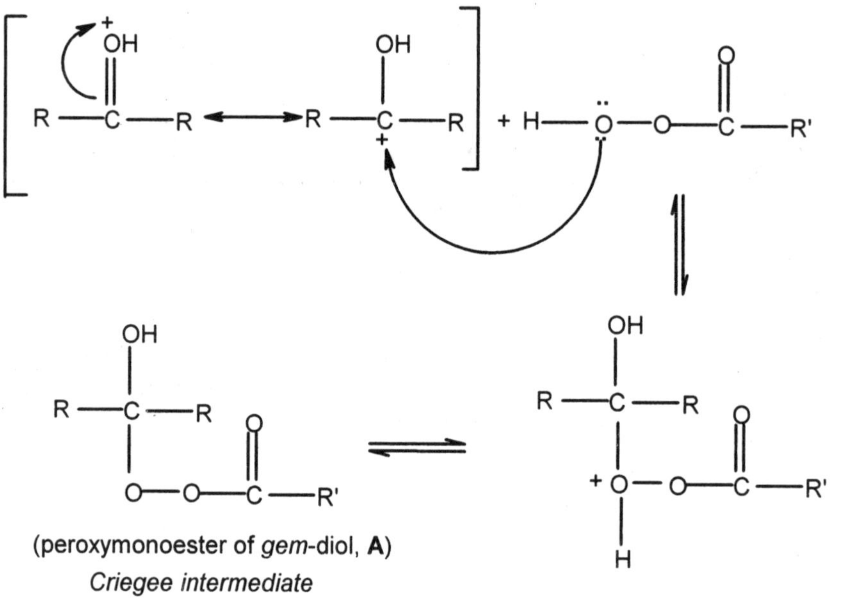

(peroxymonoester of *gem*-diol, **A**)
Criegee intermediate

Step II: The intermediate (**A**) from Step I undergoes rearrangement. Cleavage of the weak O–O bond of the peroxy ester (**A**) is assisted by migration of one of the substituents of the carbonyl group to electron-deficient oxygen. The group 'R' migrates with its pair of bonding electrons in much same way as alkyl groups migrate on carbocation rearrangements. More often, the "*Criegee intermediate*" rearranges to ester or lactone in a rate limiting *synchronous* fashion, although cleavage processes have also been noted in some cases.

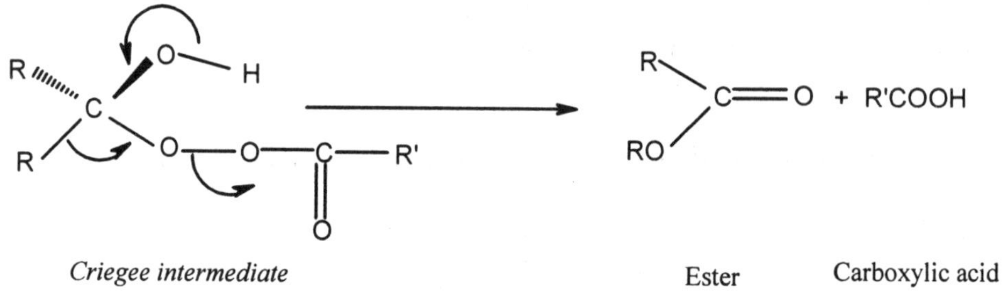

Criegee intermediate   Ester   Carboxylic acid

Scheme 1: Mechanism of the *Baeyer-Villiger oxidation* of a ketone

This mechanism was supported by the isotope labelling experiment of Doering and Dorfman[16]; the substrate, [O^{18}]- benzophenone (1), on *Baeyer-Villiger* oxidation with perbenzoic acid yielded phenylbenzoate (2) retaining all the labeled oxygen as the carbonylic oxygen of the ester, none as its alkoxyl oxygen.

Further, from the available experimental evidences it can be argued that not the formation of *Criegee intermediate*, but its destruction (*i.e.* rearrangement) is the rate-determining step of the reaction. Palmer and Fry[17] carried out *Baeyer-Villiger oxidations* of *p*-substituted acetophenone-1-C^{14} and compared these rates of oxidation with those of unlabeled ketones.

(X = H, CH_3, Cl, CN)
(*p*-substituted acetophenone-1-C^{14})

All the substituents, there is a significant ^{14}C isotope effect ($k_{12}/k_{14} > 1$) suggesting the rearrangement step as the rate-determining step; because rate-determining formation of *Criegee intermediate* (**A**) would not give an isotope effect, since this step does not involve significant bond alternation of the labeled position.

Stereochemistry

Baeyer–Villiger rearrangement proceeds with *retention of configuration*[18] and the reaction is *stereospecific* in nature — an optically active ketone leads to an optically active product. Actually retention is the general feature of this reaction, even when inversion would give rise to a more stable product.

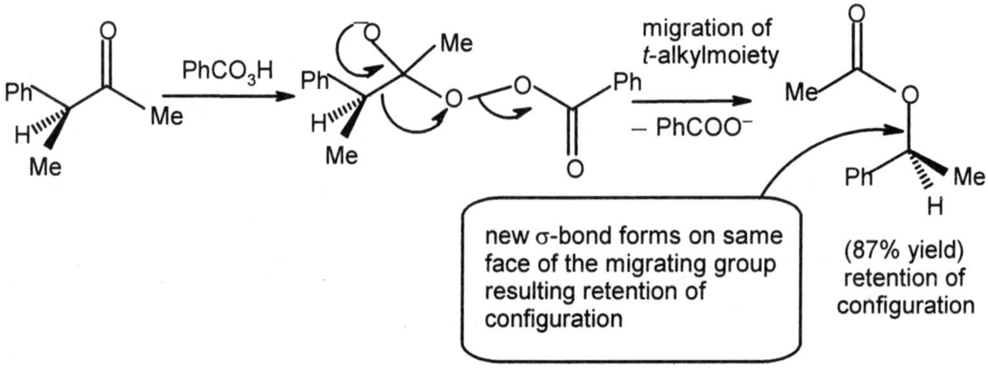

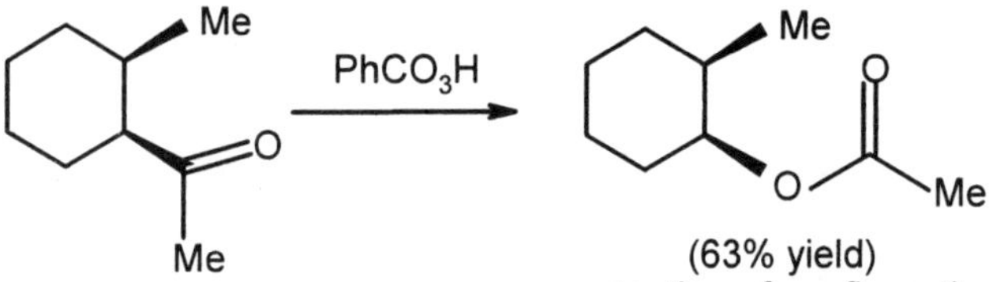

[Reaction scheme showing Baeyer-Villiger mechanism with PhCO₃H, including annotations: "migration of t-alkylmoiety", "– PhCOO⁻", "new σ-bond forms on same face of the migrating group resulting retention of configuration", "(87% yield) retention of configuration", and a cyclohexane example giving (63% yield) retention of configuration.]

A few more illustrations may be cited:

(a) *Cis*-1-acetyl-2-methylcyclopentanone (**3**) on *Baeyer–Villiger oxidation* with perbenzoic acid furnishes exclusively *cis*-2-methylcyclopentyl acetate (**4**) in 69% yield.

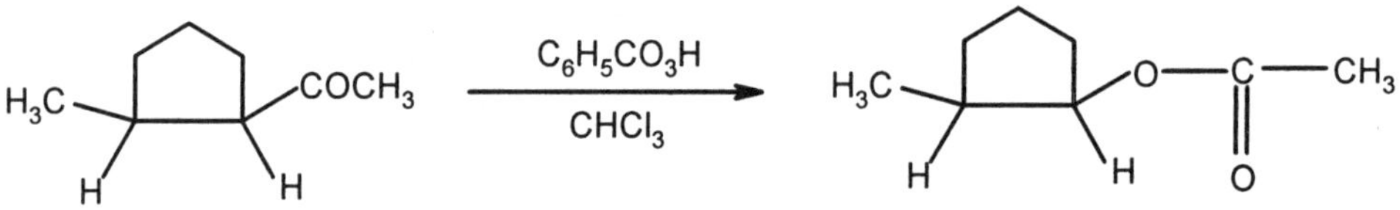

(**3**) → (**4**) (sole product), reagents: $C_6H_5CO_3H$ / $CHCl_3$

Similarly in case of the *trans*-isomer, the *trans*-acetate becomes the only product.

(b) Burson and Suzuki[19] carried out the following transformation –

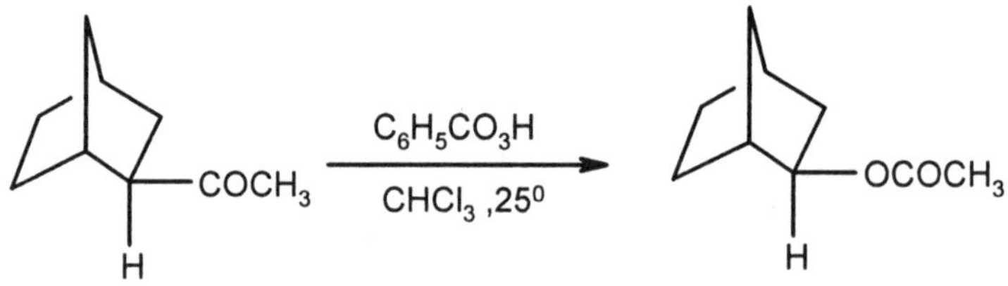

(**5**; optically active) → (**6**; 81% yield, optically active), reagents: $C_6H_5CO_3H$ / $CHCl_3$, 25°

(c) One more example from the works of House and Bare[20] is also cited here –

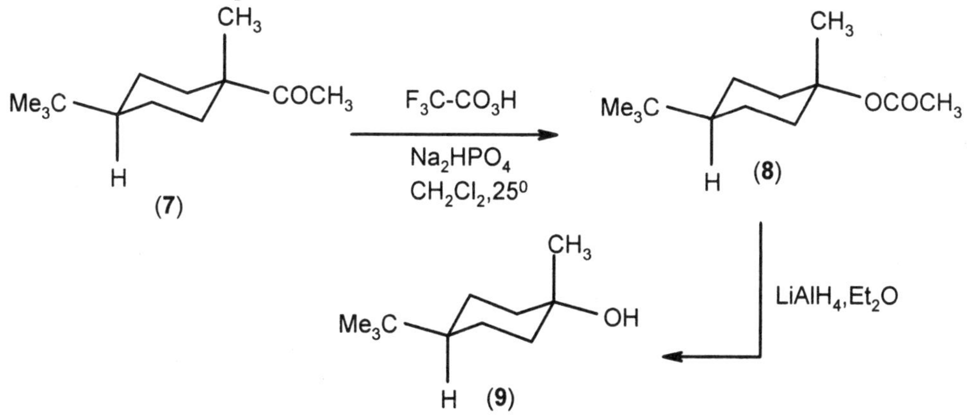

It is seen that the oxidation not only provides a useful method for the conversion of a methyl ketone to an alcohol, but offers a method for relating the stereochemistry of a ketone and an alcohol also.

Migratory Aptitude

In case of unsymmetrical ketones question of migratory aptitude of migrating groups arises. From a study of a series of alkyl aryl ketones[21,22] as well as from other studies[3,23,24], the relative ease of migration of various groups in the *Baeyer–Villiger reaction* has been found to be of the order: *tertiary* alkyl > cyclohexyl ~ *secondary* alkyl ~ benzyl ~ phenyl > *primary* alkyl > cyclopropyl > methyl. Primary groups are much more reluctant to undergo migration than secondary ones or aryl groups, and this makes *Baeyer-Villiger reactions* regioselective (experimental observation in **Scheme 2**). It is also evidenced that bridgehead *t*-alkyl group (e.g. **10**) migrates in preference to the phenyl group, as evidenced from the experimental results of Hawthorne and Emmens[21].

(**10**) (bridgehead moiety migrates)

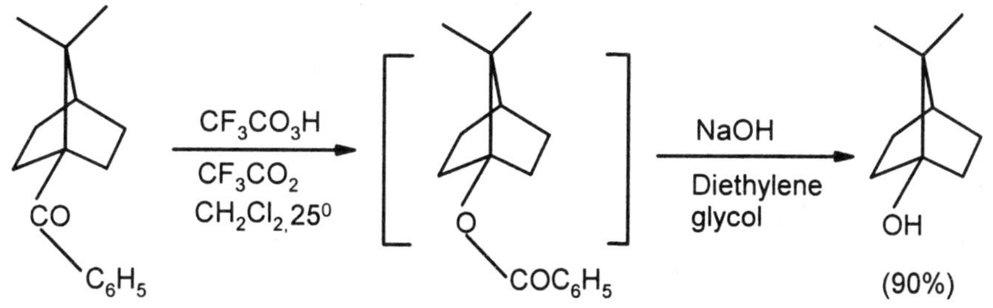

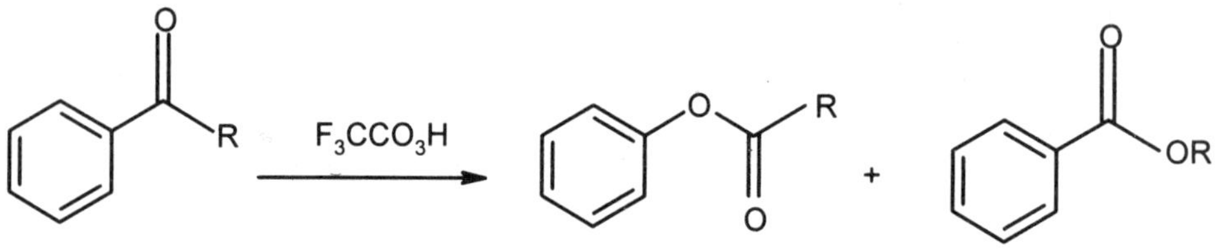

Scheme 2

Phenylalkyl ketone (11) → (by migration of Ph) (12) + (by migration of R) (13)

Reagent: F$_3$CCO$_3$H

R in 11	% Yield of 12	% Yield of 13
Me	90	0
Et	87	06
i-Pr	33	63
t-Bu	02	77

Relative migratory preference of cyclohexyl group over phenyl can be illustrated from the works of Friess and Franham[25], while that of phenyl over cyclopropyl moiety may be exemplified from the works of Sauers and Ubersax[23].

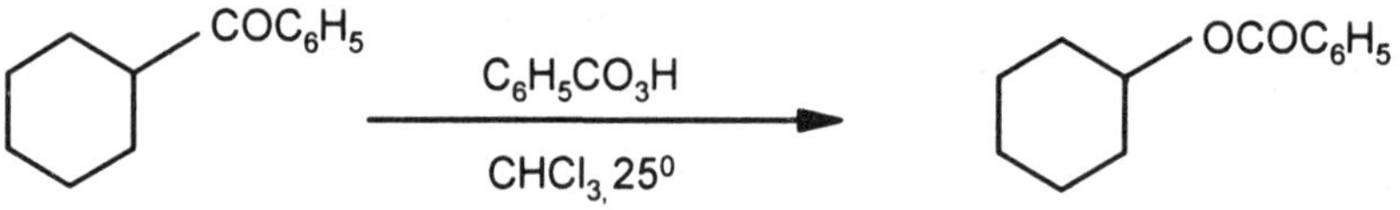

Cyclohexyl–COC$_6$H$_5$ →[C$_6$H$_5$CO$_3$H, CHCl$_3$, 25°] Cyclohexyl–OCOC$_6$H$_5$ (71%, isolated as alcohol) [Ref. 25]

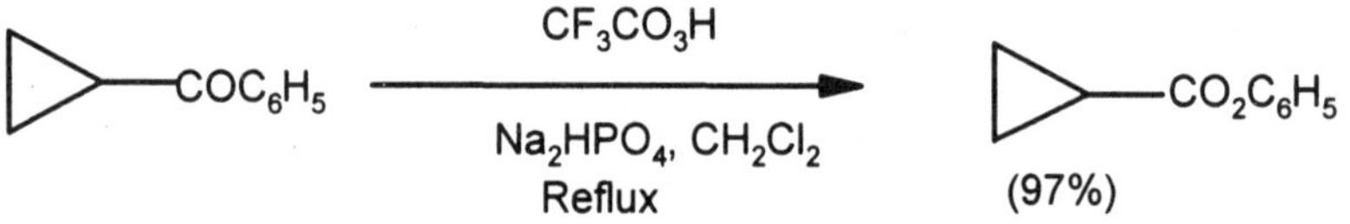

Cyclopropyl–COC$_6$H$_5$ →[CF$_3$CO$_3$H, Na$_2$HPO$_4$, CH$_2$Cl$_2$, Reflux] Cyclopropyl–CO$_2$C$_6$H$_5$ (97%) [Ref. 23]

The group, which is rich in electrons, generally migrates preferentially. The migration of a phenyl ring is facilitated by the presence of electron-donating substituents and retarded by the presence of electron-withdrawing substituents; thus the migratory preference in aromatic ketones decreases in the following order —

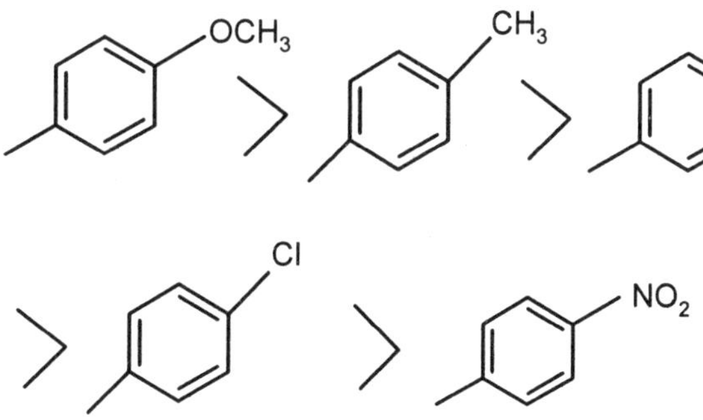

[Migratory aptitude series of aryl groups: p-MeO-C6H4 > p-Me-C6H4 > C6H5 > p-Cl-C6H4 > p-O2N-C6H4]

Migratory preference of phenyl moiety may be illustrated by the following reactions—

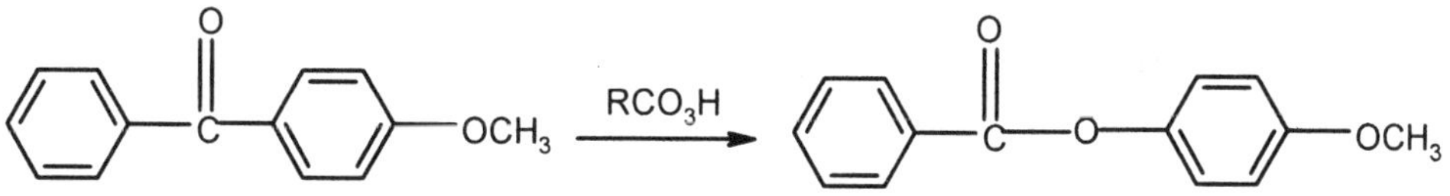

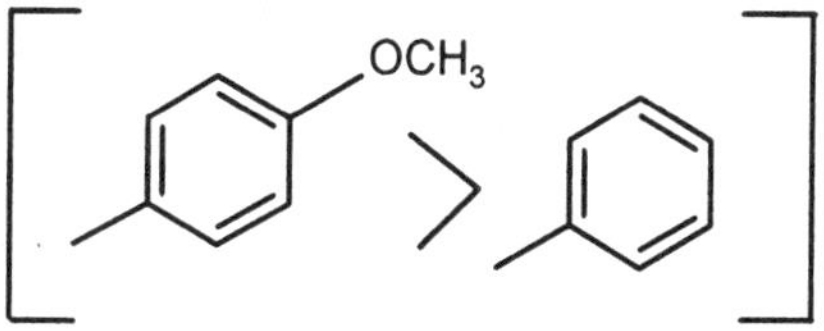

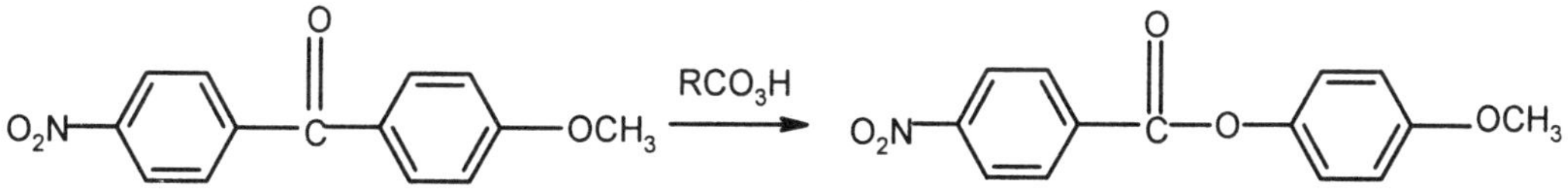

[Ph-CO-C6H4-OCH3 (para) + RCO3H → Ph-CO-O-C6H4-OCH3]

[p-MeO-C6H4 > C6H5]

[O2N-C6H4-CO-C6H4-OCH3 + RCO3H → O2N-C6H4-CO-O-C6H4-OCH3 (95%)]

[C6H5 > p-O2N-C6H4] [Ref. 4]

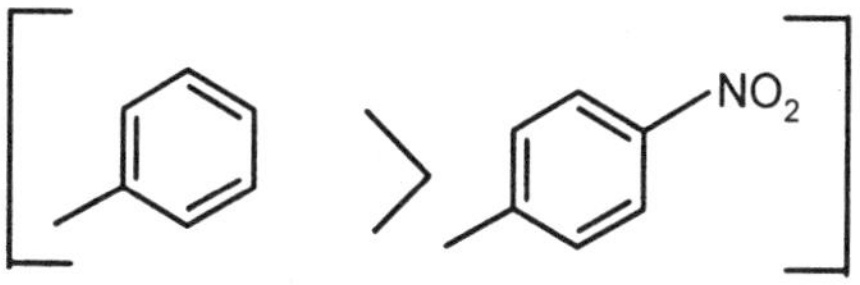

The migratory aptitude reflects the ability of a migrating group to accept a partial positive charge in the transition state. The transition state has a positive charge spread out over the molecule as the carboxylate leaves as an anion. Thus, the group that can better stabilized the positive charge developed would tend to migrate — the more stable the charge, the faster the reaction.

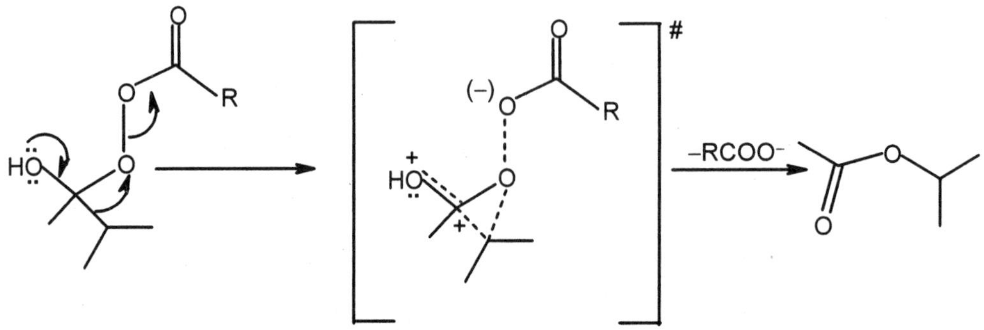

(secondary alkyl group having greater capability of stabilizing positive charge than methyl migrates)

When a benzene ring migrates, π participation is involved as the benzene ring acts as a nucleophile and the positive charge can be spread out even further. Hence, a trastition state in alkyl migration becomes an intermediate in phenyl migration.

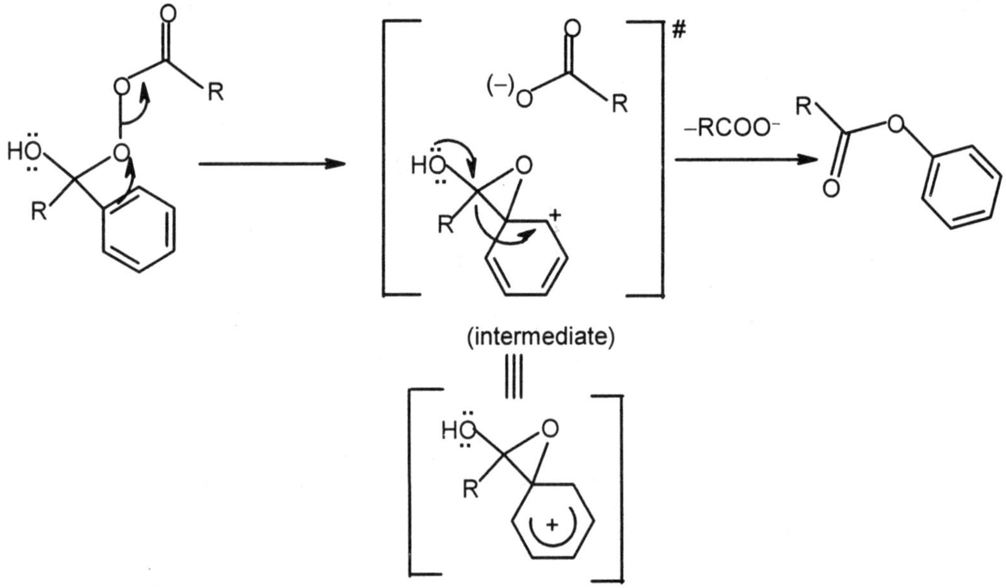

(intermediate)

But still migratory aptitude of migrating groups in *Baeyer–Villiger rearrangements* depends not only on the electronic factor alone; it may be influenced by other various factors like reaction conditions including steric effects.

More on Baeyer-Villiger reaction: Baeyer-Villiger oxidation of aldehydes

Peracids usually oxidize aldehydes to carboxylic acids; thus benzaldehyde gives benzoic acid. Aromatic aldehydes on reaction with peracids may furnish phenols also as predominant product; this has been observed to being facilitated in alkaline media. This particular phenomenon is also called as *Dakin reaction*; the mechanism is similar to that of *Baeyer-Villiger reaction*[26]. For example —

Ogata and Suzuki[27] studied the *Baeyer-Villiger oxidation* of benzaldehyde and substituted benzaldehydes kinetically in aquo-organic solvents at various pH's and reported significant results. Benzaldehydes on such oxidation yield ultimately phenols and/or benzoic acids. The rate in acid media is found to being enhanced with increasing pH of the reaction medium and exhibited general base catalysis, but not acid catalysis.

Migrating groups in the *Baeyer-Villiger* rearrangement of benzaldehydes are either hydride or aryl anion.

$$ArCHO \xrightarrow{PhCO_3H} ArCOOH + ArOH$$

Migarting ratios of Ar/H differ largely, not only by ring substituents in the aldehydes, but by the acidity of the media also. The aryl shift occurs exclusively with *p*- and *o*-hydroxybenzaldehydes leading to the formation of phenols. On the other hand, unsubstituted benzaldehyde and those with electron-withdrawing groups (like *p*-Cl, *m*- & *p*-NO_2) give solely the corresponding carboxylic acids by hydride shift regardless of the acidity of the media[27]. The oxidation of anisaldehyde has been found to be a border-line case; the hydride shift is predominant in alkaline media while the aryl shift is preferred in acid media[27]. The rate of the reaction also depends on solvent effect. The rate of aryl shift for *p*-hydroxybenzaldehyde has been found to being increased with increasing content of water; the rate in aqueous methanol is considerably faster than that in aqueous ethanol. On the contrary, the rate for hydride shift is in the order of aq. MeOH < aq. EtOH < aq. dioxane.

Enzymatic Baeyer-Villiger Oxidation

Baeyer-Villiger oxidation effected by enzyme was reported by Ryerson *et al.*[28]; they carried out the transfiormation of cyclohexanone (**14**) to the corresponding lactone (**15**) using purified Cyclohexanone oxygenase enzyme[29]. The rearrangement induced by this enzyme proceeds with complete stereochemical retention, as shown by Schwas[30], analogous to the classical *Baeyer-Villiger rearrangement* that is effected by peroxyacids.

(**14**)　　　　　　　　　　　　　　　　　(**15**)

Enzymatic *Baeyer-Villiger oxidation* is thus an interesting addition to the well-known *Baeyer-Villiger rearrangement*. The formation of chiral, non-racemic products using the enzymatic *Baeyer-Villiger reaction* would find immense applications[31] in organic syntheses as an extension of the better-known peroxide induced *Baeyer–Villiger reaction*. Two more examples are cited here —

(80% yield with > 98% ee)

Ref. 33

(Cyclohexanone oxygenase enzyme obtained from *Acinetobactor* NCIB9871)[32]

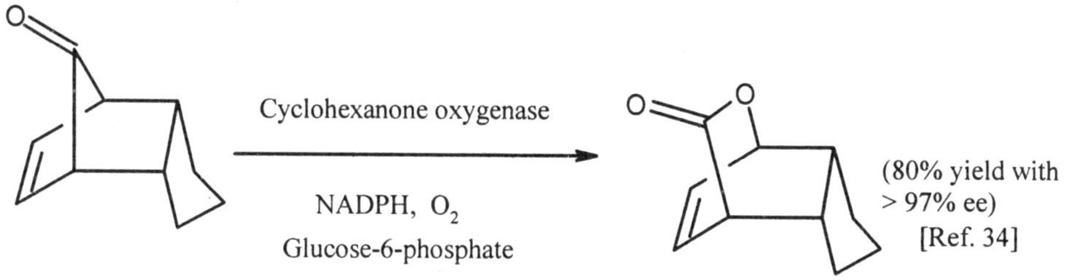

Baeyer – Villiger Oxidation of Bridged Bicyclic Ketones: 2-Oxo-bicyclic ketones

Insertion of oxygen into norbornan-2-one (**16**) with 28% peracetic acid/sulphuric acid[35] or TFPAA/Na_2PO_4 buffer[36] leads to the isolation of only bridgehead migrated lactone (**17**) in 88–100% yield. Some methylene migrated lactone (**18**) may also be obtained if the *Baeyer-Villiger oxidation* of **16** is carried out using 15% H_2O_2/NaOH (65% yield of a mixture of **17** & **18** at a ratio of 54:46).[37]

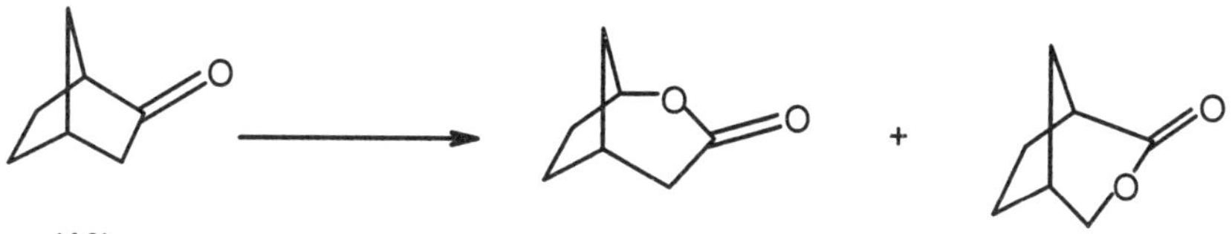

Few more examples, concerning *Baeyer-Villiger oxidations* of this class are —

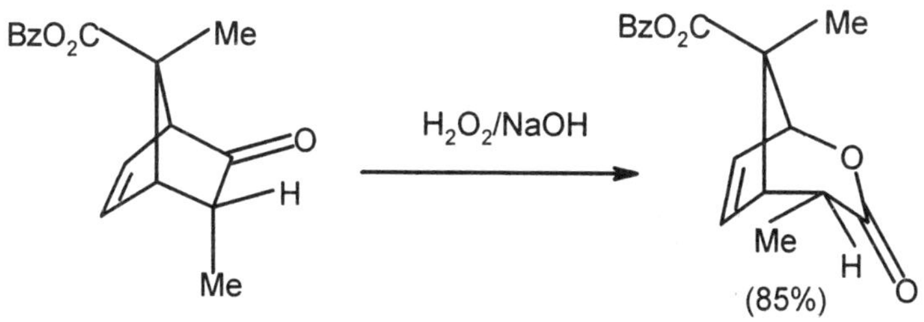

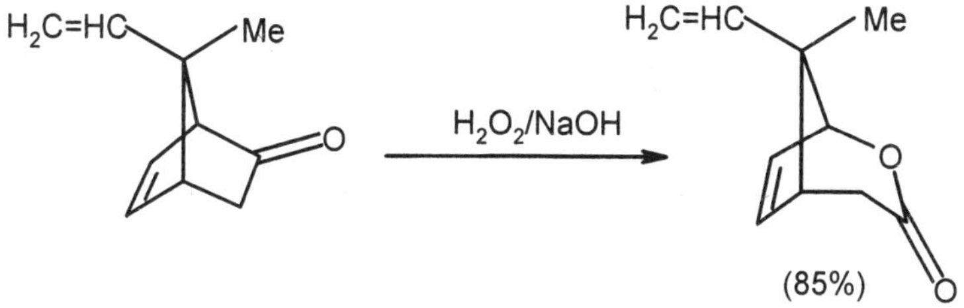

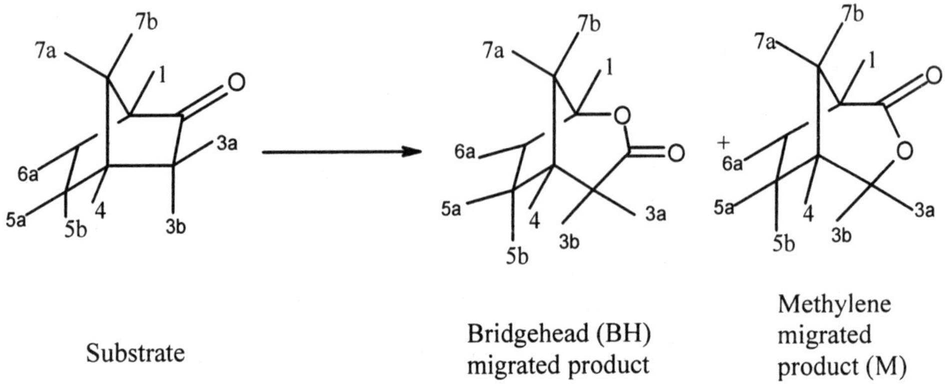

(if not mentioned otherwise, the corresponding labelled groups are hydrogens)

Substituents	Reagents	Predominant products	% Yield	Ref.
5a: Br, 7a: -CH$_2$=CCl$_2$	MCPBA	100% BH	89	38
5a: Br, 7a: -COOH	PAA	78% BH	-	39
1,7a,7b-tri-Me	40% PAA/buffer/30days	75% BH	42	40
1,7a,7b-tri-Me	H$_2$SO$_5$	100% M	50	41
3a:Cl, 5a: CN 7b: OMe	Fails to react with various peracids	-	-	42
5b: OAc, 7a: OMe	MCPBA/NaHCO$_3$	70% BH	-	42
	HCOOH/30% H$_2$O$_2$	100% BH	-	42
7b: Br	40% PAA / NaOAc	71% M	73	43
5a: Br 7a: COOMe 7b: Me	Peracids	100% M	96	44

In 1899 Baeyer and Villiger reported the isolation of campholide (**23**) formed by methylene migration upon oxidation of camphor (**19**) with Caro's acid. Latter, in

1956, Murray et al.[45] explained the preference for methylene migration in camphor (**19**) on the basis of conformational factor as shown in **Scheme 3**. During decomposition of more preferred '*Criegee intermediate*' (**20**) formed by peracid addition to the *endo*-face of camphore (**19**), migration of the *tertiary* carbon through path 'a' to give (**22**) involves a boat-like transition state while migration of methylene carbon through path 'b' to give (**23**) involves an energetically more favoured chair-like transition state.

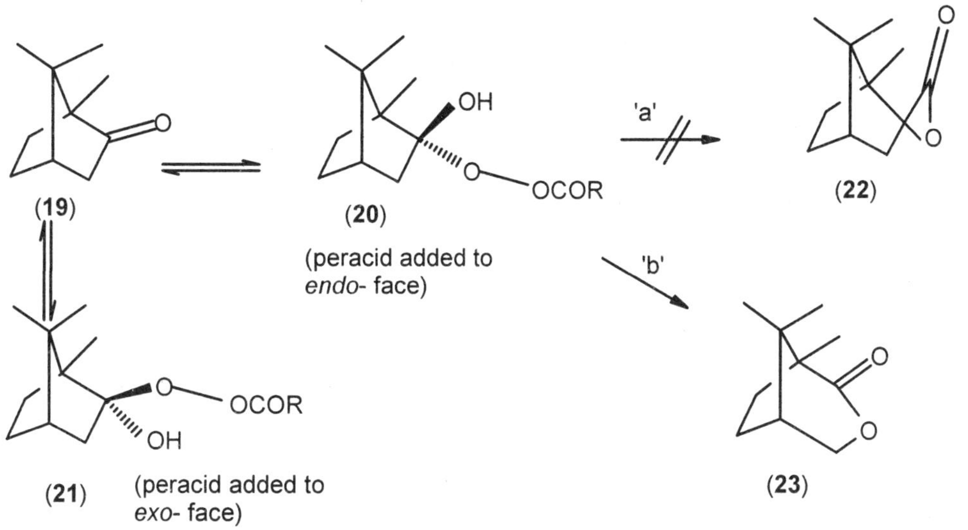

Scheme 3: The boat/chair model for the *Baeyer–Villiger rearrangement* of (**19**)

On the other hand, norcamphor (**12**) on oxidation with peracetic acid yielded solely the electronically favoured bridgehead migrated product (**17**) as reported by Meinwald and Frauenglass[35] (**Scheme 4**). The lactone (**17**) results from the decomposition of *Criegee intermediate* (**24**), formed by sterically favoured *exo*-attack of the peracid on norcamphor (**16**). Rearrangement of the tetrahedral intermediate (**24**) *via* a chair conformation affords bridgehead migrated lactone (**17**).

Sauers[40,46] restudied the peracid oxidation of camphor using PAA and found that the reaction course depends upon the acidity of the reaction medium. With 6:15:6 40% PAA/acetic acid/sulphuric acid at 25⁰, a 30% yield of α-campholide (**23**) was obtained; the product arises as a result of methylene carbon migration. But in weakly acidic medium (PAA/NaOAc/AcOH) the investigator reported to obtain a mixture of lactones **22** &**23** at 75:25 ratio (**Scheme 5**).

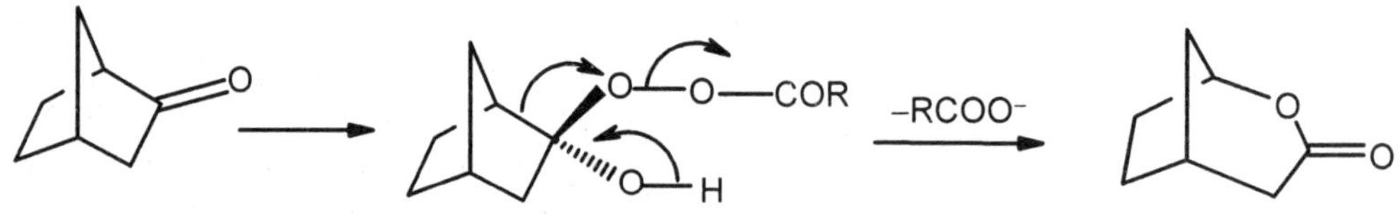

(**16**) (**24**) (**17**)

(sterically favoured;
peracid added to *exo*- face)

Scheme 4: Chair model for *Baeyer – Villiger oxidation* of norcamphor (**16**)

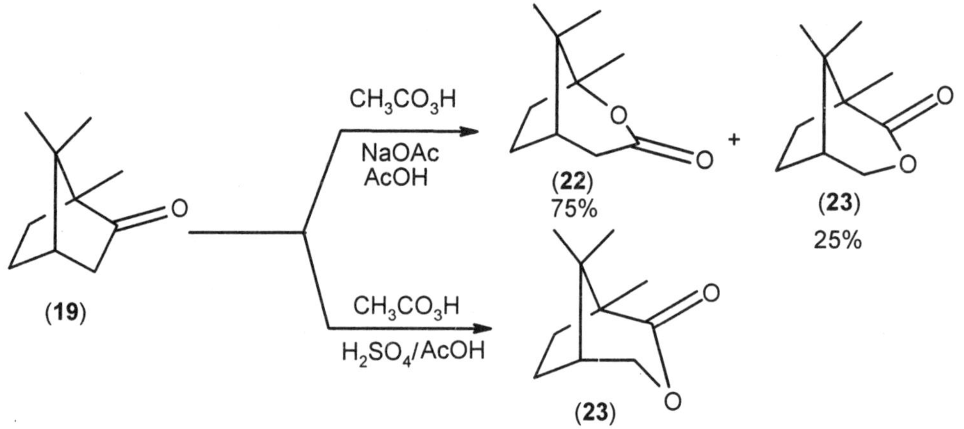

(**23**)
30%

Scheme 5: *Baeyer – Villiger oxidation* of camphor (**19**) both in strong and weak acidic medium

Sauers suggested that such product ratio (75:25) results from boat conformation steric effects (which direct methylene carbon migration) competing successfully with electronic effects (which direct bridgehead migration) in controlling reaction outcome. The investigator further suggested[43] that an additional torsional starin factor may sometimes be appeared; this stain factor, along with the electronic effects and boat/chair steric interactions in the transition states for the *Baeyer–Villiger reaction*, would influence the bridgehead *vs*. methylene carbon migration ratio. In support of his suggestion, Sauers[40] cited the results of fenchone (**25**) oxidation (with 40% PAA) giving (60:40) mixture of lactones **26** & **27**. For fenchone (**25**) electronic effects are similar for the competing *tertiary* migrating groups. Assuming rearrangement of a '*Criegee intermediate*' formed by peracid attack on the *exo*-face of **25**, competition is between a chair transition state favouring bridgehead migrated lactone (**27**) and of torsional strain relief, favouring methylene migrated lactone (**26**). The opposing factors are nearly equal, slightly favouring methylene carbon migration.

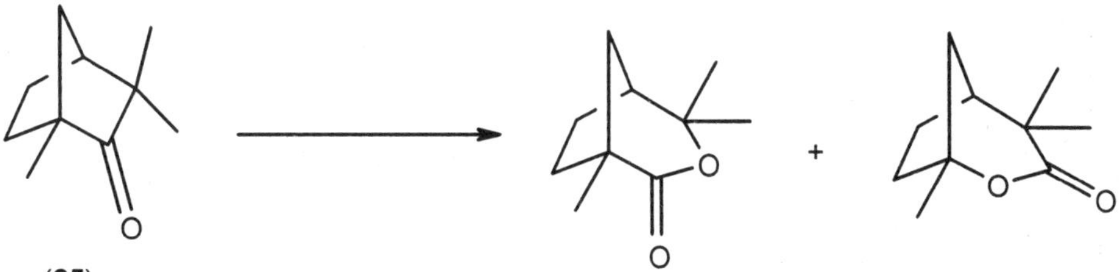

Other Bridged Cyclic Ketonic Systems

a) A novel and selective *Baeyer-Villiger reaction* of camphen-7-ene (**28**) has been effected under mildly acidic condition in dioxane contaminated with its hydroperoxides[47]. Lactones **29** & **30** were obtained in a 4:1 ratio. No double bond epoxidation was observed.

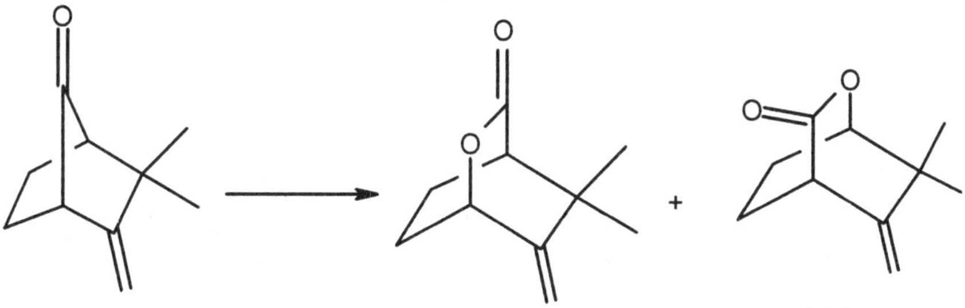

b) Mehta *et al.* reported the conversion of 1,4-bishomocubanone (**31**) into the corresponding lactone (**32**) using *Baeyer-Villiger oxidation* effected with 30% H_2O_2/AcOH in 80-85% yield[48]; ceric ammonium sulphate/aq. acetonitrile in 80% yield[49] and MCPBA/C_6H_6/TsOH in 86% yield[49].

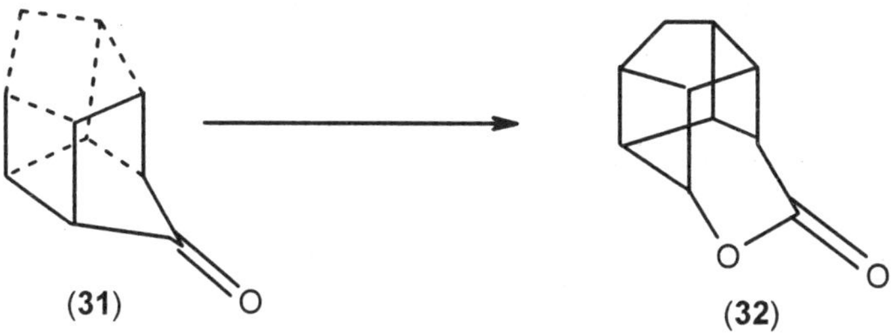

Mehta *et al.*[50] also reported the following transformation —

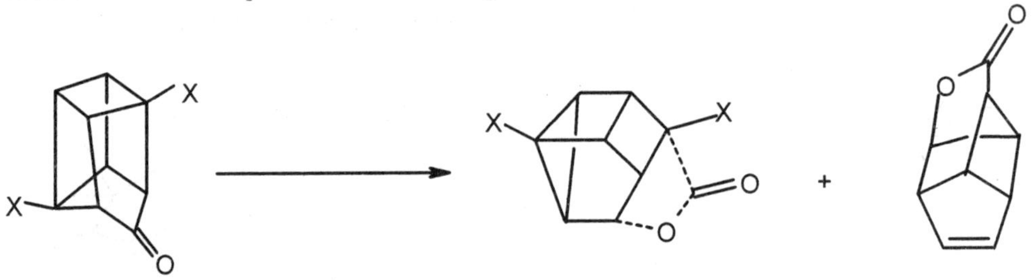

33 a: X=H
    b: X=Br

34 a (16%)
34 b (15%)

35 (6%)

c) On *Baeyer-Villiger oxidation* (effected by MCPBA) the *syn*-fused ketone (**36**) rearranges with bridgehead migration *regiospecifically* to give lactone (**37**) (80% yield). The *anti*-fused ketone (**38**), under the same conditions, affords an 8:2 mixture of lactones (**39**) & (**40**) with bridgehead migration again preferred[51].

36 → **37**

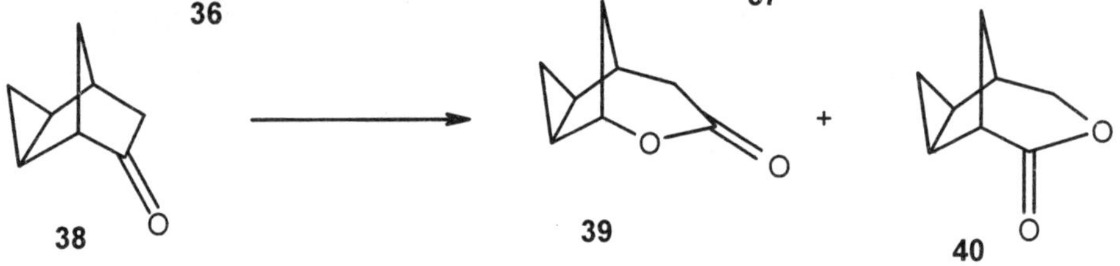

38 → **39** + **40**

But the cyclopropyl-fused ketones (**41**) & (**42**), showed interesting behaviour[51] upon NaHCO$_3$ buffered MCPBA oxidation in CH$_2$Cl$_2$. The *anti*-fused ketone (**41**) afforded bridgehead migrated lactone (**43**) solely (62% yield). This lactone rearranges in aq. perchloric acid to the *trans*-fused butyrolactone (**44**).

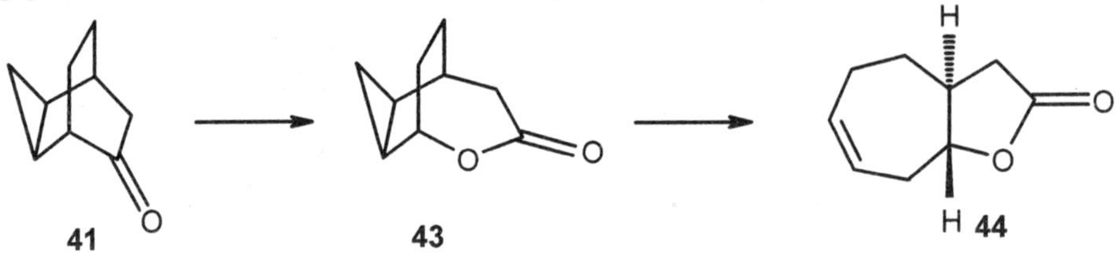

41 → **43** → **H 44**

By contrast, the *syn*-fused ketone (**42**) afforded lactones (**45**) & (**46**) in a 9:1 ratio with methylene migration preferred. The bridgehead migrated lactone (**46**) rearranges in presence of acid to a *cis*-fused butyrolactone (**47**). The preference for methylene migration in the oxidation of ketone (**42**) contrasts with favoured bridgehead migration for (**41**) and for the bicyclo[2.2.1] analogs **36** &**38**.

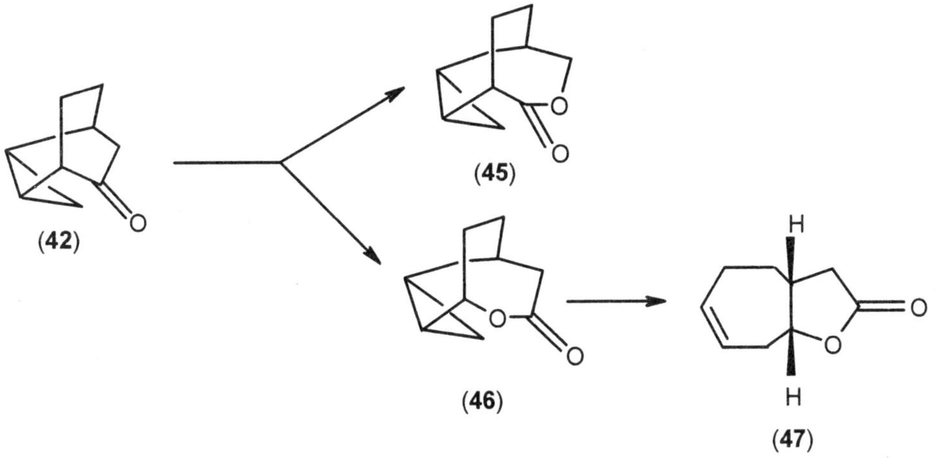

d) Oxidation of azabicyclic ketone (**48**) with peracetic acid proceeds *regiospecifically* to furnish bridgehead migrated lactone (**49**) in 65% yield[52]; but the oxidation effected with MCPBA/CH$_2$Cl$_2$ resulted in the isolation of a mixture of lacones (**49**)& (**50**) at a ratio 2:1 in 50% combined yield[52].

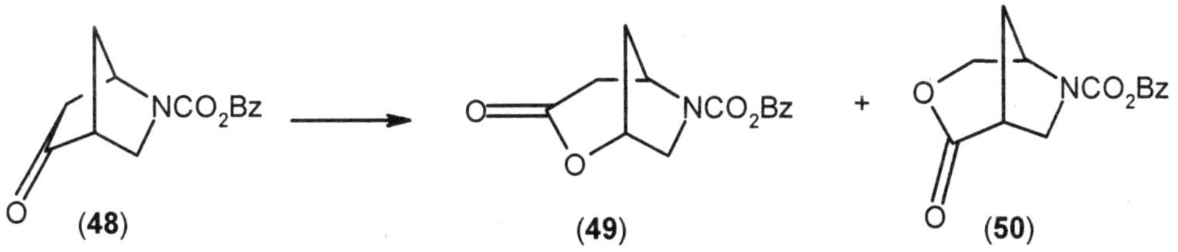

References

1 Baeyer, A. and Villiger, V. (1899), *Ber.*, **24**, 3625.
2 Sager, W. F. and Duckworth, A. (1955), *J. Am. Chem. Soc.*, **77**, 188.
3 For a study of the mechanism of this conversion, see Cullis, P.M., Arnold, J.R.P., Clarke, M., Howell, R., Demira, M., Naylor, M., Nicholls, D. (1987), *J. Chem. Soc, Chem. Commun.*, 1088.

4 Doering, W. and Speers, L., *J. Am. Chem. Soc.* (1950), **72**, 5515.
5 Emmons, W. D. and Lucas, G. B. (1965), *J. Am. Chem. Soc.*, **77**, 2287.
6 Friess, S. (1949), *J. Am. Chem. Soc.*, **71**, 2571.
7 Wiberg, K. B. and Snoonian, J. R. (1998), *J. Org. Chem.*, **63**, 1390.
8 Burton, J. W., Clarke, J. S., Derres, S., Stork, T. C., Bendall, J. G. and Holmes, A. B. (1997), *J. Am. Chem. Soc.*, **119**, 7483.
9 Krafft, G. A. and Katzenellenbogen, J. A. (1981), *J. Am. Chem. Soc.*, **103**, 5459.
10 Magnusson, G. (1977), *Tetrahedron Lett.*, 2713.
11 Bailey, W. F. and Shih, M.-J. (1982), *J. Am. Chem. Soc.*, **104**, 1769.
12 Mino, T., Masuda, S., Masayuki, N. and Yamashita, M. (1997), *J. Org. Chem.*, **62**, 2633.
13 Watanabe, A., Uchida, T., Ito, K. and Katsuki, T. (2002), *Tetrahedron Lett.*, **43**(25), 4481.
14 Plesnicar, B. (1978), *Oxidation in Organic Chemistry, Part C*, (edited by Trahanovsky, W.S.), pp. 254-62, Academic Press, New York.
15 Criegee, R. and Kasper, K. (1948), *Ann.*, **560**, 127.
16 Doering, W. E. and Dorfman, E. (1953), *J. Am. Chem. Soc.*, **75**, 5595.
17 Palmer, B. W. and Fry, A. (1970), *J. Am. Chem. Soc.*, **92**, 2580.
18 Turner, R. B., *J. Am. Chem. Soc.*, **72**, 878 (1950); Mislow, K. and Brenner, J. (1953), *J. Am. Chem. Soc.*, **75**, 2318.
19 Berson, J. A. and Suzuki, S. (1959), *J. Am. Chem. Soc.*, **81**, 4088.
20 House, H. O. and Bare, T. M. (1968), *J. Org. Chem.*, **33**, 943.
21 Hawthrone, M. F. and Emmons, W. D. (1958), *J. Am. Chem. Soc.*, **80**, 6398.
22 Smissman, E. F., Li, J. P. and Israili. Z. H. (1968), *J. Org. Chem.*, **33**, 4231;. Chambers, R. D. and Clark, M. (1970), *Tetrahedron Lett.*, No. 32, 2741.
23 Sauers, R. R. and Ubersax, R. W. (1965), *J. Org. Chem.*, **30**, 3939.
24 Robertsen, J. C. and Swelim, A. (1967), *Tetrahedron Lett.*, No. 30, 2871.
25 Friess, S. L. and Franham, N. (1950), *J. Am. Chem. Soc.*, **72**, 5518.
26 For a discussion, see Hocking, M.B., Bhandari, K., Shell, B. and Symth, T. A. (1982), *J. Org. Chem.*, **47**, 4208.
27 Ogata, Y. and Sawaki, Y. (1969), *J. Org. Chem.*, **34**(12), 3985.
28 Ryerson, C. C., Ballone, D. P. and Walsh, C. (1982), *Biochemistry*, **21**, 2644.
29 Donoghue, N. A., Norris, D. B. and Trudgill, P. W. (1976), *Eur. J. Biochem.*, **63**, 175.
30 Schwab, J. M. (1981), *J. Am. Chem. Soc.*, **103**, 1876.
31 Taschner, M. J., Chen, Q.-Z. (1991), *Bioorganic and Medicinal Chem. Lett.*, **1**, 535 (1991).
32 Schwab, J. M., Li, W. and Thomas, L. P. (1983), *J. Am. Chem. Soc.*, **105**, 4800.
33 Taschner, M. J. and Black, D. J. (1988), *J. Am. Chem. Soc.*, **110**, 6892.

34 Taschner, M. J. and Peddada, L. (1992), *J. Chem. Soc., Chem. Commun.*, 1384.
35 Meinwald, J. and Frauenglass, E. (1960), *J. Am. Chem. Soc.*, **82**, 5235.
36 Rassat, A. and Ourisson, G. (1959), *Bull. Soc. Chim. France*, 1133.
37 Deslongchamps, P. (1975), *Tetrahedron*, **31**, 2463.
38 Takano, S., Kubodera, N. and Ogasawara, K. (1977), *J. Org. Chem.*, **42**, 786.
39 Peel, R. and Sutherland, J. K. (1974), *J. Chem. Soc., Chem. Commun.*, 151.
40 Sauers, R. R. and Ahearn, G. P. (1961), *J. Am. Chem. Soc.*, **83**, 2759.
41 Clark, R. D. and Heathcock, C. H. (1976), *J. Org. Chem.*, **41**, 1396.
42 Grudezinski, Z., Roberts, S. N., Howard, C. and Newton, R. F. (1978), *J. Chem. Soc., Perkin Trans. I*, 1182.
43 Sauers, R. R. and Beisler, J. A. (1964), *J. Org. Chem.*, **29**, 210.
44 Grieco, P. A., Pogonoulski, C. S., Burke, S. D., Nishizawa, M., Miyashita, M., Masaki, Y., Wang, C.-L. J. and Majetich, G. (1977), *J. Am. Chem. Soc.*, **99**, 4111.
45 Murray, M. F., Johnson, B. A., Pederson, R. R. and Ott, A. C. (1956), *J. Am. Chem. Soc.*, **78**, 981.
46 Sauers, R. R. (1955), *J. Am. Chem. Soc.*, **81**, 925; *Chem. Abstr.* (1980), 93, 150386q.
47 Patil, D., Chawla, H. and Dev, S. (1979), *Tetrahedron*, **35**, 527.
48 Mehta, G. and Pandey, P. (1975), *Synthesis*, **6**, 404.
49 Mehta, G., Pandey, P.N., Ho, T.-L. (1976), *J. Org. Chem.*, **41**, 953; Mehta, G., Singh, N., Pandey, P.N., Chaudhury, B. and Duddeck, H. (1980), *Chem. Lett.*, 59.
50 Mehta, G. and Suri, S. C. (1980), *Tetrahedron Lett.*, 3825 (1980).
51 Marshall, J. A. and Ellison, R. H. (1975), *J. Org. Chem.*, **40**, 2070.
52 Krow, G. P. (1981), *Tetrahedron*, **37**(16), 2697.

- Krow, G. R. (1993), *Org. React.*, **43**, 251. [Review]
- Renz, M. and Meunier, B. (1999), *Eur. J. Org. Chem.*, **4**, 737.
- Crudden, C. M., Chen, A. C. and Calhoun, L. A. (2000), *Angew. Chem. Int. Ed. Engl.*, **39**, 2851.
- Fukuda, O., Sakaguchi, S. and Ishii, Y. (2001), *Tetrahedron Lett.*, **42**, 3479.
- Watanabe, A., Uchida, T., Ito, K. and Katsuki, T. (2002), *Tetrahedron Lett.*, **42**, 3479
- Kobayashi, S., Tanaka, H., Ami, H. and Uneyama, K. (2003), *Tetrahedron*, **59**, 1547.

2

Barton Reaction

Introduction

Organic nitrites are a good source of alkoxy radicals. The nitrogen-oxygen bond in organic nitrites is particularly susceptible to homolysis leading to the formation of alkoxide radicals. Alkoxide radicals may participate in the reaction by any of the following pathways[1], and several of these pathways have been observed in nitrite photolysis:

(a) association with other radicals, including dimerization
(b) addition to unsaturated compounds
(c) hydrogen abstraction from another molecules
(d) disproportionation
(e) rearrangement, including internal hydrogen abstraction, and
(f) decomposition with carbon or hydrogen elimination.

Although the synthetic utility of alkoxides is much less because of their tendency of undergoing a variety of termination reactions, yet this photolytic decomposition of organic nitrites had been made a potentially useful tool in the field of synthetic organic chemistry. Prof. D. H. R. Barton was the pioneer in this field[2]. He concentrated his thinking in selecting the starting materials, which bear certain desired structural framework. He observed that an alkoxy radical, if produced by photolysis of an organic nitrite (in solution) in which the –C– O–NO moiety and a C–H bond are oriented in potentially close proximity, brings about selective *intramolecular hydrogen abstraction* from a desired site. Moreover, the resulting free radical undergoes chain termination by recombination with NO, presumably the very NO generated by the original photolysis, to form a nitroso-alcohol, which may be isolated as dimer or may be rearranged to an oxime. Hence, direct introduction of functionality into a non-activated position is achieved; a functionality that can easily be converted to carbonyl or a variety of other chemical functionalities. Here lies the significance of *Barton reaction*.

Definition

This novel course of reaction involving photolytic decomposition of organic nitrites to alkoxy radicals, followed by a *stereoselective intramolecular hydrogen abstraction* by the latter and recombination of the resulting carbon radicals with nitric oxide produced, to form nitroso-alcohols is designated as the *Barton Reaction*, in honour of the inventor.

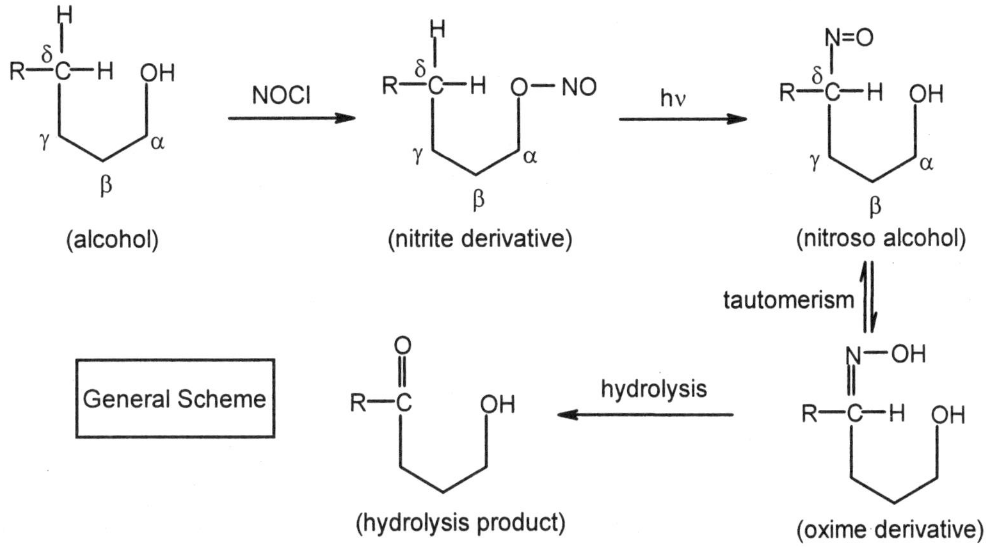

Photolysis of the nitrite results in the conversion of the nitrite group into –OH function and nitrosation of the methyl (if R = H) or methylene (if R is not H) moiety. Hydrolysis of the tautomer (oxime) yields the corresponding carbonyl derivative. Thus, this method provides a means to oxidize a carbon atom separated from an OH group by three other carbon atoms i.e. carbon atom concerned would be in the δ-position with respect to an OH group. The reaction is effected by irradiation of a solution of the nitrite in a suitable non-hydroxylic solvent with light from a high-pressure mercury arc lamp in the atmosphere of nitrogen. A Pyrex filter is usually employed to limit the radiation to wavelengths higher than 300 nm, thus avoiding undesired side reactions induced by more energetic lower wavelength radiations.

Mechanism

The mechanism of *Barton reaction* is believed to proceed through the following discrete steps:

Step 1: Homolysis of the nitrite derivative **(1)** on photo-irradiation takes place firstly. Mechanistic studies using ^{15}N have shown that photolysis occurs in reversible fashion forming alkoxy radical **(2)** and nitrogen monoxide, which are completely dissociated from each other.

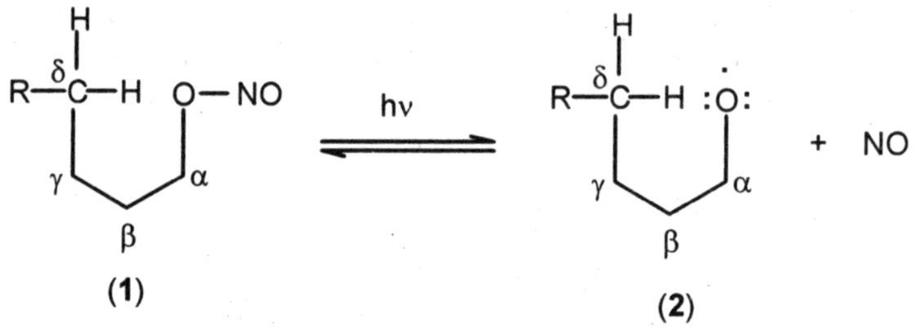

Step 2: The alkoxy radical rearranges itself rapidly, by *intramolecular hydrogen abstraction* via a *six-membered cyclic transition state* **(3)**, to a carbon radical **(4)**.

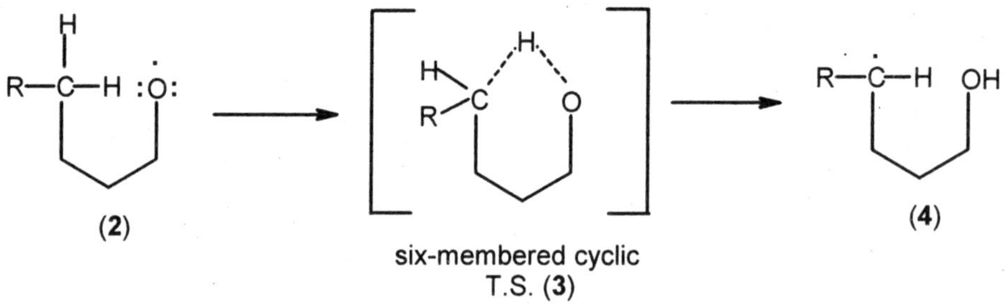

Step 3: Recombination between the carbon radical **(4)** and NO occurs forming nitroso-alcohol monomer **(5)**.

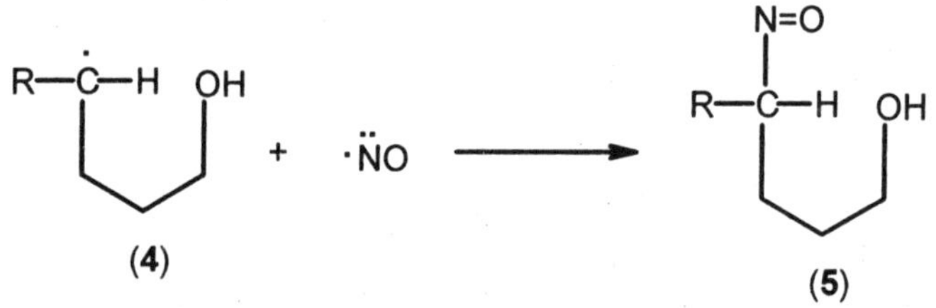

Step 4: Isomerization of the nitroso compound **(5)** to an oxime.

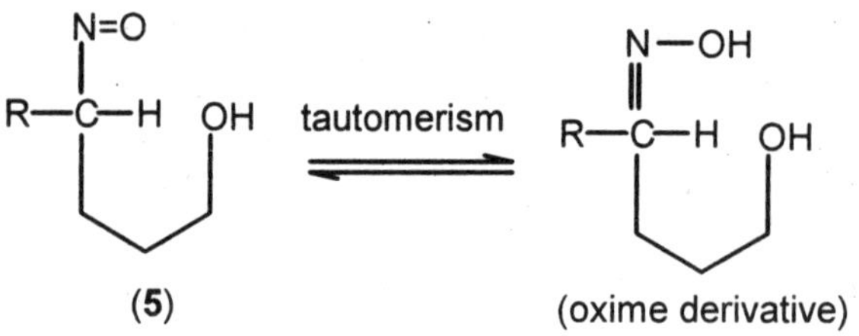

Discussion on mechanism

Barton reaction is an example of *intramolecular hydrogen abstraction*; in practice the reaction occurs almost exclusively by abstraction of a hydrogen atom from the δ-carbon atom through a six-membered cyclic transition state. This contention has already received ample experimental support. It was found that photolysis of 1-octylnitrite in benzene solution gave exclusively the dimer of 4–nitroso-1-octanol (46% yield) through the transition state (**6**).

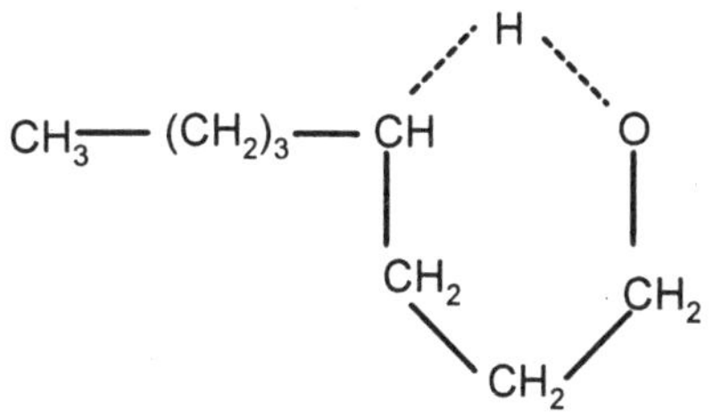

(6)

Conclusive support for the formation of six-membered cyclic transition state in the *Barton reaction* received from the works of Kabasakalian *et al.*[3,4] who studied the photolysis of aromatic alkyl nitrites in solution.

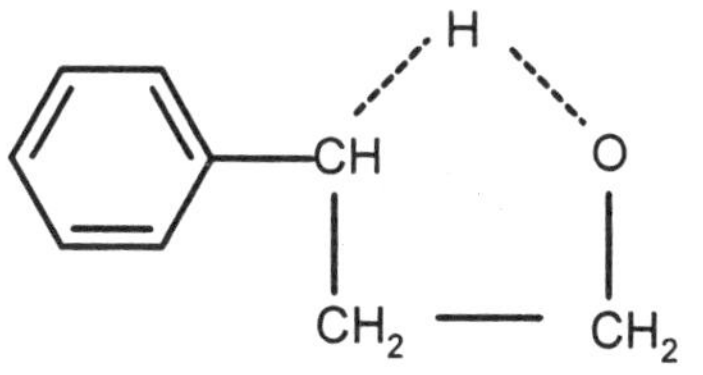

(7)

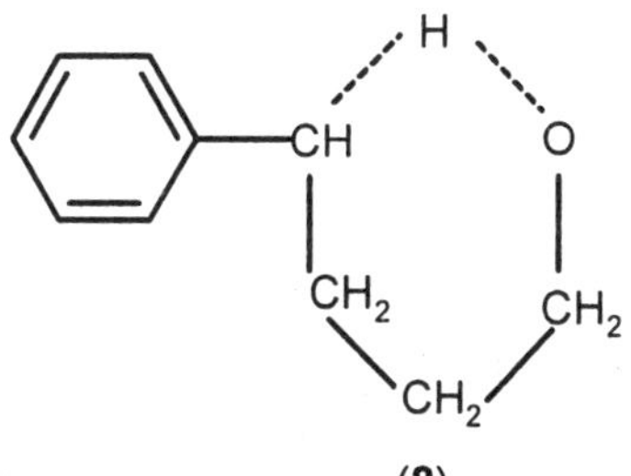

(8)

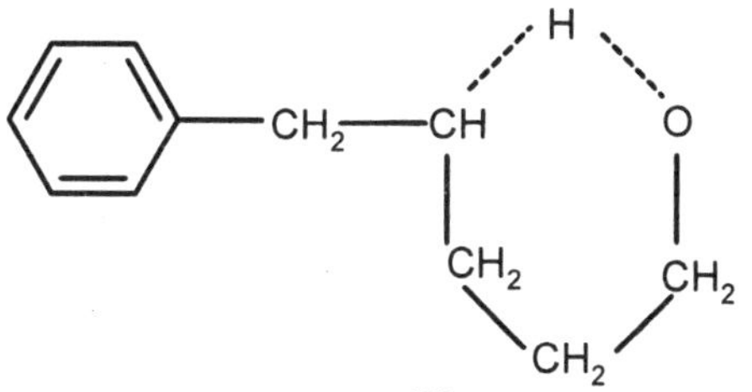

(9)

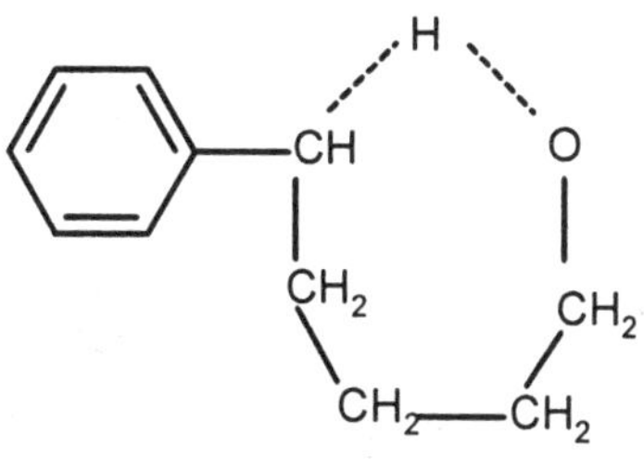

(10)

The experimental results showed that 3-phenyl-1-propylnitrite did not give rise to any nitroso dimer; the postulated five membered T.S. (**7**) apparently cannot to be formed, in spite of the predicted ease of benzyl-hydrogen abstraction. But 4-phenyl-1-butylnitrite is able to form the postulated six-membered T.S. (**8**) as evidenced from the isolation of corresponding nitroso dimer from the substrate. It is very much interesting to note that 5-phenyl-1-pentylnitrite undergoes hydrogen abstraction at C-4 (**9**) rather than at C-5 (**10**); the phenomenon may be argued by considering the advantages, to be gained from a six-membered *vs.* a seven-membered T.S., which outweigh the greater ease of benzyl-hydrogen abstraction.

Remote functionalization and Molecular modeling relating to Barton reaction

Burke *et al.* offered a predictive basis for *Barton reaction* from molecular modeling study utilizing MACROMODEL™ software[5]. The investigators showed that calculated distances in computationally minimized substrate structures correlate with experimental results. A survey of known examples of *Barton reaction* provides support for either the desired extra-annular methyl activation or the undesired functionalization of the intra-annular methyl group.

(11)   (12)

Substrate (**11**) shows a marked preference for C-18 methyl (intra-annular methyl group) involvement[6], whereas substrate (**12**) [$\Delta^{5,6}$- system] shows a 2:1 preference for the C-19 methyl group (extra-annular methyl moiety)[7,8]. Similarly, functioalization of extra-annular methyl group is also observed in the following transformation of (**13**) to (**14**).

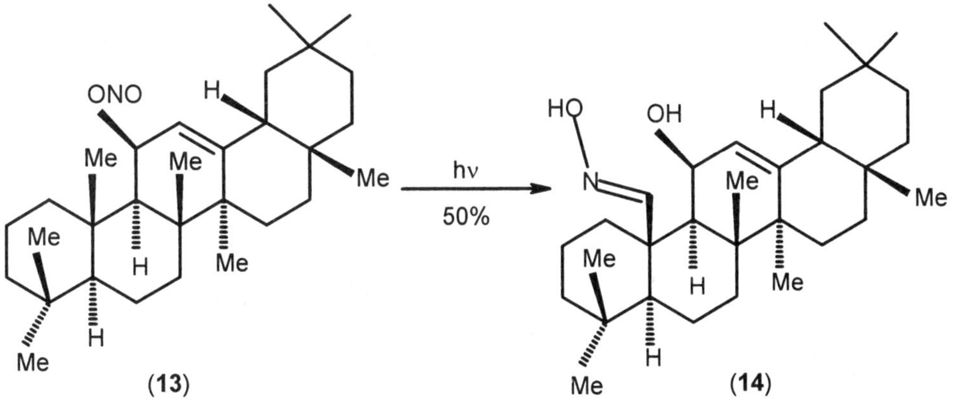

(13) → hv, 50% → (14)

The authors defined a term "Δ d" [('undesired distance') – ('desired distance')] to be positive for extra-annular methyl functionalization and negative for intra-annular methyl functioalization.

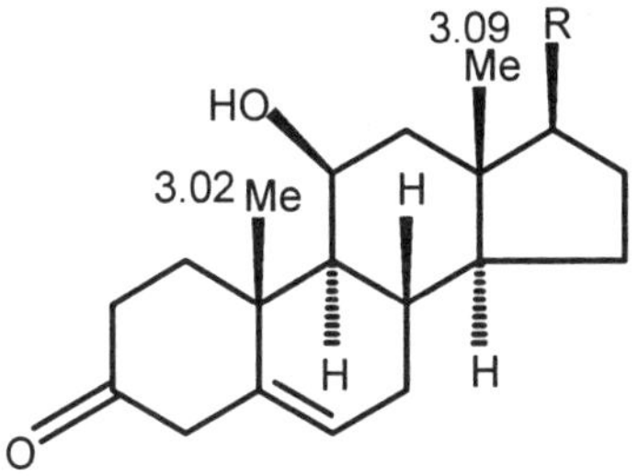

(15)

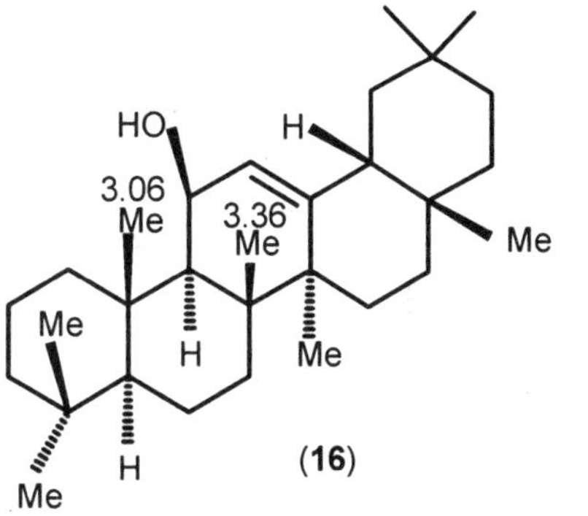

Δ d = (3.09 – 3.02) °A = + 0.07 °A

(16)

Δ d = (3.36 – 3.06) °A = + 0.30 °A

Energy minimization (MM2) of the known substrates (12) & (13) gave rise to the distances (in Angstroms) depicted in structures (15) & (16) respectively. These calculated distances are from the centres of mass of the oxygen atom to the centre of mass of methyl carbon, and are considered to be reflective of the corresponding oxygen to hydrogen distances in the six-membered transition states. The theoretical assumptions for the aforediscussed substrates are qualitatively in accord with the experimental results.

Applications

The *Barton reaction* finds immense applications in the steroidal series, particularly in the functionalization of the two non-activated C-18 and C-19 angular methyls of these molecules having suitably placed hydroxyl groups. D. H. R. Barton *et al.*[2,9] used the reaction to effect the key step in their elegent synthesis of aldosterone acetate (**18**), a biologically useful hormone of the adrenal cortex, starting from corticosterone acetate (**17**).

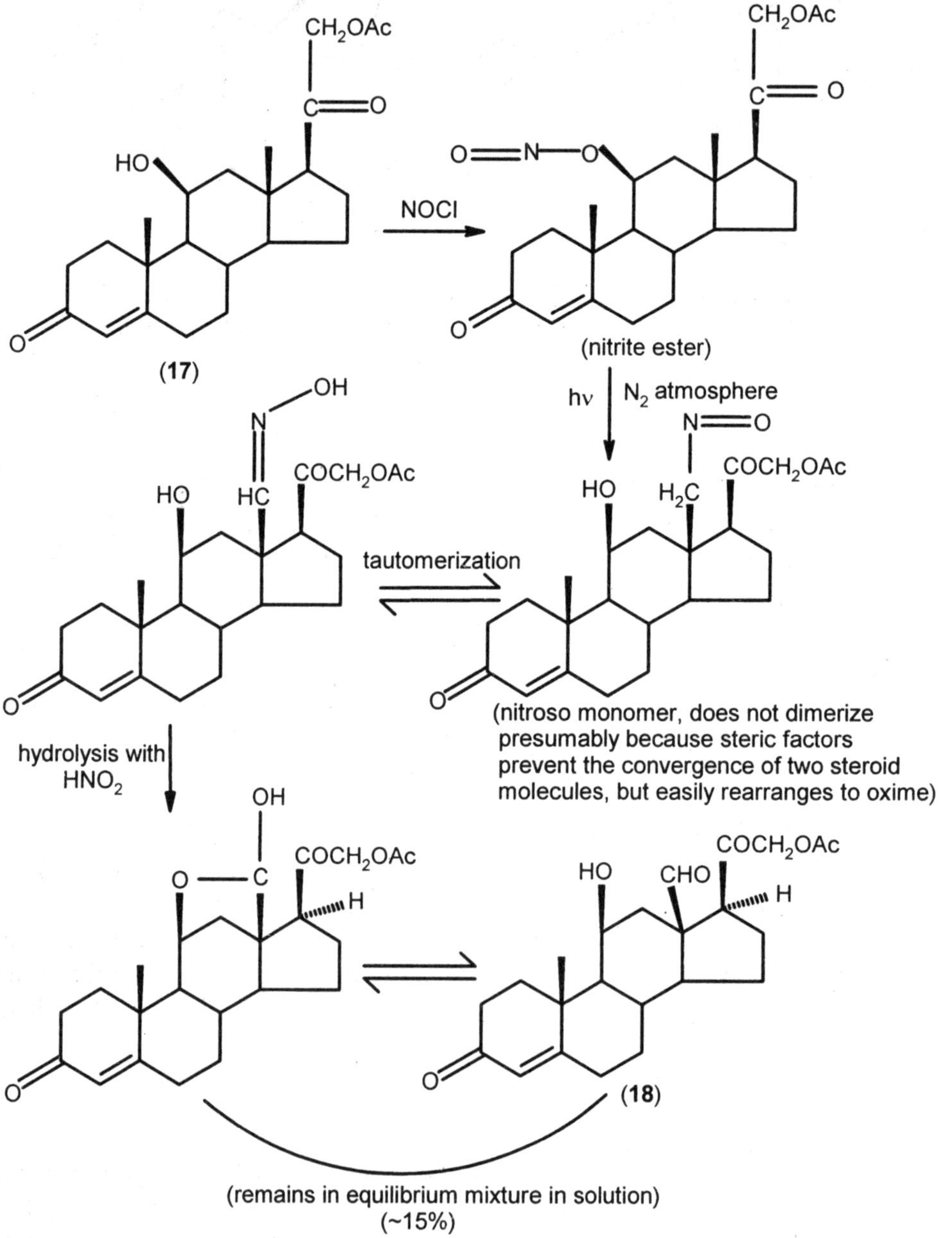

Subsequently, this general procedure of attack at C-18 *via* the 11-nitrite was broadened and resulted in the synthesis of 21-desoxyaldosterone[10] (**19–24**) and 19-noraldosterone[9,10] (**25–27**). Both the syntheses (**Schemes 1 & 2**) are outlined in flow-sheet form:

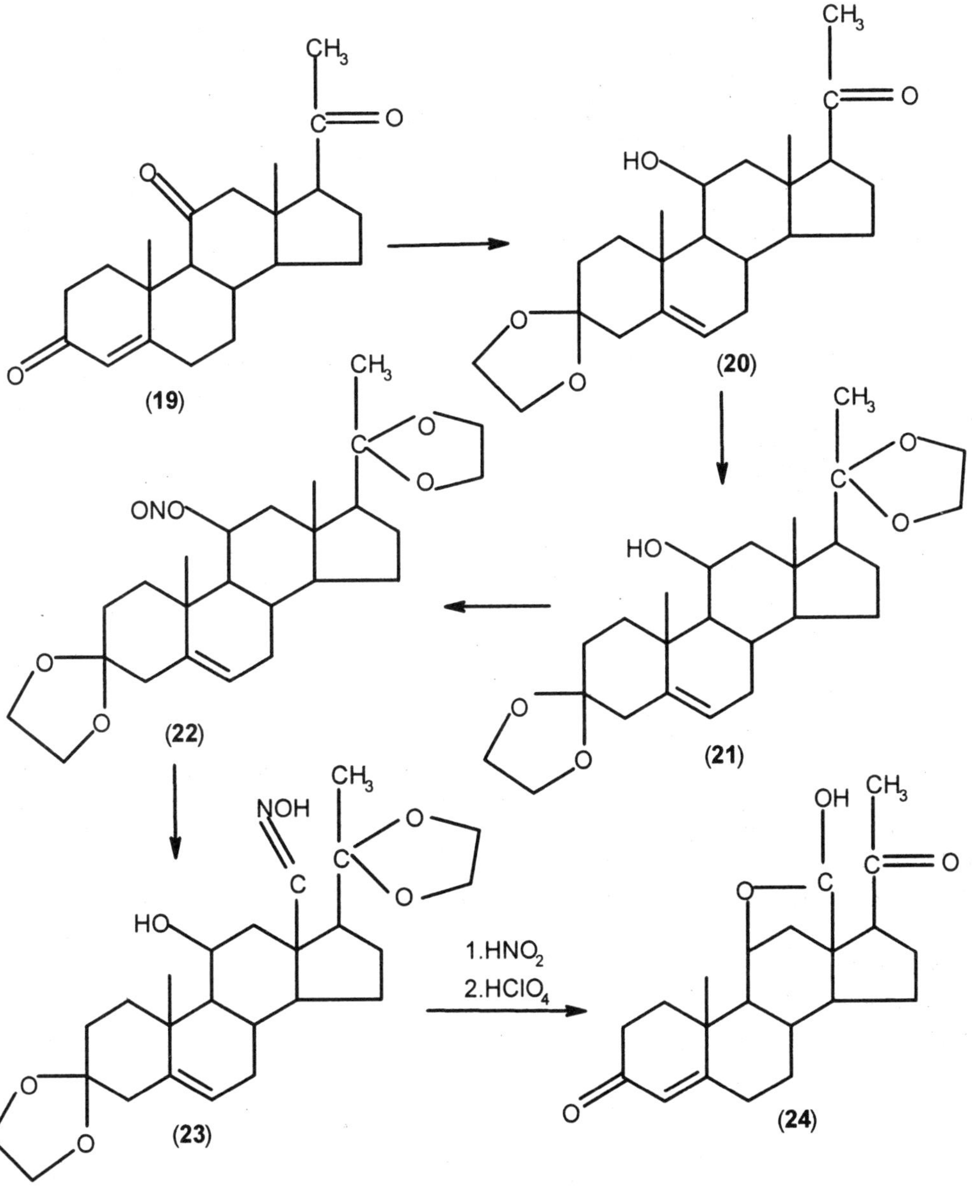

Scheme 1: Conversion of **19** to **24**

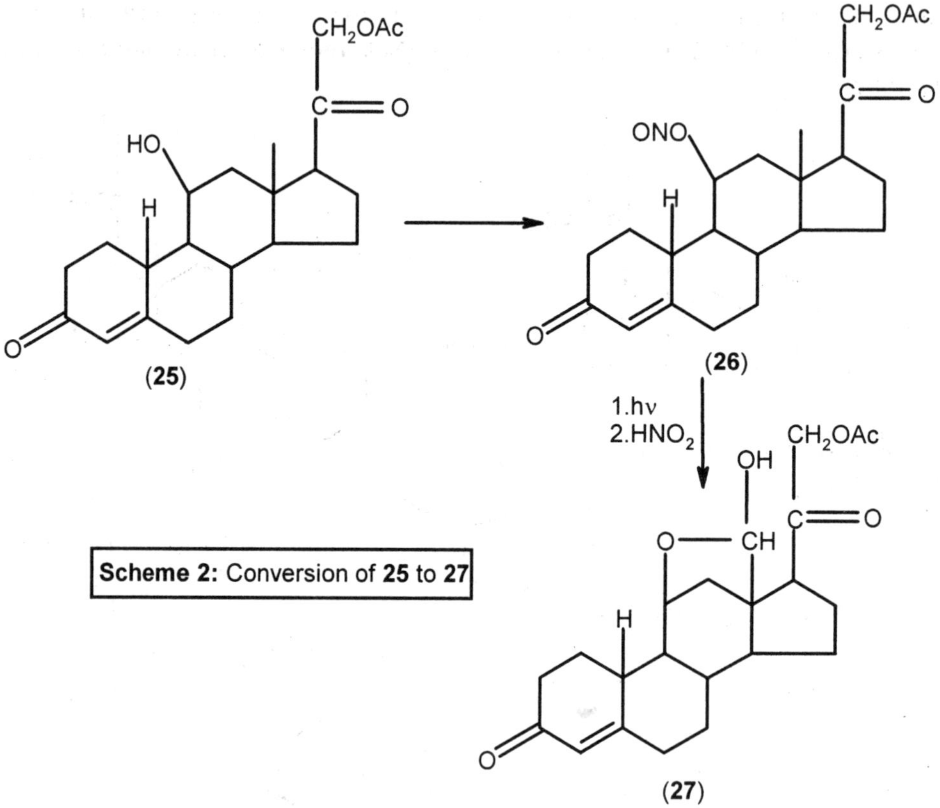

Scheme 2: Conversion of 25 to 27

Sometimes, more than one site in a molecule may favourably be disposed for attack by the newly formed alkoxy radical and in such cases a mixture of products would result. In course of the reaction leading to the conversion of (17) to (18), along with the attack at C-18, attack at C-19 can also be occurred giving rise to a mixture of (28) and (29) (**Scheme 3**) as identified by Barton et al.[10,11].

Bosworth et al.[12] and latter Hobbs et al.[13] successfully utilized *Barton reaction* as a convenient method for functionalizing the unactivated bridge methyl groups of pinane structure (**Scheme 4**).

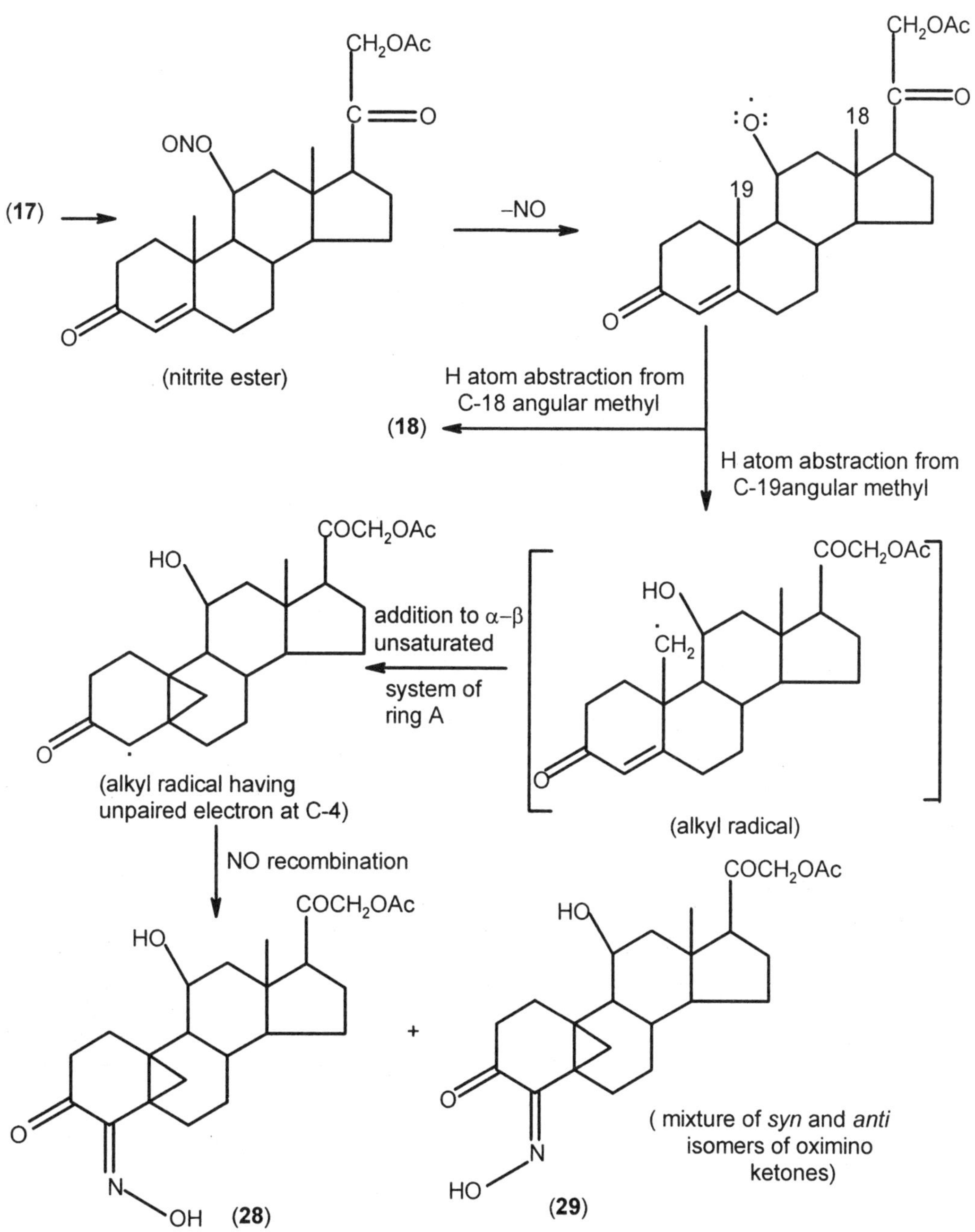

Barton Reaction

(17) → (nitrite ester) → −NO →

H atom abstraction from C-18 angular methyl → (18)

H atom abstraction from C-19 angular methyl → (alkyl radical)

addition to α–β unsaturated system of ring A → (alkyl radical having unpaired electron at C-4)

NO recombination → (28) + (29)

(mixture of *syn* and *anti* isomers of oximino ketones)

Scheme 3

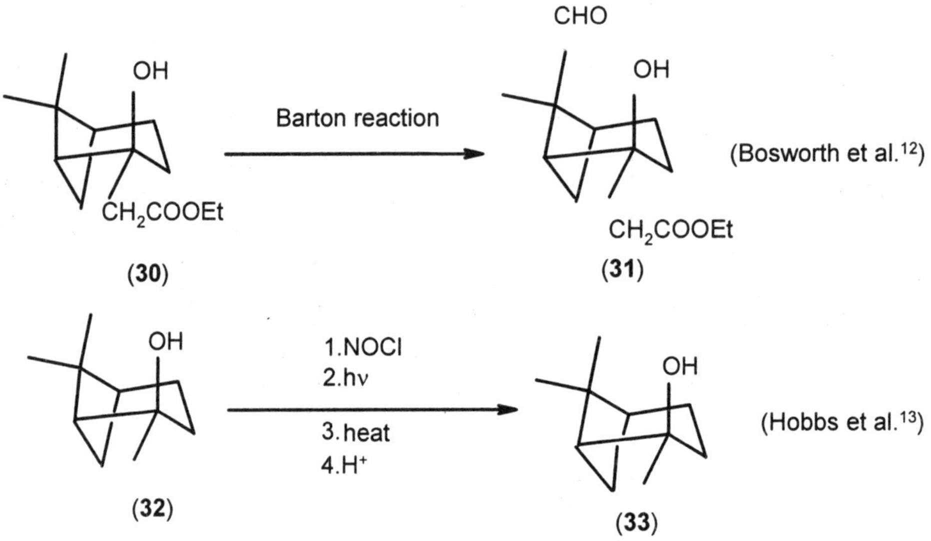

Abnormal Barton reaction: *Barton fragmentation*

In cases of certain substrates it has been observed that instead of intramolecular H-abstraction by the intermediate alkoxy radical produced from corresponding nitrite ester under Barton condition, cleavage of α,β- C–C bond occurs. Such reactions are termed as *Barton fragmentations*. Baldwin and Blomquist[14] synthesized a number of cyclohexanone derivatives from tertiary alcohols using *Barton fragmentation* as the key step.

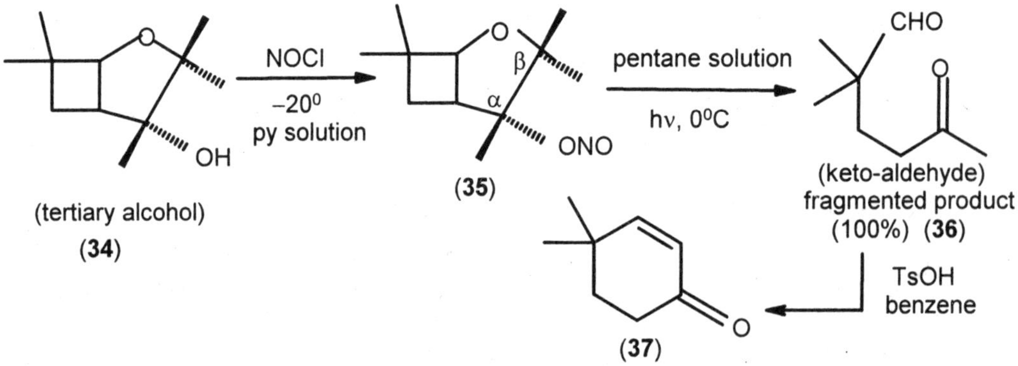

Even, extremely hindered alcohols were observed to undergo the nitrosation/fragmentation process with ease. For instance —

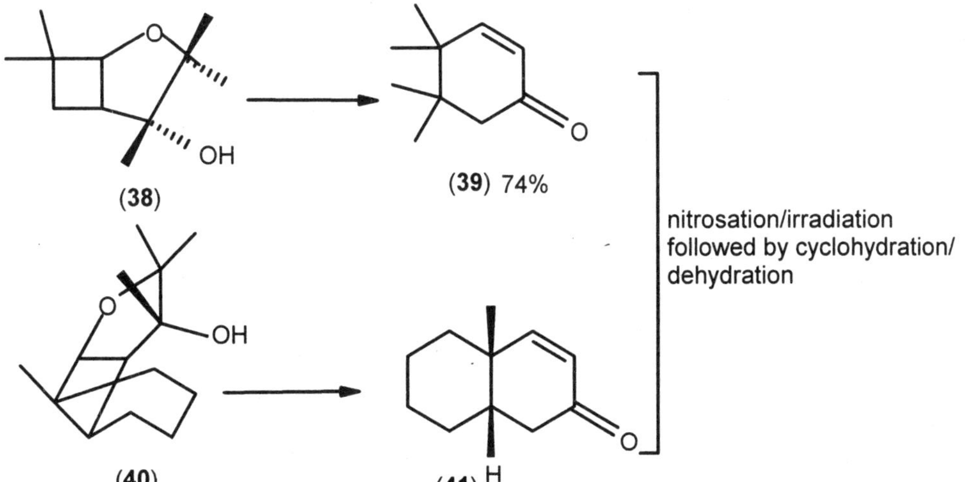

Factors that facilitate *Barton fragmentation*[15,16] include stabilization of the derived carbon radical, particularly by oxygen[17-19], as well as strain or steric compression in the initial alkoxy radical[20,21]. Baldwin and Blomquist[14] suggested the following mechanism for the reaction leading to the conversion of **35** to **36**.

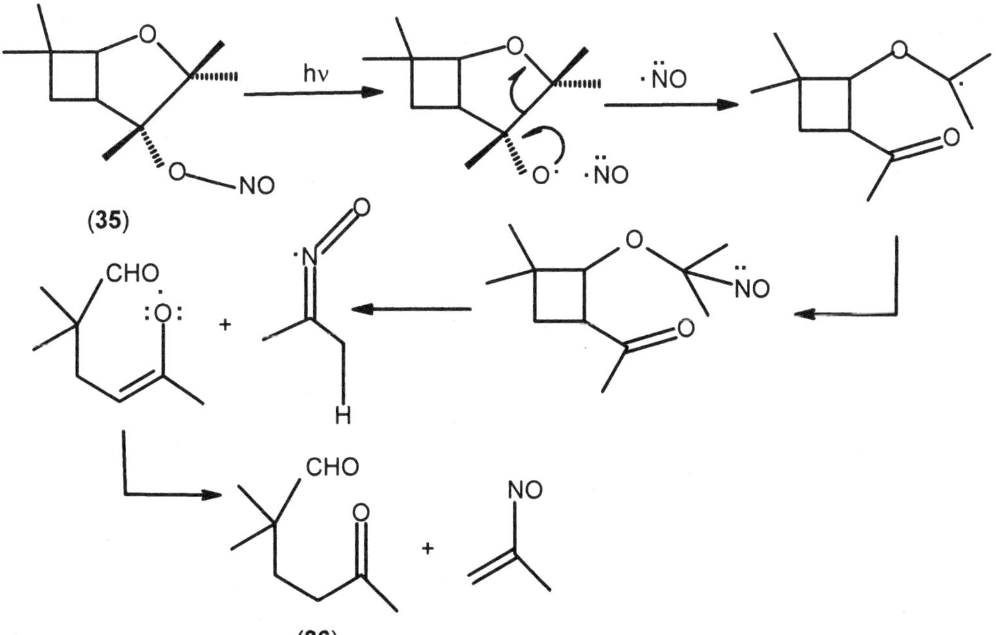

Abnormal Barton reaction: Cyclization via five membered transition state

It is the general observation that the *Barton reaction* requires a *six-membered transition state*. But the experimental observations by Rieke and Moore[22] revealed that the majority of the 4-pentenoxyl radicals cyclize *via* a *five-membered transition state* to give the thermodynamically less stable primary radicals. One possible explanation for this observation is that entropy factors are governing the mode of cyclization. The preference of a *five-membered transition state* over a *six-membered transition state* due to entropy factors has frequently been observed[23,24]. For instance, the following scheme (**42** to **43**) of Rieke and Moore may be cited.

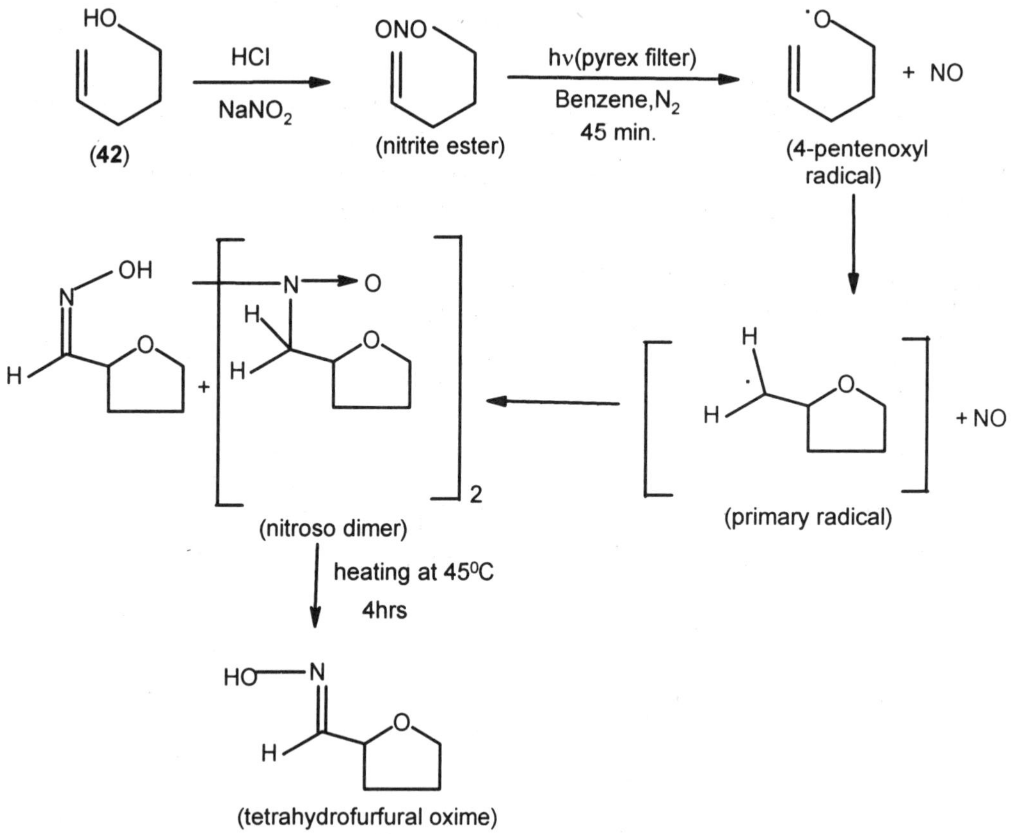

(tetrahydrofurfural oxime)
(**43**)

References

1. Gray, P. and Williams, A. (1959), *Chem. Rev.*, **59**, 239. Barton, D. H. R., Beaton, J. M., Geller, L. E. and Pechet, M. M. (1960), *J. Am. Chem. Soc.*, **82**, 2640.
2. Gray, P. and Williams, A. (1959), *Chem. Rev.*, *59*, 276.
3. Kabasakalian, P., Townley, E. R. and Yudis, M. D. (1962), *J. Am. Chem. Soc.*, **84**, 2716.
4. Kabasakalian, P. and Townley, E. R. (1962), *J. Am. Chem. Soc.*, **84**, 2724.
5. Burke, S. D., Silks III, L. A. and Strickland, S. M. S. (1968), *Tetrahedron Lett.*, **29**(23), 2761.
6. Akhtar, M., Barton, D. H. R., Beaton, J. M. and Hortmann, A. G. (1963), *J. Am. Chem. Soc.*, **85**, 1512.
7. Barton, D. H. R. and Beation, J. M. (1962), *J. Am. Chem. Soc.*, **84**, 199.
8. Hesse, R. H. and Pechet, M. M. (1965), *J. Org. Chem.*, **30**, 1723.
9. Barton, D. H. R. and Beaton, J. M. (1961), *J. Am. Chem. Soc.*, **83**, 750.
10. Nussbaum, A. L. and Robinson, C. H. (1962), *Tetrahedron*, **17**, 85.
11. Barton, D. H. R. and Beaton, J. M. (1960), *J. Am. Chem. Soc.*, **82**, 2641.
12. Bosworth, N. and Magnus, P. D. (1972), *J. Chem. Soc., Perkin I*, 943.
13. Hobbs, P. D. and Magnus, P. D. (1976), *J. Chem. Soc.*, **98**, 4594.
14. Baldwin, S. W. and Blomquist, H. R. (1982), *J. Am. Chem. Soc.*, **104**, 4990.
15. Beckwith, A. L. J. and Gold, K. U. in "*Rearrangements in Ground and Excited States*", deMayo, P. (Editor) (1980), Academic Press: New York, vol. 1, P. 260.
16. Hesse, R. H. (1969), *Adv. Free Radical Chem.,* **3**, 83.
17. Nickon, A., Mcguire, F. J., Mahajan, J. R., Umezawa, B. and Narang, S. A. (1964), *J. Am. Chem. Soc.*, **86**, 1437.
18. Nickson, A., Ferguson, R., Bosch, A. and Iwadare, T. (1947), *J. Am. Chem. Soc.*, **99**, 4518.
19. Nussbaum, A. L., Yuan, E. P., Robinson, C. H., Mitchell, A., Oliveto, E. P., Beaton, J. M. and barton, D. H. R. (1962), *J. Org. Chem.*, **27**, 20.
20. Kabasakalian, P. and Townley, E. R. (1962), *J. Org. Chem.*, **27**, 2918.
21. Depuy, C. H., Joney, H. L. and Gibson, D. H. (1972), *J. Am. Chem. Soc.*, **94**, 3924.
22. Rieke, R. D. and Moore, N. A. (1969), *Tetrahedron* Lett., No. 25, 2035.
23. Freudlich and Kroeplin, Z. (1926), *Phys. Chem.*, **122**, 39.
24. Capon, B. (1964), *Quarterly Reviews*, **18**, 45 and references therein.

- McBay, H. C. and Tucker, O. (1954), *J. Org. Chem.*, **19**, 869.
- Davis, W. (Jr) and Noyes, W.A. (1947), *J. Am. Chem. Soc.*, **69**, 2153.
- Yang, N. C. and Yang, D. D. H. (1958), *J. Am. Chem. Soc.*, **80**, 2913.

- Nussbaum, A. L., Carlon, F. E., Oliveto, E. P., Townley, E., Kabasakalian, P. and Barton, D.H.R. (1960), *J. Am. Chem. Soc.*, **82**, 2973.
- Kabasakalian, P., Townley, E.R. and Yudis, M. D. (1962), *J. Am. Chem. Soc.*, **84**, 2716.
- Bernstein, S. and Lenhard, R. H. (1955), *J. Am. Chem. Soc.*, **77**, 2331.
- Zaffaroni, A., Ringold, H. J., Rosenkranz, G., Sondheimer, F., Thomas, G. H. and Djerassi, C. (1958), *J. Am. Chem. Soc.*, **80**, 6110.
- Suginome, H., Sato, N. and Masamune, T. (1971), *Tetrahedron*, **27**(20), 4863.
- Burke, S. D., Silks III, L. A. and Strickland, S. M. S. (1988), *Tetrahedron Lett.*, **29**, 2761.
- Green, M. M., Boyle, B. A., Vairamani, M., Mukhopadhyay, T., Saunders Jr., W. H., Bowen, P., Allinger, N. L. (1986), *J. Am. Chem. Soc.*, **108**, 2381.
- Barton, D. H. R. (1990), *Aldrichimica Acta*, **23**, 3. [Review]
- Majetich, G. and Wheless, K. (1995), *Tetrahedron*, **51**, 7095
- Herzog, A., Knobler, C. B. and Hawthorne, M. F. (1998), *Angew. Chem. Int. Ed. Engl.*, **37**, 1552.

3

Beckmann Rearrangement

Introduction

The conversion of ketoximes into *N*-substituted amides in acid catalyzed reaction is widely known as *Beckmann rearrangement*. Although certain aldoximes have been found to undergo this reaction, it is not generally a means of converting aldoximes to unsubstituted amides. It may not be out of point to mention here that in most of the cases, esters of ketoximes with many acids, organic and inorganic, undergo the very facile *Beckmann rearrangement*. This rearrangement reaction is an example of 1,2-shift in which migration origin is carbon and the migration terminus is nitrogen.

$$RR'C=NOH \xrightarrow{H^+} RCONHR' \text{ or } R'CONHR$$

The reaction is catalyzed by a wide variety of acidic reagents, *e.g.* conc. H_2SO_4, HCOOH, BF_3, liq. SO_2, $SOCl_2$, $PhSO_2Cl$, SO_3, PCl_5, silica gel, P_2O_5-methanesulphonic acid, polyphosphoric acid, etc. Polyphosphoric acid (PPA) is regarded as such a catalyst and solvent system that suppresses abnormal rearrangements[1] and also resists ready isomerization of oximes[2].

General scheme

$$\underset{R'}{\overset{R''}{>}}C=N^{\cdots OH} \xrightarrow[2. H_2O]{1. PCl_5/\text{ether}} R-\underset{\|}{\overset{O}{C}}-NHR' \quad (N\text{-substituted amide})$$

$$\underset{R'}{\overset{R''}{>}}C=N_{\searrow OH} \xrightarrow[2. H_2O]{1. PCl_5/\text{ether}} R'-\underset{\|}{\overset{O}{C}}-NHR \quad (N\text{-substituted amide})$$

(R/R' *anti* to oxime OH migrates)

The scope of the reaction is quite broad. The migrating group may be alkyl, aryl, alicyclyl or may also be hydrogen in some cases. The migrating aptitude does not depend on the nature of the group, but on its stereochemical arrangement in the oxime; the migrating group generally attains the *anti*-orientation with respect to the leaving group in the oxime molecule.

Examples

(a) 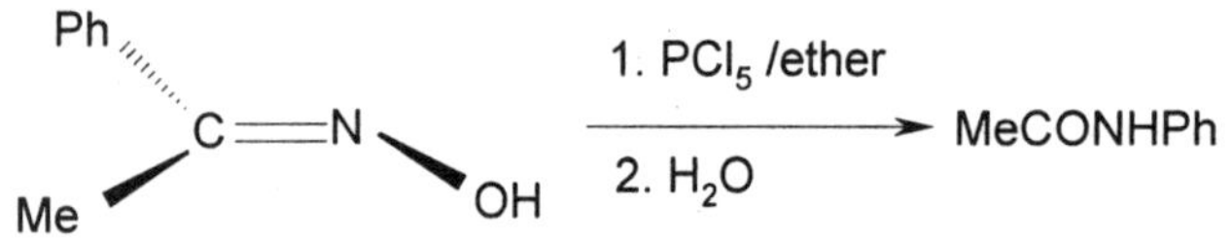 [Structure: Ph and Me on C=N-OH, Ph anti to OH] → MeCONHPh
   Reagents: 1. PCl$_5$/ether 2. H$_2$O
   (Ph *anti* to the oxime OH)

(b) 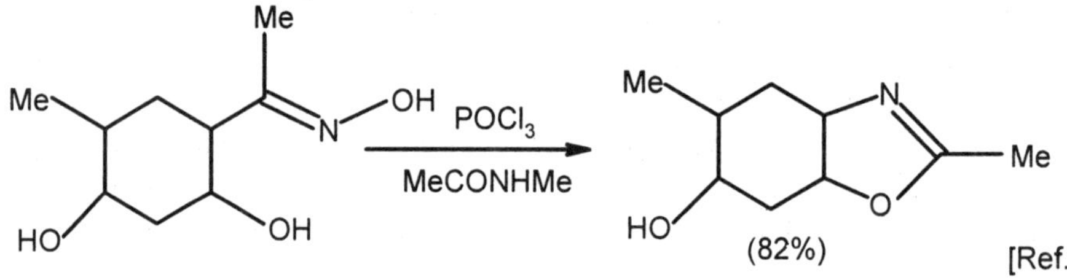 [Cyclohexane structure with oxime] → [bicyclic oxazoline product] (82%) [Ref. 3]
   Reagents: POCl$_3$, MeCONHMe

(c) 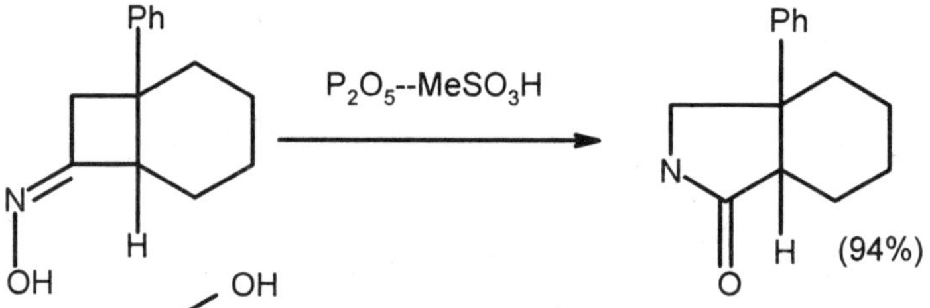 [Bicyclic oxime structure] → [lactam product] (94%) [Ref. 4]
   Reagents: P$_2$O$_5$–MeSO$_3$H

(d) 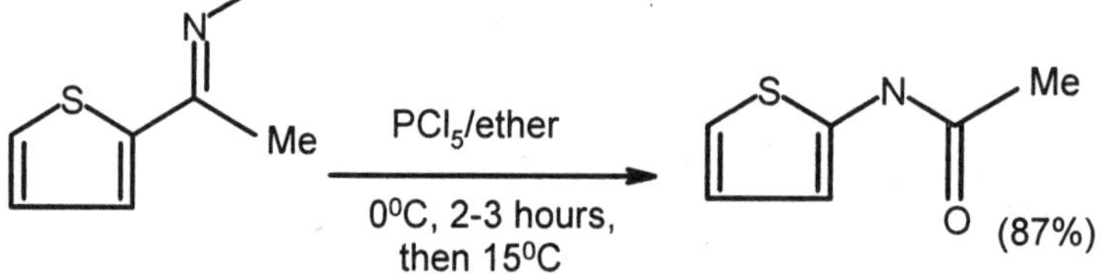 [Thiophene oxime structure] → [thiophene amide product] (87%) [Ref. 5]
   Reagents: PCl$_5$/ether, 0°C, 2-3 hours, then 15°C

(e) 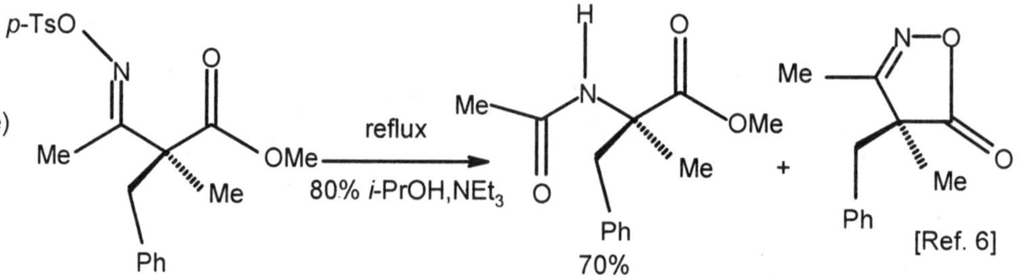 [p-TsO oxime structure] → [amide product] 70% + [isoxazoline product] [Ref. 6]
   Reagents: reflux, 80% *i*-PrOH, NEt$_3$

Discussion on mechanistic path

The mechanism of *Beckmann rearrangement* is a subject of much discussion. Different workers at different times proposed different mechanisms for this rearrangement reaction. It was proposed that this rearrangement reaction involves a bridged-ion. This proposal was suggested by Kenyon *et al*. (1946) who observed that when (+)-α-phenylethylmethylketoxime is treated with sulphuric acid, the product obtained [*N*-(α-phenylethylmethy)acetamide] is almost 100 percent optically pure. Thus, the migrating group never detaches itself during the rearrangement. This retention of optical activity supports the formation of a bridged-ion during the migration. Another evidence in favour of this bridged-ion intermediate formation is that when the migrating group is aryl and contains an electron-releasing functionality (e.g. Me) in the *para*-position, the rate of the rearrangement is accelerated. Thus, at least for migrating aryl groups, the oxime substrates proceed through bridged-ion intermediates.

In 1964, Grob *et al*. suggested that the rearrangement proceeds *via* nitrilium salt intermediate on the basis of their studies on the *Beckmann rearrangements* of various tosylates. The existence of nitrilium ion was detected by infrared spectroscopy. According to them, the mechanism of this rearrangement follows the following path (below). This view of Grob *et al*. was latter supported by Schofield *et al*. (1970) who studied the *Beckmann rearrangement* of *ortho*-substituted acetophenone oximes in presence of sulphuric acid and detected the nitrilium ion (by means of infrared spectroscopy) as an intermediate.

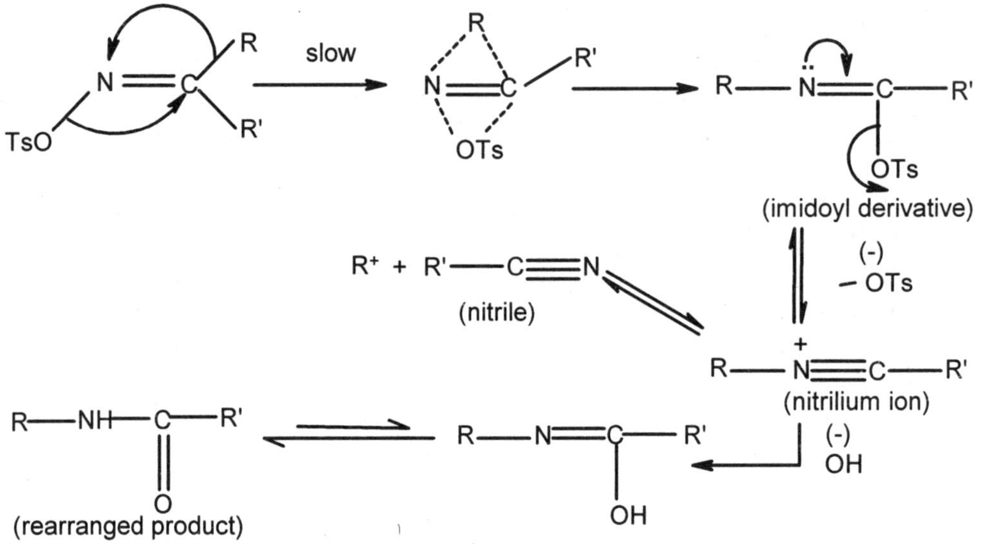

Hill *et al.* have also proposed an *intermolecular mechanism* in 1962. They reported an example in which the migrating group attained the inverted configuration in the rearranged product. The authors studied the rearrangement of 9-acetyl-*cis*-decalin oxime (**1**) and suggested the following mechanism.

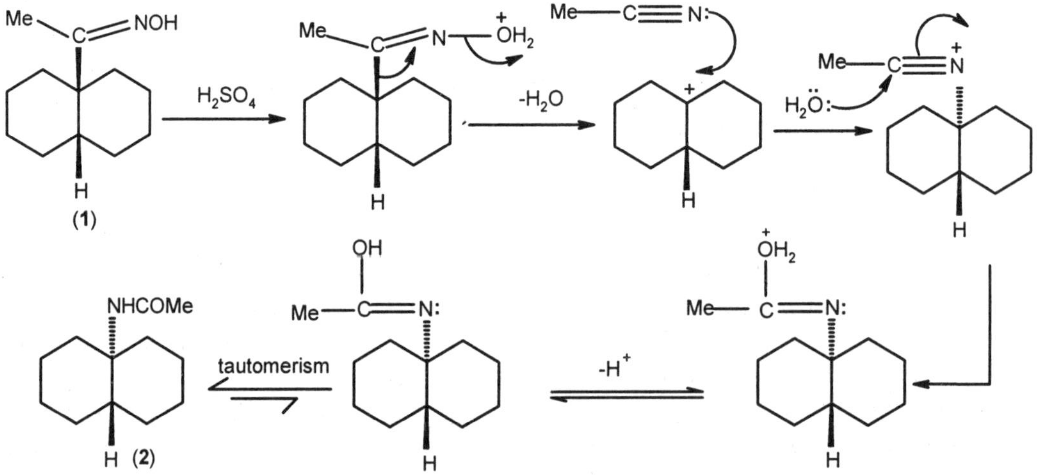

The authors identified methyl cyanide as an intermediate product and also showed that methyl cyanide and *cis*-decalol in sulphuric acid yielded the compound (2). The *intermolecular* mechanism has been further evidenced from the work of Conley (1963) who carried out the *Beckmann rearrangement* of a mixture of phenyl-2-phenylisopropylketoxime and *tert*-butylmethylketoxime in polyphosphoric acid and isolated the product as a mixture of four secondary amides. The formation of crossed-product from the mixture of oximes reveals a *fragmentation-recombination* mechanism.

$$Ph(Me)_2C\!\!-\!\!C(=N\!\!-\!\!OH)\!\!-\!\!Ph \;+\; Me_3C\!\!-\!\!C(=N\!\!-\!\!OH)\!\!-\!\!Me \xrightarrow{PPA} PhCONHCMe_2Ph + MeCONHCMe_3 + PhCONHCMe_3 + MeCONHCMe_2Ph$$

From the studies of various groups of workers, the most general mechanistic pathway that emerges is as discussed below:

Step I: Formation of a better leaving group.

$$R\!\!-\!\!C(=N\!\!-\!\!OH)\!\!-\!\!R' \xrightarrow{H_2SO_4} R\!\!-\!\!C(=N\!\!-\!\!\overset{+}{O}H_2)\!\!-\!\!R' \text{ (protonation)}$$

$$\xrightarrow[\text{ether}]{PCl_5} R\!\!-\!\!C(=N\!\!-\!\!OPCl_4)\!\!-\!\!R'$$

$$\xrightarrow{PhSO_2Cl} R\!\!-\!\!C(=N\!\!-\!\!OSO_2Ph)\!\!-\!\!R'$$

] esterification

Step II: Migration of the *anti* group (with respect to the leaving group) and loss of the leaving group occur simultaneously, *i.e.* it is believed that the breakage of C-C bond and the formation of new C-N bond take place *synchronously*. This step is the ionization step and is the rate-determining step. It has been observed that the rate of the reaction of various oxime esters depend on the strength of the esterifying acids which is found to be in the order — $PhSO_3H > ClCH_2COOH > CH_3COOH$. The rate of the reaction also increases as the solvent polarity increases.

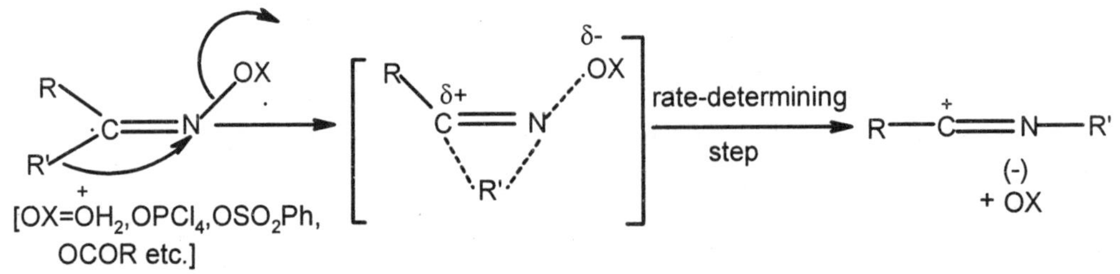

$$\underset{\substack{[OX=OH_2, OPCl_4, OSO_2Ph,\\ OCOR\ etc.]}}{\overset{R}{\underset{R'}{>}}C=N\overset{OX}{\curvearrowleft}} \longrightarrow \left[\overset{R}{\underset{R'}{>}}\overset{\delta+}{C}\cdots\cdots\overset{\delta-}{\underset{}{N}}\overset{OX}{}\right] \xrightarrow{\text{rate-determining step}} R-\overset{+}{C}=N-R'\quad (-)+OX$$

In this step the newly formed carbocation may take up the leaving group.

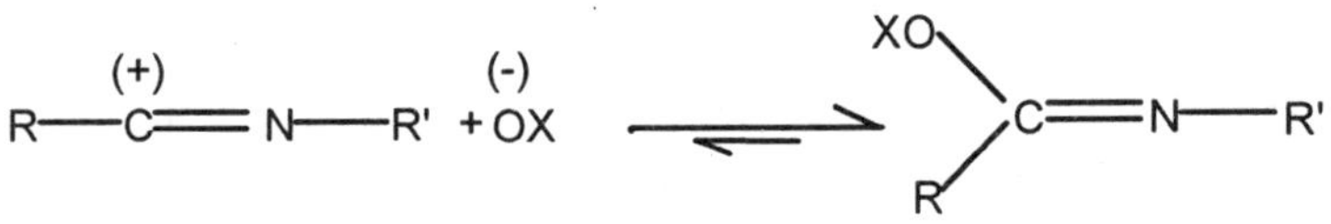

$$R-\overset{(+)}{C}=N-R'+\overset{(-)}{OX} \rightleftharpoons \underset{R'}{\overset{XO}{>}}C=N-R'$$

Step III: Addition of water molecule to the carbocation to form the enol of the amide, which rapidly tautomerizes to yield an *N*-substituted amide.

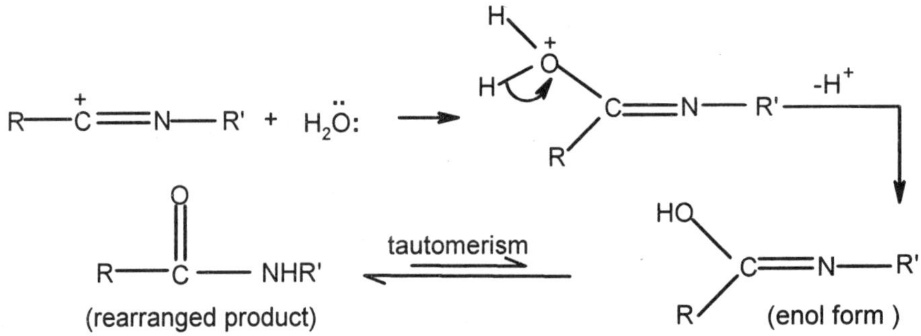

$$R-\overset{+}{C}=N-R' + H_2\ddot{O}: \longrightarrow \underset{R'}{\overset{H\overset{+}{O}H}{>}}C=N-R' \xrightarrow{-H^+}$$

$$\underset{\text{(rearranged product)}}{R-\overset{O}{\overset{\|}{C}}-NHR'} \xleftarrow{\text{tautomerism}} \underset{\text{(enol form)}}{\overset{HO}{\underset{R'}{>}}C=N-R'}$$

In support of the above mechanistic pathway of this reaction, the following experimental evidences are being cited:

(a) Evidence for Step-I

Kuhara and latter Chapman established that the rearrangement does not only take place in the oxime itself, but in its acetyl derivatives also. Subsequently, a number of intermediates of the general formula RR'-C=NOX (X=PCl$_4$, SO$_2$Ph, COCH$_2$Cl, etc.) have been prepared and found to yield expected amides in the neutral medium in absence of catalysts through the rearrangement reactions. Thus, the species formed in the Step-I are the probable intermediates of the *Beckmann rearrangement* reaction.

(b) *Evidence for Step-II (evidence for anti-group migration)*

The exclusive migration of *anti* group has been confirmed in a number of cases. The well-known example is the *Beckmann transformation* of 2-bromo-5-nitro-acetophenone oxime (Meisenheimer *et al.*, 1925). It can exist in two forms (**3** & **4**).

Configuration (**3**) was established from the fact of its easy cyclization to the nitro-substituted methyl benzisoxazol (**5**) with base even in the cold, showing that the aromatic moiety lies on the same side of the C=N linkage as the OH group. On *Beckmann rearrangement*, this isomer furnishes *N*-methyl derivative of 2-bromo-5-nitrobenzamide (**6**), formed due to the migration of Me group that is *anti* with respect to oxime OH group. On the other hand, isomer (**4**) — whose configuration is ascertained from the phenomenon that under the same condition (cold alkali treatment) it does not undergo cyclization — yields *N*-(2-bromo-5-nitrophenyl) acetamide (**7**) formed due to the migration of aromatic moiety that is *anti* to the leaving functionality. Thus, it can be concluded that the *anti* group migrates in a *Beckmann rearrangement*.

Another method used for examining which group migrates is the method of dipole measurements. In 1931, Sutton *et al.* measured the dipole moments of two isomeric *N*-methyl ethers of *p*-nitrobenzophenone oxime and obtained the values as given. The values clearly indicate the configurations (**8** & **9**).

[Structures: nitrone (8) μ=6.60D → Beckmann rearrangement → p-nitrobenzanilide (10); nitrone (9) μ=1.09D → Beckmann rearrangement → benzo-p-nitranilide (11)]

The oxime corresponding to the nitrone (**8**) with higher dipole moment on undergoing the *Beckmann rearrangement* yields *p*-nitrobenzanolide (**10**), while the other isomer (**9**) gives benzo-*p*-nitranilide (**11**).

(c) *Evidence for synchronous migration of anti group and loss of leaving functionality*

Mixtures of two ketoximes when subjected to *Beckmann rearrangement*, no cross products are formed. For example, on treatment of a solution of benzophenone oxime and acetophenone oxime with acid furnishes *N*-phenylbenzamide and *N*-methylacetamide only, neither *N*-methylbenzamide nor *N*-phenylacetamide has been isolated. This phenomenon establishes that the migrating group never detaches itself from the substrate — its migration and loss of the leaving functionality occur in a *synchronous* manner.

[Scheme: Ph₂C=N-OH + Me₂C=N-OH → 1. H₂SO₄ 2. H₂O → PhCONHPh + MeCONHMe]

(d) *Evidence for the formation of ionic intermediate*

Beckmann rearrangement does not proceed by an *intramolecular* exchange of the leaving group and the migrating group between N and C atoms. This has been demonstrated by carrying out the rearrangement of benzophenone oxime in an aqueous medium containing $H_2^{18}O$. The product, benzanilide isolated therefrom contains PhCONHPh along with PhC^{18}ONHPh. So, the rearrangement must involve loss of the oxime OH group and subsequent re-introduction of oxygen from a water molecule. The main function of the acidic catalyst is to convert the oximino-hydroxyl group into a better leaving functionality by protonation, esterification, etc. The resulting oxime derivative subsequently undergoes ionisation to form a carbonium ion intermediate (e.g. $RC^+=NR'$) and a leaving anion in a *synchronous* fashion. Formation of such type of carbonium ion intermediates has been detected by NMR and UV spectroscopy[7]. This ionisation step is possibly the rate-controlling step. The contention is verified from the fact that the rate of reaction increases as the solvent polarity increases and also from the observation that the rate of the rearrangement is enhanced with increasing stability of the leaving group anions.

Stereochemistry

The *Beckmann rearrangement* is highly stereoselective concerning the *anti*-migration of group (with respect to the leaving functionality) from carbon to nitrogen atom in the oxime. Moreover, the migration of the group occurs with retention of its configuration during the rearrangement reaction. Kenyon et al. (1946) found that when (+)-α-phenylethylmethylketoxime is subjected to *Beckmann rearrangement* with sulphuric acid, the product obtained is almost 100 percent optically pure (retention).

Limitations

Certain abnormalities in some cases have been observed during carrying out the *Beckmann rearrangement* reaction. For instances ——

(a) A side reaction may also take place as noticed in case of many substrates leading to the formation of nitriles. This happens due to the fragmentation of carbocation intermediates as formed during the course of the *Beckmann rearrangement* reactions. This phenomenon is often called *"abnormal"* or *"second-order Beckmann rearrangement"*[8]. Fragmentation is a side reaction even with ordinary ketoximes[9] and the cases, where a particularly stable carbocation can be cleaved from the parent carbocation intermediate, may be the main reaction[10].

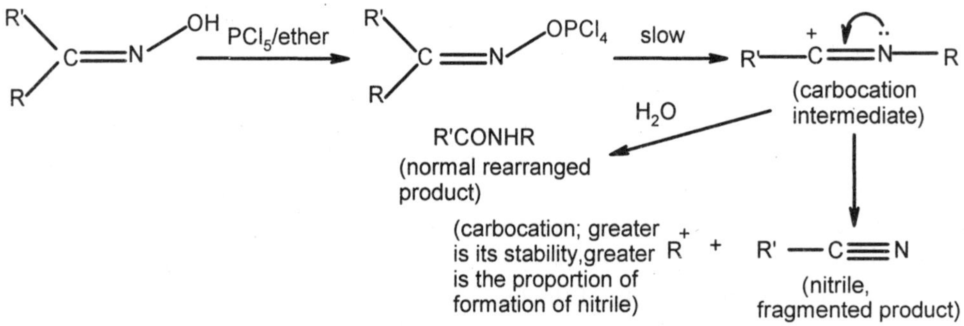

Beckmann fragmentation is a likely competing process if one of the centres adjacent to the oxime is secondary, tertiary or quaternary. For examples —

(norcamphoroxime) (unsaturated nitrile)

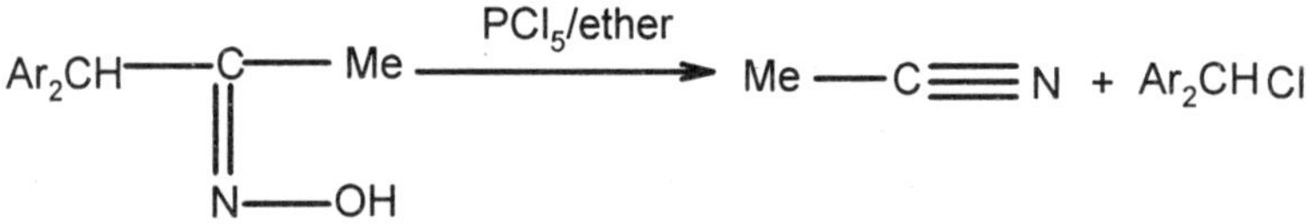

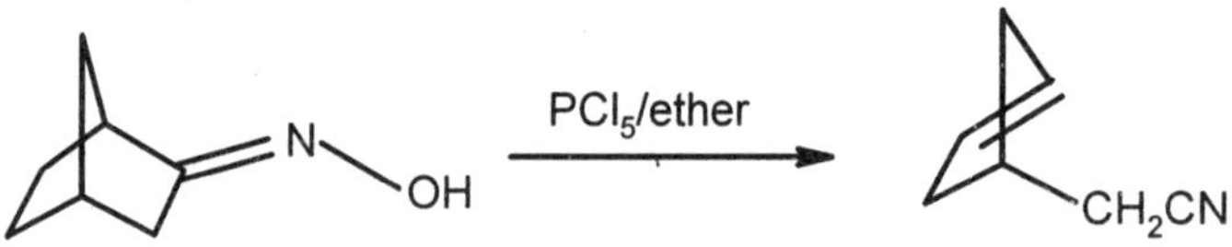

Lansbury *et al.*[11] carried out the similar type of reaction. They prepared Balduilin stereoselectively using *Beckmann fragmentation* technique. Oxime derivative of 9-acetoxy-camphor (**12**) undergoes *Beckmann fragmentation* to provide nitrile acetate intermediate (**13**), which takes part in the stereoselective synthesis of Balduilin.

(**12**) (**13**)

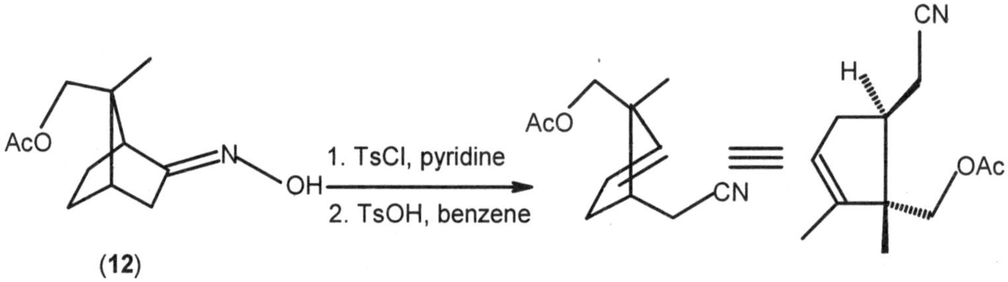

(b) It was also shown by Lansbury and Mancuso[12] that some *Beckmann rearrangement* reactions are authentically *non-stereospecific*. They carried out the *Beckmann transformations* of the oximes of the following ketones with PPA at 110-120^{0}C and noticed the migration aptitudes as:

(**14**) (**15**) (**16**)

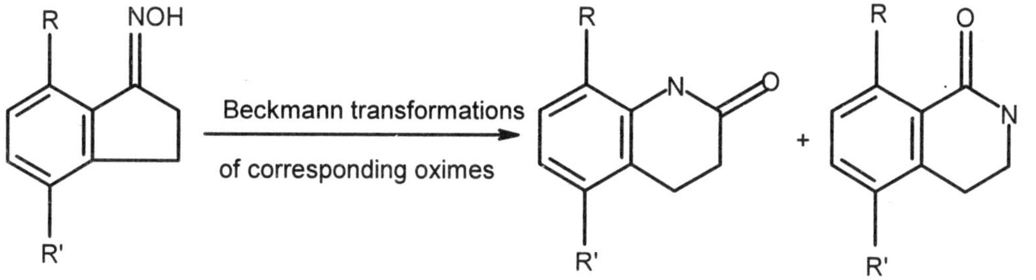

14 Substrate	**15** %Aryl migration	**16** %Alkyl migration
14a: R=R'=H	90	10
14b: R=R'=Me	34	66
14c: R=Bu^t, R'=Br	19	81

As the bulkiness of the substituents (R & R′) increases, alkyl-migrated products become more predominant. In this connection it may pointed out that the oximes of ketones (**14**b-**14**c) prefer to attain the *anti*-aryl orientation because in *syn*-aryl configuration, the oximes would be too sterically compressed to exit relative to *anti*-isomer.

(*syn*-aryl orientation) (*anti*-aryl orientation)

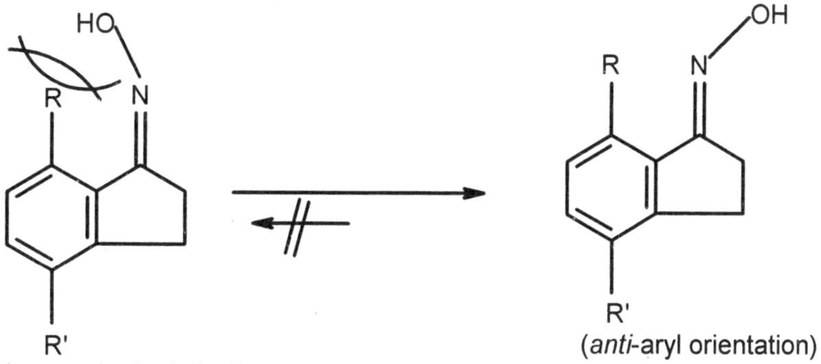

Hence, the *Beckmann rearranged* products as obtained indicate the occurrence of predominant *syn*- migration (*i.e.* alkyl migration) instead of expected *trans*-migration.

Thus, the generalization of *trans*-stereochemistry in *Beckmann reaction* must be used with caution in predicting rearrangement products and assigning oxime configurations.

Extensions

A lot of extensions have been made to *Beckmann rearrangement reaction*. Suginome et al.[13-16] carried out *photo-induced Beckmann transformations* of a number of acyclic ketone oximes, particularly of steroidal skeleton. They noticed that in each case the chirality of migrating groups is retained. Hattori et al.[17,18] studied the *Beckmann rearrangement* of oxime sulphonates with Grignard reagents which provides an efficient and general entry to α-alkyl and α,α- dialkylamines in good yields. Besides these, a number of other significant developments have also been done. Just one or two instances are cited:

(a) Chattopadhyaya and Rama Rao[19] extended the *Beckmann rearrangement* by introducing silica gel-induced conversion of aldoximes to corresponding amides. This method is found to be much better than other methods for simplicity and high yields. Unlike acid-catalyzed isomerization, the nitrile is not an intermediate in this isomerization. It has been observed that the reaction occurs satisfactorily even in compounds having *o*-hydroxyl group or double bond conjugated with the oxime without causing interference in the smooth conversion to the amide. Electron-donating or electron-withdrawing substituents on the aromatic aldoximes do not affect the yields, and in all cases as they studied the conversions are much better when silica gel is employed as catalyst compared with nickel acetate. *This reaction is only useful for converting aldoximes to the corresponding amides and not applicable for the conversion of ketoximes.* Thus, acetophenone oxime, when subjected to undergo the silica gel induced conversion for the same length of time, has been recovered quantitatively. Similarly, actaldoxime has yielded exclusively acetamide and no traces of *N*-methylformide.

The mechanism of this isomerization is not clearly understood. The possibility of a nitrile intermediate, as in all acid-catalyzed isomerizations[20-22], is ruled out because at no stages of the reaction, the presence of nitrile could be detected. The authors suggested the possibility of *intramolecular* cyclization through a transitional intermediate **B** that might have arisen from a nitrone-type tautomer, A which possibly becomes dominant in presence of silica gel catalyst. They also supposed that the possibility of an intramolecular cyclisation involving six-membered cyclic intermediate can't be ruled out.

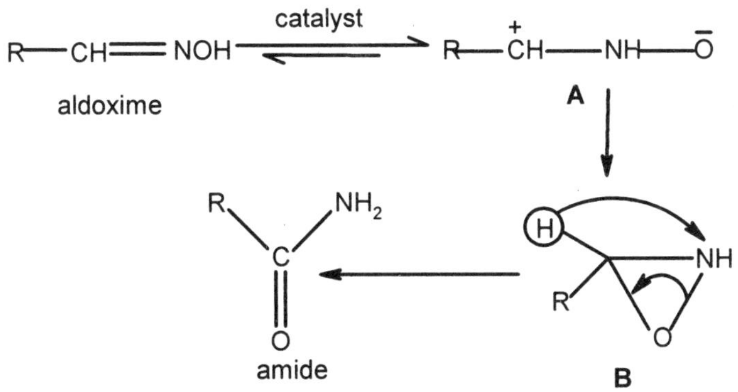

Beckmann Rearrangement

e.g.

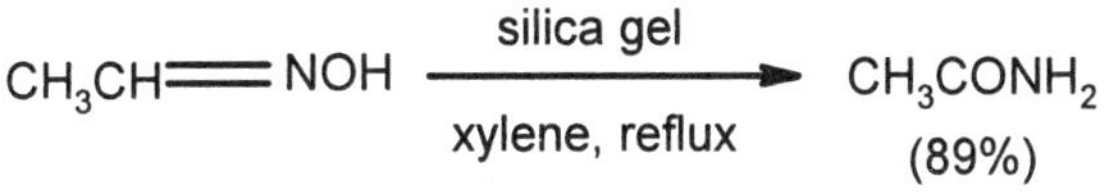

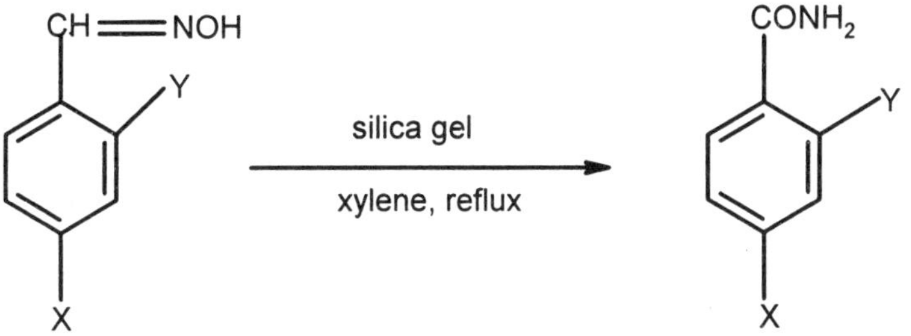

18a: X = OH, Y = H; yield 84%
18b: X = OMe, Y = H; yield 81%
18c: X = Y = OH;   yield 84%

(b) R.T. Conley and L.J. Frainier[23] carried out *Beckmann rearrangement* of a series of 2,2-disubstituted-1- indanone, tetralone and benzosuberone oximes using PPA as the catalyst and solvent.

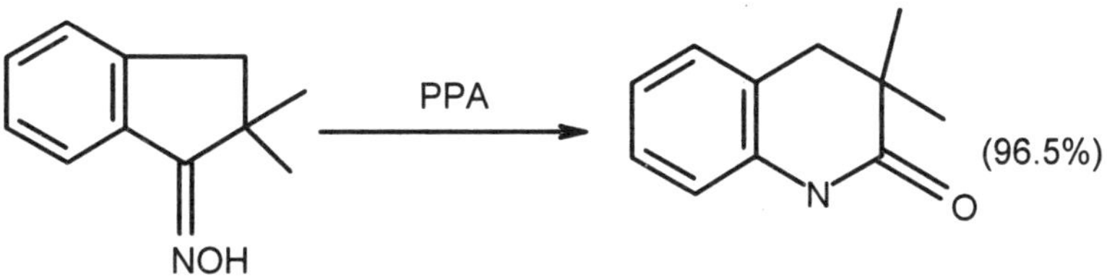

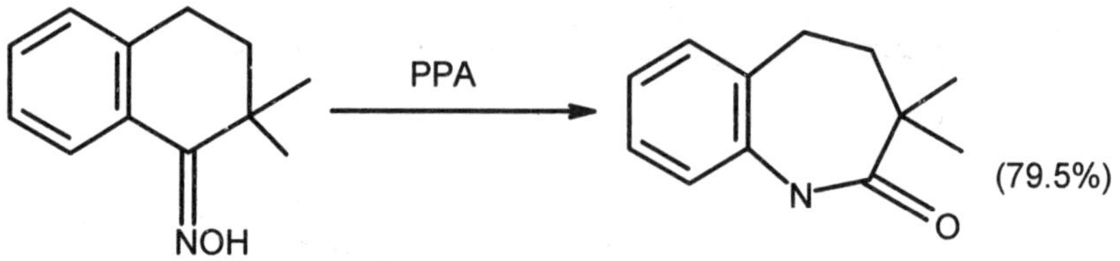

This dissociation by protonation or scission of the N — O bond of the oxime-polyphosphate ester intermediate would give rise to a non-stereoselective positively charged nitrogen ion which would be stabilized through bridged ion and hence group migration would be controlled by migratory aptitude due primarily to transition state stabilization. The authors proposed an alternate pathway for the rearrangement of such type of substrates as follows:

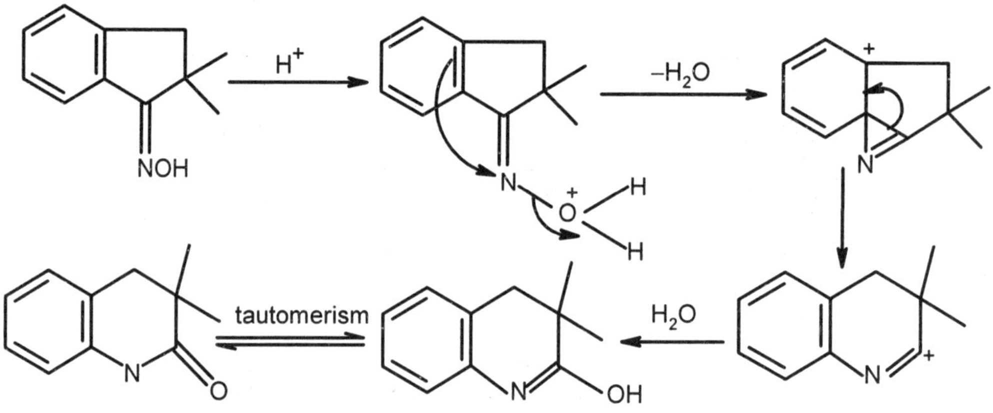

Applications

Beckmann rearrangements have various applications along with organic synthesis. Some of them are illustrated here.

(a) *Configuration of ketoxime:*

Configurations of ketoximes can be assigned by studying their *Beckmann rearrangements*. In general, *anti*-group with respect to the leaving functionality in the oxime derivative migrates during rearrangement. The reaction product (amide) on hydrolysis yields amine —— structural analysis of which clearly suggests the configuration of the parent oxime.

$$RR'C=NOH \xrightarrow{H^+} R'CONHR \xrightarrow{hydrolysis} R'COOH + RNH_2$$

Formation of RNH_2 indicates the migration of the 'R' group to the nitrogen atom. The groups 'R' and 'OH' are, therefore, *anti* to each other. Thus, the structure of the oxime is:

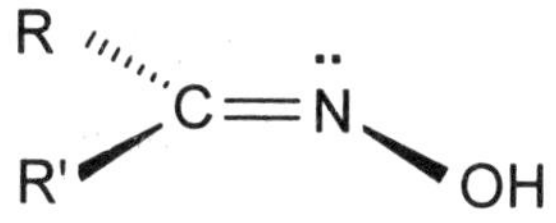

R─C(R')=N─OH (with R and OH anti)

(b) *Synthesis of heterocyclic rings:*

An interesting application of this reaction is in the synthesis of heterocyclic rings; *e.g.*

(i) Bamberger and Goldschmidt (1894) prepared isoquinoline from cinnamaldehyde oxime by treating it with phosphorous pentaoxide.

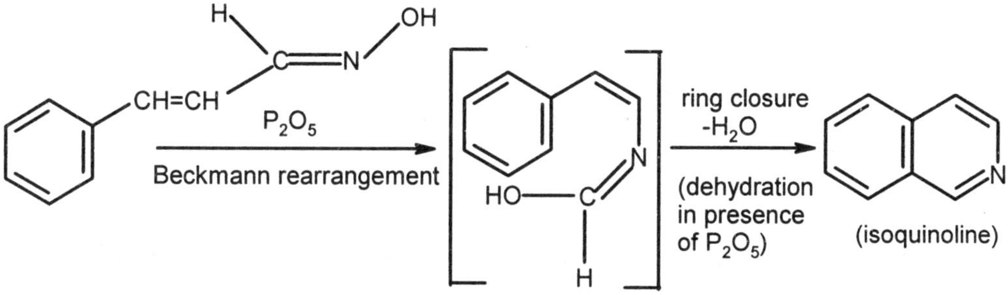

PhCH=CH−CH=N−OH $\xrightarrow[\text{Beckmann rearrangement}]{P_2O_5}$ [intermediate] $\xrightarrow[-H_2O]{\text{ring closure}}$ isoquinoline
(dehydration in presence of P_2O_5)

(ii) Cyclopentanone oxime on *Beckmann rearrangement* forms 2-piperidone.

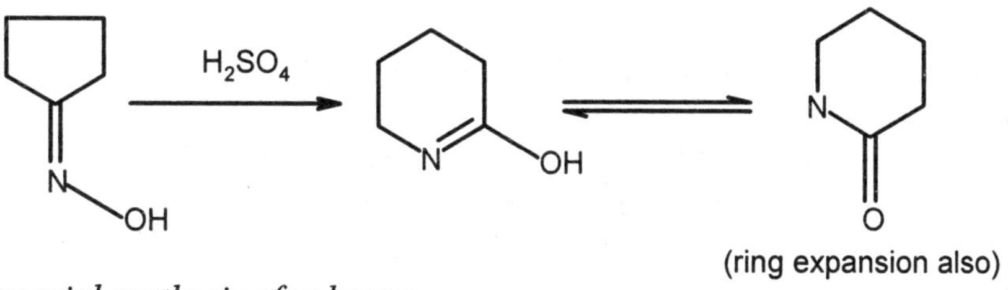

cyclopentanone oxime $\xrightarrow{H_2SO_4}$ 2-hydroxy tetrahydropyridine ⇌ 2-piperidone
(ring expansion also)

(c) *Commercial synthesis of polymer:*

A product of considerable industrial importance is Nylon-6, which is a worthful textile polymer. It is synthesized in large amount from ε-caprolactum obtained by the

Beckmann rearrangement of cyclohexanone oxime. This oxime can be prepared from phenol as shown:

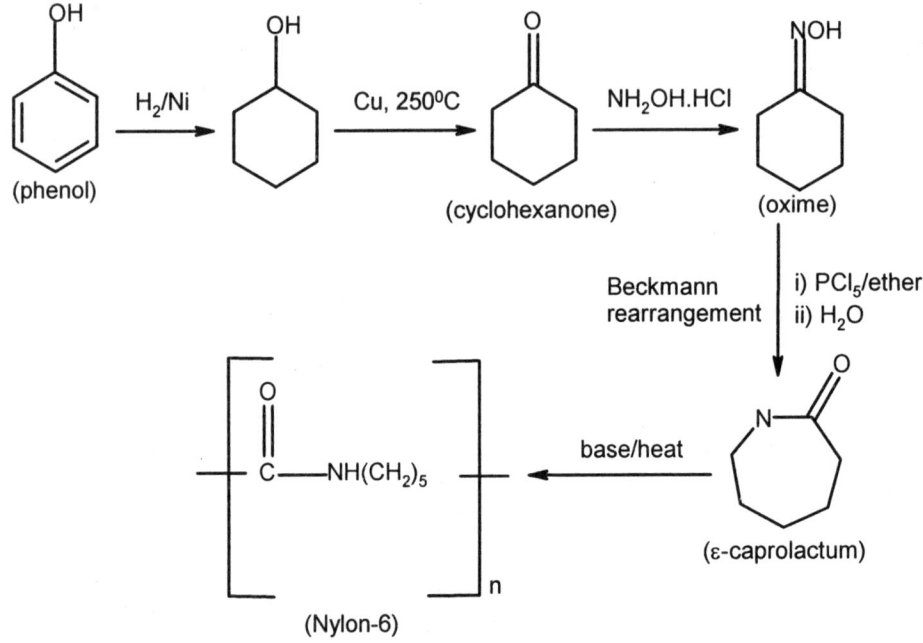

References

1. Horning, E. C., Stromberg, V. L. and Lloyd, H. A. (1952), *J. Am. Chem. Soc.*, **74**, 5153.
2. Smith, N. H. P. (1961), *J. Chem. Soc.*, 4209.
3. Fujita, S., koyona, K. and Inagaki, Y. (1982), *Synthesis*, 68.
4. Jeffs, P. W., Molina, G., Cass, M. W. and Cortese, N. A. (1982), *J. Org. Chem.*, **47**, 3871.
5. Meth-Cohen, O. and Narina, B. (1980), *Synthesis*, 133.
6. Frutos, R. P. and Spero, D. M. (1998), *Tetrahedron Lett.*, **39**, 2475.
7. Gregory, Moodie and Schofield (1970), *J. Chem. Soc.* B, 338.
8. See the discussion in Ferris (1960), *J. Org. Chem.*, **25**, 12.
9. See, for example, Hill and Conley (1960), *J. Am. Chem. Soc.*, **82**, 645.
10. Hassner and Nash (1965), *Tetrahedron Lett.*, 525.
11. Lansbury, P. T., Mazur, D. J. and Springer, J. P. (1985), *J. Org. Chem.*, **50**(10), 1632.
12. Lansbury, P. T. and Mancuso, N. R. (1965), *Tetrahedron Lett.*, No.29, 2445.

13 Suginome, H. and Yagihashi, F.(1977), *J. Chem. Soc., Perkin I,* 2488.
14 Suginome, H. and Takahashi, T. L. (1975), *Bull. Chem. Soc., Japan,* **48**, 582; Suginome, H. and Takahashi, T. L. (1975), *Tetrahedron Lett.,* 5119.
15 Suginome, H. and Uchida, T. (1973), *Tetrahedron Letters,* 2293; *Bull. Chem. Soc., Japan* (1974), **47**, 687.
16 Suginome, H. (1976), *Kagakno Ryoiki,* **30**, 578.
17 Hattori, K., Matsumura, Y., Miyazaki, T., Manruoka, K. and Yamamoto, H. (1981), *J. Am. Chem. Soc.,* **103**, 7368.
18 Hattori, K., Manruoka, K. and Yamamoto, H. (1982), *Tetrahedron Lett.,* **23** (33), 3395.
19 Chattophadhyaya, J. B. and Rama Rao, A. V. (1974), *Tetrahedron,* **30**, 2899.
20 Hoffenberg, D. S. and Hauser, C. R. (1955), *J. Org. Chem.,* **20**, 1496.
21 Horning, E. C. and Stromberg, V. L. (1952), *J Am. Chem. Soc.,* **74**, 5151.
22 Dunsta, N. R. and Dymbond, T. S. (1894), *J. Chem. Soc.,* 106.
23 Conley, R. T. and Frainier, L. J. (1962), *J. Org. Chem.,* **27**, 384.

- Padwa, A. and Albrecht, F. (1974), *J. Am. Chem. Soc.,* **96**, 4849.
- Schmitz, E. and Ohme, R. (1963), *Angew. Chem. Internat. Edn.,* **2**, 157.
- Saski, T., Eguchi, S. and Toru, T. (1970), *Chem. Comm.,* 1239.
- Kusama, H., Yamashita, Y and Naradaka, K. (1995), *Bull. Chem. Soc. Japan,* **68**, 373.
- Cunningham, M., Lim, L.S.N. and Just, G (1971), *Canad. J. Chem.,* **49**, 2891.
- Taylor, T.T., Douek, M. and Just, G.(1966), *Tetrahedron Letters,* 4143.
- Kobayashi, Y. (1973), *Bull. Chem. Soc., Japan,* **46**, 3467.
- Just, G. and Cunningham, M. (1972), *Tetrahedron Letters,* 1151.
- Eaton, P.E., Carlson, G.R. and Lee, J.T. (1973), *J. Org. Chem.,* **38**(23), 4071.
- Chatterjea, J. N., Singh, K. R. R. P. (1982), *J. Indian Chem. Soc.,* **59**, 527.
- Gawley, R. E. (1988), *Org. React.,* **35**, 1. [Review]
- Catsoulacos, P. and Catsoulacos, D. (1993), *J. Heterocycl. Chem.,* **30**, 1.
- Anilkumar, R. and Chandrasekhar, S. (2000), *Tetrahedron Lett.,* **41**, 7235.
- Khodaei, M. M., Meybodi, F. A., Remai, N. and Salehi, P. (2001), *Synth. Commun.,* **31**, 2047.
- Sharghi, H. and Hosseini, M. (2002), *Synthesis,* 1057.
- Chandrasekhar, S. and Copalaiah, K. (2003), *Tetrahedron Lett.,* **44**, 715.

4

Claisen Rearrangement

Introduction

The *Claisen rearrangement*[1] is an example of pericyclic reactions, and belongs to the category of [3,3]-*sigmatropic rearrangements*. The *Claisen rearrangement* involves intramolecular thermal conversion of allyl aryl ethers to allylphenols. The allyl group in the substrate migrates from the ethereal oxygen to the ring carbon *ortho* to it. But when both the *ortho*-positions are blocked, migration occurs at the respective *para*-position.

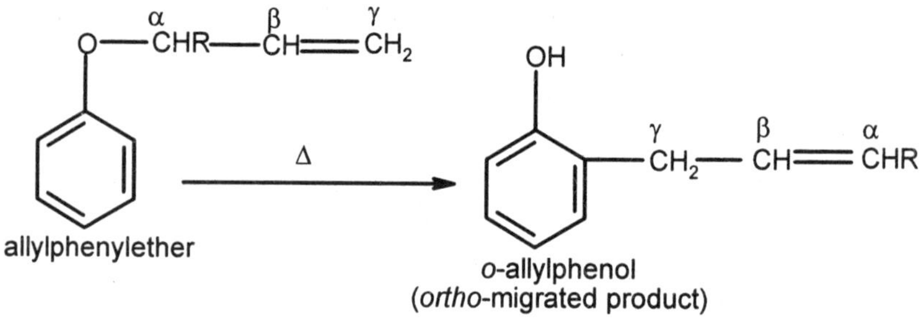

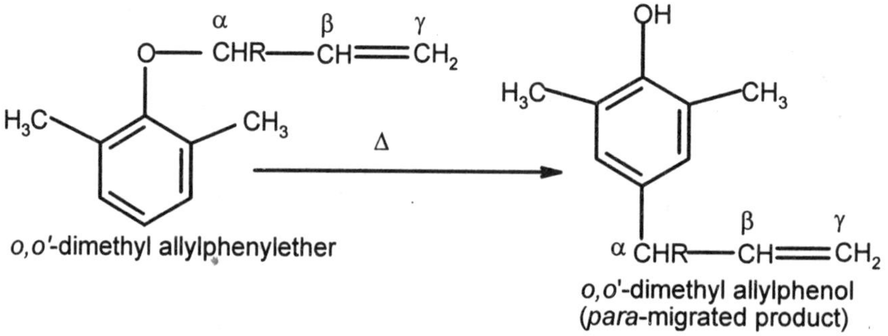

During *ortho*-migration the allyl group always undergoes an allylic shift — the carbon atom α to the ethereal oxygen atom in the substrate becomes γ to the ring in the product. However in *para*-migration, the allylic group is found exactly as it was in the starting ether. When the *para-* and both *ortho*-positions are filled, there is no reaction at all — migration to the *meta*-position has not been observed.

Mechanism

The *Claisen rearrangement* follows the first order kinetics. The rearrangement is strictly intramolecular and the mechanism is a concerted pericyclic [3,3]-sigmatropic rearrangement[2]. The reaction proceeds through a cyclic six-membered transition state in which the rupture of the oxygen-allyl bond is *synchronous* with the formation of a carbon–carbon bond at an *ortho*-position. A cyclohexadienone intermediate (1) is thus formed, which subsequently undergoes rapid tautomerism to yield the more stable aromatic compound (2).

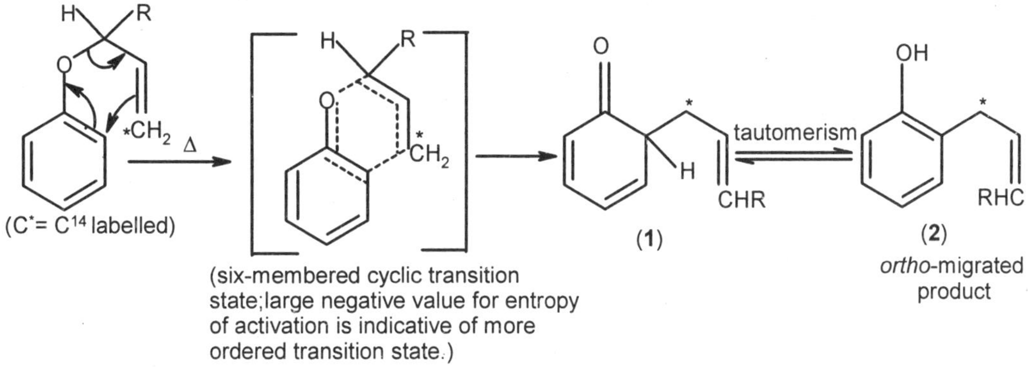

Supportive evidence for the migration of *o*-hydrogen in the dienone (1) may be drawn from the experimental work of Kistiakowsk *et al.* (1942).

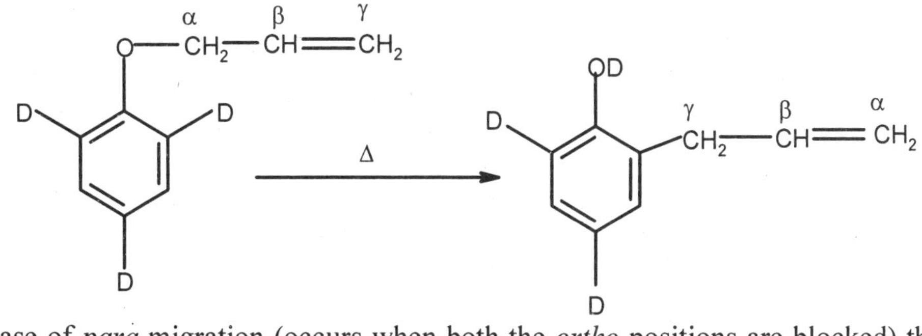

In case of *para*-migration (occurs when both the *ortho*-positions are blocked) the first step leads to the formation of *ortho*-substituted cyclohexadienone (3), but the absence of hydrogen atom at the *ortho*-position prevents aromatization. It undergoes further rearrangement involving migration of the allyl group, again through cyclic six-membered transition state (**4**) yielding (**5**) that finally tautomerizes to the *para*-substituted product (**6**).

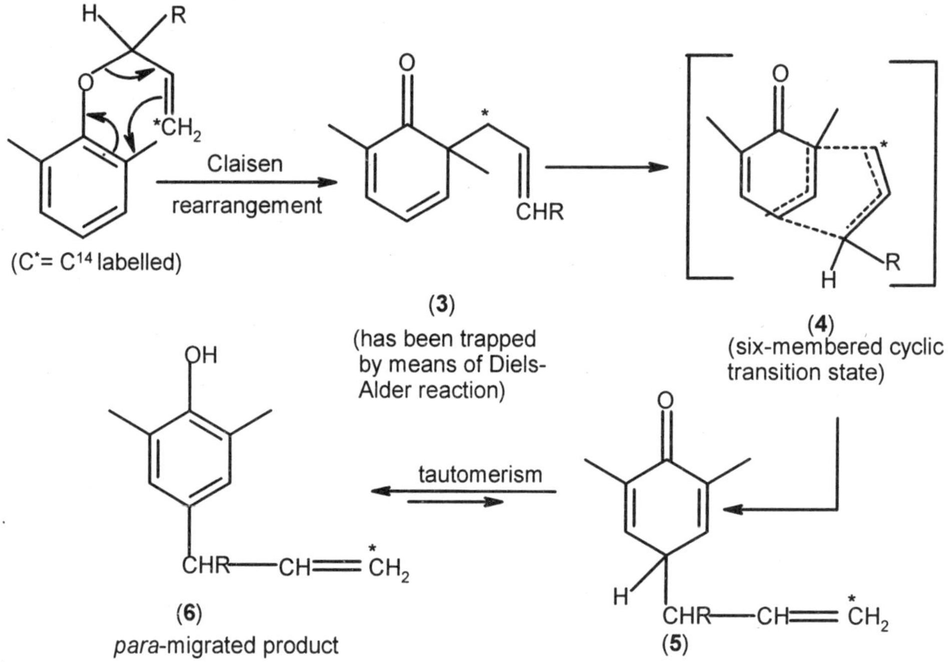

(**3**)
(has been trapped by means of Diels-Alder reaction)

(**4**)
(six-membered cyclic transition state)

(**6**)
para-migrated product

(**5**)

The *intramolecular mechanistic approach* is evidenced from the following facts that the reaction —
(i) requires no catalyst,
(ii) is of first order kinetics,
(iii) yields no crossover products when the reaction is carried out with a mixture of two different substrate molecules,
(iv) gives *ortho*-migrated product with allylic shift of the allylic group, while the attachment of the migrated moiety at *para*-position remains the same, due to double inversion, as derived from ^{14}C-labelling experiments.

It is also interesting to note that the six-membered cyclic transition state of this rearrangement usually prefers to attain chair-like conformation[3].

Discussion

Some major points regarding this type of rearrangement are being cited:
(a) The electronic nature of substituents, if present on the ring, has no so much influence on the rate of the *Claisen rearrangement*. This is because of the fact that the reaction mechanism does not involve ions. Although the effect is small, electron-

releasing substituents have been found to enhance the reaction rate, and electron-withdrawing ones to decrease the rate at a minor scale.

However, the reaction rate has been observed to become greatly influenced by solvent effect; trifluoroacetic acid is usually employed as a good solvent for carrying out the reaction at room temperature.

(b) *"Abnormal" Claisen rearrangement* [4]

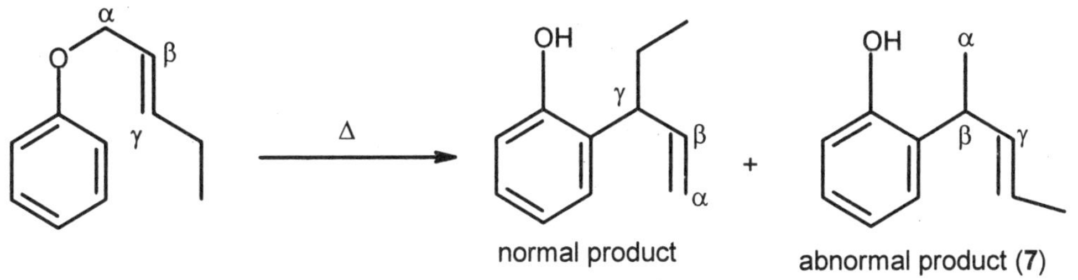

In case of ethers of the type, ArO–C–C=C–R, having an alkyl group at γ-position, an *'abnormal product'* (**7**) is sometimes obtained —— the β–carbon instead of γ-carbon attaches itself to the ring. It has been established that such *'abnormal products'* are not formed directly from the starting ether, but from the initially formed 'normal product' through a further rearrangement —

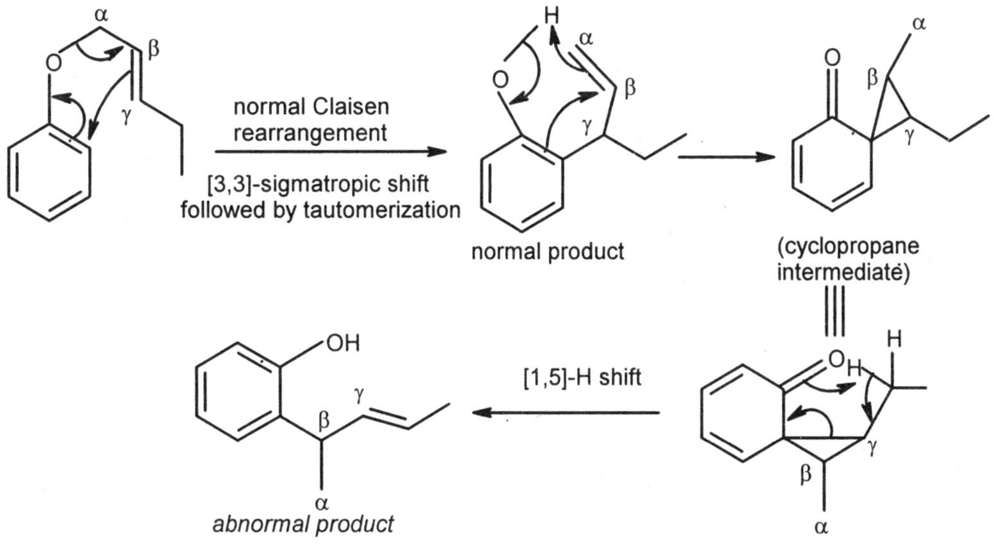

(c) Allyl vinyl ethers[5] also undergo the *Claisen rearrangement* in similar fashion via concerted step six-membered cyclic transition state. In such cases there is no energetic driving force for tautomerism to restore aromaticity, and thus enone is the end product.

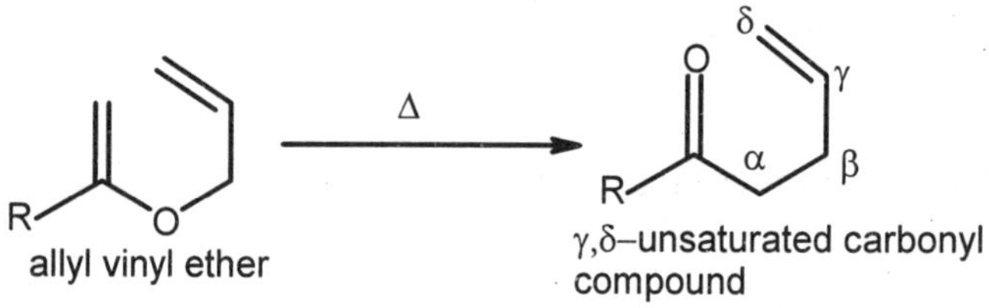

allyl vinyl ether → γ,δ–unsaturated carbonyl compound

The reaction offers an excellent stereoselective route to γ,δ-unsaturated carbonyl compounds (aldehydes, ketones, esters and amides) from allyl alcohols.

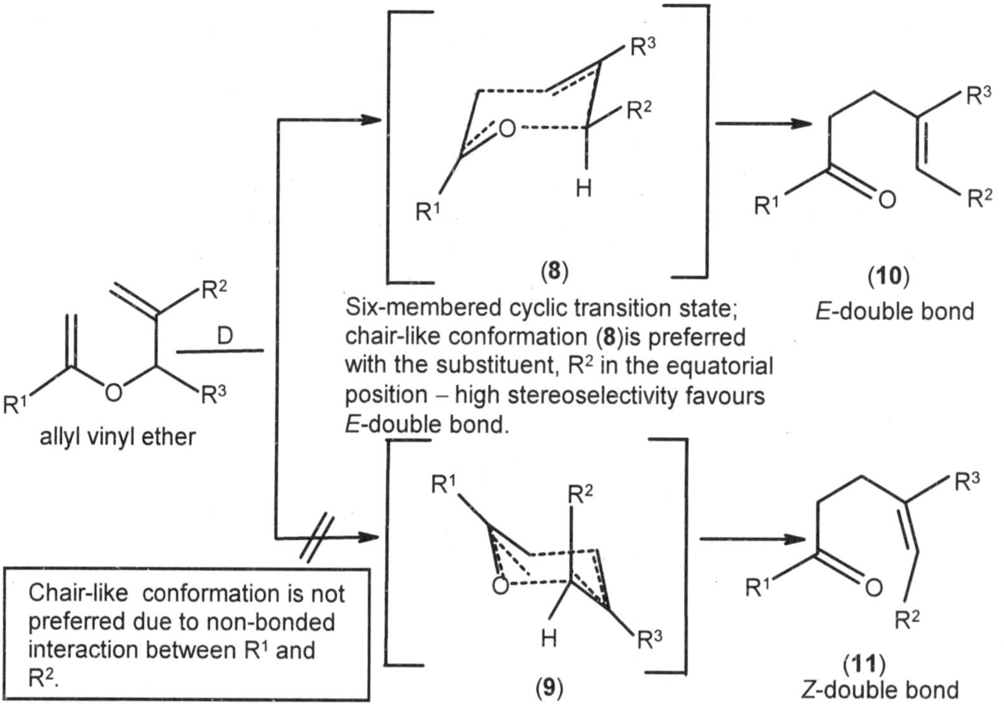

Six-membered cyclic transition state; chair-like conformation (**8**) is preferred with the substituent, R^2 in the equatorial position – high stereoselectivity favours *E*-double bond.

(**8**) → (**10**) *E*-double bond

Chair-like conformation is not preferred due to non-bonded interaction between R^1 and R^2.

(**9**) → (**11**) *Z*-double bond

(d) *Claisen-Ireland Rearrangement*[6]

A variation of *Claisen rearrangement* using silylketene acetals as substrates is known as *Claisen-Ireland rearrangement*; this reaction obviously occurs under much milder conditions (even at room temperature) than the classical *Claisen rearrangement*. The one major cause for this feasibility is the product development control (PDC) — the C=O bond in the product (α-allylated silyl ester) is stabilized by ester resonance (~14kcal/mole). This additional driving force corresponds to a lowered activation barrier leading to the increase in reaction rate (Hammond postulate).

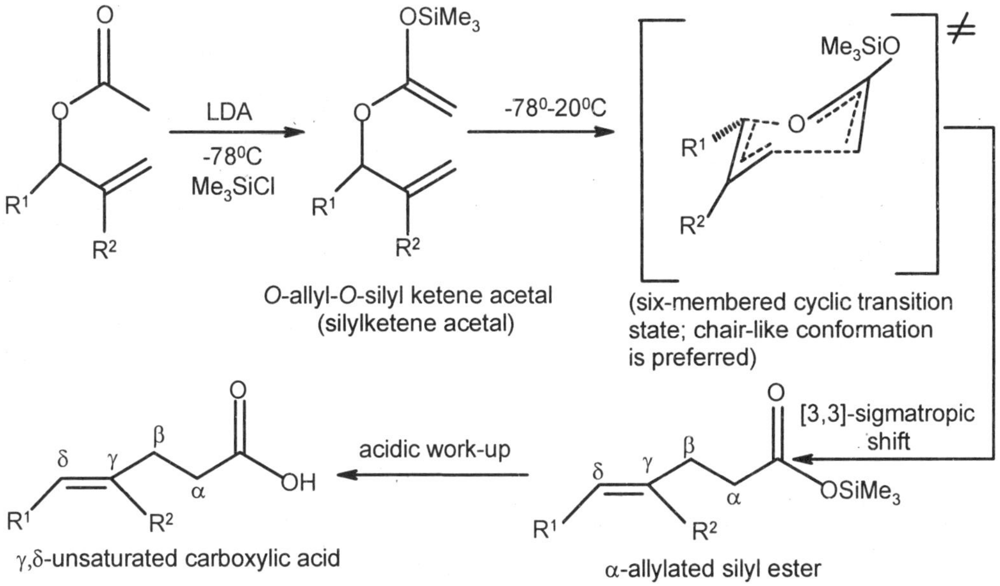

Since silyl esters are very sensitive toward hydrolysis, attempt is usually made to hydrolyze the silyl ester (as formed) during workup. Thus, in actual practice the *Claisen - Ireland rearrangement* affords γ,δ–unsaturated carboxylic acids.

Applications

The classical *Claisen rearrangement* and its extensions are of tremendous utility in organic syntheses. Some are mentioned here:

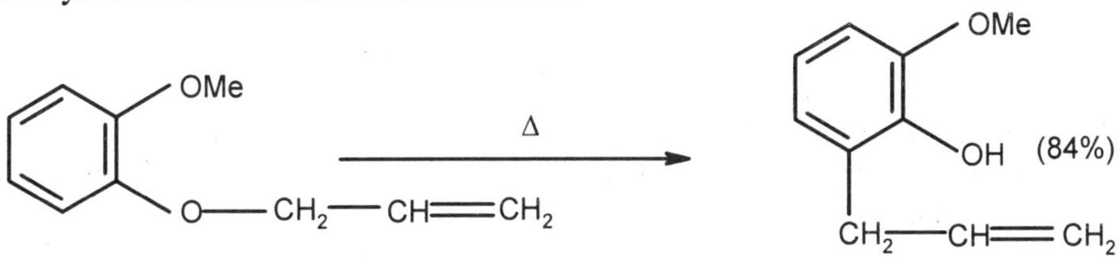

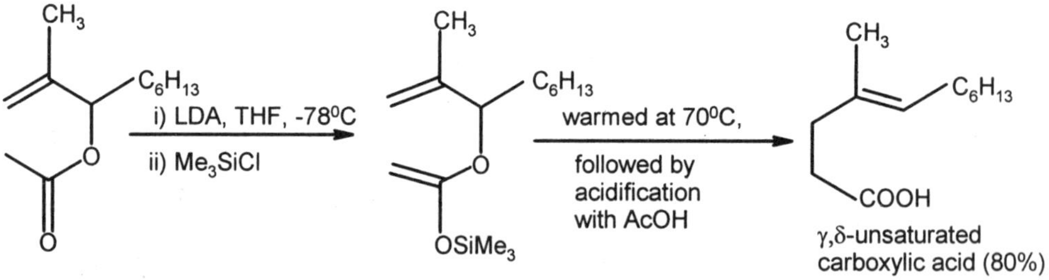

5) The thermal [3,3]-sigmatropic rearrangement of *N*-allyl-*N*-arylamines is called *aza-Claisen* (or *amino-Claisen*) *rearrangement*[7] — it is the nitrogen analogue of simple *Claisen rearrangement*.

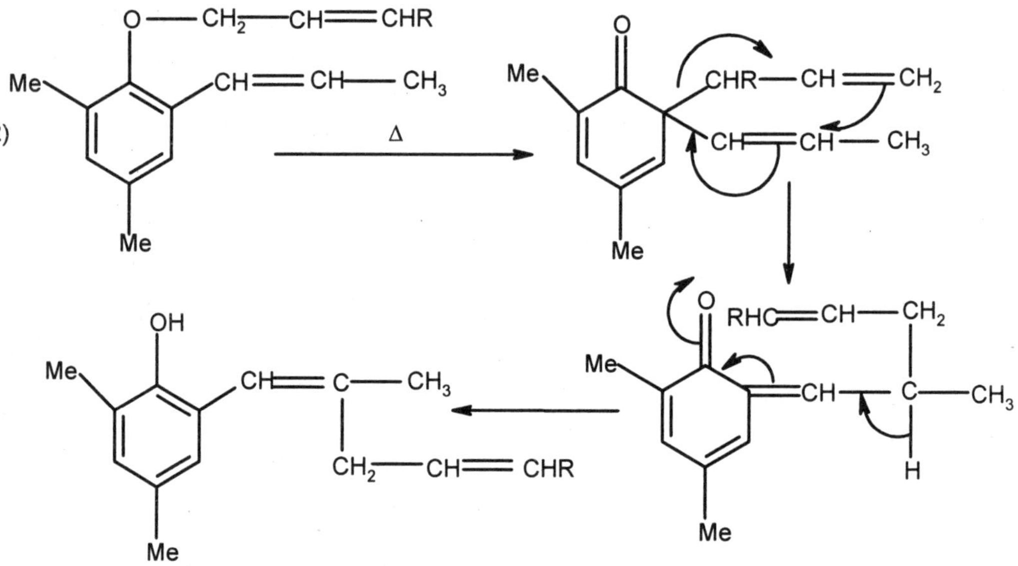

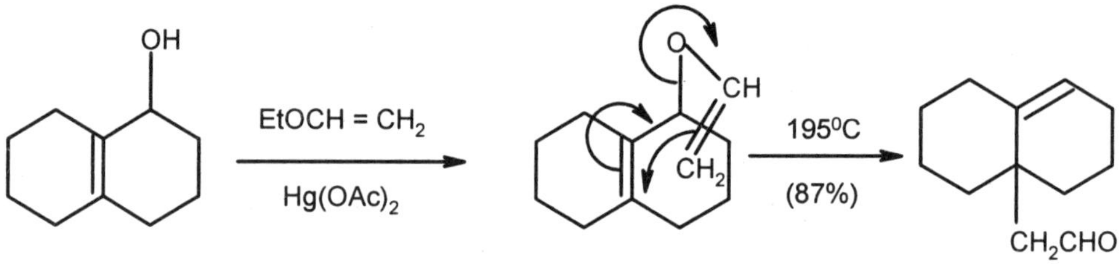

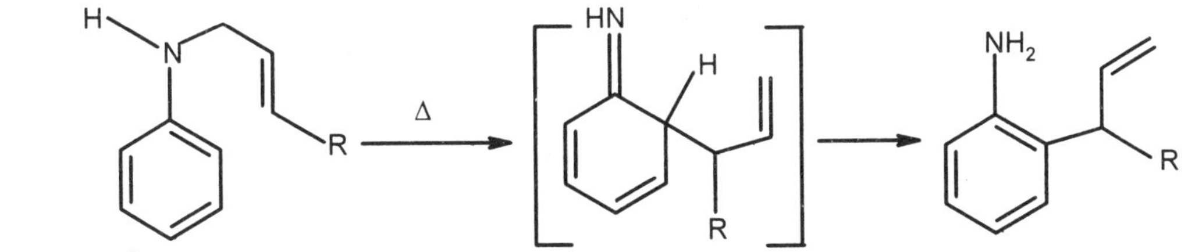

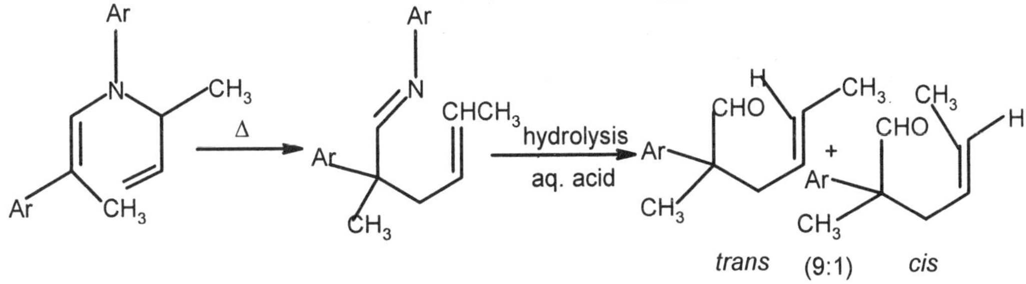

Claisen Rearrangement

The formation of *trans* and *cis* isomers in a roughly 9:1 ratio suggested a *concerted* cyclic process[8].

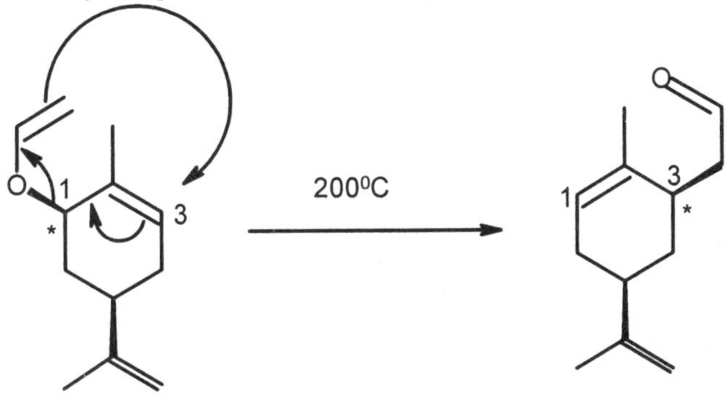

Hence, the *Claisen rearrangement* occurs with complete transfer of stereochemical information from the substrate to the product —— the stereocentre at C-1 is completely transferred to the newly generated stereocentre at C-3. Such type of stereocontrolled transformation of a stereocentre into a new one is termed as *chirality transfer*, and in this case it is *1,3-chirality transfer*.

References

1. Claisen, L (1912), *Ber. Dtsch. Chem. Ges.*, **45**, 3157; for reviews, see Fleming, I. (1999), *Pericyclic Reactions*, Oxford Univ. Press, Oxford, p. 71; Bennett, G. B.

(1977), *Synthesis,* 589; Jefferson, A. and Scheinmann, F. (1968), *Q. Rev. Chem. Soc.*, **22**, 391.

2. McMichael, K. D. and Korver, G. L. (1979), *J. Am. Chem. Soc.*, **101**, 2746; Gajewski, J. J. and Conrad, N. D. (1979), *J. Am. Chem. Soc.*, **101**, 2747; Kupczyk – Subotkowska, L. (1988), Saunders Jr., W. H. and Shine, H. J., *J. Am. Chem. Soc.*, **110**, 7153.

3. Kupczyk – Subotkowska, L., Saunders Jr., W. H. and Shine, H. J. (1992), *J. Am. Chem. Soc.*, **114**, 3441; Copley, S. D. and Knowles, J. R. (1985), *J. Am. Chem. Soc.*, **107**, 5306; Yoo, H. Y. and Houk, K. N. (1994), *J. Am. Chem. Soc.*, **116**, 12047.

4. Habich, A., Barner, R., Roberts, R. and Schmid, H. (1962), *Helv. Chim. Acta.*, **45**, 1943; Marvell, E. N. and Schatz, B. (1967), *Tetrahedron Lett.*, 67; Hansen, H. (1971), *Mech. Mol. Migr.*, **3**, 177 (Review); Watson, J. M., Irvine, J. L. and Roberts, R. M. (1973), *J. Am. Chem.. Soc.*, **95**, 3348.

5. Claisen, L. (1912), *Berechtt*, **45**, 3157; Ziegler, F. E. (1988), *Chem. Rev.*, **88**, 1423 (Review).

6. Ireland, R. E. and Mueller, R. H. (1972), *J. Am. Chem. Soc.*, **94**, 5897; Ireland, R. E., Mueller, R. H. and Willard, A. K. (1976), *J. Am. Chem. Soc.*, **98**, 2868; Ireland, R. E., Wipf, P. and Armstrong III, J. D. (1991), *J. Org. Chem.*, **56**, 650; Dell, C. P., Khan, K. M. and Knight, D. W. (1994), *J. Chem. Soc., Perkin Trans. 1*, 341.

7. Heimgartar, H., Hansen, J. and Schmid, H. (1979) in *"Iminium Salts in Organic Chemistry"*, eds. Bohme, H. and Viehe, H., Wiley-Interscience, New York, Part-2, p. 655; Marcinkiewicz, S., Green, J. and Mamalis, P. (1961), *Tetrahedron*, **14**, 208; Jolidon, S. and Hansen, H. (1977), *Helv. Chim. Acta*, **60**, 978; Majumdar, K. C. and Bhattacharya, T. (2002), *J. Indian Chem. Soc.*, **79**(2), 112 (Review).

8. Hill, R. K. and Gilman, N. W. (1967), *Tetrahedron Lett.*, **15**, 1421.

- Saidi, M. R. (1982), *Heterocycles*, **19**, 1473.
- Corey, E. J., Danheiser, R. L., Chandrasekaran, S., Siret, P., Keck, G. E. and Gras, J.-L. (1978), *J. Am. Chem. Soc.*, **100**, 8031.
- Mohamed, M. and Brook, M. A. (2001), *Tetrahedron Lett.*, **42**, 191.
- Hong, S.-p., Lindsay, H. A., Yaramasu, T., Zhang, X. and McIntosh, M. C. (2002), *J. Org. Chem.*, **67**, 2092.
- Chai, Y., Hong, S.-p., Lindsay, H. A., McFarland, C. and McIntosh, M. C. (2002), *Tetrahedron,* **58**, 2905.
- Khaledy, M. M., Salani, M. Y. S., Khuong, K. S., Houk, K. N., Aviyente, V., Neier, R., Soldermann, N. and Velker, J. (2003), *J. Org. Chem.*, **68**, 572.

5

Diels – Alder Reaction

Introduction

In 1928 Otto Diels and his research student Kurt Alder presented a convenient route[1] to the formation of six-membered rings through pericyclic process, categorically a [4 + 2]π-cycloaddition reaction. The reaction is well-known as *Diels-Alder reaction* in honour of the inventors. The usefulness of this reaction in synthesis arises from its versatility and also from its remarkable stereoselectivity[2]. Besides six-membered carbocyclic rings, a variety of six-membered heterocyclic compounds may also be synthesized by using this reaction. The *Diels-Alder reaction* involves a *concerted* step cycloaddition between a 1,3-conjugated diene (4π-electron system) and an alkene (2π-electron system, usually referred to as *dienophile*); during the course of the reaction two new σ-bonds are formed at the cost of two π-bonds. The reaction may be prototypically represented as:

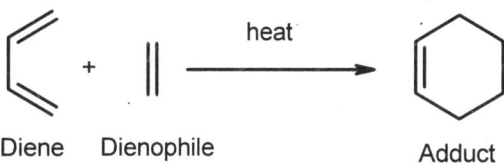

Diene   Dienophile   Adduct

Diels-Alder reaction takes place simply by mixing the components at room temperature or by gentle warming in suitable solvent — however, more vigorous conditions may be necessary in some cases where dienes or dienophiles are not so reactive. Since *Diels-Alder reaction* is reversible, heating at high temperature is disadvantageous because it facilitates adduct to decompose into its components; that's why *Diels-Alder reaction* is generally carried out at low temperature. The forward reaction can be facilitated by using an excess of one of the components, or a solvent from which the adduct separates out readily. It has also been observed that Lewis acid catalysts accelerate many *Diels-Alder reactions*. By varying the nature of diene and dienophile different types of ring structures can be built up.

The Diene

The diene component in the *Diels-Alder reaction* may either be open-chain or cyclic; the diene must essentially have *cisoid* (or *s-cis*) conformation or must be able to adopt it during the reaction before undergoing cycloaddition with dienophiles — this is required so that the ends can interact with the dienophile simultaneously. No reaction takes place in the *transoid* (or *s-trans*) conformation of the diene. Butadiene itself prefers the transoid conformation with the two double bonds as far as away from each other for steric reasons; since the rotational energy barrier (about the central σ-bond) separating the two conformations is small (about 3.9 kcal/mole)[3], rotation to the less favourable but reactive *cisoid* conformation becomes rapid.

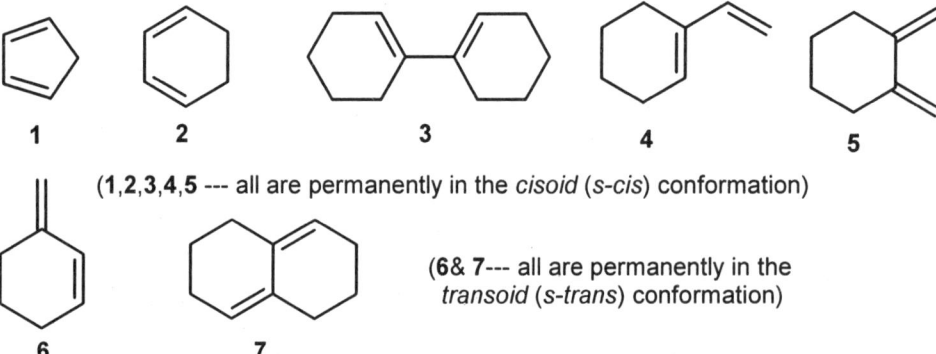

Cyclic dienes that are permanently in the *cisoid* conformation are exceptionally good toward *Diels-Alder reaction* — cyclopentadiene is a classic example — but cyclic dienes that are frozen into the *transoid* conformation and cannot adopt the *cisoid* form will not undergo the reaction at all. Thus, the cyclic dienes (**1** to **5**) can undergo the reaction while **6** and **7** cannot.

It is observed that electron-donating substituents (*e.g.* –Me, -OMe, -NMe$_2$, etc.) often enhance the reaction-rate while electron-withdrawing groups have the opposite effect. As expected bulky substituents, which prevent the diene from adopting the *cisoid* conformation, retard the reaction. Thus, 2-methyl, 2,3-dimethyl and 2-*t*-butylbutadienes react normally with maleic anhydride but 2,3-diphenyl-1,3-butadiene is less reactive while 2,3-di-*t*-butyl-1,3-butadiene is completely unreactive.

(2-methyl-1,3-butadiene)   (2,3-dimethyl-1,3-butadiene)   (2-*t*-butyl-1,3-butadiene)

(2,3-diphenyl-1,3-butadiene)   (2,3-di-*t*-butyl-1,3-butadiene)

The bulky *t*-butyl groups situated at both the C-2 & C-3-positions prevent the molecule from attaining planar cisoid conformation by imposing steric effects

An *E*-substituted 1,3-butadiene reacts with dienophiles much more readily than the *Z*-isomer — because *Z*-alkyl or aryl substituents in the C-1 position of the diene undergoes non-bonded interaction with a hydrogen atom at C-4, thus, hindering the *cisoid* conformation. It is reflected from the fact that while *Z*-1-methyl-1,3-butadiene (**7**) gives only 4% yield of adduct when heated with maleic anhydride at 100^0C, the *E*-isomer (**8**) forms an adduct in almost quantitative yield in benzene even at 0^0C.

(7)   (8)

Similarly, *E,E*-1,4-dimethyl-1,3-butadiene reacts with many dienophiles rapidly, but the *Z,E*-isomer forms an adduct only when the components are heated in benzene at 150^0C; in an extreme case, *i.e.* the *Z,Z*-isomer does not undergo the reaction.

Aromatic compounds can also behave as dienes[4]. Benzene is very unreactive toward dienophiles — very few ones (e.g. benzyne) have been reported to give *Diels-Alder* adduct with it[5]. Naphthalene has been reported to undergo the reaction at high pressures[6]. However, anthracene and other compounds with at least three linear benzene rings give *Diels-Alder reaction* readily. For example:

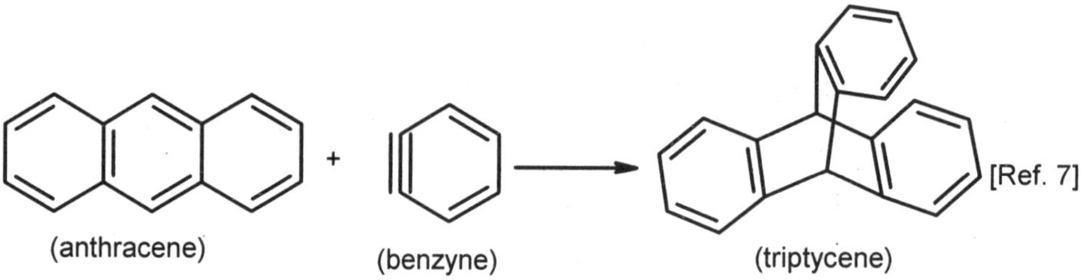

anthracene + benzyne → triptycene [Ref. 7]

Very few examples of a variety of diene systems are presented below:

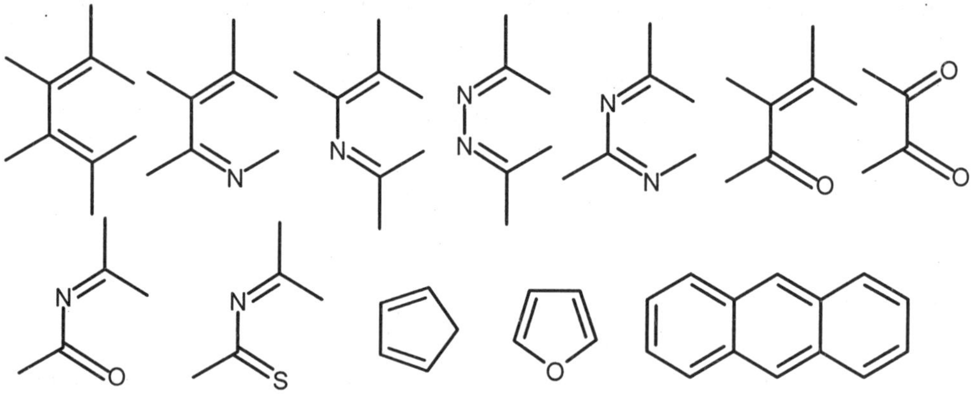

The Dienophile

In *Diels-Alder reaction* the dienophile component comprises of a large variety of ethylenic and acetylenic derivatives or reagents in which one or both the reacting atoms is a heteroatom. Dienophioles may vary in reactivity — the reactivity depends upon the structural pattern. Presence of electron-withdrawing substituents (e.g CO, COOR, CN, NO_2, etc.) on the double or triple bond of dienophiles enhances the reactivity; it is attributed to the lowering of the energy of the lowest unoccupied molecular orbital (LUMO) of the dienophile by the strong electron-attracting substituents. At room temperature, reaction between 1,3-butadiene and maleic anhydride is slow; for quantitative yield it is required to carry out the reaction in boiling benzene, but the highly activated tetracyanoethylene reacts with the diene extremely rapid even at 0^0C.

Reactivity of dienophiles is also influenced by steric effect exerted by the substituents. The experimental observation reveals that the reactions of butadiene and 2,3-dimethylbutadiene with methylmaleic anhydride requires more vigorous conditions than those with maleic anhydride itself.

Very few examples of a variety of dieneophile systems are presented below:

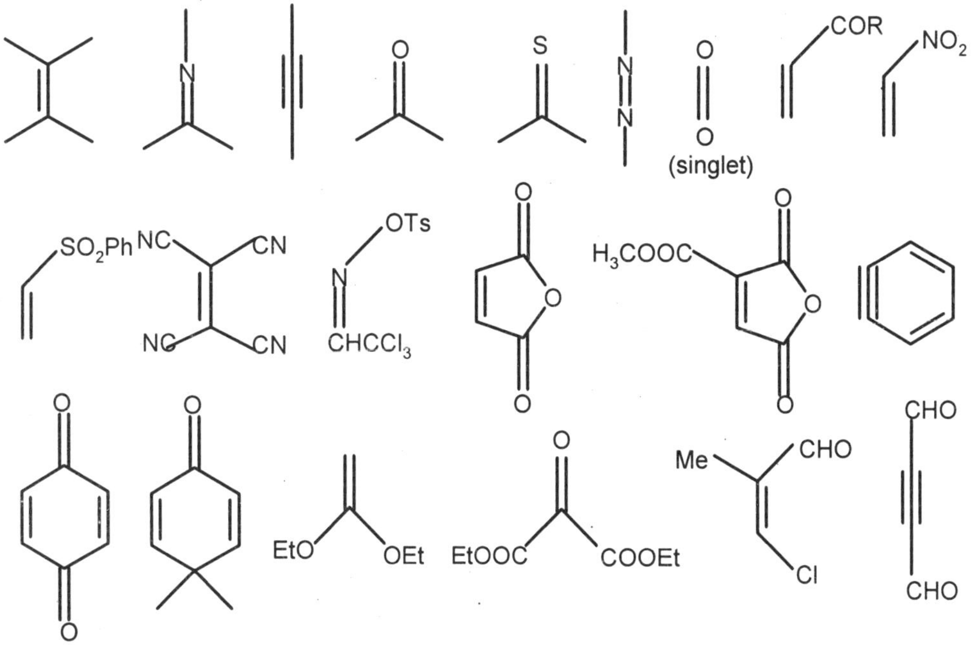

Few Examples of Diels-Alder Reaction

Enormous chemical reactions involving varieties of dienes and dienophiles are known; a very few representative examples are cited below:

(a)

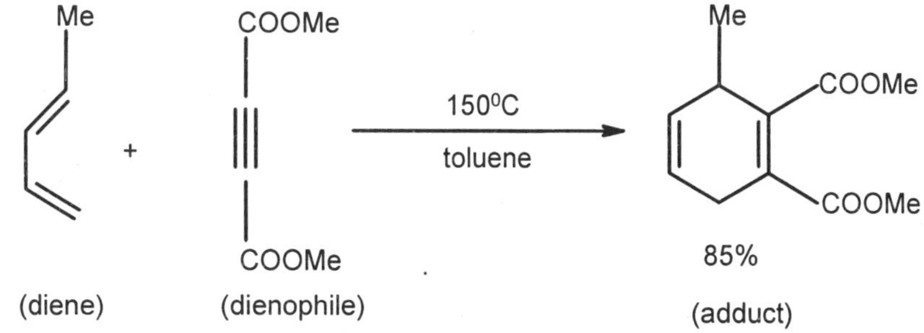

(b), 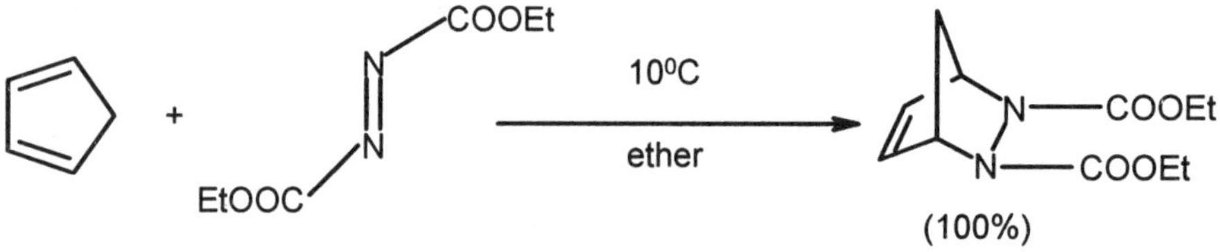 (c), 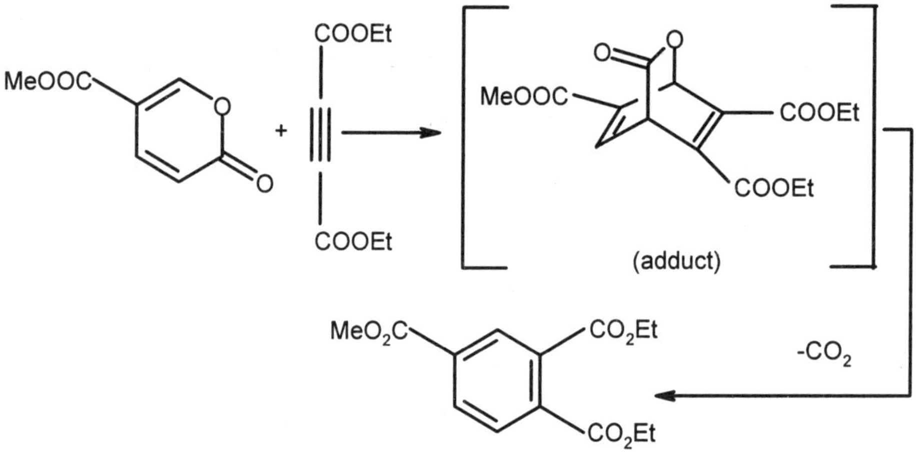 (d), 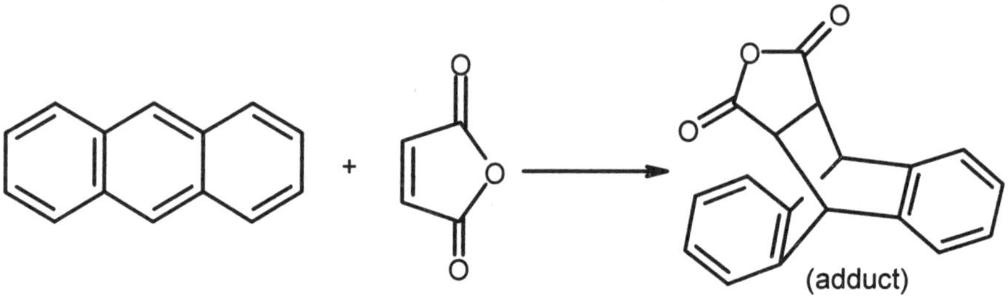 (e) 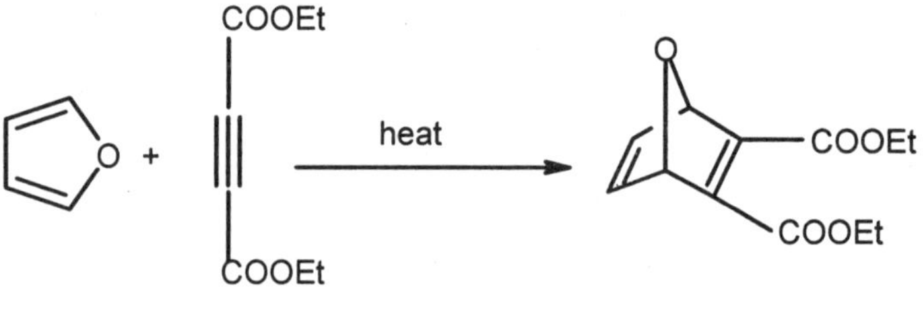 [reaction schemes]

(f) Fulvenes react preferentially as good dienophiles in the *Diels-Alder reaction* with dienes. However, mode of cycloaddition varies with the electronic property of dienes — with electron-poor dienes like enones a [4+2]π cycloaddition takes place, whereas a [6+4]π cycloaddition occurs with electron-rich dienes like 1-amino-butadiene[8].

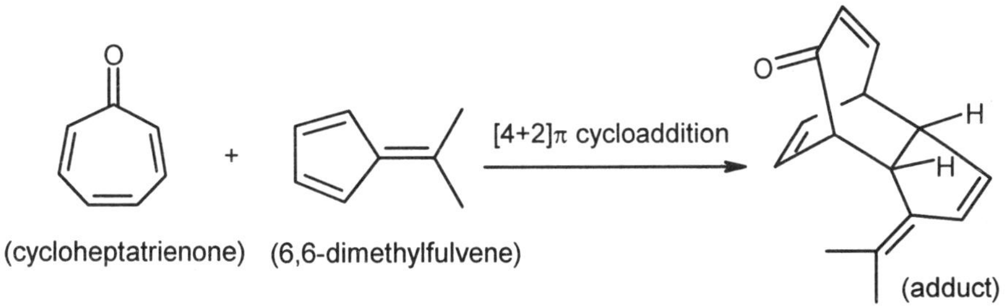

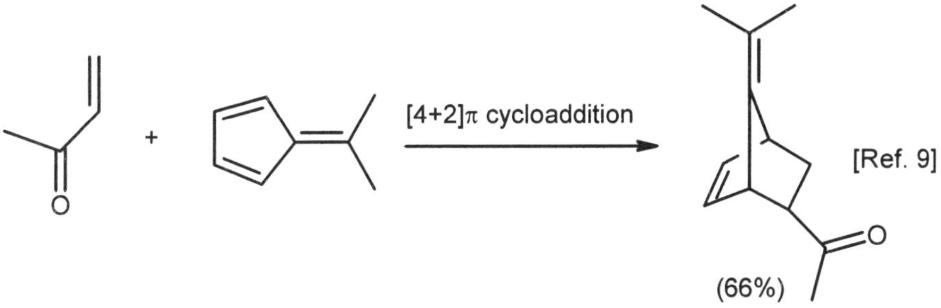

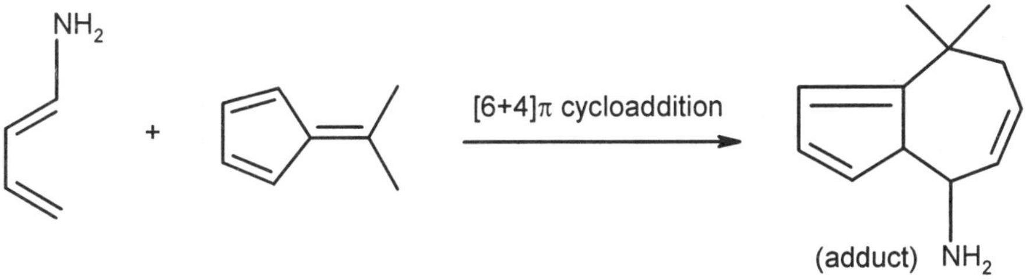

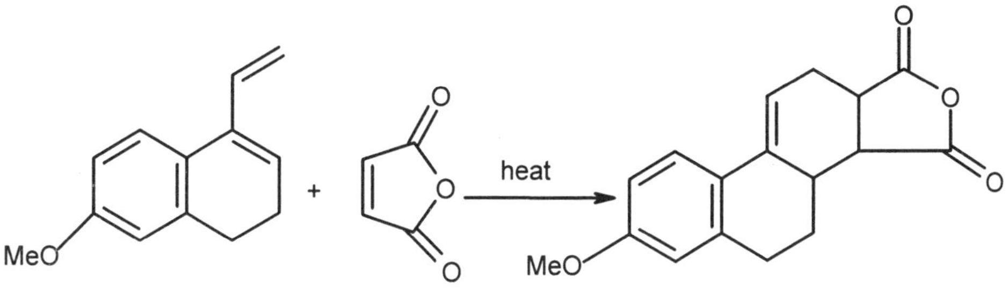

(h) *Diels-Alder reactions* may take place in water medium also[10]; in this case water-soluble dienes are used. Sodium salts of carboxylic acids and protonated amines both behave well under these conditions. Presumably, the soluble tail is in the water but the diene itself is inside the oily drops with the dienophile. For example —

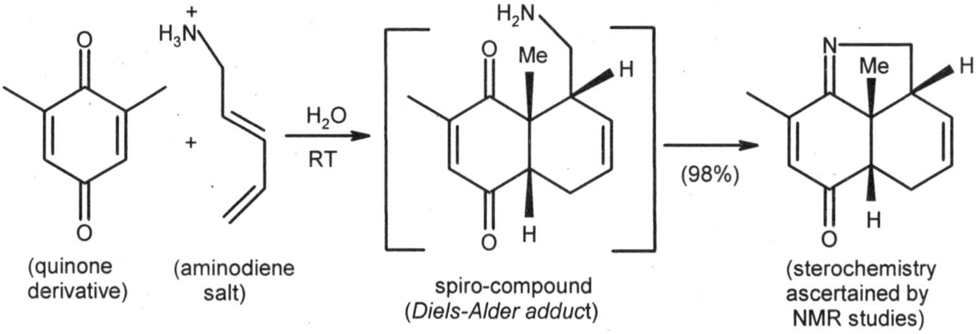

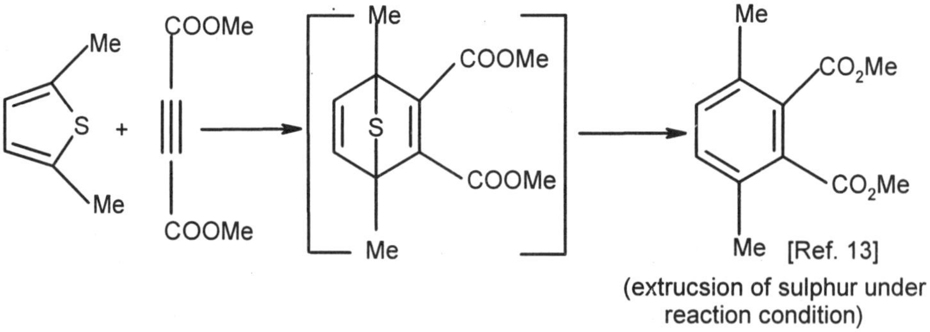

Mechanism

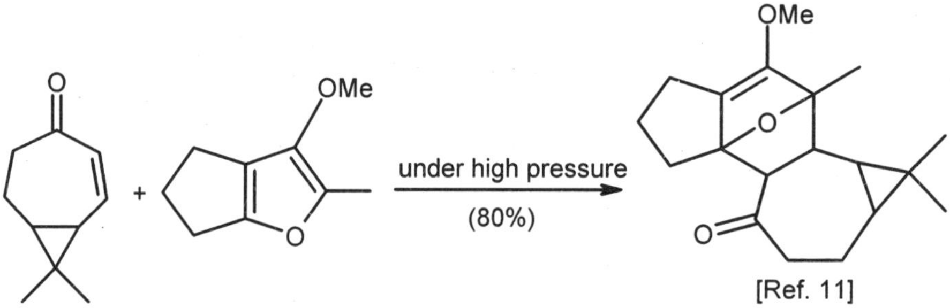

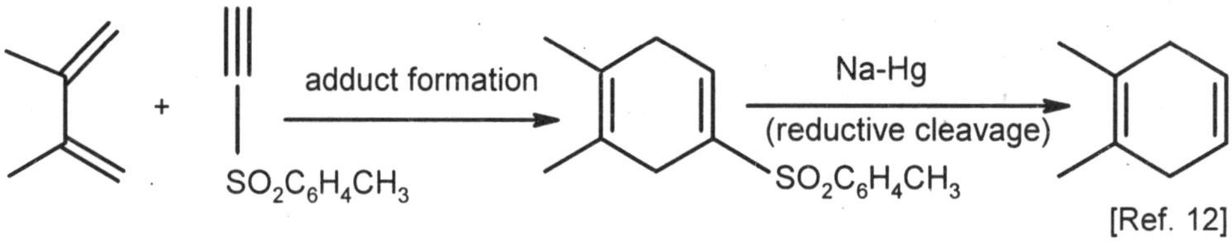

The true mechanism of the *Diels-Alder reaction* has long been a matter of much debate and controversy. The rate-determining step in adduct formation is bimolecular and formation of two new σ-bonds occurs by overlap of meolecular π-orbitals in a direction corresponding to endwise overlap of atomic p-orbitals. Two different mechanisms have been proposed for the *Diels-Alder reaction* between a diene and dienophile —— *concerted* and *nonconcerted*. In a *concerted* mechanism, all bonds that are to be formed or broken during the reaction are being so formed or broken at the same time; therefore, no intermediates are involved and it takes place through a cyclic transition state (**fig.1**). The *nonconcerted* mechanism occurs in two steps —— the first step becomes the rate controlling one. The mechanism involves either a zwitterionic or diradical intermediate (**fig.2**).

(Fig. 1)

(zwitterion)   (diradical)

(Fig. 2)

The high stereoselectivity of *Diels-Alder reaction* favours the *concerted* path; however a two-step mechanism can't be ignored completely by this evidence if rotation about carbon-carbon single bonds in the diradical intermediates is slow compared with the rate of ring closure. A vast amount of works was carried out according to which it has generally been accepted that major *Diels-Alder reactions* are *concerted*[14]. Addition of *cis* and *trans*-dichloroethylene to cyclopentadiene has been found to be completely stereospecific, that ruled out the two-step mechanism (because long-lived biradical intermediate in this case do not give expected mixture of products). Recently S. Sakai[15] established by using *ab initio MO* methods and examining the energy barrier of the transition states that *Diels-Alder reaction* between butadiene and ethylene is more energetically favourable for a *concerted* mechanism by 2.7 kcal/mole to take place rather than a step-wise mechanism. As the diene and dienophile become more substituted in different *Diels-Alder reactions*, the concerted mechanism becomes much more energetically favourable than the nonconcerted one. Further evidence of a concerted mechanism includes the fact that no one has ever been able to isolate a product created by the intermediate of diradical intermediates in a step-wise mechanism. Similarly, the kinetic effects of *para* substituents in 1-phenylbutadiene are very small for a rate-determining transition state corresponding to a zwitterion intermediate. However, certain *Diels-Alder reactions* are likely to take place nonconcertedly because they are interestingly noted to be affected by Lewis acid catalysts[16].

The generally accepted concerted mechanism for *Diels-Alder reaction* can either have *synchronous* (involves equal distribution of charges of the three π bonds with an arene-like property) or *asynchronous* (involves unequal distribution of charge of the transferred π bonds between the diene and dienophile) transition states. The issue whether the *Diels-Alder reaction* mechanism is of *synchronous* or *asynchronous concerted* one has been heavily disputed among scientists[17].

Normally, dienophiles with electron-withdrawing groups are more reactive because such groups lower the LUMO energy, thereby enhancing interaction with the diene (bearing electron-releasing groups) HOMO. This complementary nature of electron-rich diene with electron-deficient dienophile affecting greater reactivity is very much general, and that's why such *Diels-Alder reactions* are termed as *normal Diels-Alder reactions*. However, some electron-deficient dienes (*e.g.* hexacholoro-cyclopentadiene) are found to undergo this reaction, and in that case it is observed that electronic requirements of the dienophile become reversed —— the electron-donating substituents favour the reaction. These reactions are often termed as *Diels-Alder reactions* with *reversed* or *inverse* electron demand.

Stereochemistry

Diels-Alder reaction is stereospecific. The stereochemistry of an adduct can be assumed on the basis of an empirical rule "*cis* principle" formulated by Alder and Stein in 1937, according to which the stereochemistry of all the substituents located at the C-1 and C-4 positions on the diene and at the C-1 and C-2 positions on the dienophile remains preserved. Thus —

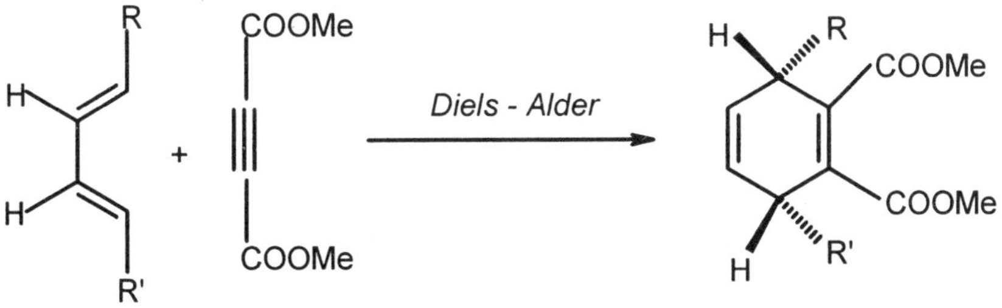

Diels – Alder Reaction

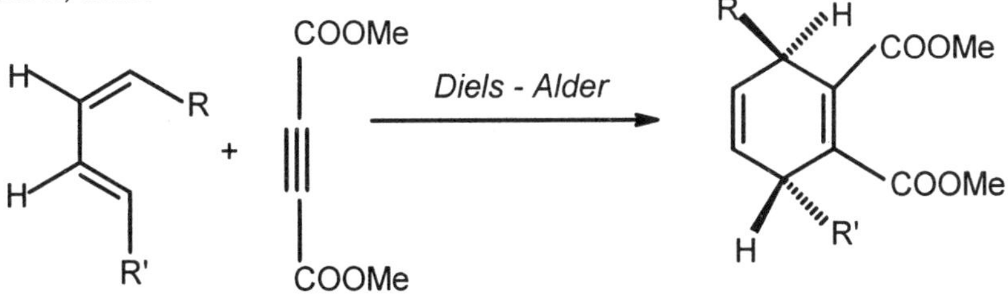

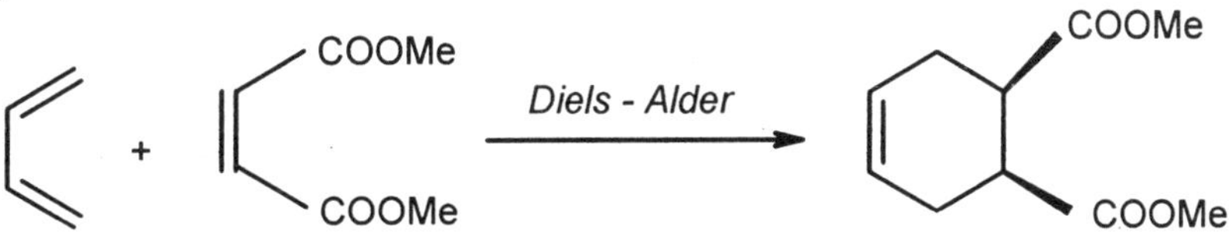

Similarly,

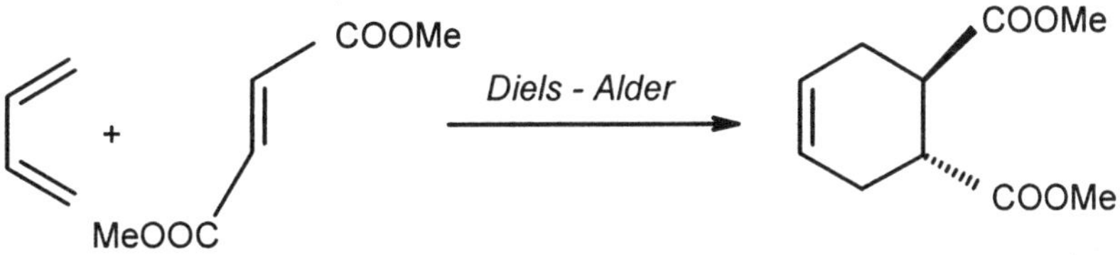

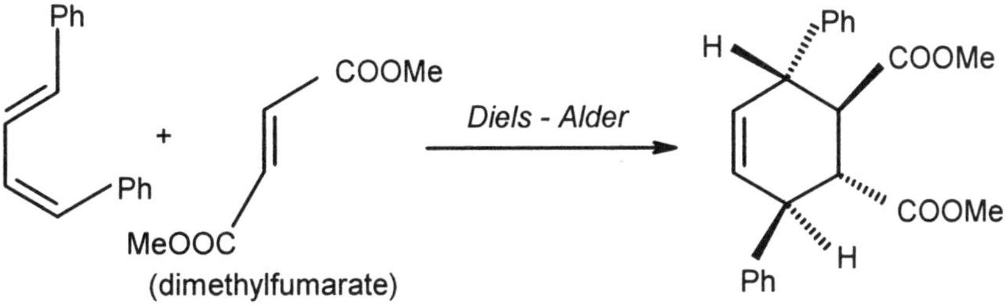

Thus,

Diels-Alder reaction maintains regioselectivity during product formation — 1-substituted dienens form *ortho*-products while 2-substituted ones give para-products when add to dienophiles with an electron-withdrawing group at one end. Hence,

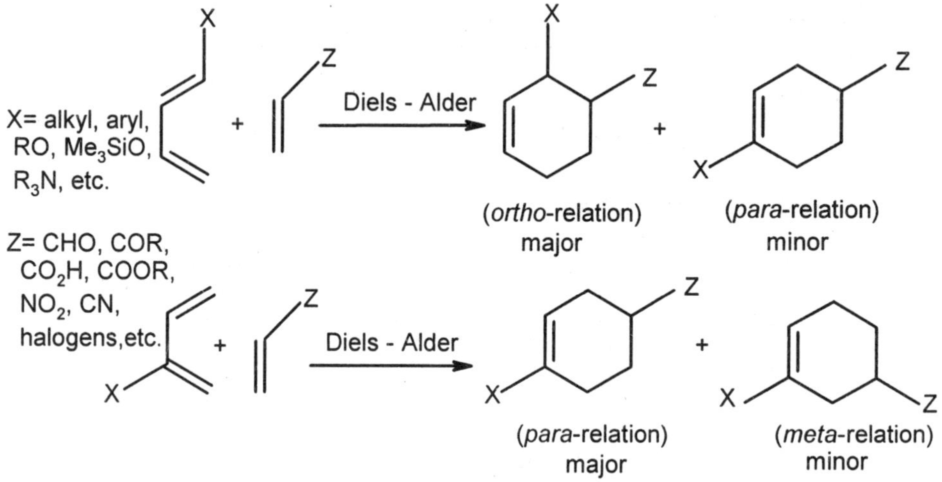

X= alkyl, aryl, RO, Me$_3$SiO, R$_3$N, etc.

Z= CHO, COR, CO$_2$H, COOR, NO$_2$, CN, halogens, etc.

(*ortho*-relation) major + (*para*-relation) minor

(*para*-relation) major + (*meta*-relation) minor

Application of such regioselectivity, as for example, has been used up in the synthesis of limonene (a natural odour principle found in most citrus plants):

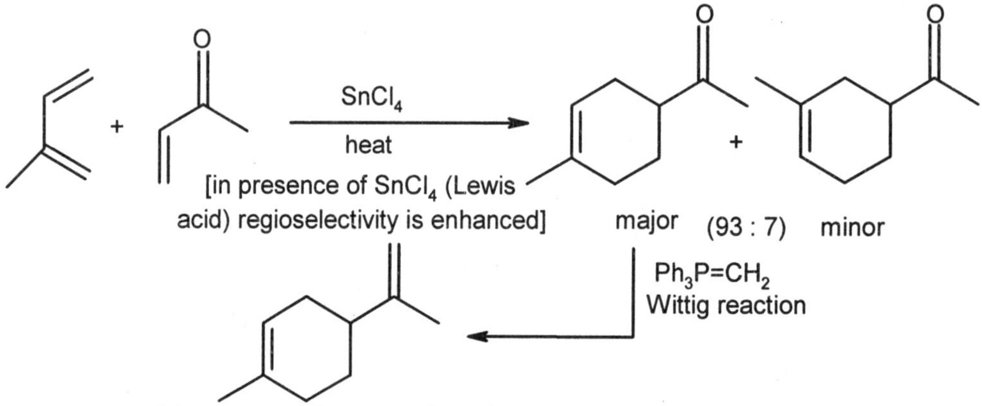

SnCl$_4$ / heat [in presence of SnCl$_4$ (Lewis acid) regioselectivity is enhanced]

major (93 : 7) minor

Ph$_3$P=CH$_2$ Wittig reaction

(Limonene, a citrus constituent)

The stereochemistry of the resultant ring is assigned by the manner in which the diene and dienophile approach each other — either described as '*endo*' or '*exo*'. The '*endo*' approach can be defined as orientation in which the diene and dienophile come to have their ends match up, while achieving maximum 'overlap' of their faces. The '*exo*' approach is, in turn, described as an orientation in which the two faces have the least interaction, while still maintaining the proper alignment of the ends of the diene and dienophile. To illustrate the phenomenon let us consider the *Diels-Alder reaction* between maleic anhydride and cyclopentadiene — the two possible products ('*endo*' and '*exo*') may be shown as:

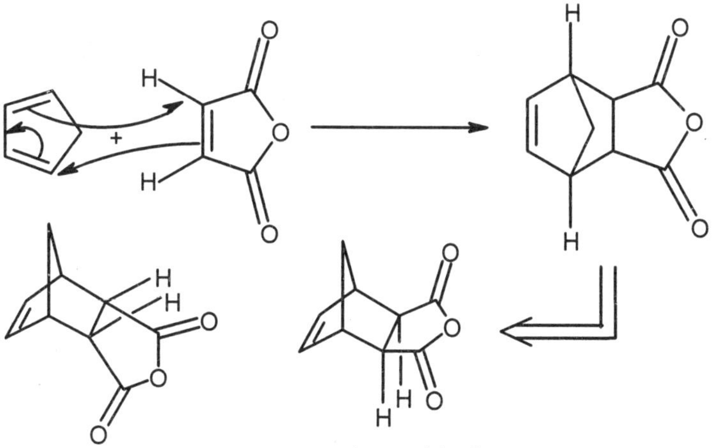

(*endo*-adduct)   (*exo*-adduct)
(two possible products)

Thermodynamically an *endo*-adduct becomes less stable than *exo*-adduct, but the interesting fact is that *endo*-products normally predominate, often almost to the exclusion of *exo*-products. In irreversible *Diels-Alder reaction endo*-adduct must be kinetic product of the reaction. In case of reversible reactions initially formed *endo*-adducts may, however, isomerize to the less sterically hindered more stable *exo*-adducts by a *retro-Diels-Alder reaction* followed by recombination, as observed in:

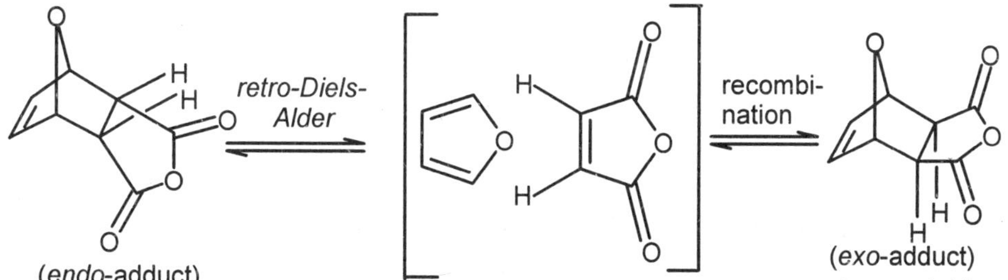

(*endo*-adduct)
(less stable; kinetically controlled product)

(*exo*-adduct)
(more stable; thermodynamically controlled product)

A recent experimental proof in favour of *endo*-approach has been demonstrated by Houk and Lech (2001)[18]. Woodward and Hoffmann explained the matter on the basis of greater stabilization of the *endo*-transition state by means of *secondary orbital interactions* between appropriate highest occupied and lowest unoccupied vacant orbitals (acts across the space between orbitals even though no bonds are formed)[19]. Such favourable interactions (represented by dashed lines) lower the *endo*-transition state relative to that of the *exo*-transition state, where these *secondary orbital interactions* are absent; hence the *endo*-adduct is formed under kinetically controlled conditions.

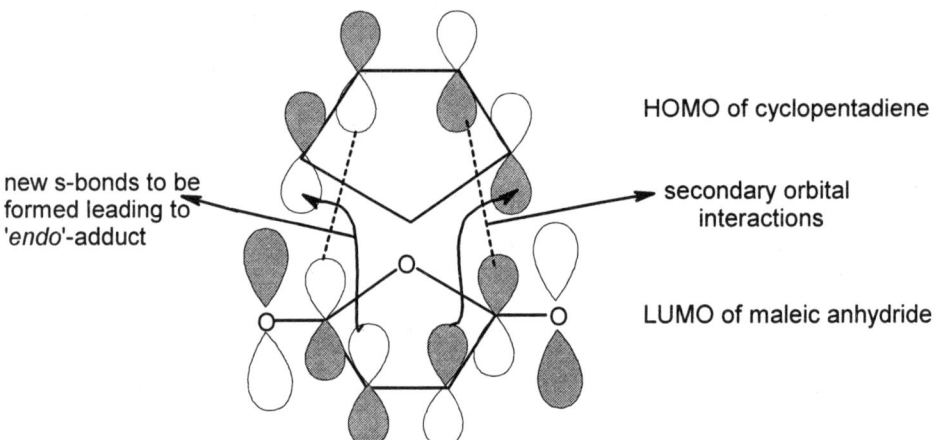

Preference for *endo*-selectivity depends upon a number of factors; hence other explanations are also available. Herndon and Hall (1967) argued that such addition results from more efficient geometrical overlap in comparison to that occurred in *exo*-addition[20]. The cause of greater aptitude for *endo*-addition may also be the dipole-induced dipole or charge transfer interactions between diene and dienophile[21]. The charge transfer interactions are described as involving p-electrons facilitating exchanges between electrons in an atom's 'd' orbital[22].

Intramolecular Diels-Alder Reaction

When a molecule bears the diene and dienophile part within it in proper orientation for cyclization, the two components undergoes ring-closure in an intramolecular fashion; the reaction is said to be *intramolecular Diels-Alder reaction*. This reaction is very much useful for the synthesis of important natural products belonging to the series of alkaloids, steroids, terpenoids, etc.[23] It is highly regio- and stereoselective. Thus, *cis*-dienyl acrylic ester (**9**) gives the *cis*-bicyclic product (**10**) while the *trans*-isomer (**11**) offers the *trans*-product (**12**).[24]

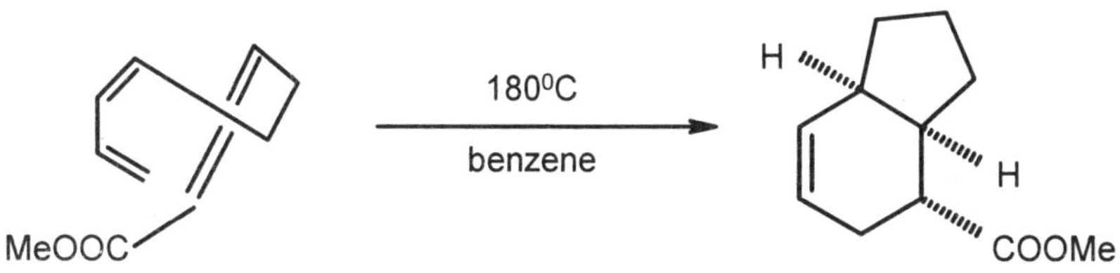

Few more examples of the intramolecular reactions of synthetic utility are cited below:

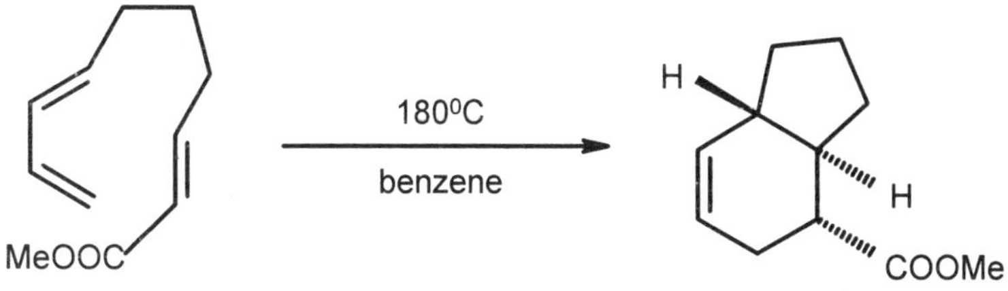

(a) The starting compound (**13**) affords the bicyclic lactum (**14**) on pyrolysis, which is subsequently converted into the alkaloid d-coniceine (**15**)[25].

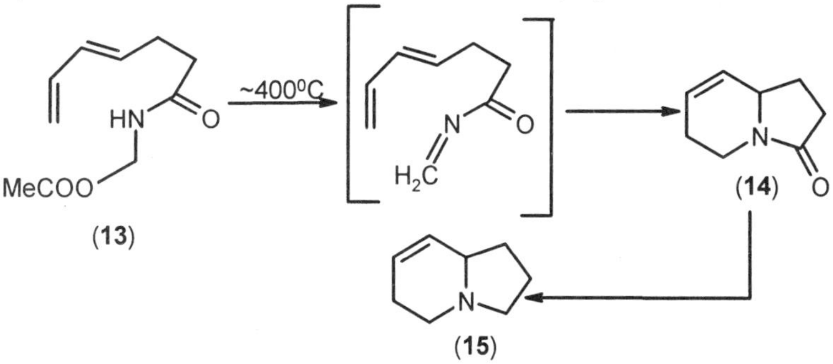

[Ref. 26]

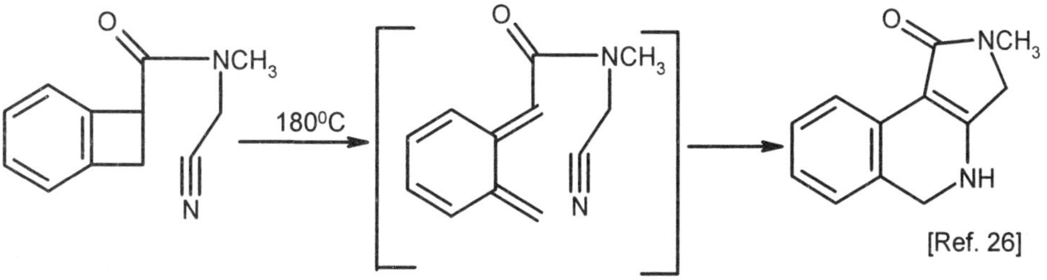

(c) 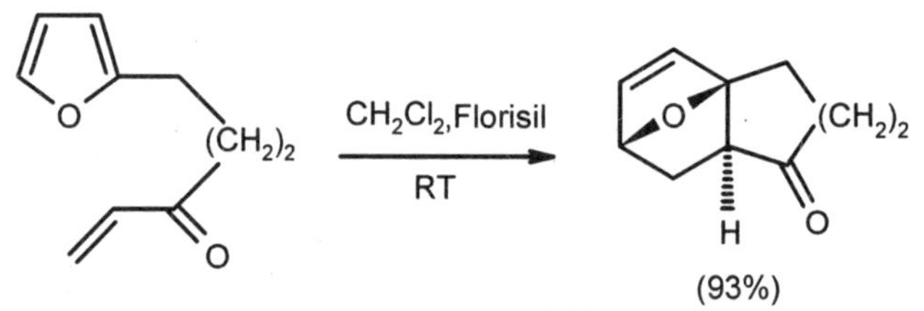 [Scheme showing furan-tethered enone cyclizing with CH₂Cl₂, Florisil, RT to tricyclic ketone (93%)] [Ref. 27]

(d) 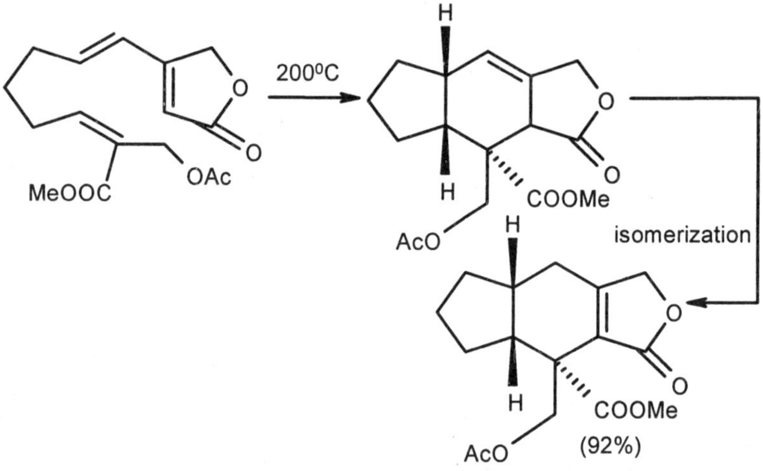 [Scheme showing intramolecular Diels-Alder at 200°C followed by isomerization to give lactone product (92%)] [Ref. 28]

Homo-Diels-Alder Reaction

For undergoing *Diels-Alder reaction* the diene component (both all-carbon and hetero systems) is usually a conjugated (1,3) enyne; but reports are available where the diene can even be nonconjugated if the geometry of the molecule is suitable. In this particular case 1,5-addition of dienophiles to 1,4-dienes takes place and is known as *homo-Diels-Alder reaction*. This is a useful synthetic tool for the preparation of certain tetracyclic compounds.

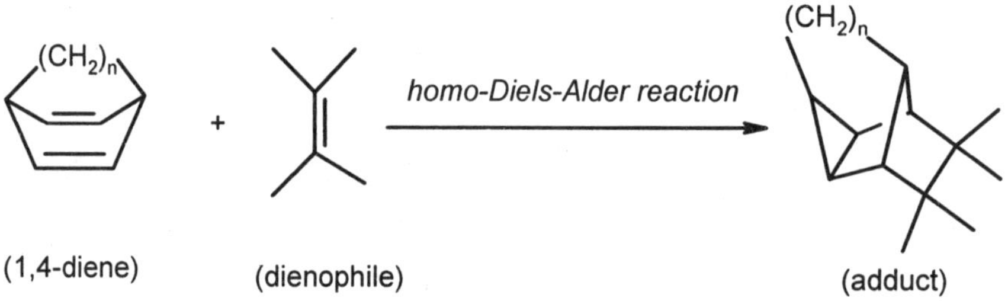

[Scheme: (1,4-diene) + (dienophile) → homo-Diels-Alder reaction → (adduct)]

Reports have already established that norbornadiene (**16**) exhibits unusual chemical reactivity as compared with ordinary nonconjugated diene[29]. Of particular interest are the observations that norbornadiene gave 1:1 adduct with dienophiles by a *homo-conjugated Diels-Alder type addition* across the 2,6-position.

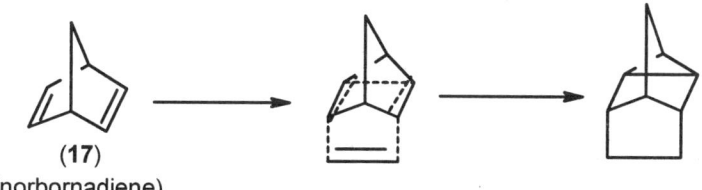

(1:1 adduct)
8,8,9,9-tetracyano-quadricyclo-
[2,2,1,0^{2,3},2^{3,5}]nonane (**17**)
crystalline solid, m.p. 186-88°C

A number of studies with norbornadiene have been concerned with the stereochemistry of the reaction[31], a competing ionic reaction[32], and the effects of Ni(0) catalyst on the reaction[33]. In 1988 Paquette *et al.*[34] gave an insight into the mechanistic pathway of this reaction depending on the observations of their studies on secondary deuterium isotope effects of transtition state structural features —— norbornadien-2-d undergoes cycloaddition to 2π addends (dienophiles) of differing reactivity to give products wherein the distribution of deuterium is commensurate with the operation of *concerted* processes but with *asynchronous* capture of the attacking reagent.

Fig. 3

The *homo-Diels-Alder reactivity* exhibited by norbornadiene is not commonly found in other bridged dienes because of the strong dependency on through-space interaction[35] and spatial proximity of its π bonds[36]. The reaction involves *endo-*selectivity[37]; and widely interpreted as proceeding *via concerned* six-electron transition state, provided that addend polarity does not induce zwitterions formation[38].

Other examples
(a) Zimmerman and Grunewald[39] carried out *homo-Diels-alder reaction* between barrelene (**18**) and addends, dimethyl acetylenedicarboxylate and dicyanoacetylene;

the corresponding 1:1 adduct on subsequent heating yielded 1,2-disubstituted naphthalene. This is a remarkable transformation of barrelene to naphthalene derivatives.

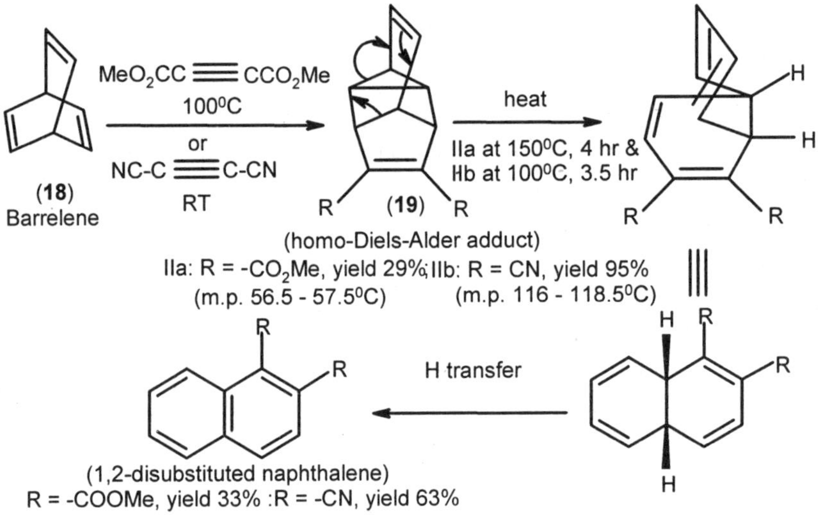

(**18**) Barrelene

MeO$_2$CC≡CCO$_2$Me, 100°C
or
NC-C≡C-CN, RT

(**19**) (homo-Diels-Alder adduct)
IIa: R = -CO$_2$Me, yield 29%; IIb: R = CN, yield 95%
(m.p. 56.5 - 57.5°C)   (m.p. 116 - 118.5°C)

heat
IIa at 150°C, 4 hr &
IIb at 100°C, 3.5 hr

H transfer

(1,2-disubstituted naphthalene)
R = -COOMe, yield 33% : R = -CN, yield 63%

(b) H. K. Hall[40] used this reaction to synthesize two atom-bridged tetracyclic ketones (**20**) and (**21**). Addition of acrylonitrile to norbornadiene gave 6-cyanotetracyclo-[3,2,1,1^{3,8},0^{2,4}]nonane (*homo-Diels-Alder adduct*), which was subsequently converted to tetracyclo[3,2,1,1^{3,8},0^{2,4}]nonan-6-one (**20**) and tetracyclo[3,2,1,1^{3,9},0^{2,4}]decan-6-one (**21**).

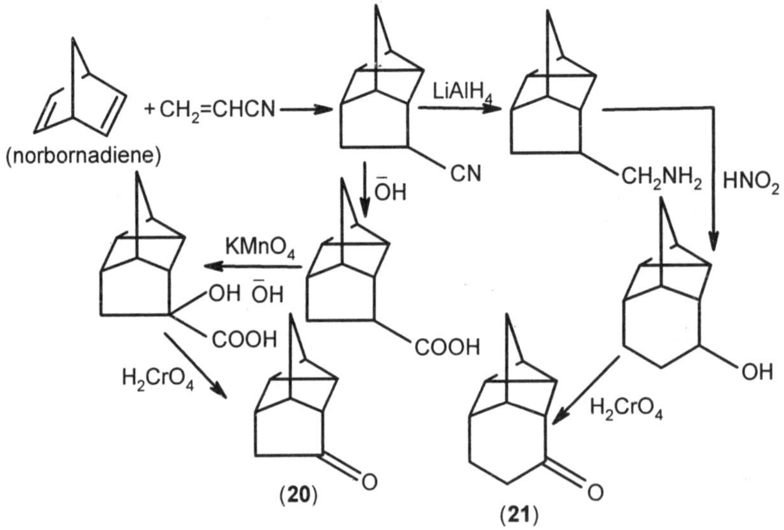

(norbornadiene) + CH$_2$=CHCN → ... LiAlH$_4$ → ... CH$_2$NH$_2$ | HNO$_2$

$\bar{O}$H ↓

KMnO$_4$
OH ŌH
COOH

H$_2$CrO$_4$

COOH

OH

H$_2$CrO$_4$

(**20**)   (**21**)

(c) Fickes and Metz[41] extended the scope of the reaction ——

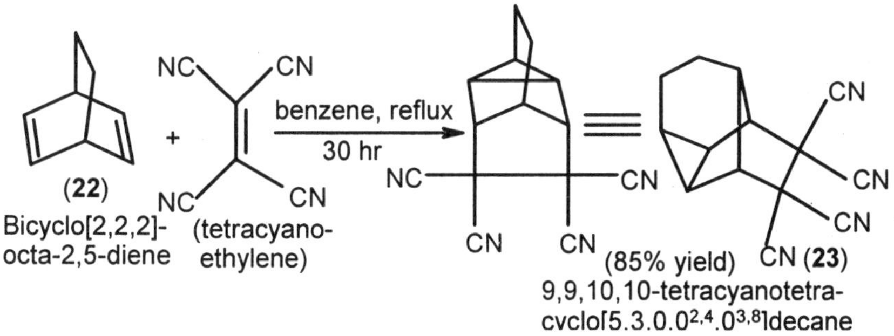

(22) Bicyclo[2,2,2]-octa-2,5-diene + (tetracyanoethylene) →(benzene, reflux, 30 hr)→ (85% yield) (23) 9,9,10,10-tetracyanotetracyclo[5.3.0.0^{2,4}.0^{3,8}]decane

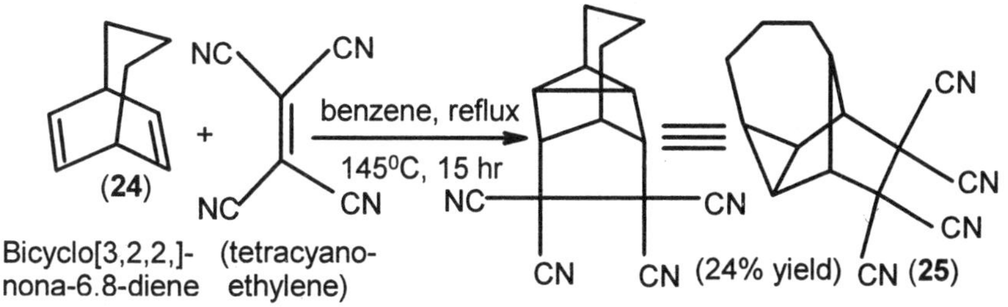

(24) Bicyclo[3,2,2,]-nona-6,8-diene + (tetracyanoethylene) →(benzene, reflux, 145°C, 15 hr)→ (24% yield) (25)

Catalysis in Diels-Alder reactions

Usually no catalyst is required for carrying out *Diels-Alder reactions*; however, it has been observed that Lewis acids catalyze some *Diels-Alder reactions* effecting a remarkable acceleration in the reaction-rate, particularly those in which the dienophiles bear oxygen or nitrogen functional groups[42]. A large number of Lewis acids with varying reactivity have been used[42c,43] —— aluminium chloride is the most common one. For example —— benzene does not react with dicyanoacetylene under normal condition but the reaction occurs in presence of AlCl$_3$.

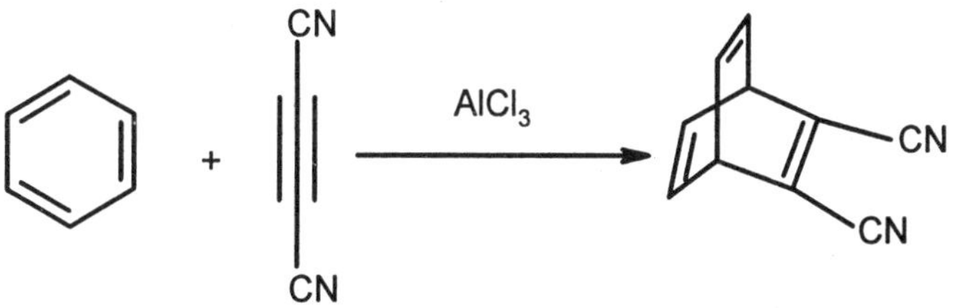

Similarly, adduct formation between anthracene and maleic anhydride is very slow, but considerable acceleration in the reaction-rate can be achieved using Lewis acid catalyst. A Lewis acid catalyst usually enhances both the regioselectivity of the reaction and the extent of *endo* addition[44]. An example is cited:

Catalyst	Temp.	Endo	Exo
Nil	0°C	84%	16%
47% AlCl$_3$. OEt$_2$	0°C	93%	7%
47% AlCl$_3$. OEt$_2$	–70°C	97%	3%

It is supposed that the catalysts form complexes with the polar groups (carbonyl or cyano) of the dienophiles —— such coordination between the non-bonding electrons in a dienophile and a Lewis acid lowers the energies of the frontier orbitals of the dienophile and alters the distribution of atomic orbital coefficients. The energy difference between the HOMO of the diene and LUMO of the dienophile is thus reduced —— the smaller is the energy difference between these orbitals the better is the overlap and the more readily the reaction occurs[46].

Besides Lewis acids, *tris* (4-bromophenyl) aminium hexachloroantimonate (Ar$_3$N$^+$. SbCl$_6^-$) is also used[47]. Bronsted acids have also been used to accelerate the rate of *Diels-Alder reactions*[48]. Singh et al.[49] showed that micellar catalysts can be effective in enhancing the reaction-rate in some cases; thus, the following *Diels-Alder reaction* takes palce in presence of the micellar catalyst, cetyltrimethylammonium bromide (CTAB).

(5-cyclopropylcyclopentadiene) + (*p*-benzoquinone) ⟶ (endo-adduct)

Use of microwave oven[50], ultrasound[51], and light[52] to catalyze *Diels-Alder reaction-rate* in some cases are also known. Alumina has been used to promote the reaction[53].

Synthetic Applications of Diels-Alder Reaction

A variety of organic compounds including natural molecules can successfully be synthesized using *Diels-Alder reaction* as one of the key steps. A very few illustrations are cited here:

(a) Conversion of azodicarboxylic ester to a useful reductant, diimide involves *Diels-Alder reaction*.

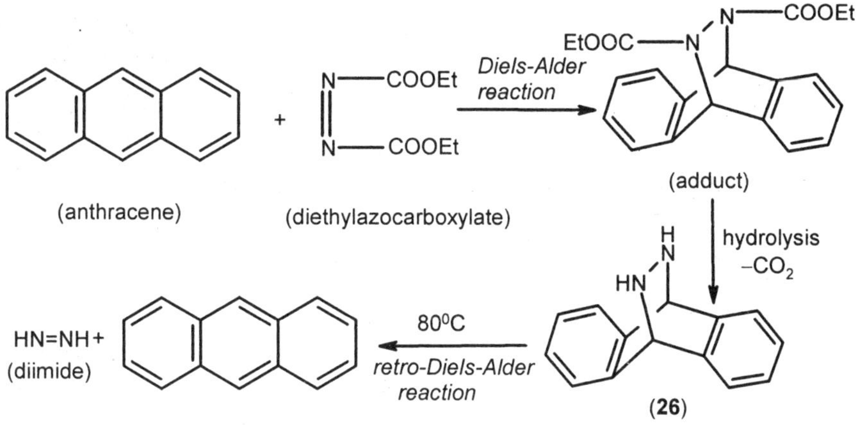

Reductions of substrates can, therefore, be carried out conveniently with (26) in refluxing ethanol[54].

(b) Vogel *et al.*[55] utilized both the *Diels-Alder* and *retro-Diels-Alder reactions* to synthesize a useful molecule, benzocyclopropene (28).

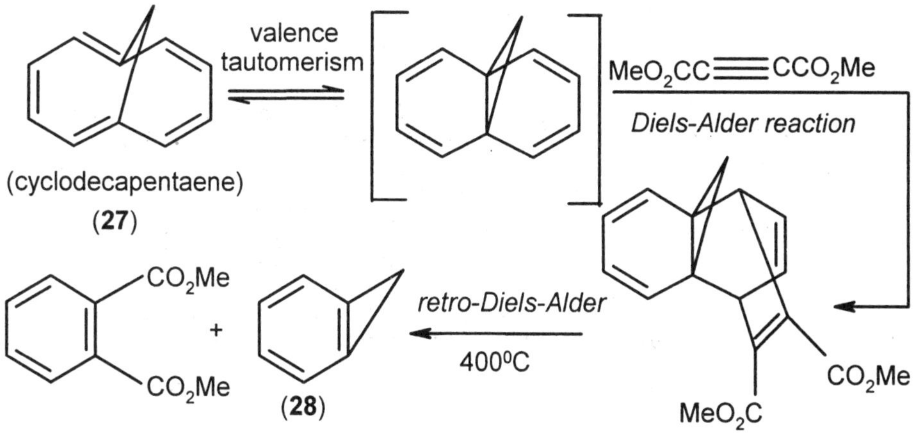

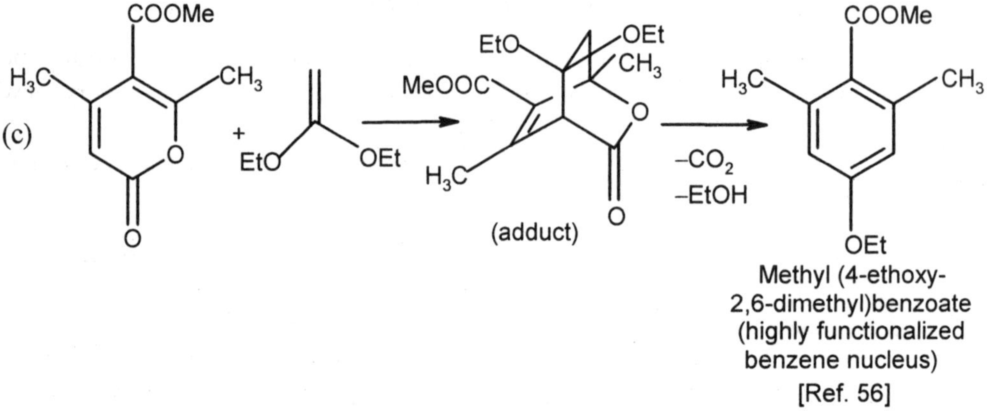

(d) Heterocyclic ring systems can be generated from the corresponding adduct.

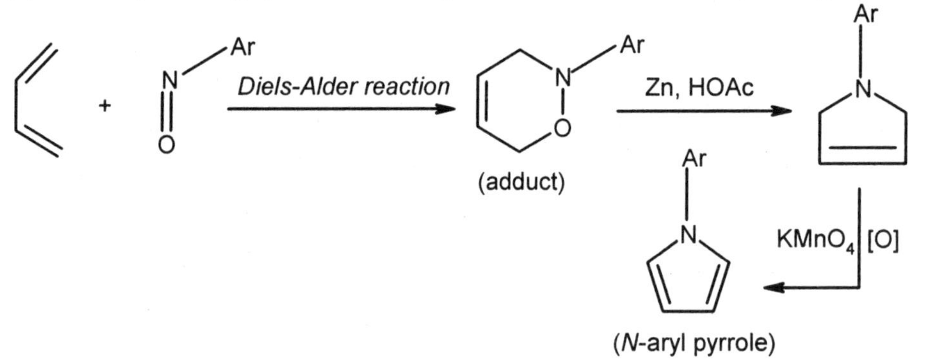

(e) Diels-Alder adducts formed on cycloaddition of cyclopentadienes with Fisher carbene complexes on subsequent treatment yield a variety of functional group derivatives[57].

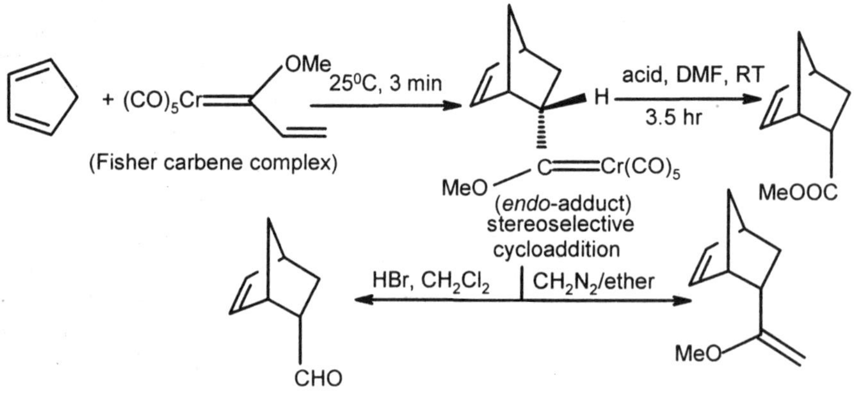

Diels – Alder Reaction

(f) Captan, an agricultural fungicide is synthesized industrially using the following reaction scheme where *Diels-Alder addition* is the key step.

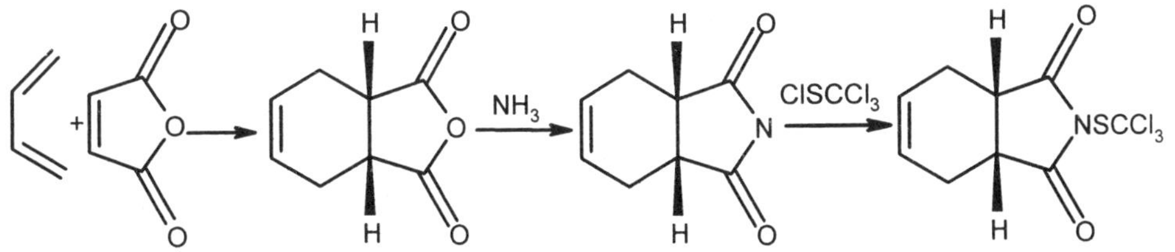

(g) Diels-Alder recaction involving *ortho*-quinodimethanes is a useful method for heterocyclic ring synthesis.

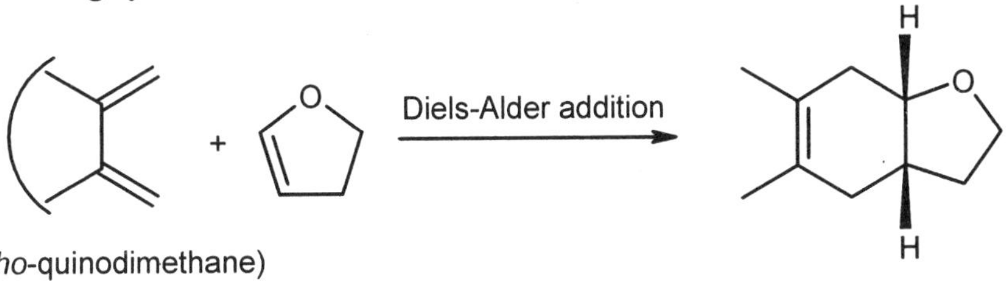

[Ref. 58]

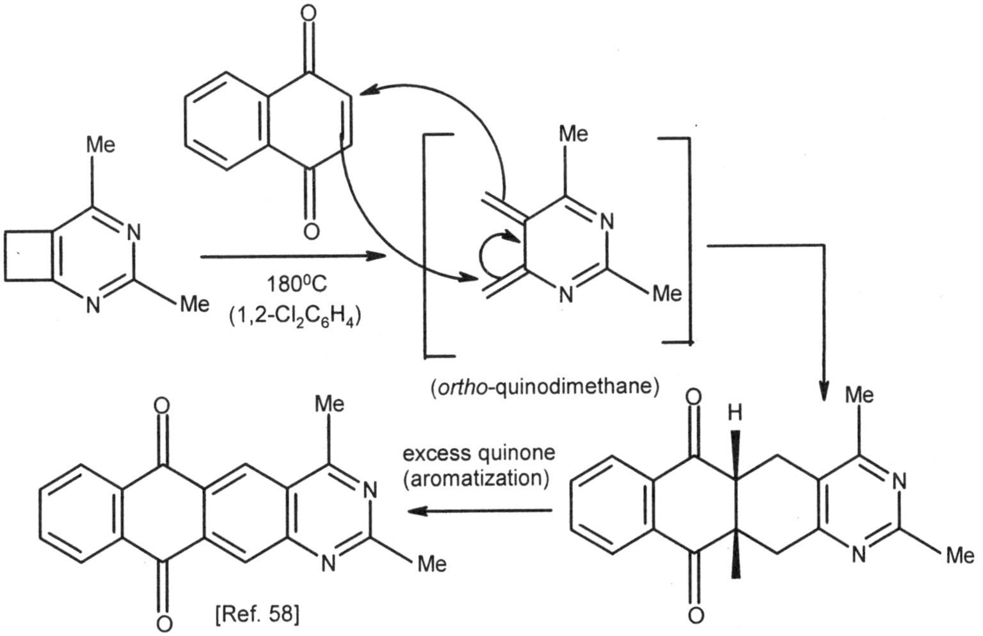

(h) Conversion of the initial adducts into the desired ketones:

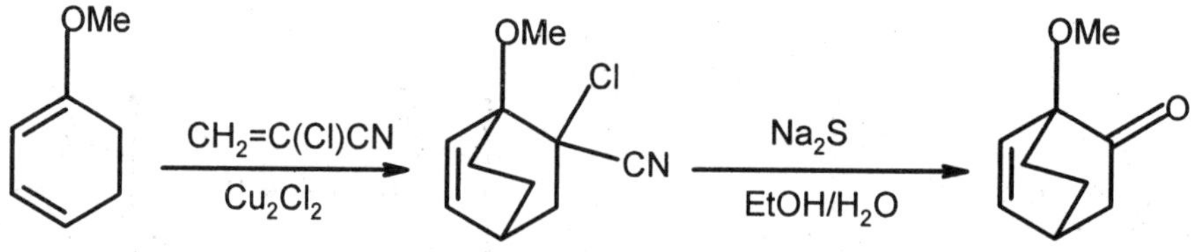

[Ref. 59]

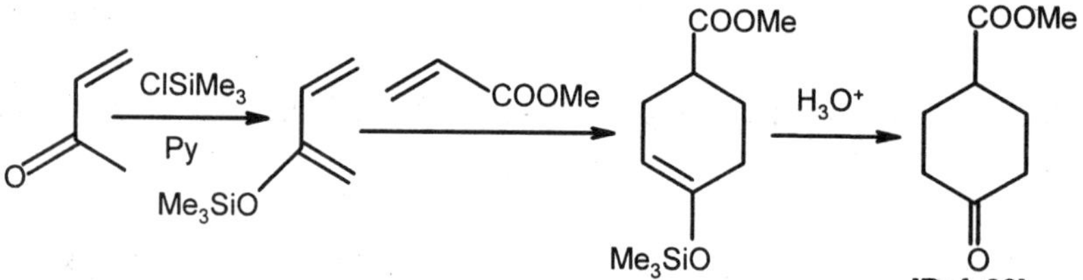

[Ref. 60]

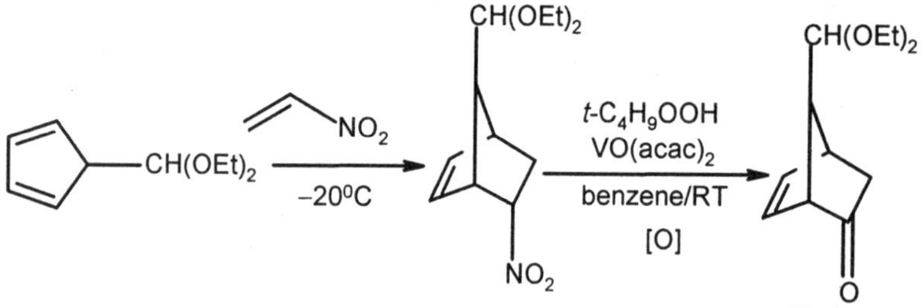

[Ref. 61]

References

1. Diels, O. and Alder, K. (1928), *Justus Liebigs Ann. Chem.*, **460**, 98.
2. Konovalov, A. I. (1983), *Russ. Chem. Rev.*, **52**, 1064. [Review]
3. Squillacote, M. E., Sheridan, R. S. and Chapman, O. L. (1979), *J. Am. Chem. Soc.* **101**, 3657.
4. Wagner-Jauregg, J. (1980), *Synthesis*, **165**, 769.
5. Friedman, L. (1967), *J. Am. Chem. Soc.*, **89**, 3071; Krespan, C. G. (1969), *J. Org. Chem.*, **34**, 1271.
6. Plieninger, H., Wild, D. and Westphal, J. (1969), *Tetrahedron*, **25**, 5561.
7. Wittig, G. and Niethammer, K. (1960), *Chem. Ber.*, **93**, 944.
8. Houk, K. N., George, J. K. and Duke Jr., R. E. (1974), *Tetrahedron*, **30**, 523; Houk, K. N., Luskus, L. J. and Bhacca, N. S. (1970), *J. Am. Chem. Soc.*, **92**, 6392; Sasaki, T., Kanamatsu, K. and Kotooka, T. (1975), *J. Org. Chem.*, **40**, 1201.

9 Ichihata, A., Kimura, R., Yamada, S. and Sakamura, S. (1980), *J. Am. Chem. Soc.*, **102**, 6353.
10 Rideout, D. C. and Breslow, R. (1980), *J. Am. Chem. Soc.*, **102**, 7816; Breslow, R. (1991), *Acc. Chem. Res.*, **24**, 159 (Review); Furlani, T. R. and Gao, J. (1996), *J. Org. Chem.*, **61**, 5492.
11 Smith, A. B., Liverton, N. J., Hrib, N. J., Sivaramakrishnan, H. and Winzenberg, K. (1986), *J. Am. Chem. Soc.*, **108**, 3040.
12 Davis, P. and Whitham, G. H. (1980), *J. Chem. Soc. Chem. Commun.*, 639.
13 Kuhn, H. J. and Gollnick, K. (1972), *Tetrahedron Lett.*, 1909.
14 Woodward, R. B. and Hoffmann, R. (1970), *The Conservation of Orbital Symmetry* (New York, Academic Press); Sauer, J. and Sustmann, R. (1980), *Angew. Chem. Internat. Edn.*, **19**, 779; Seltzer, S. (1965), *J. Am. Chem. Soc.*, **87**, 1534.
15 Sakai, S. (2000), *J. Physical Chemistry A*, **104**(5), 922.
16 Kiselev, V. D. and Konovalov, A. I. (1989), *Russ. Chem. Rev.*, **58**, 230 [Review].
17 Dewar, M. J. S and Pierini, A. B. (1984), *J. Am. Chem. Soc.*, **106** (1), 203; Leach, A. G. and Houk, K. N. (2001), *J. Org. Chem.*, **66**(15), 5192.
18 Leach, A. G. and Houk, K. N. (2001), *J. Org. Chem.*, **66**(15), 5192
19 Woodward, R. B. and Hoffmann, R. (1969), *Angew. Chem. Int. Edn. Engl.*, **8**, 781.
20 Herndon, W. C. and Hall, L. H. (1967), *Tetrahedron Lett.*, 3095.
21 Williamson, K. L., Hsu, Y. F. L., Lacko, R. and Youn, C. H. (1969), *J. Am. Chem. Soc.*, **91**, 6129; Houk, K. N. (1970), *Tetrahedron Lett.*, 2621.
22 Fringuelli, F. and Taticchi, A., *Dienes in the Diels-Alder Reaction*, 1990, John Wiley and Sons, New York.
23 Briger, G. and Bennett, J. N.(1980), *Chem. Rev.*, **80**, 63; Kametani, T. and Nemoto, H. (1981), *Tetrahedron*, **37**,3; Shea, K. J. and Wada, E. (1982), *J. Am. Chem. Soc.*, **104**, 5715); Shea, K. J. and Gilman, J. W. (1983), *Tetrahedron Lett.*, 657; Rousch, W. R. and Gillis, H. R. (1980), *J. Org. Chem.*, **45**, 4267.
24 Fullis, A.G. (1984), *Canadian J. Chem.*, **61**, 183 (Rview); Oppolzer, W. (1977), *Angew. Chem. Int. Edn. (Engl.)*, **16**, 10.
25 Schmittenner, H. F. and Weinreb, S. M. (1980), *J. Org. Chem.*, **45**, 3372; Weinreb, S. M., Khatri, N. A. and Shringarpure, J. (1979), *J. Am. Chem. Soc.*, **101**, 5073.
26 Oppolzer, W. (1972), *Angew. Chem. Int. Edn. (Engl.)*, **11**, 1031.
27 Van Royan, L. A., Mijngheer, R. and DeClercq, P. J. (1982), *Tetrahedron Lett.*, 3283.
28 Boeckmann Jr., R. K. and Kao, S. S. (1980), *J. Am. Chem. Soc.*, **102**, 7149.

29 Winsteil, S. and Shatavski, M. (1956), *J. Am. Chem. Soc.*, **78**, 592; Schmerling, L., Luviei, J. P. and Welch, R. W. (1956), *J. Am. Chem. Soc.*, **78**, 2819; Cristol, S. J., Brindell, G. D. and Reeder, J. A. (1958), *J. Am. Chem. Soc.*, **80**, 635; Krespan, C. J., McKusick, B. C. and Cairns, T. L. (1961), *J. Am. Chem. Soc.*, **83**, 3428.
30 Blomquist, A. T. and Meinwald, Y. C. (1959), *J. Am. Chem. Soc.*, **81**, 667.
31 Kobuke, Y., Sugimoto, T, Furukawa, J. and Tueno, T. (1972), *J. Am. Chem. Soc.*, **94**, 3633; Tabushi, I., Yamamura, K., Yoshida, Z. and Togashi, A. (1975), *Bull. Chem. Soc. Jpn.*, **48**, 2922.
32 Sasaki, T., Eguchi, S., Sugimoto, M. and Hibi, F. (1972), *J. Org. Chem.*, **37**, 2317.
33 Noyori, R., Umeda, I., Kawauchi, H. and Takaya, H. (1975), *J. Am. Chem. Soc*, **97**, 812; Yoshikawa, S., Aoki, K., Kiji, J. and Furukawa, J. (1975), *Bull. Chem. Soc. Jpn.*, **48**, 3239.
34 Paquette, L. A., Kessalmayer, M. A. and Kunzer, H. (1988), *J. Org. Chem.*, **53**, 5185.
35 Hoffmann, R. (1971), *Acc. Chem. Res.*, **4**, 1.
36 Goldstein, M. J., Natowasky, S., Heilbronner, E. and Hornung, V. (1973), *Hel. Chim. Acta*, **56**, 294.
37 Tabushi, I., Yamamura, K. and Yoshida, Z. (1972), *J. Am. Chem. Soc.,* **94**, 787.
38 le Noble, W. J. and Mukhtar, R. J. (1974), *J. Am. Chem. Soc.,* **96**, 6191.
39 Zimmerman, H. E. and Grunewald, G. L. (1964), *J. Am. Chem. Soc.*, **86**, 1434.
40 Hall, H. K. (1960), *J. Org. Chem.*, **25**, 42.
41 Fickes, G. N. and Metz, T. E. (1978), *J. Org. Chem.*, **43** (21), 4057.
42 (a) Yates, P. and Eaton, P. (1960), *J. Am. Chem. Soc.*, **82**, 4436; (b) Sbai, A., Branchadell, V., Ortuno, R. M. and Oliva, A. (1997), *J. Org. Chem.*, **62**, 3049; (c) Bonnesen, P. V., Puckett, C. L., Honeychuk, R. V. and Hersh, W. H. (1989), *J. Am. Chem. Soc.*, **111**, 6070.
43 Narasaka, K., Iwasawa, N., Inoue, M., Yamada, T., Nakashima, M. and Sugimori, J. (1989), *J. Am. Chem. Soc.*, **111**, 5340.
44 (a) Houk, K. N. and Strozier, R. W. (1973), *J. Am. Chem. Soc.*, **95**, 4094; (b) Alston, P. V. and Ottenbrite, R. M. (1975), *J. Org. Chem.*, **40**, 1111.
45 Sauer, J. and Kreel, J. (1966), *Tetrahedron Lett.*, 731, 6359.
46 (a) Houk, K. N. (1975), *Acc. Chem. Res.*, **8**, 361; (b) Sustmann, R. (1974), *Pure and Appl. Chem.*, **40**, 569.
47 (a) Gassman, P. G. and Singleton, D. A. (1984), *J. Am. Chem. Soc.*, **106**, 7993; (b) Pabon, R. A., Bellville, D. J. and Bauld, N. L. (1983), *J. Am. Chem. Soc.*, **105**, 5158; (c) Bauld, N. L. (1989), *Tetrahedron*, **45**, 5307.

48 Ishihara, K., Kurihara, H. and Yamamoto, H. (1996), *J. Am. Chem. Soc.*, **118**, 3049.
49 Singh, V. K., Raju, B. N. S. and Deota, P. T. (1988), *Synthetic Commun.*, **18**, 567.
50 (a) Berlan, J., Giboreau, P., Lefeuvre, S. and Marchand, C. (1991), *Tetrahedron Lett.*, **32**, 2363; (b) DaCunha, L. and Garrigues, B. (1997), *Bull. Soc. Chim. Belg.*, **106**, 817.
51 Raj, C. P., Dhas, N. A., Cherkinski, M., Gedanken, A. and Braverman, S. (1998), *Tetrahedron Lett.*, **39**, 5413.
52 Pandey, B. and Dalvi, P. V. (1993), *Angew. Chem. Int. Ed. Engl.*, **32**, 1612.
53 Pagni, R. M., Kabalka, G. W., Hondrogiannis, G., Bains, S., Anosike, P. and Kurt, R. (1993), *Tetrahedron,* **49**, 6743.
54 Corey, E. J. and Mock, W. L. (1962), *J. Am. Chem. Soc.*, **84**, 685.
55 Vogel, E., Grimme, W. and Korte, S. (1965), *Tetrahedron Lett.*, 3625.
56 (a) Jung, M. E. and Hagenah, J. A. (1989), *J. Org. Chem.*, **52**, 1889; (b) Boger, D. and Mulhcan, M. D. (1984), *J. Org. Chem.*, **49**, 4033.
57 (a) Wulff, W. D. and Yang, D. C. (1983), *J. Am. Chem. Soc.*, **105**, 6726; (b) Dotz, K. H., Noack, R., Harmas, K. and Moller, G. (1990), *Tetrahedron*, **46**, 1235.
58 Herrera, A., Martinez, R., Gonzalez, B., Illescas, B., Martin, N. and Seoane, C. (1997), *Tetrahedron Lett.*, **38**, 4873.
59 Madge, N. C. and Holmes, A. B. (1980), *J. Chem. Soc. Chem. Commun.*, 956.
60 Jung, M. E., McCombs, C. A., Takeda, Y. and Pan, Y. G. (1981), *J. Am. Chem. Soc.*, **103**, 6677.
61 Bartlett, P. A., Green, F. R., Webb, T. R. (1977), *Tetrahedron Lett.*, 331; Ranganathan, D., Rao, C. B., Ranganathan, S., Mehrotra, A. K. and Iyengar, R. (1980), *J. Org. Chem.*, **45**, 1185; Kornblum, N., Erickson, A. E., Kelly, W. J. and Henngler, B. (1982), *J. Org. Chem.*, **47**, 4534.

- Dave, P. R., Duddu, R., Surapaneni, R. and Gilardi, R. (1999), *Tetrahedron Lett.*, **40**, 443.
- Mehta, G. and Uma, R. (2000), *Acc. Chem. Res.*, **33**, 278. [Review]
- Jorgensen, K. A. (2000), *Angew. Chem. Int. Edn. Engl.*, **39**, 3558.
- Evans, D. A., Johnson, J. S. and Olhava, E. J. (2000), *J. Am. Chem. Soc.*, **122**, 1635.
- Hung, Y. and Rawal, V. H. (2000), *Org. Lett.*, **2**, 3321.
- Deyle, M. P., Phillips, I. M. and Hu, W. (2001), *J. Am. Chem. Soc.*, **123**, 5366.
- Richter, F., Bauer, M., Perez, C., Maichle-Mossmer, C. and Maier, M. E. (2002), *J. Org. Chem.*, **67**, 2474.
- Gainelli, G., Galletti, P., Giacomini, D. and Quintavalla, A. (2003), *Tetrahedron Lett.*, **44**, 93.

- Osborn, H. M. I. and Coisson, D. (2004), *Mini-Reviews in Organic Chemistry*, **1**, 41. [Review].
- Sunakawa, T. and Kuroda, C. (2005), *Molecules*, **10**, 244.
- Branowska, D. (2005), *Molecules*, **10**, 265.
- Branowska, D. (2005), *Molecules*, **10**, 274.

6

Favorskii Rearrangement

Introduction

F*avorskii rearrangement* is regarded as one of the most useful synthetic tools in organic chemistry. The base-induced rearrangement of α-haloketones to carboxylic acid derivatives is widely known as the *Favorskii rearrangement*[1]. The use of hydroxide ions or alkoxide ions or amines as bases leads to the formation of free carboxylic acid (salt) or ester or amide, respectively, containing the same number of carbon atoms. At the same time, concurrent formation of hydroxyketones and some solvolysis products is also observed[2]. With cyclic α-haloketones, the *Favorskii rearrangement* leads to a ring contraction by one carbon atom. A. E. Favorskii[3,4] observed such a ring contraction reaction when 2-chlorocyclohexanone was treated with alcoholic alkali.

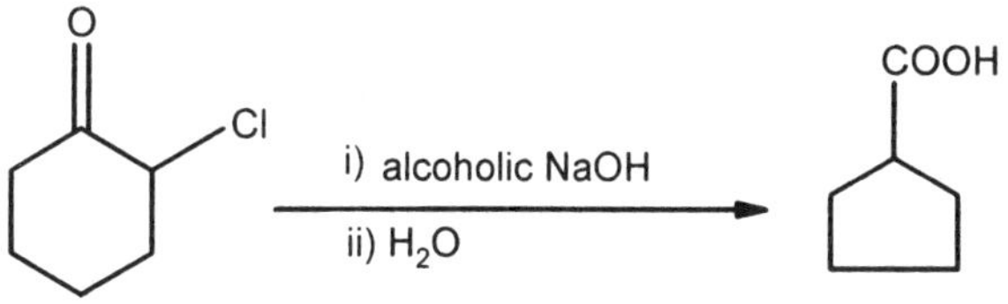

Favorskii rearrangement is akin to *Wallach degradation*[5] which involves the rearrangement of α,α' (or 2,6)-dibromocyclohexanones to 1-hydroxycyclopentane-carboxylic acids, followed by oxidation to the corresponding ketones.

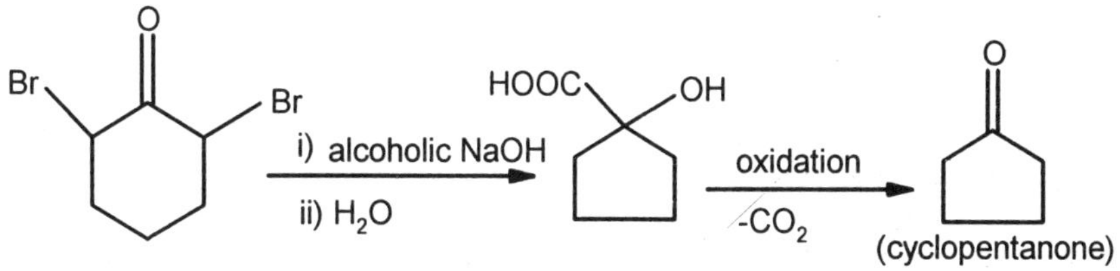

The *Favorskii rearrangement* had already been found to be applicable to a wide variety of substrates, and is regarded as a versatile and generalized reaction of α-haloketones.

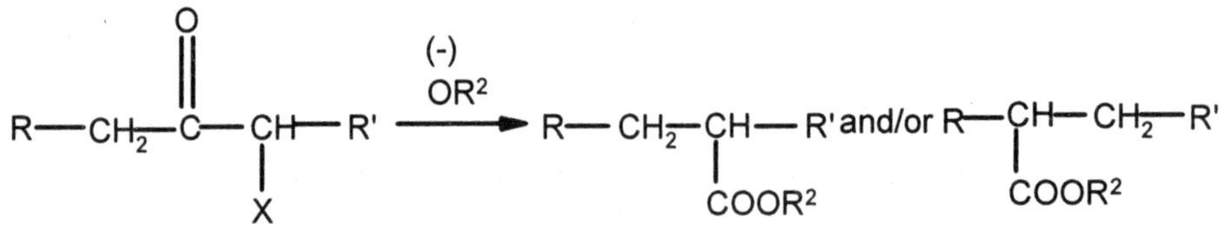

$$R-CH_2-\overset{O}{\underset{\|}{C}}-\underset{X}{\overset{|}{CH}}-R' \xrightarrow{\overset{(-)}{OR^2}} R-CH_2-\underset{COOR^2}{\overset{|}{CH}}-R' \text{ and/or } R-\underset{COOR^2}{\overset{|}{CH}}-CH_2-R'$$

Examples

A few examples of reactions undergoing *Favorskii rearrangement* are cited below.

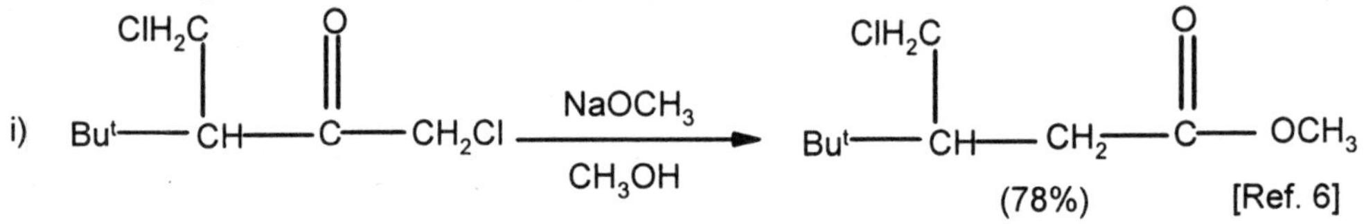

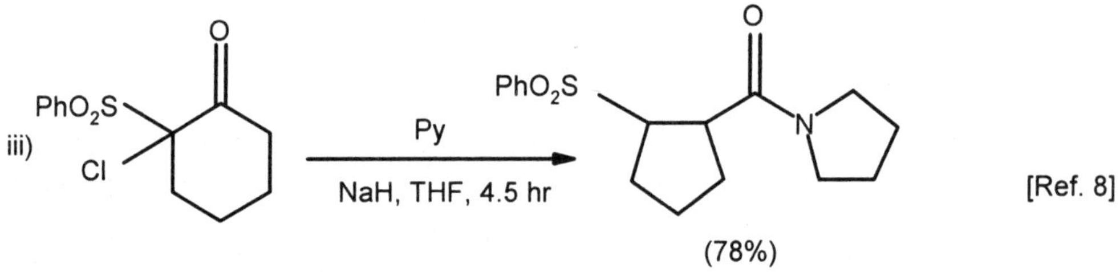

i) Bu^t—CH(CH$_2$Cl)—C(=O)—CH$_2$Cl $\xrightarrow[\text{CH}_3\text{OH}]{\text{NaOCH}_3}$ Bu^t—CH(CH$_2$Cl)—CH$_2$—C(=O)—OCH$_3$ (78%) [Ref. 6]

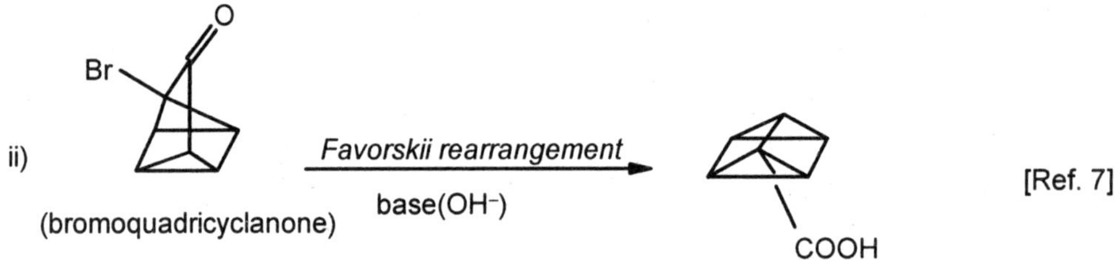

ii) (bromoquadricyclanone) $\xrightarrow[\text{base(OH}^-\text{)}]{\textit{Favorskii rearrangement}}$ [quadricyclane-COOH] [Ref. 7]

iii) [2-chloro-2-(PhO$_2$S)cyclohexanone] $\xrightarrow[\text{NaH, THF, 4.5 hr}]{\text{Py}}$ [2-(PhO$_2$S)cyclopentyl pyrrolidinyl ketone] (78%) [Ref. 8]

Discussion on mechanism

For a long period, the mechanism[9] of *Favorskii rearrangement* has been the subject of much investigation and from time to time a number of different mechanisms were proposed for this rearrangement reaction. The *Favorskii rearrangement* was believed to involve a cyclopropanone[10] or mesomeric zwitterion[10,11]. However, Turro and Hammond[12] gave the first direct evidence for the long-standing hypothesis that cyclopropanones are readily cleaved to Favorskii products by strong base. The possible intermediacy of a cyclopropanone hemiketal (as its anion) was also suggested[13]. Besides, a *"semi-benzilic like"* mechanism[14] had also been proposed for certain ketonic substrates.

Mechanistic pathway involving cyclopropanone intermediate

It has been observed that the stereochemistry and structure of most of the Favorskii products are best reasonably explained in term of a *cyclopropanone* intermediate. The well-accepted general mechanistic scheme involving *cyclopropanone* intermediate for *Favorskii rearrangement* is given below:

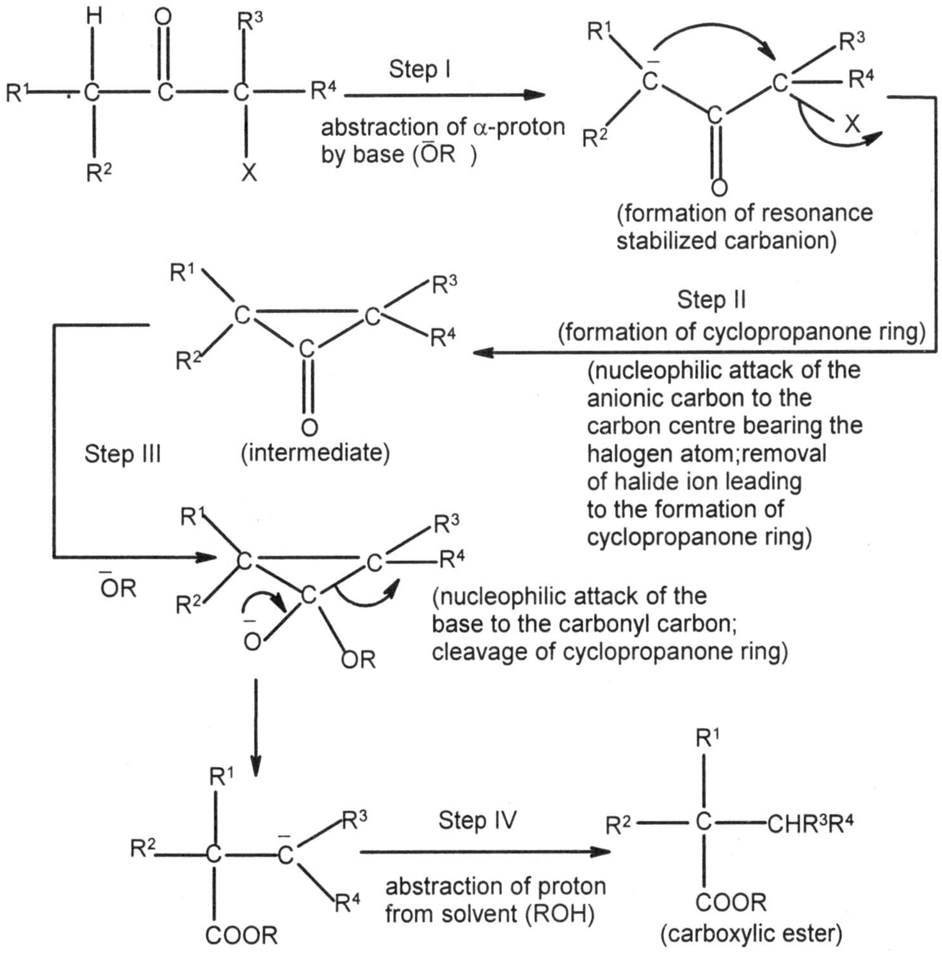

Experimental proof in favour of cyclopropanone intermediate

Several experimental evidences in favour of the formation of cyclopropanone ring as the reaction intermediate were reported:

a) Bordewell et al.[15] observed that the same product (**4**) is obtained from two different α-chloroketones (**1** & **2**) as given in the following reaction scheme. This can most reasonably be explained by assuming that both the substrates undergo the rearrangement through common *cyclopropanone* intermediate (**3**).

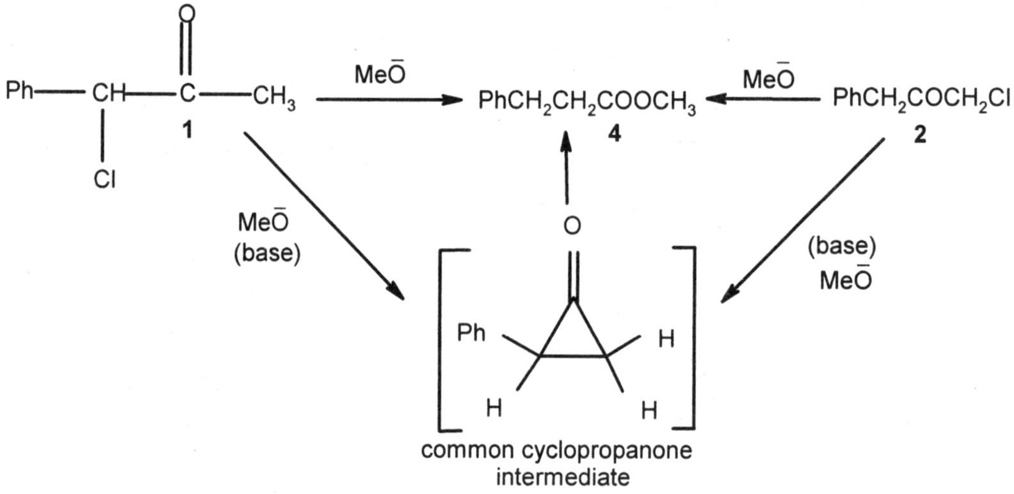

Ph—CH(Cl)—C(=O)—CH₃ (**1**) $\xrightarrow{\text{MeO}^-}$ PhCH₂CH₂COOCH₃ (**4**) $\xleftarrow{\text{MeO}^-}$ PhCH₂COCH₂Cl (**2**)

$\downarrow$ MeŌ (base) $\qquad\qquad$ $\uparrow$ $\qquad\qquad$ $\swarrow$ (base) MeŌ

common cyclopropanone intermediate (**3**) — cyclopropanone with Ph, H, H, H substituents

b) Another important experiment was carried out with the help of radioactive labeling by Lotfield[16]. The author obtained convincing results from ^{14}C-labelling studies on 2-chlorocyclohexanone (**5**); an even distribution of ^{14}C between the two carbon centers (C-1 & C-2) was found in the product formed.

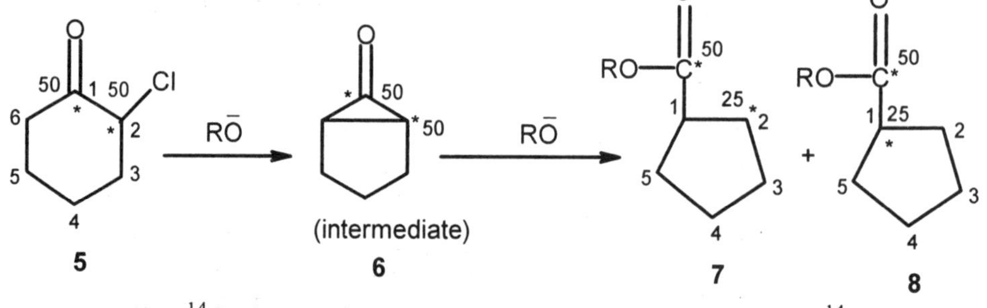

5 (2-chlorocyclohexanone, C-1: 50, C-2: 1, labelled) $\xrightarrow{\text{RO}^-}$ **6** (intermediate) $\xrightarrow{\text{RO}^-}$ **7** (RO—C*(=O), C-1: 50, C-2: 25*) + **8** (RO—C*(=O), C-1: 25, C-2: *)

(* = ^{14}C, the number nearly stands for the percentage of ^{14}C)

Although C-2 & C-6 of (**5**) are not equivalent, but it is the fact that C-1 & C-2 in the products (**7** & **8**) were found to be equally labelled (25% each). This phenomenon can be understood from the formation of a symmetrical intermediate[16]. The type of intermediate that fits the circumstances is a *cyclopropanone* (**6**). A preliminary migration of the chlorine atom from C-2 was ruled out by the fact that recovered product (**5**) had the same isotopic distribution as the starting (**5**).

c) Although cyclopropanones are very reactive compounds, several of them have been isolated and trapped[17]; *e.g.*

PhCH₂COCHPh | Cl →(base)→ Ph,H / Ph,H cyclopropanone with =O (intermediate) →(furan)→ Ph-[O-bridged bicyclic with =O]-Ph (adduct)

Buᵗ—CH₂—CO—CH(Cl)—Buᵗ →(base)→ Buᵗ,H / Buᵗ,H cyclopropanone with =O (cyclopropanone intermediate isolated and trapped [18,19])

Besides, cyclopropanones synthesized by other methods also have been shown to give Favorskii products on treatment with NaOMe or other bases[20].

Direction of Cyclopropanone ring opening

In general, it is thought that the direction of ring opening of cyclopropanone intermediates may be predicted by considering the stability order of the two possible carbanions produced by base attack on the cyclopropanone; formation of more stable carbanion directs the path of the ring opening. For example, treatment of 2,2-dimethylcyclopropanone (**9**) with NaOMe in CH_2Cl_2 or MeOH followed by hydrolysis leads to the formation of methyl trimethylacetate (**11**) in quantitative yield[21]. The absence (<0.5%) of methyl isopropylacetate (**13**) is consistent with the hypothesis that the lower energy of the transition state leading to **11** relative to **15** is in tune with the lower energy of primary carbanion (**10**) in comparison to **12**. The rearrangement of 3-bromo-3-methyl-2-butanone (**14**) with methoxide in ether was found to result exclusively in the formation of (**11**)[20b,22]

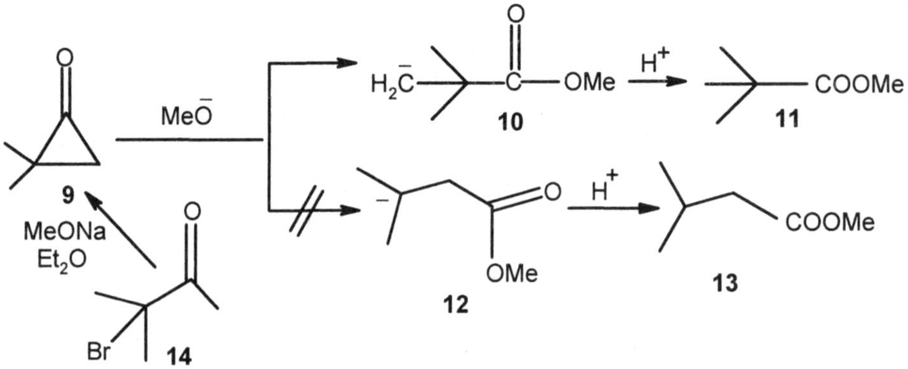

But in case of the rearrangement of α-bromoisopropyl ketones (**15**), which can be assumed to proceed *via* the *cyclopropanone* intermediate (**22**), was observed to yield a mixture of esters (**17**) & (**18**)[23].

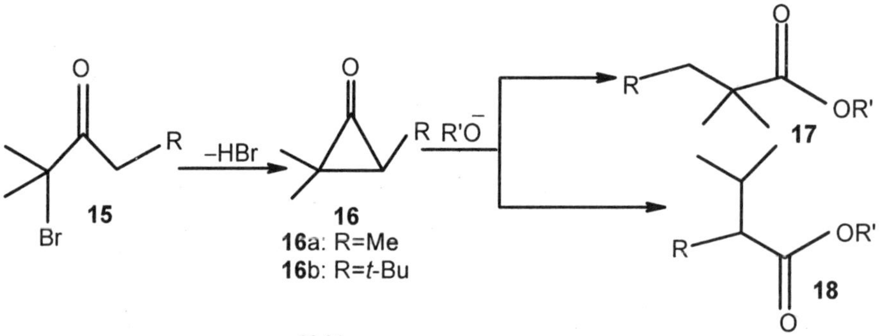

On the basis of later works[23,24], however, it is argued that this cleavage of the cyclopropanone may occur from both directions depending upon the carbanion stabilities as well as steric and other factors that play a significant role in determining the energy of the transition state for cleavage.

An interesting observation relating to the reverse result in case of **19** was observed since the carbanion generated from cleavage of less substituted bond is stabilized by the phenyl group as anticipated [17b,25].

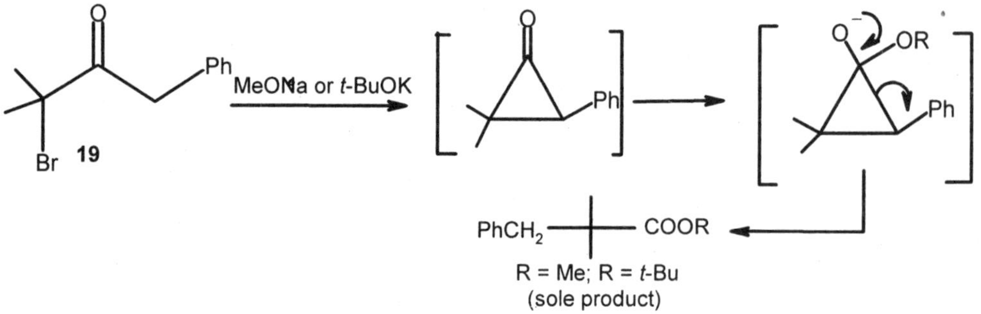

Favorskii rearrangement involving semi-benzilic like mechanism

It is also observed that α-halo ketones which do not possess an α-hydrogen atom, still undergo this reaction to yield the same type of rearranged product. This is usually called the *quasi-Favorskii rearrangement*. In 1939, Tchoubar and Sackur[14a] experimentally observed that base-catalyzed rearrangement of α-chlorocyclohexyl-phenyl ketone (**20**) yields 1-phenyl-1-cyclohexane carboxylic acid.

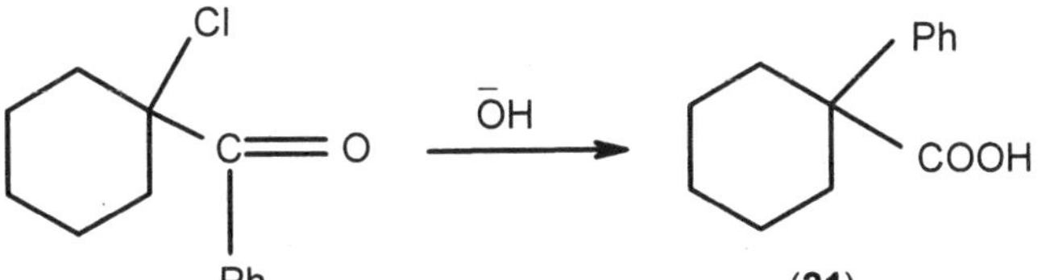

Such type of rearrangement obviously cannot take place by the *cyclopropanone mechanism*. Hence, an alternative pathway was proposed by Tchoubar and Sackur[14a] for the above reaction —

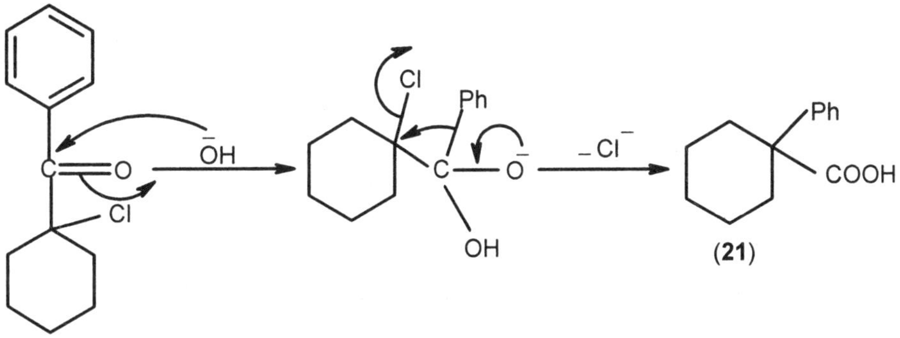

This mechanism is very similar to the *benzilic acid rearrangement*; this mechanism was latter (1952) supported by Stevens and Farkas[26] and termed as *semi-benzilic like* mechanism. Thus, a general scheme for *Favorskii rearrangement* ionvolving *semi-benzilic like* mechanism may be given as —

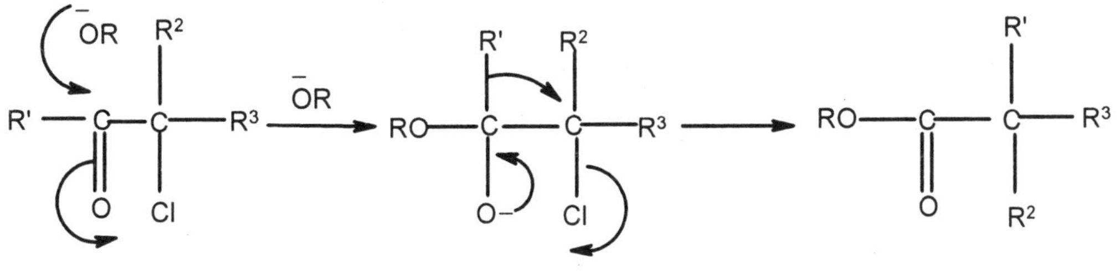

From the above scheme it is seen that this mechanism requires inversion at the migration terminus and in practice it has been found to be so[27].

One more example of *quasi-Favorskii rearrangement*, reported in recent times, is cited here. Harmata et al.[28] has reported that compound **(22)** bearing a bromo substituent at a bridge-head position undergoes a straightforward *quasi-Favorskii rearrangement* in near quantitative yield on reaction with lithiumaluminiumhydride (LAH). The implications of this process for the synthesis of angular triquinanes are obvious and are being pursued.

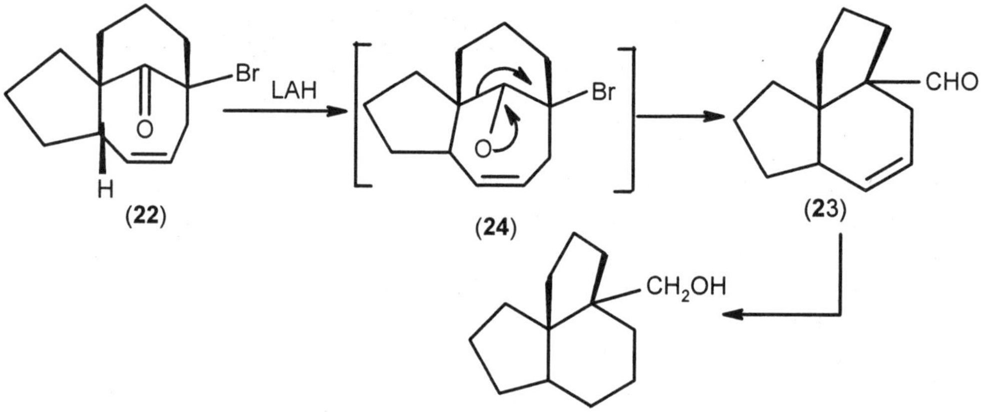

It has been shown that even when there is a suitably placed α-hydrogen, the *semi-benzilic like* mechanism may still operate[29]. Thus, 2-bromocyclobutanone undergoes *Favorskii rearrangement* when treated with water as the base. When treated with D_2O, no deuterium is incorporated in the ring. However, *semi-benzilic like* mechanism can explain the formation of the actual product without deuterium insertion in the ring.

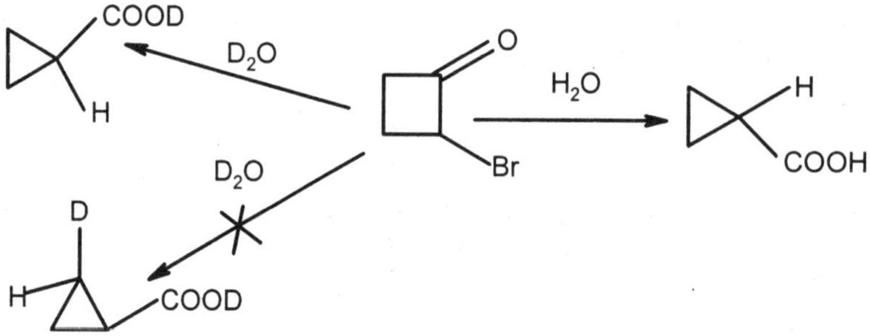

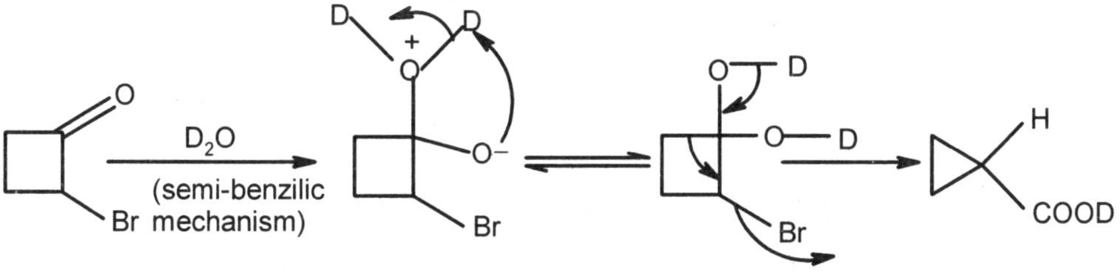

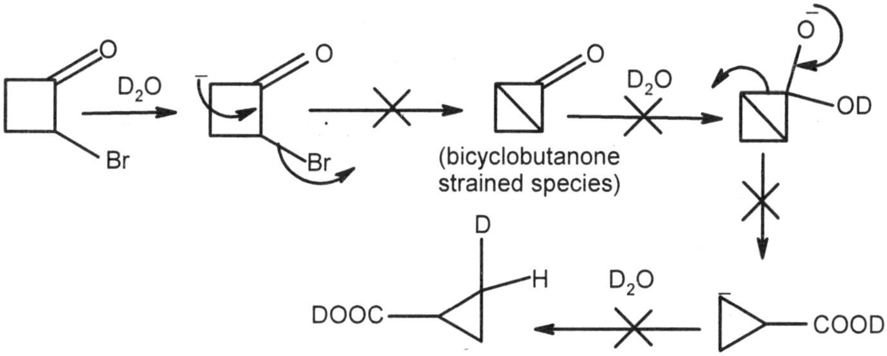

(Probably, the strain in the bicyclobutanone ring restricts the operation of cyclopropanone mechanism.)

Applications

A very few illustrations are mentioned below:

(a) The most extensive use of *Favorskii rearrangement* is by far in ring contractions. An on-top example is the synthesis of Eaton's cubane (**28**) *via* a double *Favorskii rearrangement*[30] as depicted below:

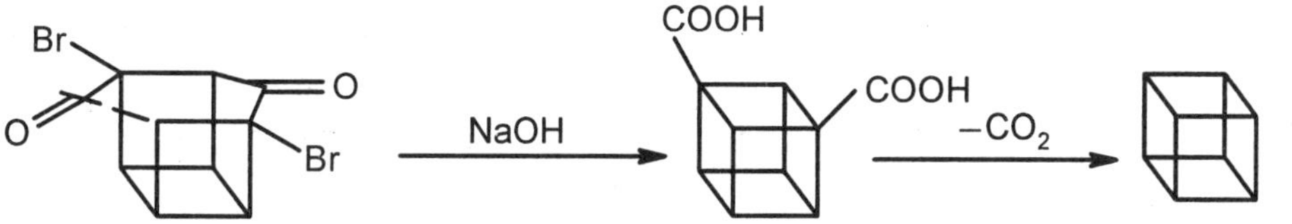

(b) The application of *quasi-Favorskii rearrangement* of α-haloketones having no α-hydrogen, in the preparation of two analgesics, Demerol (ethyl 1-methyl-4-phenyl-4-piperidine carboxylate, **32**) and β-Pethidine (ethyl 1-methyl-3-phenyl-3-piperidine carboxylate, **33**) were reported by Smissman and Hite[31]. This constitutes a novel synthesis for analogs of Demerol-type compounds.

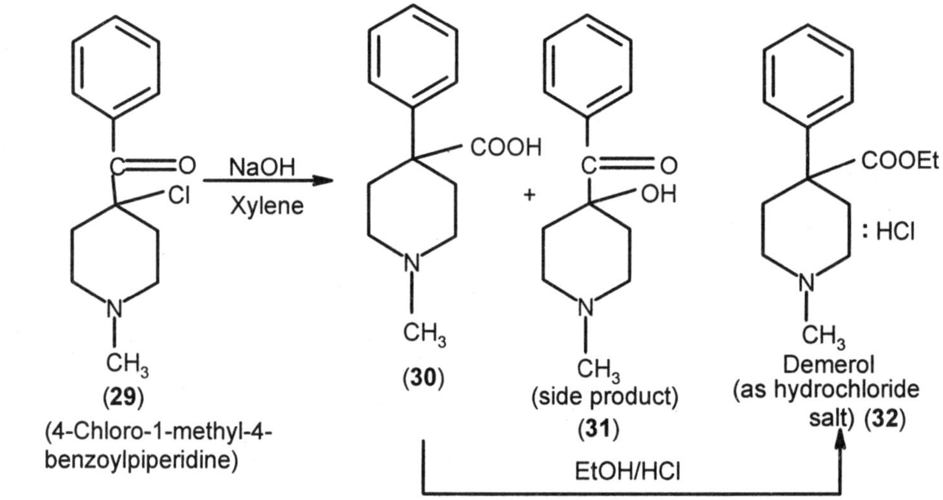

[Scheme showing conversion of (29) 4-Chloro-1-methyl-4-benzoylpiperidine with NaOH/Xylene to (30) plus side product (31), then EtOH/HCl to give Demerol (as hydrochloride salt) (32)]

They prepared β-Pethidine (**33**) with the help of same sequence of reactions using nicotinic acid as the starting material.

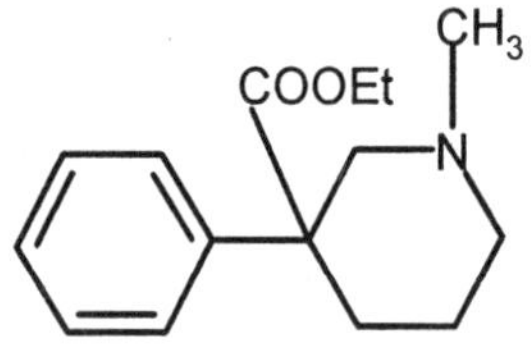

(β-Pethidine)
(**33**)

(c) In 1993, Satoh *et al.*[32] reported the application of *Favorskii rearrangement* as the key step in the formation of optically active α-alkyl amides from optically active α-chloro, α-sulfonyl ketones.

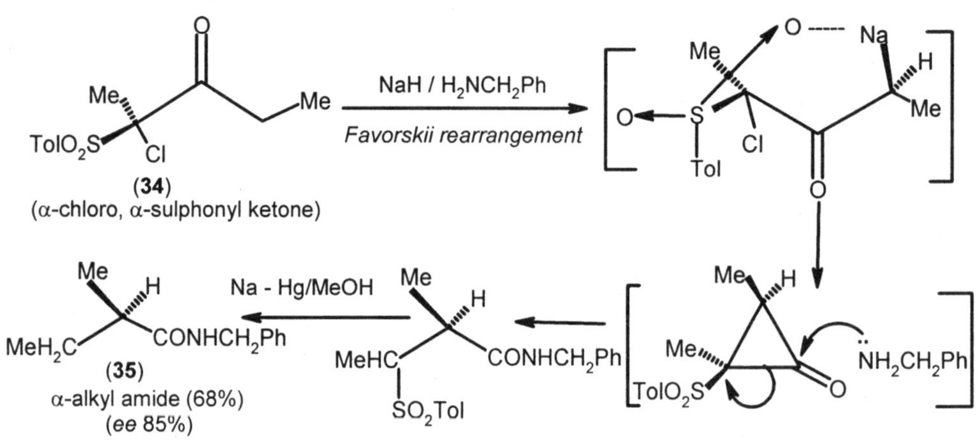

[Scheme showing (**34**) (α-chloro, α-sulphonyl ketone) reacting with NaH / H$_2$NCH$_2$Ph via Favorskii rearrangement through cyclopropanone intermediate, then Na-Hg/MeOH to give (**35**) α-alkyl amide (68%) (ee 85%)]

i) Instead of alkoxide ions, the reaction of α-haloketones is also reported to take place in presence of BF_3 – MeOH and Ag^+. The first time reporting on such silver assisted rearrangement for 1^0 and 2^0 α-bromo-alkylarylketones was made by Giordano et al.[33]

$$Ar-\underset{\underset{R}{|}}{\overset{\overset{O}{\|}}{C}}-CH-Br \xrightarrow[Ag^+(Ag_2CO_3)]{BF_3, 2MeOH} Ar-\underset{\underset{R}{|}}{CHCOOMe}$$

(rearranged product)

Ar	R	%yield
6'-Methoxy-2'-naphthyl	Me	85
4'-Methoxy-2'-phenyl	Me	78
Phenyl	H	8
4'-Chlorophenyl	H	4

It has been found that electron-releasing groups favour aryl migration; hence an electrophilic character of the reaction in the transition state may be assumed.

References

1. Favorskii, A.E. (1894), *J. Russ. Phys. Chem. Soc.*, **26**, 559; *ibid.* (1895), *J. Prakt. Chem.*, **51**, 533.
2. House, H.O. and Frank, G.A. (1965), *J. Org. Chem.*, **30**, 2948.
3. Favorskii, A.E (1913)., *J. Prakt. Chem.*, **88**, 658.
4. Favorskii, A.E. and Bozhovskii, V.N. (1914), *J. Russ. Phys. Chem. Soc.*, **46**, 1097.
5. Wallach, O. (1918), *Ann.*, **414**, 296.
6. Schamp, N., Dekimpe, N. and Coppens, W. (1975), *Tetrahedron*, **31**, 2081.
7. Boyer, L.E., Brazzilo, J., Forman, M.A. and Zanoni, B. (1996), *J. Org. Chem.*, **61**(21), 7611.
8. Lee, E. and Yoon, C.H.J. (1994), *J. Chem. Soc., Chem. Commun.*, 479.
9. For a review of the mechanism, see Baretta, A., Waegell, B. (1982), *React. Intermed.* (Plenum), **2**, 527.
10. (a) Kenede, A.S. (1960), *Org. Reactions*, **11**, 261; (b) Loftfield, R.B. (1950), *J. Am. Chem. Soc.*, **72**, 632.
11. (a) McPhee, W. D. and Klingberg, E. (1944), *J. Am. Chem. Soc.*, **66**, 132; (b) Fort, A.W. (1962), *ibid.*, **84**, 2620, 2625, 4979; (c) Aston, J.G. and Newkirk, J. D.

(1951), *ibid.,* **73**, 3900; (d) Burr, J.G. and Dewar, M.J.S. (1954), *J. Chem. Soc.*, 1201; (e) Cookson, R.C. aand Nye, M. J. (1963), *Proc. Chem. Soc.*, 123; *ibid.* (1965), *J. Chem. Soc.*, 2009.

12 Turro, N.J. and Hammond, W.B. (1965), *J. Am. Chem. Soc.*, **87**(14), 3258.

13 (a) House, H.O. and Gilmore, U.F. (1961), *J. Am. Chem. Soc.*, **83**,. 3980; (b) House H.O. and Thompson, H.W. (1963), *J. Org. Chem.*, **28**, 164.

14 (a) Tchoubar, B. and Sackur, O. (1939), *Compt. Rend. Acad. Sci.*, **208**, 1020; (b) Stevens, C.L. and Farkas, E. (1952), *J. Am. Chem. Soc.*, **74**, 5352; (c) Rappe, C. and Knutson, L. (1967), *Acta Chem. Scand.*, **21**, 163; (d) Conia, J.M. and Salaum, J. (1963), *Tetrahedron Lett.*, 1175.

15 Bordwell, F.G. and Strong, J.G. (1973), *J. Org. Chem.*, **38**, 579.

16 Lotfield, R.B. (1951), *J. Am. Chem. Soc.*, **73**, 4707.

17 (a) Wasserman, H.H., Clark, G.M. and Turley, P.C. (1974), *Top. Curr. Chem.*, **47**, 73 ; (b) Turro, N.J. (1969), *Acc. Chem. Res.*, **2**, 25.

18 Pazos, J.F., pacifici, J.G., Pierson, G.O., Sclove, D.B. and Greene, F.D. (1974), *J. Org. Chem.*, **39**, 1990.

19 (a) Fort, A.W. (1962), *J. Am. Chem. Soc.*, **84**, 4979; (b) Cookson, R.C. and Nye, M.J. (1963), *Proc. Chem. Soc.*, 129; (c) Breslow, R., Posner, J. and Krebs, A. (1963), *J. Am. Chem. Soc.*, **85**, 234; (e) Baldwin, J.E. and Cardellina, J.H.I. (1968), *Chem. Commun.,* 558.

20 (a) Crandell, J.K. and Machleder, W.H. (1968), *J. Org. Chem.*, **90**, 7347; (b) Turro, N.J., Gagosian, R.B., Rappe, C. and Knutsson, L. (1969), *Chem. Commun.*, 270; (c) Wharton, P.S., Fritberg, A.R. (1968), *J. Org. Chem.*, **37**, 1899.

21 Hammond, W.B. and Turro, N.J. (1966), *J. Am. Chem. Soc.*, **88**, 2880.

22 Aston, J.G. and Greenburg, R.B. (1940), *J. Am. Chem. Soc.*, **62**, 2590.

23 Rappe, C. and Knutson, L. (1967), *Acta Chem. Scand.*, **21**, 2205.

24 Rappe, C., Knutsson, L., Turro, N.J. and Gagosian, R.B. (1970), *J. Am. Chem. Soc.*, **92**(7), 2032.

25 Cram, D.J. (1965), *"Fundamentals of Carbanion Chemistry"*, Academic Press, New York, N.Y., p.243.

26 Stevens, C.L. and Farkas, E. (1952), *J. Am. Chem. Soc.,* **74**, 618, 5352.

27 Baudry, D., Begue, J. and Charpentier – Morize, M. (1971), *Bull. Soc. Chim. Fr.*, 1416; *ibid.* (1970)*Tetrahedron Lett.*, 2147.

28 Harmata, M., Bohnert, G., Kurti, L. and Barnes, C.L. (2002), *Tetrahedron Lett.*, **43**, 2347.

29 Salaun, J.R., Garnier, B. and Conia, J.M. (1973), *Tetrahedron*, **29**, 2895; (b) Warnhoff, E.W., Wong, C.M. and Tai, W.T. (1968), *J. Am. Chem. Soc.*, **90**, 514.

30 Barborak, J.C., Watts, L. and Petit, R. (1966), *J. Am. Chem. Soc.*, **88**, 1328.

31 Smissman, E.E. and Hite,, G. (1959), *J. Am. Chem. Soc.*, **81**, 1201.

32 Satoh, T., Motohashi, S., Kimura, S., Tokutaka, N. and Yamakawa, K. (1973), *Tetrahedron Lett.*, **34**(30), 4823.
33 Giordano, C., Castaldi, G., Casagrande, F. and Abis, L. (1982), *Tetrahedron Lett.*, **23**, 1385.

- Gambacorta, A., Turchetta, S., Bovivelli, P. and Botta, M. (1991), *Tetrahedron*, **47**, 9097.
- Satoh, T., Oguro, K., Shishikura, J-i., Kanetaka, N., Okada, R. and Yamakawa, K. (1993), *Bull. Chem. Soc., Japan*, **66**, 2339.
- Satoh, T., Oguro, K., Shishikura, J-i., Kanetaka, N., Okada, R. and Yamakawa, K. (1993), *Bull. Chem. Soc., Japan*, **66**, 2339.
- Braverman, S., Cherkinsky, M., Kumar, E. V. K. S. and Gottlieb, H. E. (2000), *Tetrahedron*, **56**, 4521.
- Mamedov, V. A., Tsuboi, S., Mustakimova, L. V., Hamamoto, H., Gubaidullin, A. T., Litvinov, I. A. and Levin, Y. A. (2001), *Chem. Heterocycl. Compd.*, **36**, 911.
- Muldgaard, L., Thomsen, I. B., Hazell, R. G. and Bols, M. (2002), *J. Chem. Soc., Perkin I,* 1297.
- Harmata, M. and Bohnet, G. (2003), *Org. Lett.*, **5**, 59.

7

Fries Rearrangement

Introduction

Fries rearrangement* is a synthetically useful reaction involving the conversion of phenolic esters to acylated phenols (or may say rearrangement of aryl esters to aryl ketones) in presence of Lewis acid as a catalyst. A phenolic ester when heated with a Lewis acid in an inert solvent, it undergoes a rearrangement reaction to yield a mixture of both *ortho-* and *para-*acylphenols on subsequent hydrolysis. This is usually known as *Fries rearrangement*. Besides Lewis acids, light has also been found to act as catalyst in certain cases.

Catalyst, solvent and reaction temperature

The common Lewis acid used as catalyst in *Fries rearrangement* reaction is anhydrous aluminium chloride/bromide, and the common solvents used are carbondisulphide, nitrobenzene, chlorobenzene, etc. The *ortho-* and *para-*isomers can be separated from their mixture by means of steam distillation or fractional crystallization, as the case may be. However, it is often possible to select conditions so that either one of the isomers predominates.

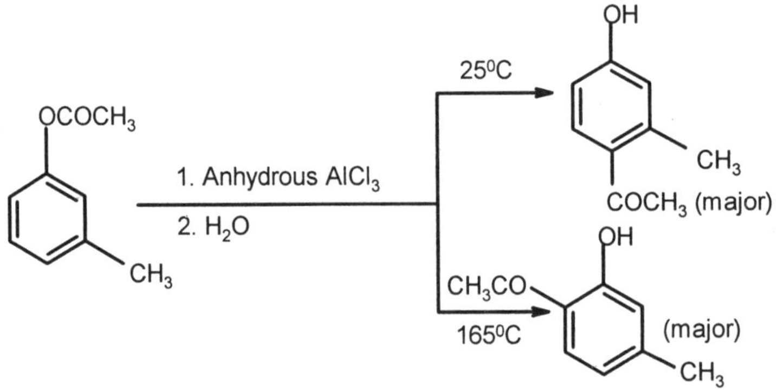

The *ortho/para* ratio is largely dependent on reaction temperature, solvents used and catalyst concentration. Generally, it has been observed that at low temperature the *para*-isomer predominates, while at higher temperature the *ortho*-isomer becomes the major product.

The *para*-product is appeared to be *kinetically controlled*, whereas the *ortho*-isomer is a *thermodynamically controlled* product. Perhaps, owing to steric hindrance the *ortho*-isomer can't be formed at a low temperature. Greater stability of the *ortho*-product may be ascribed considering the existence of *intramolecular* hydrogen bonding in the molecule (**1**). The *ortho*-isomer also gets stabilized in its aluminium complex (product before hydrolysis) due to the union of the two oxygen atoms through aluminium (**2**).

Several workers[1-4] reported that the *ortho/para* ratio increases with increasing reaction temperature. Ogata and Tabuchi[5] (1964) carried out *Fries reaction* with the reactants composed of equimolar phenylacetate and aluminium chloride in nitrobenzene solution at varying temperatures and found that the *ortho/para* ratio of hydroxyacetophenones obtained as 0.25 at 40^0C, 0.42 at 60^0C and 0.64 at 80^0C. The *ortho/para* ratio is dependent not only on reaction temperature, but also on the variety of catalysts used and their concentrations along with on the variation of solvents. The wise selection of solvent in *Fries rearrangement* leads to the formation of desired rearranged product(s) in many cases, as reported. For instance, the side products, *ortho*- and *para*-chlorobenzophenones, as also are obtained along with the desired *ortho*- and *para*-hydroxybenzophenones in course of the *Fries rearrangement* of phenylbenzoate[6] using chlorobenzene as solvent (substrate/catalyst molar ratio 1:2), but in nitrobenzene solvent[7] the products are the only *ortho*- & *para*-hydroxybenzophenones (with both $AlCl_3$ and $TiCl_4$ as catalysts). Again, a number of workers found that phenylbenzoate, when heated with anhydrous $AlCl_3$ in the absence of a solvent at 140^0C and above, it rearranges exclusively[8-11] or predominantly[7,12] to *para*-hydroxybenzophenone. Titanium tetrachloride[7] and anhydrous ferric chloride[13] in the absence of a solvent also direct the transformation predominantly to *para*-rearrangement. Similarly, hydrogen fluoride[14] and also boron trifluoride[15] were reported to convert phenylbenzoate to *para*-hydroxybenzophenone.

Examples

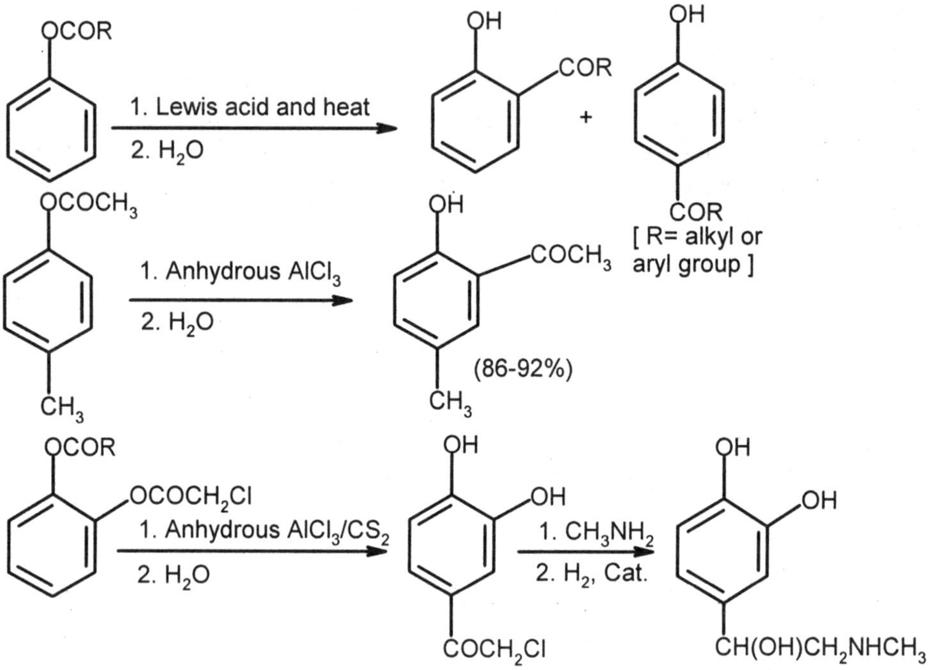

[Scheme: Aryl ester OCOR → (Lewis acid and heat; H₂O) → ortho-hydroxyaryl ketone (2-hydroxy COR) + para-hydroxyaryl ketone (4-hydroxy COR). R = alkyl or aryl group]

[Scheme: 4-methylphenyl acetate (OCOCH₃, para-CH₃) → (1. Anhydrous AlCl₃; 2. H₂O) → 2-hydroxy-5-methylacetophenone (86–92%)]

[Scheme: 2-(OCOR)-phenyl chloroacetate (OCOCH₂Cl) → (1. Anhydrous AlCl₃/CS₂; 2. H₂O) → 3,4-dihydroxy-ω-chloroacetophenone (COCH₂Cl) → (1. CH₃NH₂; 2. H₂, Cat.) → 3,4-dihydroxy-CH(OH)CH₂NHCH₃ (adrenaline-type product)]

Abnormalities

Certain abnormal findings relating to *Fries rearrangement* were reported. The most notable exception is the unusual formation of *meta*-product in some cases. T. Reichstein (1927)[16] for the first time recognized the formation of a *meta*-isomer in the rearrangement of acetylguaiacol under *Fries* condition. Latter, in 1961 Paolo and Cimatoribus[22b] showed that in the acylation of creosol by boron trifluoride/carboxylic acids method involving the *Fries rearrangement*, it is possible to obtain selectively, by varying temperature, *ortho* or *meta* migration of the acyl group. The authors extensively studied the *Fries rearrangement* reaction on acetylcreosol in a varying range of temperatures from 25^0 to 160^0C. Between 25^0 to 80^0, only the *meta*-isomer was obtained with decreasing yield from 90 to 56%, while between 10^0-120^0 the two isomers (*ortho* & *meta*) were formed in a ratio (*ortho/meta*) that varies from 1:4.5 to 4:1. Above 120^0, only the *ortho*-isomer was isolated. Thus, by saturating a creosol (3) and acetic or propionic acid mixture (molar ratio 1:2) with boron at ice bath temperature and leaving the reaction mixture at room temperature for two days, the *meta*-migrated products — 2-methyl-4-methoxy-5-hydroxyacetophenone (4) and respectively propiophenone (5) were obtained.

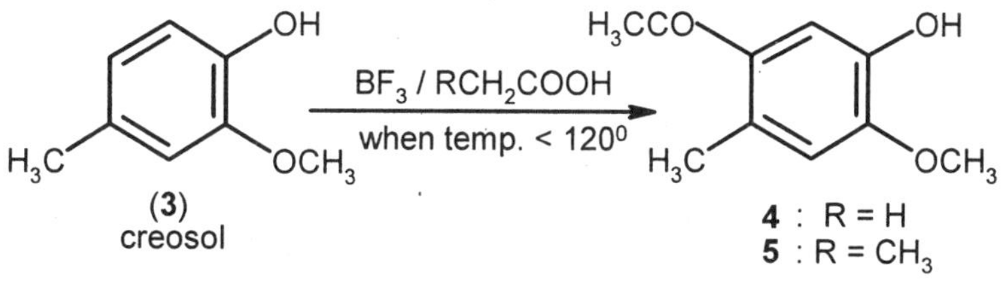

1.7 Fries Rearrangement

The results as obtained from the experimental observations in carrying out the *Fries rearrangement* of acetylcreosol at varying range of temperatures are summarized below:

Temperature (^{0}C)	*ortho*-isomer (%)	*meta*-isomer (%)
25	-	90
40	-	75
60	-	68
80	-	56.6
100	18	40
120	25	10
140	60	-
160	78	-

Hence, it is evident that a temperature of 25^{0}C is sufficient to direct the migrating group exclusively in the *meta*-position with a maximum of 90% yield, while the *ortho*-isomer becomes the exclusive product (maximum of 78% yield) at 160^{0}C.

In 1982, Chorn *et al.*[18] found that 1-acetoxy-4-hydroxy-5-methoxynaphthalene (6) underwent an unusual "*meta- Fries rearrangement*" leading to the formation of 3-acetyl-1,4-dihydroxy-5-methoxylnaphthalene (7).

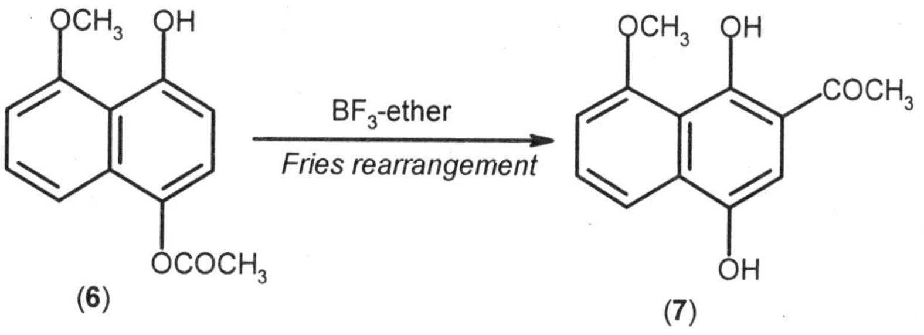

The acetyl group of compound (**6**) migrates to C-3 (*meta*-position) on treatment with boron trifluoride etherate. The authors supposed that the *abnormal Fries rearrangement* reactions of this type of substrates is induced by the *peri*-methoxyl group and proposed that the *meta*-migration of the migrating moiety proceeds presumably through *intermolecular path*. Besides, in case of arylbenzoates treated with F$_3$CSO$_3$H (trifluoromethanesulphonic acid) the reversible *Fries rearrangement* was observed to maintain an equilibrium[19]. *Fries rearrangement* has also been found not to work well with *meta*-directing substituents on the aromatic ring.

Discussion on Mechanism

The mechanism of *Fries rearrangement* is a matter of much controversy. The exact mechanism has still not been completely worked out. The question is whether the rearrangement is completely *intramolecular* or completely *intermolecular* or partially *intra-* and *intermolecular*. In this connection it may be mentioned that a number of *cross-over* experiments have been carried out; sometimes *cross-over* products have been found, sometimes not.

Recent workers have expressed their views on each of the possible mechanisms as proposed prior to 1940's[1,20] or combinations of them. Thus, *intermolecular* mechanism as proposed by Skraup and Poller[21] and Cox[22] was supported by Ralston et al.[23], Martin et al.[24], Hauser and Man[25] (established by exchange experiments) and also by others. On the contrary, Ogata and Tabuchi[5] supported the *intramolecular* proposal of Von Auwers[25], while a number of research groups[26] argued in support of the combination of *inter-* and *intramolecular* mechanism for *Fries rearrangement*. Amin and Shah[27] for the first time expressed their view that proton plays an important role in the *Fries rearrangement*. The contention was supported latter by Gerecs et al.[9,28]. Their findings also suggested that proton catalysis plays an essential part in this rearrangement reaction; the *ortho-* and *para*-rearrangement being affected to different degrees by altering the proton concentration. Gerecs[28] proved that in the case of lower proton concentration, an *ortho*-directed *intramolecular* rearrangement of catalyst-ester complex [e.g. ArOC(R)=O-AlCl$_3$] would occur. The formation of a protonated catalyst-ester complex [ArO$^+$HC(R)=O-AlCl$_3$] would favour *intermolecular* reaction leading to *para*-migration. Dewar and Hart[6], for the first time, established the duality in mechanism (may proceed through normal π-complex or ion-pair type intermediate) in case of *ortho*-migration. In our present discussion, we would consider the works of Ogata and Tabuchi[5] and Dewar and Hart[6], in detail.

Mechanism as offered by Ogata and Tabuchi[5]

The authors studied on the mechanism of *Fries rearrangement* with phenylacetate and their experimental findings enabled them to suggest an *intramolecular* migration of acetyl group to both *ortho-* and *para*-positions, involving a normal *π-complex* intermediate. The schematic representation as offered by them is cited:

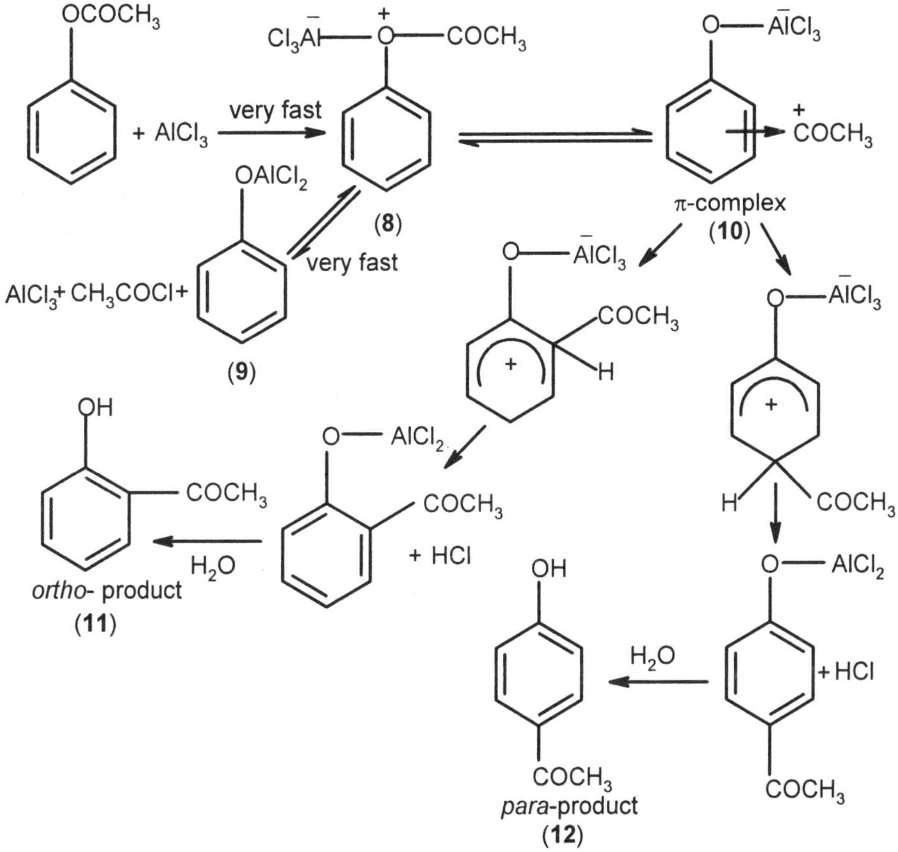

Regarding to their proposed mechanistic pathway they inferred that the formation of coordination complex (8) is very rapid and the step from (8) to *π-complex* (10) is probably reversible, while the steps (10) to hydroxyacetophenones (11) & (12) are irreversible since the ketones can't be converted to the ester.

Mechanism as offered by Dewar and Hart[6]

Dewar and Hart selected phenylbenzoate as the reacting substrate for *Fries rearrangement* in chlorobenzene solution using aluminium bromide as catalyst. They studied the reaction in two different conditions involving the molar proportions of catalyst and substrate as (a) 1:1 (designated as *1:1 rearrangement*) and (b) 2:1 (designated as *2:1 rearrangement*). From the detailed analysis of the experimental results as obtained, they proposed two different mechanistic schemes, for the two categories, as depicted separately.

(a) Proposed mechanism for 1:1 rearrangement

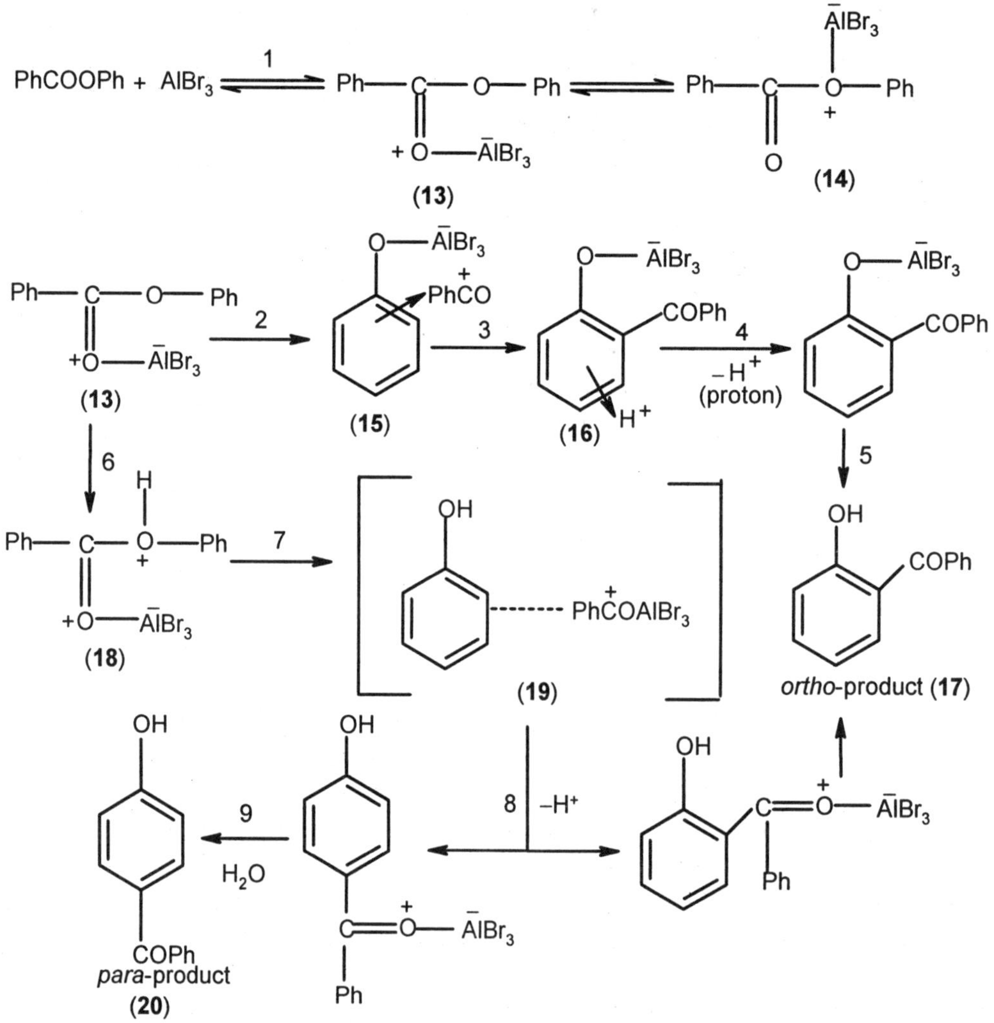

The first step (reaction 1) is very rapid and reversible[29] formation of a catalyst-ester complex (**13**)[4,7,29,30] [the complex (**13**) was originally shown with catalyst attached to the carbonyl oxygen[30,31]]. This complex then undergoes an exclusively *ortho*-directed rearrangement (*via* reaction 2,3,4 & 5; the process termed as *'first stage'* reaction), consistent with an *intra-molecular* migration involving a π-complex intermediate (**15**).

During *ortho*-rearrangement, a proton is liberated and this proton is responsible for an alternative and faster rearrangement mechanism to operate. The complex (**13**) accepts this proton at its second nucleophilic centre[28] and the protonated complex (**18**) then undergoes rearrangement to both *ortho*- and *para*-hydroxybenzophenones (**17** & **20** respectively). This process (6 to 9) is termed as *'second stage'* reaction — which may not be truly *intramolecular*, but an ion-pair type intermediate (**19**) may plausibly be involved in this process. Though the solvent (chlorobenzene) used does not take part into the reaction, it is not the sufficient proof of the *intramolecularity* of the reaction and may be merely a consequence of its lower reactivity in comparison to phenol moiety in its ion-pair type intermediate (**19**) toward the acylating species (PhCOAlBr$_3$) in (**19**). Hence, Dewar and Hart established a duality in mechanism relating to the *ortho*-migrating rearrangement.

The *ortho/para* ratio in the *1:1 rearrangement* was observed to decrease with increasing time; it is due to the fact that the *'second stage'* reaction is faster than the *'first stage'* reaction, because there is a better leaving group in the protonated catalyst-ester complex (**18**) than in the catalyst-ester complex (**13**). The rate of the *'first stage'* reaction will naturally decrease with increasing time since along with the exclusive conversion to *ortho*-rearranged product (*'first stage process'*) the catalyst-ester complex (**13**) also undergoes a simultaneous conversion to protonated catalyst-ester complex (**18**) — consumption of which (through *'second stage'* process) is faster than that of (**13**), leading to the formation of *ortho*- and *para*-hydroxybenzophenones. This production of both *ortho*- and *para*-product, also produces more protons and that is why the *'second stage'* reaction might be regarded as *auto-catalytic*.

(b) Proposed mechanism for 2:1 rearrangement:
The *2:1 rearrangement* proceeds too quickly under the reaction conditions used to allow the investigators to undertake whether there is a similar duality in mechanism as in the case of *1:1 rearrangement* reaction or not. Probably, steric effects play a greater part in the 2:1 rearrangement; steric influences become more important in breaking the complex (**21**) to (**22**), and the active cationic part (Ph$^+$COAlBr$_3$) of (P') is very susceptible to nucleophilic attack leading to the formation of a mixture of hydroxybenzophenones (**17** & **20**), chlorobenzophenones and ketoester (**23**). The ketoester should probably be considered as one of the true intermediates in the *2:1 rearrangement* reaction, as its quantity in the reaction products gradually diminishes

with increasing reaction time, because of its subsequent conversion to *para*-hydroxybenzophenone.

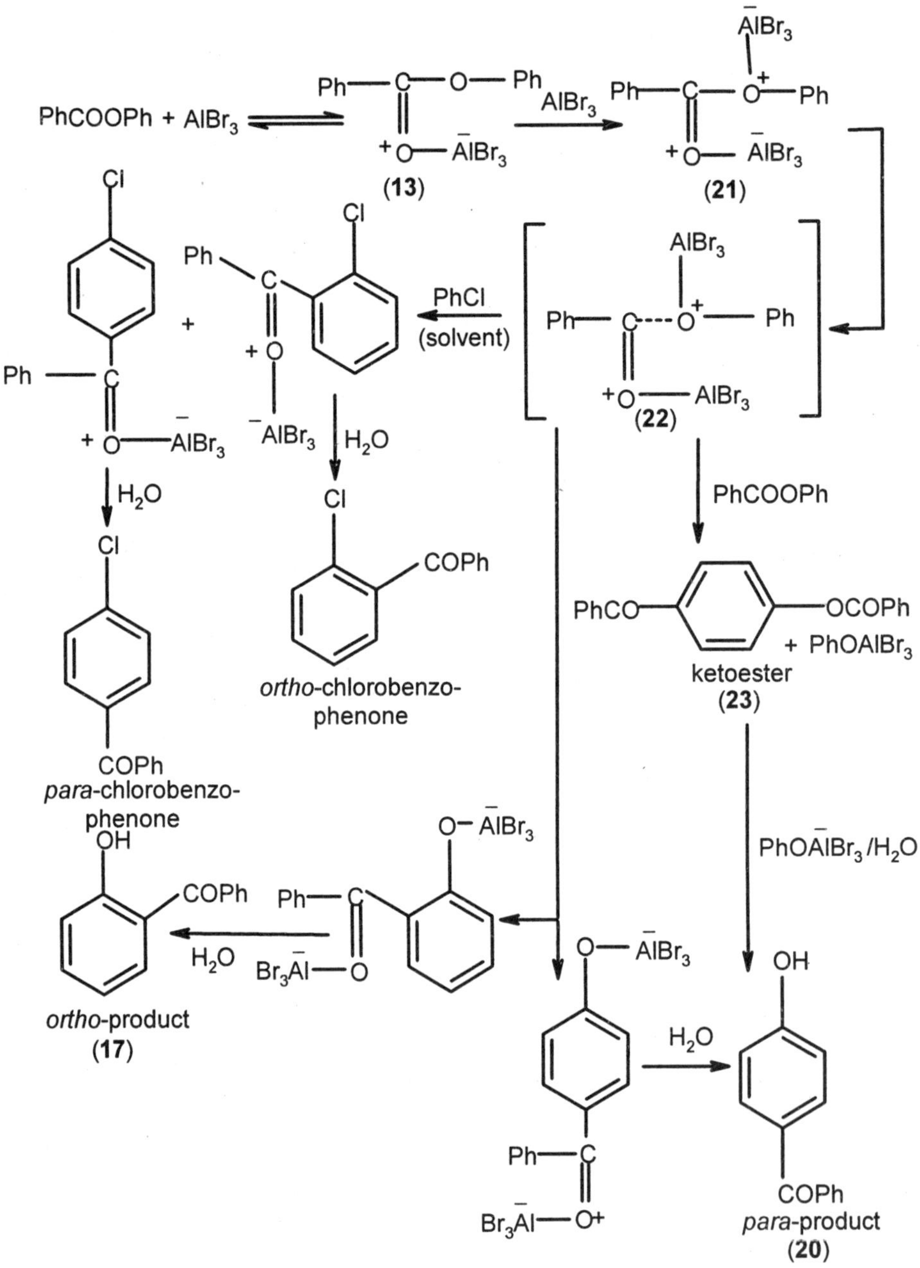

Hence, an inference regarding the mechanistic pathway of the *Fries rearrangement* may be drawn like that the rearrangement can proceed through both *intramolecular* and *intermolecular* path or may follow a simultaneous *intra-* and *intermolecular* way. In case of a particular reaction system, the actual course of the reaction is dependent on a number of factors like nature of substrates, solvent and catalysts as well as catalyst-concentration and reaction temperature.

Extensions

(a) *Photo-Fries Rearrangement*

The classical *Fries rearrangement* was reported to have a photochemical analogue. This analogous rearrangement reaction catalyzed by light is known as *photo-Fries rearrangement*. This photochemical rearrangement was first reported in 1960 by Anderson and Reese[31-34]. A variety of arylesters[35] were found to undergo *photo-Fries rearrangement* resulting in the formation of both *ortho-* and *para*-migrated products along with phenol (ArOH) as a side product[36].

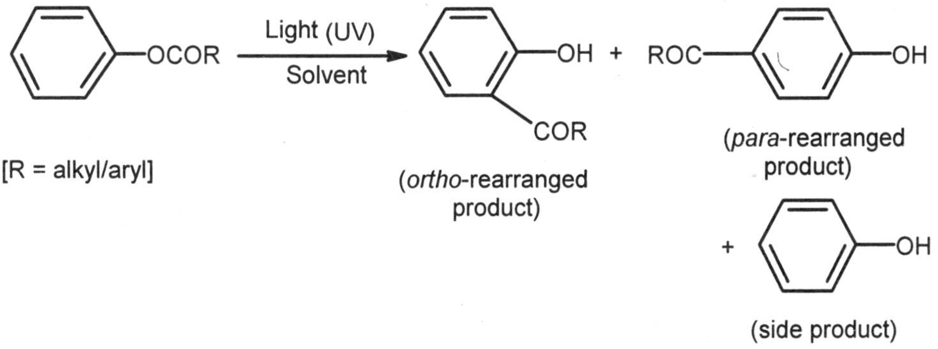

Examples

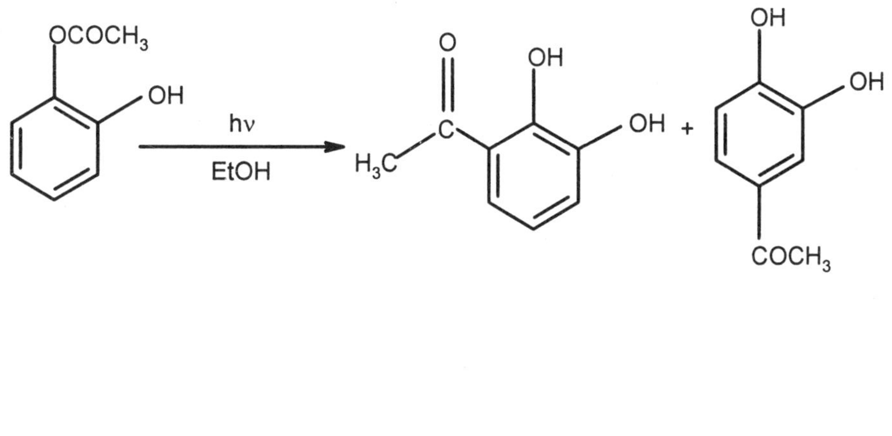

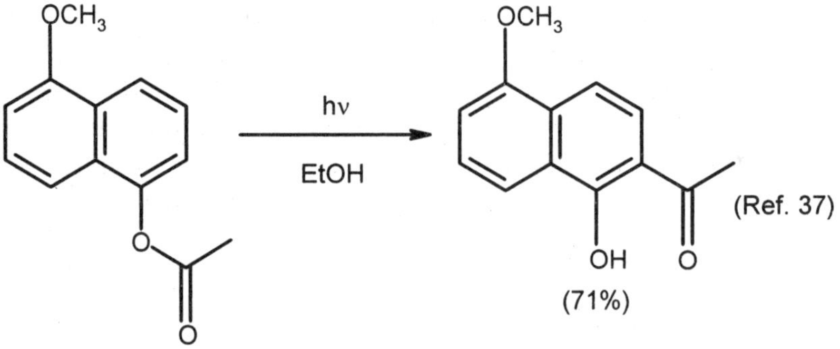

Mechanistic pathway

The *intramolecular* nature of the rearrangement appeared to be well established. Regarding the mechanistic pathway of *photo-Fries rearrangement*, the proposal advanced by Kosba is preferred. His view was latter substantiated by many workers (Anderson and Reese[33], Finnegan and Mattice[38], Kalmus and Hercules[39] and others). In Kosba's view[40] the rearrangement proceeds through isobutane photo-cleavage of ArO-COR bond (acyl-oxygen fission) leading to the formation of a pair of radicals, which remain in a solvent cage for a time sufficient enough to allow their recombination with the formation of dienones. Subsequent enolization leads to *ortho-* and *para*-products. Furthermore, migration of radicals from the solvent cage [preferably to say leakage of some ARO· radicals] accounts for the formation of corresponding phenols by hydrogen abstraction from neighbouring molecules. In this connection, it may be pointed out that when *photo-Fries reaction* was performed on phenyl acetate in absence of solvent, but in gas phase using isobutane as a source of abstractable hydrogen, the side product phenol was reported to come out as the chief product; virtually formation of no *ortho-* or *para*-hydroxy acetophenone[41]. Hence, the phenomenon of recombination of the radicals inside the solvent cage has received further support.

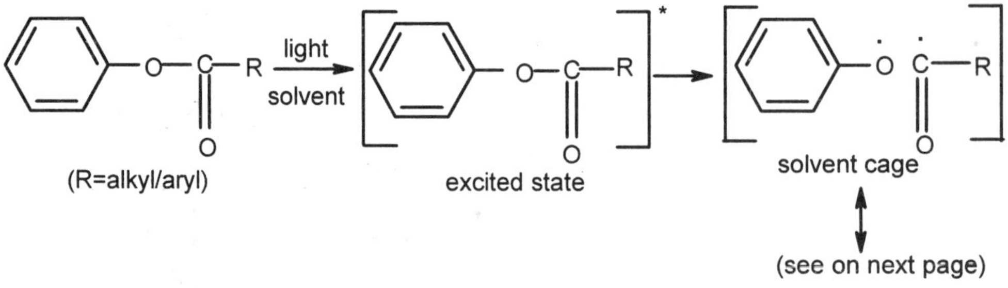

(see on next page)

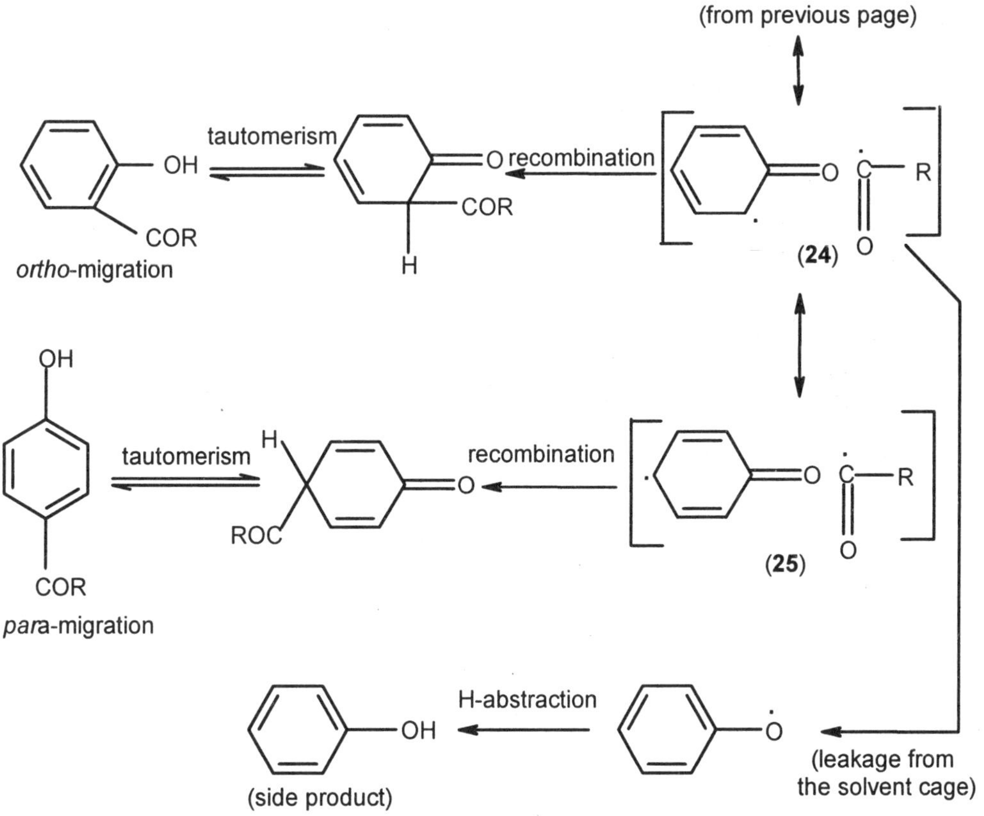

The *photo-Fries rearrangement* is supposed to follow the pathway of singlet excited state, caged radical pair mechanism as evidenced from the response of CIDNP effects in the rearrangement of *p*-tolyl-*p*-chlorobenzoate that was found to be consistent with a singlet-state reaction[42,43]. Latter on, Shine and Subotkoski[44] (1987) studied the kinetic (KIE) and Magnetic (MIE) effects, in detail, on the *photo-rearrangement* of 4-methoxy phenyl acetate and concluded on the basis of their experimental observations that the rearrangement follows the path of recombination of a caged radical pair which originates from an excited singlet state; KIE evidence for a *concerted pathway* could not be found. Evidence in support of radical-recombination mechanism arises from the detection of aryloxy radical (ArO·) by flash photolysis[45] and Raman spectroscopy[46]. Further evidence for the recombination of radicals in a cage has also been obtained with the use of cyclodextrins. Thus, the ratio of *ortho/para-* products from the rearrangement of phenyl acetate was found to alter from 4.0 in water to 1.89 in presence of α-cyclodextrin and 1.79 in presence of β-cyclodextrin. The change in the *ortho/para* ratio is due to the inhibition of recombination of radicals at the *ortho*-position and this phenomenon arises as a result of inclusion of acyl radical and *ortho*-radical (**24**) in the cavity of a cyclodextrin[47]. As the time passes away, the recombination of acyl radical occurs with more stable *para*-radical (**25**), enhancing the *ortho/para* ratio.

(b) 7-Acetoxy-4-methylcoumarin (**26**) was also found to undergo *Fries rearrangement* by Kravtchenko et al.[48].

(**26**) → AlCl$_3$/heat → 8-acetyl-7-hydroxy-4-methylcoumarin (**27**) + 6-acetyl-7-hydroxy-4-methylcoumarin (**28**)

Isomer (**27**) becomes the predominant product of this reaction at any temperature from 85^{0}C to 155^{0}C, even though the content of the other isomeric product (**28**) increases definitely at higher temperature. The ratio [(**27**) : (**28**)] was found to be 200:1 at 85^{0}C and 2.5:1 at 155^{0}C.

(c) Traven et al.[49] studied the *Fries rearrangement* of 7-acetoxy-4-methylquinolin-2-one (**29**) at different temperature and established a significant and unusual temperature dependence of *Fries rearrangement* for this substrate. Under *Fries condition*, two isomers — 8-acetyl-4-methylquinolin-2-one (**30**) and 6-acetyl-4-methylquinolin-2-one (**31**) — were detected in the reaction mixture.

(**29**) → AlCl$_3$/hea → 8-acetyl-7-hydroxy-4-methylquinolin-2-one (**30**) + 6-acetyl-7-hydroxy-4-methylquinolin-2-one (**31**)

The authors reported that the compound (**30**) predominates in the reaction mixture when the rearrangement was carried out at 85^0C, while the other isomer (**31**) predominates (yield becomes about 94%) at 155^0C. The predominant content of (**31**) in the final reaction mixture at higher temperature is due to a higher thermodynamic stability of this isomer in comparison to (**30**). The authors showed that it is more than 8Kcal/mole more stable than the isomer (**30**). On the contrary, the preferable formation of compound (**30**) at lower temperature seems to be due to kinetically controlled reaction. An *intermolecular* mechanism for these transformations was proposed by the authors as in the case of their acylhydroxycoumarin analogs[48].

References

1. Blatt, A.H. (1940), *Chem. Rev.*, **27**, 413.
2. Blatt, A.H. (1942), *Organic Reactions*, **1**, 342.
3. Ralston, A.W., McCorkle, M.R. and Segebrecht, E.W. (1941), *J. Org. Chem.*, **6**, 750.
4. Baltzly, R., Ide, W.S. and Phillips, A.P. (1955), *J. Am. Chem. Soc.*, **77**, 2522.
5. Ogata, Y. and Tabuchi, H. (1964), *Tetrahedron*, **20**, 1661.
6. Dewar, M.J.S. and Hart, L.S. (1970), *Tetrahedron*, **26**, 973.
7. Culinane, N.M. and Edwards, B.F.R. (1958), *J. Chem. Soc.*, 2926.
8. Rosenmund, K.W. and Schnurr, D. (1928), *Liebigs Ann.*, **49**, 1609.
9. Gerecs, A. and Windholz, M. (1955), *Naturwiss.*, **42**, 414; and earlier papers in *Acta Chim. Acad. Sci. Hung.*, 1935.
10. Blicke, F.F. and Weinkauff, O.J. (1932), *J. Am. Chem. Soc.*, **54**, 330.
11. Pieroni, A. and Lenghini, S. (1932), *Gazzetta*, **62**, 387; *Chem. Abstr.* (1932), **26**, 4802.
12. Cullinane, N.M., Morgan, N.M.C. and Plummer, C.A.J. (1937), *Rec. Trav. Chim.*, **56**, 627; *Chem. Abstr.* (1937), **31**, 6238.
13. Huber, H. and Brunner, K. (1930), *Monatsh.*, **56**, 322.
14. Wiechert, K. (1948), *Newer Methods of Preparative Organic Chemistry*, p.315-68, 344, 351, Interscience, New York.
15. Kindler, K., Oelschlager, H. and Henrich, P. (1954), *Arch. Pharm.*, **287**, 210.
16. Reichstein, T. (1927), *Helv. Chim. Acta*, **10**, 392.
17. DaRe, P. and Cimatoribus, L. (1961), *J. Org. Chem.*, **26**, 3650.
18. Chorn, T.A., Giles, R.G.F., Green, I.R., Hugo, V.P. and Mitchell, P.R.K., *Tetrahedron Lett.*, **23**(32), 3299.
19. Effenberger et al. (1982), *Chem. Ber.*, **115**, 1089.
20. Prajer, L. (1957), *Wiadomosci Chemiczne*, **11**, 177.

21 Skraup, S. and Poller, K. (1924), *Ber. Dtsh. Chem. Ges.*, **57**, 2033.
22 Cox, E.H. (1930), *J. Am. Chem. Soc.*, **52**, 352.
23 Martin et al. (1986), *Bull. Soc. Chim. Fr.*, 659.
24 Hauser, C.R and Man, E.H. (1952), *J. Org. Chem.*, **17**, 390.
25 Von Auwers, K. and Mauss, W. (1928), *Liebigs Ann.*, **464**, 56; *Ber. Dtsch. Chem. Ges.*, (1928) **61**, 1495.
26 Munavilli (1972), *Chem. Ind. (London)*, 293; Warshawsky et al. (1928), *J. Am. Chem. Soc.*, **100**, 4544; Dawsen et al. (1985), *J. Chem. Soc., Perkin Trans. 2*, 1601.
27 Amin, G.C. and Shah, N.M. (1948), *J. Univ. Bombay*, **17A**, 5; *Chem. Abstr.* (1949), **43**, 6593.
28 Gerecs, A. (1964), *Fridel-Crafts and Related Reactions*, (Edited by G.A. Olah), Vol.III, Part-II, Chap. XXXIII, p.499-533: *The Fries Reaction*, Interscience, New York, London, and Sydney.
29 Gohring, J. and Susz, B.P. (1966), *Helv. Chim. Acta.*, **49**, 486.
30 Furka, A. and Szell, T. (1960), *J. Chem. Soc.*, 2312; Szell, T. and Furka, A. (1960), *ibid.*, 2321.
31 Bystrov, D.S. and Filimonov, V.N. (1960), *Dokl. Akad. Nauk. USSR*, **131**, 338; *Chem. Abstr.* (1961), **55**, 11083.
32 Anderson, J.C. and Reese, C.B. (1960), *Proc. Chem. Soc.*, 217.
33 Anderson, J.C. and Reese, C.B. (1963), *J. Chem. Soc.*, 1781.
34 Klinger, H. and Standke, O. (1891), *Ber. Dtsch. Chem. Ges.*, **24**, 1340.
35 Bellus, D. (1971), *Advances in Photochemistry*, **8**, 109.
36 Kharasch, M.S., Stampa, G. and Nudenberg, W. (1952), *Science*, **116**, 309; Kelley, D.P., Pinhey, J.T. and Rigby, R.D.C. (1966), *Tetrahedron*, **48**, 5953.
37 Crouse, D.J., Hurlbut, S.L. and Wheeler, D.M.S. (1981), *J. Org. Chem.*, **46**, 374.
38 Finnegan, R.A. and Mattice, J.J.(1965), *Tetrahedron*, **21**, 1015.
39 Kalmus, C.E. and Hercules, D.M. (1972), *Tetrahedron Lett.*, No.16, 1575.
40 Kosba, H. (1962, *J. Org. Chem.*, **27**, 2293.
41 Meyer and Hammond (1970), *J. Am. Chem. Soc.*, **92**, 2187; (1972), **94**, 2219.
42 Adam, W. (1974), *J. Chem. Soc., Chem. Commun.*, **96**, 449.
43 Nakagaki, R., Hiramatsu, M., Watanable, T., Tanimoto, Y. and Nagakura, S. (1985), *J. Phys. Chem.*, **89**, 3222.
44 Shine, H.J. and Subotkaowski, W. (1987), *J. Org. Chem.*, **52**, 3815.
45 Kalmus, C.E. and Hercules, D.M. (1974), *J. Am. Chem. Soc.*, **96**, 449.
46 Beck, S.M. and Brus, L.E. (1982), *J. Am. Chem. Soc.*, **104**, 1805.
47 Chinevert, R. and Voyer, N. (1984), *Tetrahedron Lett.*, 5007.
48 Kravtchenko, D. V., Chibisova, T. A. and Traven, V. F. (1999), *Zhuranal Organicheskoi Khimii*, **35**, 924.

49 Traven, V. F., Podhaluzina, N. Y., Vasilyev, A. V. and Manaev, A. V. (2000), *ARKIVOC*, **Vol. 1, Part 6**, 931-938.

- Martin, R. (1992), *Org. Prep. Proced. Int.*, **24**, 369.
- Trehan, I. R., Brar, J. S., Arora, A. K. and Kad, G. L. (1997), *J. Chem. Educ.*, **74**, 324.
- Harjani, J. R., Nara, S. J. and Salunkhe, M. M. (2001), *Tetrahedron Lett.*, **42**, 1979.
- Kozhevnikova, E. F., Derouane, E. G. and Kozhevnikov, I. V. (2002), *Chem. Commun.*, 1178.
- Clark, J. H., Dekamin, M. G. and Moghaddam, F. M. (2002), *Green Chemistry*, **4**, 366.
- Sriraghavan, K. and Ramakrishnan, V. T. (2003), *Tetrahedron*, **59**, 1791.

8

Norrish Type I and Type II Reactions

Introduction

There are many carbonyl group reactions initiated by n-π* excitation, which involve photochemical bond cleavage. Reaction can occur from both the singlet and triplet excited states. A large number of these ketone and aldehyde photochemical reactions may be viewed within the framework of two general photochemical processes,[1] termed as *Norrish Type I Cleavage* and the *Norrish Type II Cleavage*. In Norrish Type I process fission of carbon–carbon single bond α to the carbonyl group takes place, whereas Norrish Type II reaction involves intramolecular abstraction of γ-hydrogen followed by cleavage of the resulting diradical to give an enol that tautomerizes to the lower carbonyl compound (aldehyde or ketone).

General scheme and mechanistic pathway

(a) Norrish Type I Cleavage:

The bond dissociation energy of a carbonyl–carbon bond is comparatively small. Hence photochemical excitation of ketones usually results in the homolytic rupture of the carbon–carbon single bond α to the carbonyl group to afford initially an acyl and an alkyl radical. This primary photochemical process is termed as *Norrish Type I cleavage*.

$$R_2CH-\underset{\underset{O}{\|}}{C}-R' \xrightarrow{h\nu} R_2CH-\underset{\underset{O}{\|}}{C}\cdot + \cdot R' \quad \text{(primary photochemical reaction)}$$

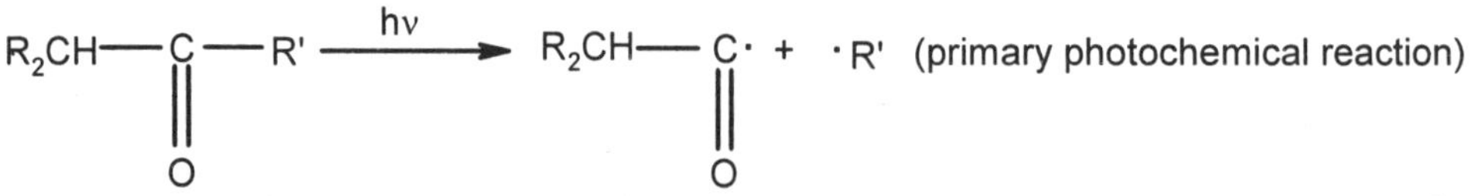

The initially formed radicals get stabilization through secondary process by one of the following routes —

Route a: α-Hydrogen abstraction by the alkyl radical to form a ketene and an alkane. The presence of a ketene as a reactive intermediate has been demonstrated by spectroscopic methods.

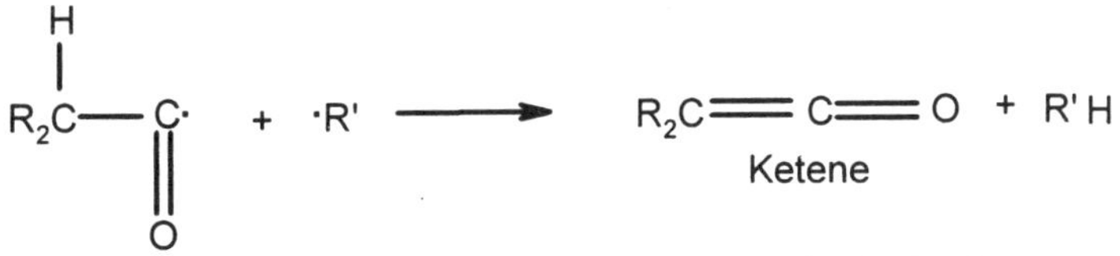

$$R_2C-\overset{H}{\underset{\underset{O}{\|}}{C}}\cdot \;+\; \cdot R' \longrightarrow R_2C=C=O \;+\; R'H$$
$$\text{Ketene}$$

Route b: The acyl radical can lose CO to give alkyl radical; the latter may recombine to form an alkane or may undergo intermolecular hydrogen abstraction to yield an alkene and an alkane.

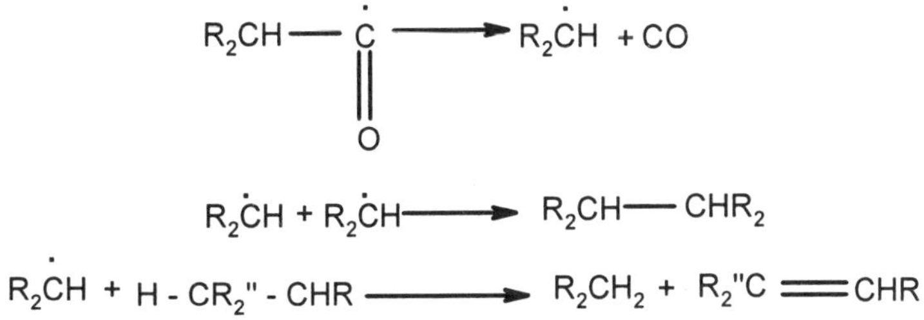

$$R_2CH-\underset{\underset{O}{\|}}{\overset{\cdot}{C}} \longrightarrow R_2\overset{\cdot}{C}H + CO$$

$$R_2\overset{\cdot}{C}H + R_2\overset{\cdot}{C}H \longrightarrow R_2CH-CHR_2$$

$$R_2\overset{\cdot}{C}H + H-CR_2''-\overset{\cdot}{C}HR \longrightarrow R_2CH_2 + R_2''C=CHR$$

Route c: Intermolecular hydrogen abstraction by the acyl radical from the alkyl radical to give an aldehyde and an alkene.

$$R_2CH-\underset{\underset{O}{\|}}{\overset{\cdot}{C}} + H-CR_2''-\overset{\cdot}{C}HR \longrightarrow R_2CH-CHO + R_2''C=CHR$$

(b) Norrish Type II Cleavage:

Carbonyl compounds (aldehydes or ketones) possessing accessible γ-hydrogen atom(s), when photoirradiated, may undergo photoelimination reaction effecting an intramolecular fission of the bond α–β to the carbonyl to give an alkene and a short-chain carbonylic compound. This reaction is known as *Norrish Type II cleavage*[2].

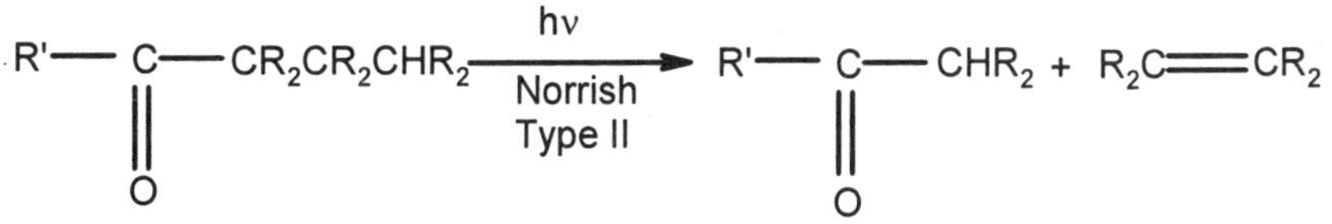

$$R'-\underset{\underset{O}{\|}}{C}-CR_2CR_2CHR_2 \xrightarrow[\text{Norrish Type II}]{h\nu} R'-\underset{\underset{O}{\|}}{C}-CHR_2 + R_2C=CR_2$$

The *Norrish Type II cleavage* involves intramolecular hydrogen transfer from the γ–carbon to the carbonyl oxygen (*i.e.* 1,5-shift) resulting initial formation of a 1,4-diradical likely through a six-membered cyclic transition state[3]; the resulting diradical then undergoes a secondary reaction[4] cleaving the α,β–carbon-carbon bond to give an alkene and an enol that tautomerizes to the corresponding short-chain aldehyde or ketonic product. The intermediate diradical can also cyclizes to a cyclobutanol derivative, which is often a side product.

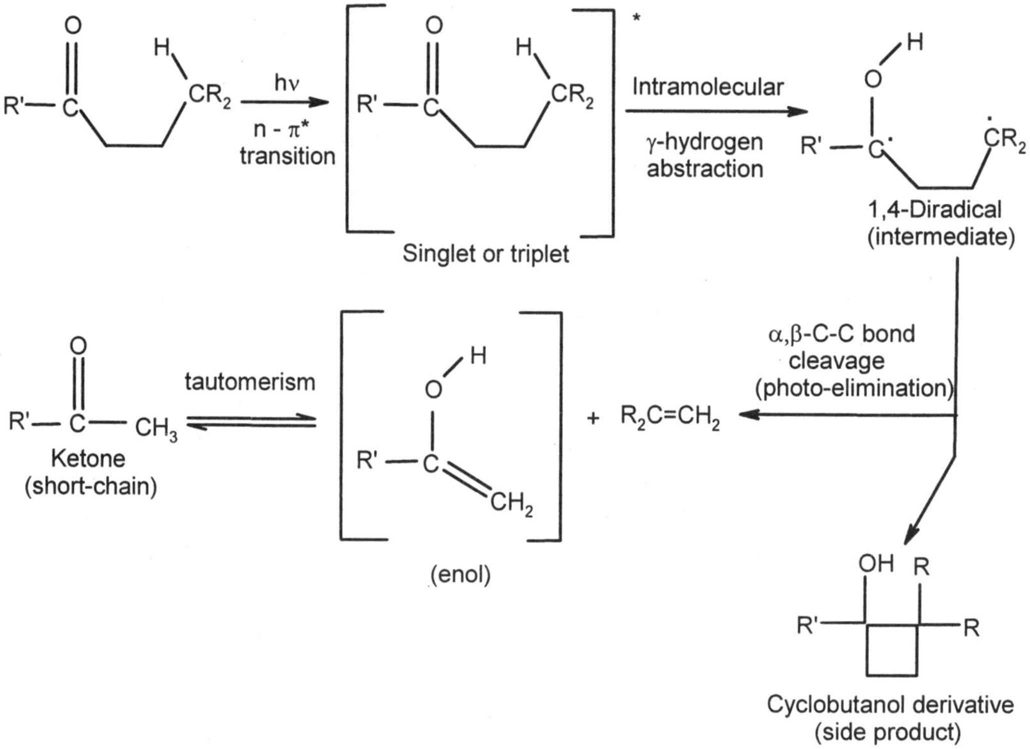

Both singlet and triplet n,π* states undergo the reaction[5]. Carboxylic esters, anhydrides, and other carbonyl compounds can also give this reaction[6].

Critical discussion on Norrish Type I and Norrish Type II cleavage

The extent of Type I cleavage for ketones is quite dependent upon the substitution at the α-carbon, and may occur from either the excited triplet or singlet state[7]. Thus for a diaryl ketone, such as benzophenone, the Type I cleavage assumes no importance. If the ketone is unsymmetrically substituted, preference for cleavage of the weakest C–C bond is noticed. For example, ethylmethylketone shows a predominant cleavage to yield an acetyl radical and an ethyl radical at 3130Å^0, although this selectivity diminishes as the energy input increases[8] —

Norrish Type I and Type II Reactions

$$CH_3-\overset{\overset{O}{\|}}{C}-CH_2CH_3 \xrightarrow{h\nu} \begin{array}{c} \Phi_a \nearrow \\ \\ \Phi_b \searrow \end{array} \begin{array}{l} CH_3-\overset{\overset{O}{\|}}{C}\cdot + \cdot CH_2CH_3 \\ \\ CH_3CH_2-\overset{\overset{O}{\|}}{C}\cdot + \cdot CH_3 \end{array}$$

λ	Φ_a/Φ_b
3130 A°	40
2654 A°	5.5
2537 A°	2.4

The Type I process is favoured by photolysis in the vapour phase and is less pronounced for photolysis in inert solvent, where the solvent cage facilitates recombination of the initially generated radical pair. Although it has been observed that generation of sufficiently stable alkyl radicals, by the Type I process, may be efficiently produced in solution. Thus Yang and Feit[7c] found that *tert*-butyl alkyl ketones undergo efficient Type I cleavage even though a Type II split is possible —

$$\underset{CH_3}{\overset{CH_3}{CH_3-\underset{|}{\overset{|}{C}}-\overset{\overset{O}{\|}}{C}-CH_2CH_2CH_3}} \xrightarrow{h\nu} \underbrace{\underset{CH_3}{\overset{CH_3}{CH_3-\underset{|}{\overset{|}{C}}-\overset{\overset{O}{\|}}{C}-CH_3}} + CH_2=CH_2}_{\Phi_{II} = 0.3 \text{ (Type II products)}}$$

$$+$$

$$\underbrace{CH_3-\underset{|}{\overset{CH_3}{C}}=CH_2 + CH_3-\underset{|}{\overset{CH_3}{\underset{|}{C}}}-H}_{\Phi_I = 0.57 \text{ (Type I products)}}$$

This is dramatic contrast to straight chain aliphatic ketones with γ-hydrogens, which undergo predominantly Type II processes. Quenching studies in this system already established that the Type I process occurs from both singlet and triplet excited states, while the Type II occurs predominantly from the singlet-excited state. Calvert and Nicol[9] in an extensive study of competing Type I and II processes in a series of *n*-propyl alkyl ketones in the gas phase observed that Q_I increases and Q_{II} decreases as the alkyl group is changed from methyl through *t*-butyl. The increasing amount of Type I cleavage is probably related to the decreasing strength of the acyl-carbon in proceeding along the series.

An especially informative investigation in establishing the presence of rotational equilibrium in a gas phase decarbonylation reaction was carried out by Alumbaugh *et al.*[10]. Photolysis of either *cis* or *trans*-2,6-dimethyl cyclohexanone yielded the same major products in the same product ratios.

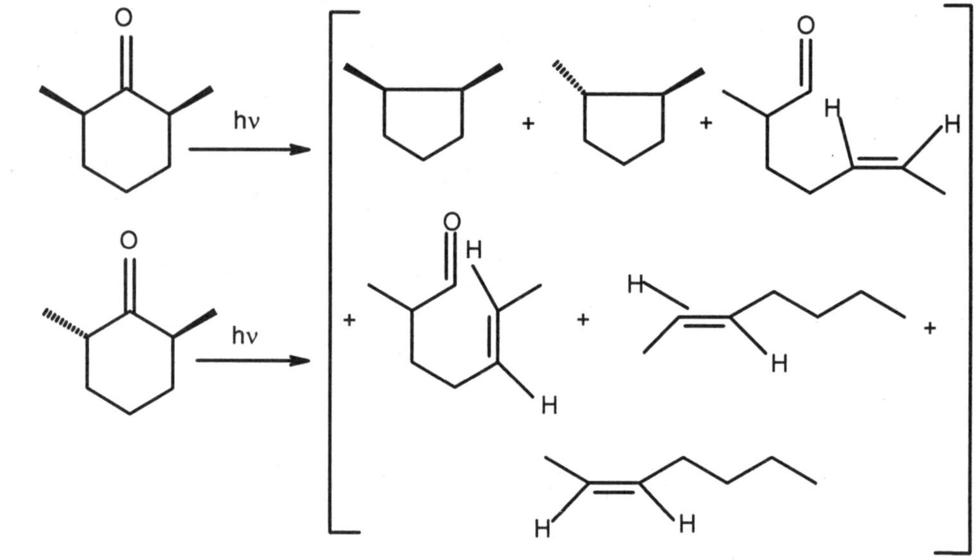

The observed lack of stereospecificity rules out a concerted decarbonylation mechanism for this system and strongly supports the intermediacy of diradicals whose rate of rotation is faster than ring closure.

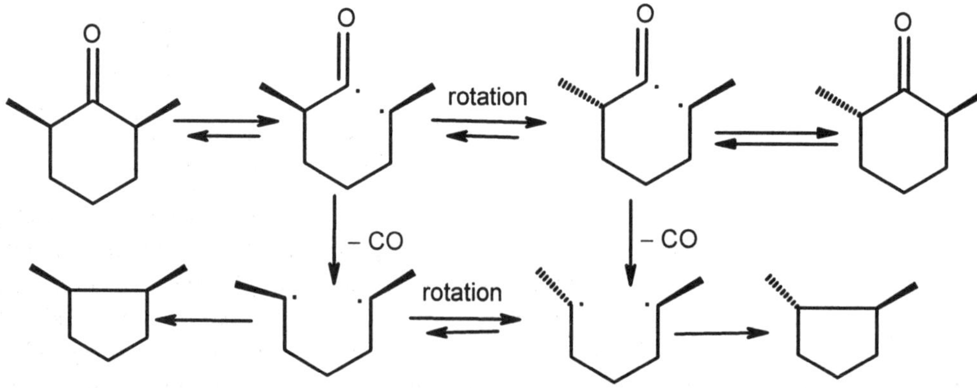

The question of the multiplicity of the excited state of aliphatic ketones in Type II processes has been complicated by different results recorded by different investigators. In one study it was observed that gas phase photoelimination from 2-hexanone at 3130A^0 is not affected by even 560 mm of oxygen[11a]. In another gas phase study of methoxy acetone at 3130 A^0, Srinivasan found that the Type II split is also unaffected by biacetyl, nitric oxide or oxygen[11b].

$$CH_3COCH_2OCH_3 \xrightarrow{h\nu} CH_3COCH_3 + HCHO$$
$$\Phi = 0.32$$

Since the triplet quenchers do not affect these processes, the reaction appeared to derive from the singlet manifold. On the other hand, Ausloos et al. found a progressive decrease in the quantum yield (Φ) for the Type II process in 2-pentanone with addition of increasing amount of oxygen[12a,b] and latter on Ausloos and Robbert[13] have shown that 2-pentanone undergoes photoelimination and cyclobutanol formation from its triplet state.

The apparently conflicting results recordred in the case of 2-pentanone and 2-hexanone stem from the fact that both singlet and triplet excited states may be responsible for the Type II split. Thus quenching data on the Type II split of 2-pentanone, 2-hexanone and 2-octanone in solution indicates that two species are reactants[5a,14]. The Type II process for these compounds is quenched by low concentrations of piperylene but the quenching effect levels off at piperylene concentrations around 0.5M as is evident from the table below; 2-pentanone reacts largely from the triplet state in agreement with the vapour-pahse results of Ausloos et al.[12,13], while 2-hexanone reacts an appreciable portion of the time for the singlet state (table below).

Quantum Yields from Disappearance of Ketones (Type II process)

Ketone	$\Phi_{overall}$	$\Phi_{singlet}$	$\Phi_{triplet}$
2-Pentanone	0.44	0.05	0.39
2-Hexanone	0.50	0.21	0.29

The rate constants for internal hydrogen abstraction depends on electronic configuration, on C – H bond strength or inductive substituent effect, and also on conformational factors. It is well known that the electronic configuration of an n,π* state allows hydrogen abstraction processes of low activation energy to take place. In aliphatic ketones, n,π* state singlets interact with C – H bonds as rapidly as do triplets because ISC is so slow, while aryl ketones generally undergo ISC so rapidly that most of their photoreactions are triplet derived. It is also important to note that typical rate constants for γ–hydrogen abstraction by n,π* triplets of acyclic ketones are 1×10^7, 1×10^8 and 5×10^8 sec^{-1} for primary, secondary and tertiary C – H bonds, respectively[15].

Stereochemistry of Norrish Type II Process

The vapour-phase photolysis result[16] of the epimeric *cis*- and *trans*-2-propyl-4-*t*-butylcyclohexanones established that Norrish Type II cleavage maintains a specific stereochemical requirement for γ-hydrogen migration. The *cis*-isomer undergoes Type II cleavage to give 4-*t*-butylcyclohexanone and it is observed that the triplet quencher, penta-1,3-diene, has no effect on the reaction. On the contrary, the *trans*-isomer undergoes Type I (effecting isomerization to more stable *cis*-isomer) instead of Type II reaction and this conversion is markedly quenched by penta-1,3-diene. Thus, it may be assumed that the former reaction occurs through singlet-excited state and the latter through triplet excited state.

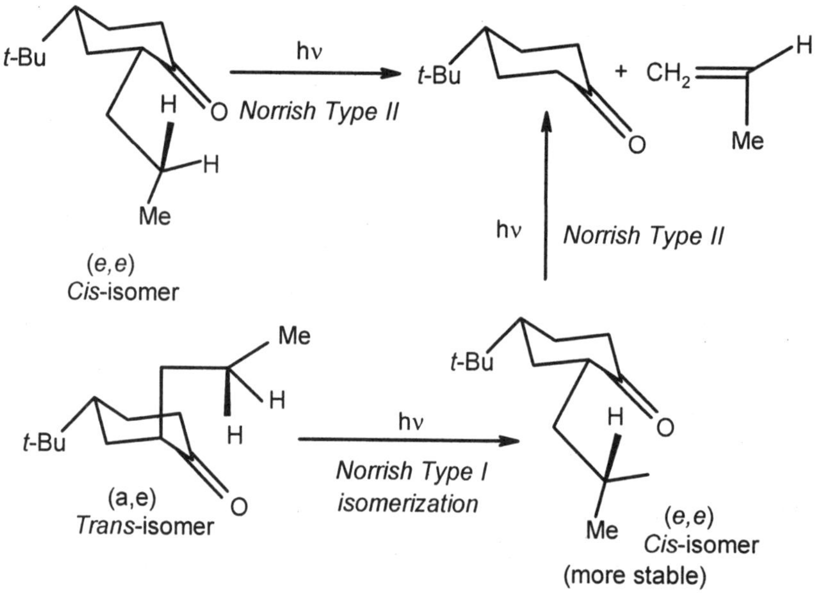

This difference in behaviour of the two epimeric ketones offers a strong evidence for a specific stereochemical requirement in the Type II process.

Molecular models indicate that for the *cis*-isomer with the propyl group at equatorial position, there is ready formation of a six-membered cyclic transition state with the γ-C – H bond axis in the plane of the half-filled n-orbital. The reaction can proceed more feasibly than that for the axial propyl ketone which would require cyclohexyl ring inversion to a boat conformation to orient the γ–C – H bond into a suitable position for hydrogen migration. This steric requirement is probably necessary for all photochemical intramolecular γ-hydrogen abstraction reactions that involve the n,π* excited singlet state[16].

T. Y. Kim *et al.*[17] offered another example of photoreactions of 5-(*o*-tolyl)-5-cyano-4,4-dimethyl-2-pentanone (OTCMP) in solution and solid state, which are governed by conformational preference. On irradiation in solution or solid state, OTCMP produces 2-(*o*-tolyl)-3-methyl-2-butenenitrile and acetone *via* intramolecular hydrogen abstraction followed by elimination (**Scheme 1**). No cyclization product is observed. OTCMP has two different types of γ-hydrogens. But the abstraction reaction occurs only at a carbon containing tolyl and cyano group. It has been observed that a phenyl group activates such process by 40 times, while the cyano group deactivates it by a half. However, the tertiary C – H bearing a phenyl and a cyano group would be more activated than the primary C-H by three fold (taking all the factors into account). The regioselectivity may also be explained by the entropy effect, which can be viewed as the probability that a molecule will have to attain the correct geometry for undergoing the reaction. The interacting groups must be able to rotate within a minimal distance of each other for proper orbital overlap to serve the purpose.

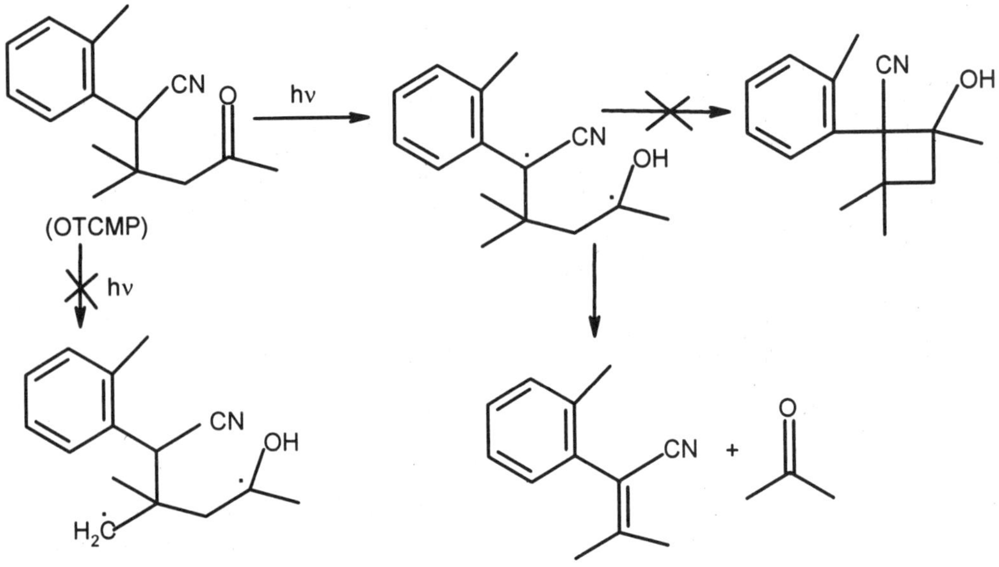

Scheme 1

More recently, J. K. Arnold et al.[18] developed a probe of conformational mobility for *Norrish Type II* photochemical reaction from the photolysis results of an aryl ketone (mercaptoundecanophenone) carried out for the first time on a monolayer-protected gold naocluster.

Thorough examples and synthetic applicability
Norrish Type I cleavage

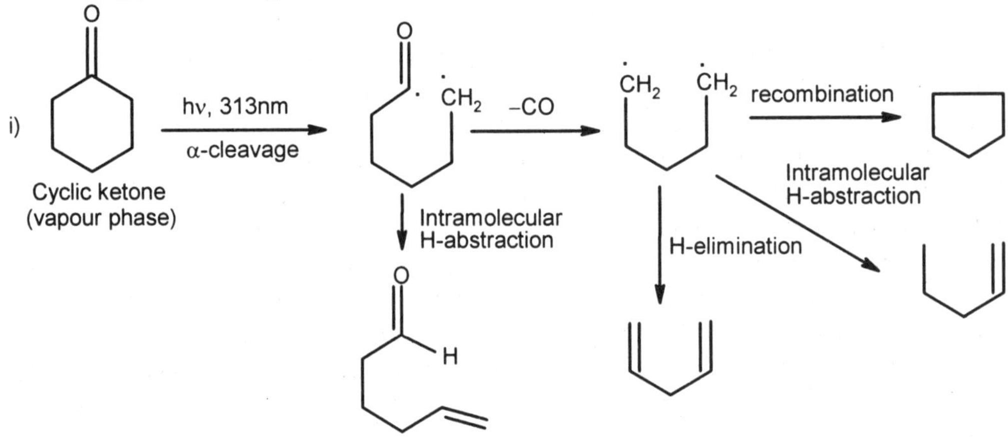

ii) In protic solvents, cyclic ketones on *Norrish Type I reaction* furnish ketene intermediates, which in turn undergo solvent addition to yield carboxylic acids or their derivatives —— as observed in the case of carvone.

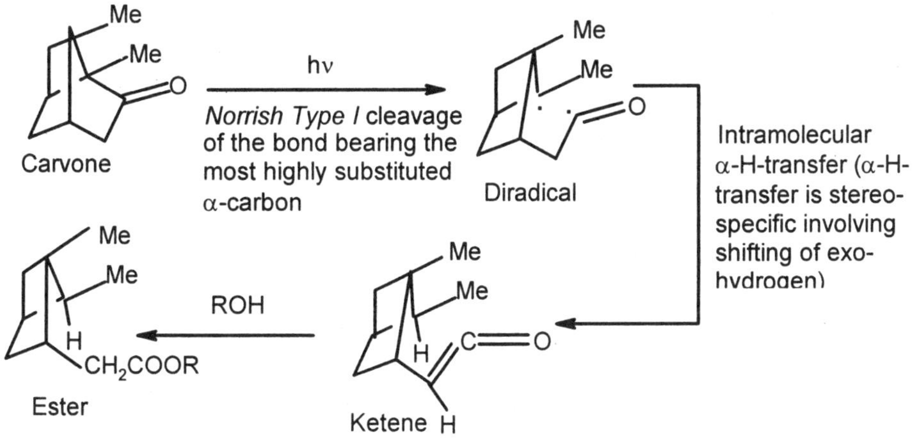

Carvone → (hv, *Norrish Type I* cleavage of the bond bearing the most highly substituted α-carbon) → Diradical → (Intramolecular α-H-transfer (α-H-transfer is stereospecific involving shifting of exo-hydrogen)) → Ketene → (ROH) → Ester (CH₂COOR)

iii) 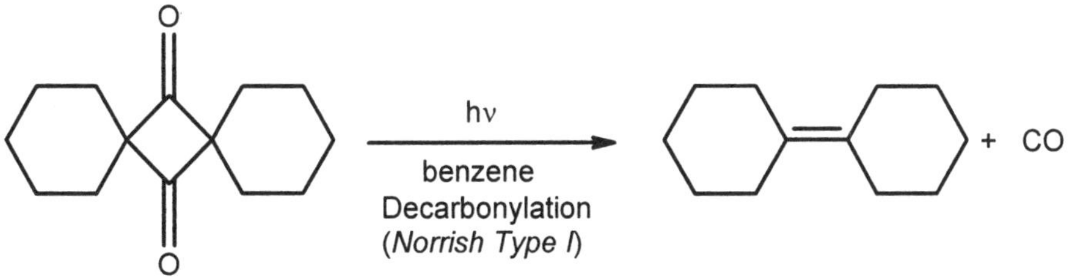 [spiro diketone] →(hv, benzene, Decarbonylation (*Norrish Type I*)) → [cyclohexylidenecyclohexane] + CO

iv) 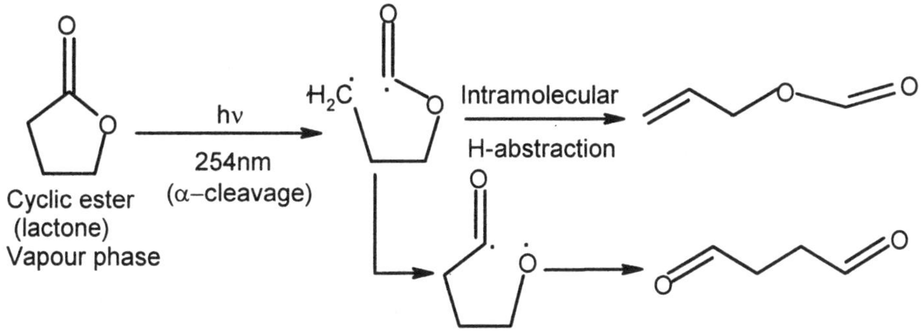 Cyclic ester (lactone), Vapour phase →(hv, 254nm, α-cleavage) → [diradical] →(Intramolecular H-abstraction) → allyl formate / succinaldehyde

v) Cyclic β,γ-unsaturated ketones are observed to undergo a photochemical rearrangement in a similar manner by initial α-cleavage, forming a diradical intermediate that in turn takes part in radical rearrangement.

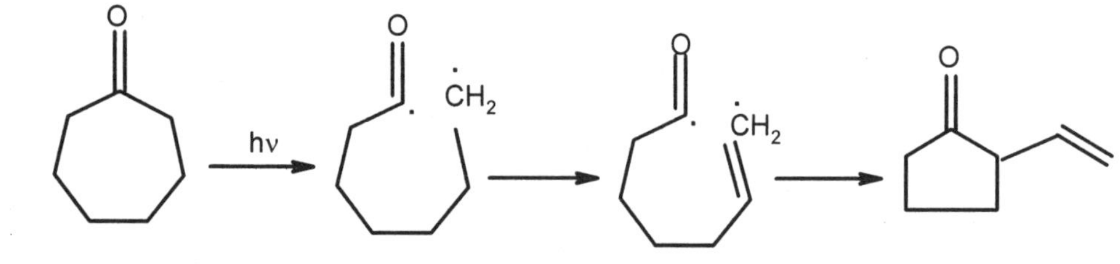

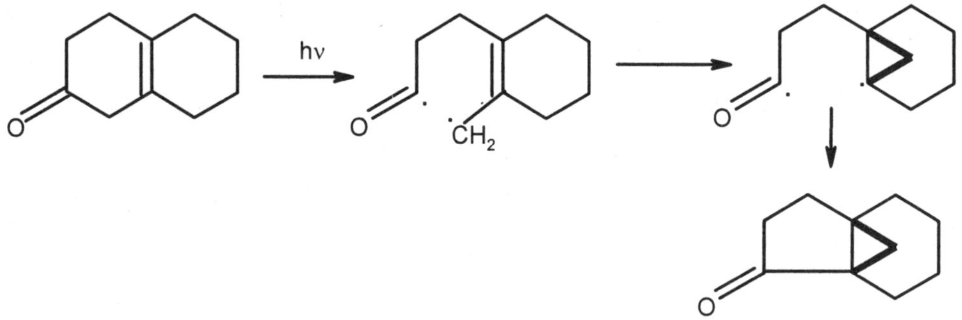

Norrish Type II cleavage

i) A synthetic example of *Norrish Type II cleavage* is the photolysis of (**1**) to form (**2**) that undergoes ring closure to alcohol (**3**).

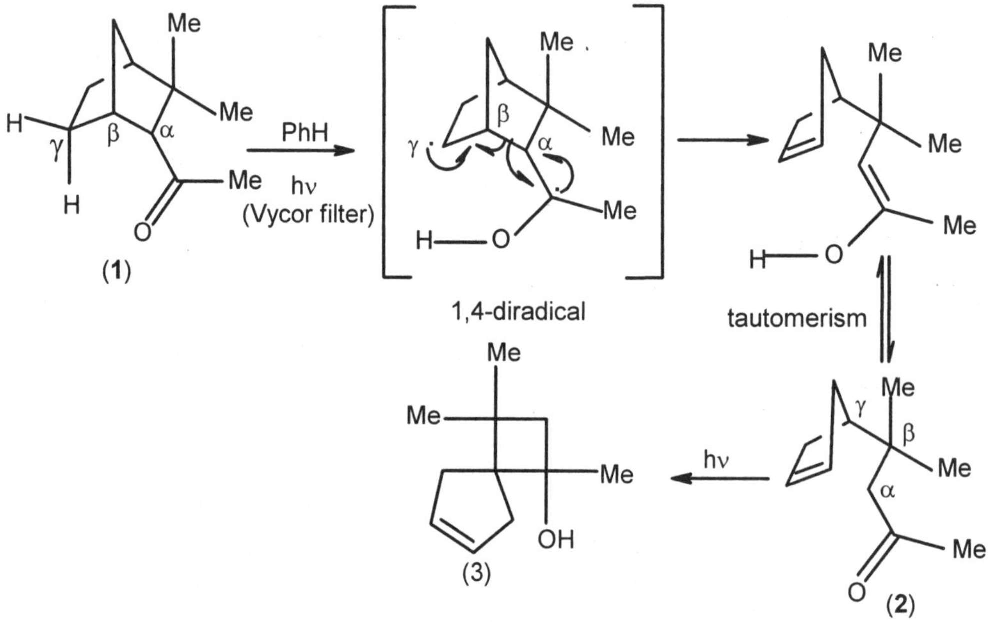

ii) Aryl alkyl ketones on photo-irradiation generate a triplet 1,4-diradical, which then undergoes cleavage to form alkene and enol (which tautomerizes to a methylketone) and at the same time may also yields cyclobutanol derivative.

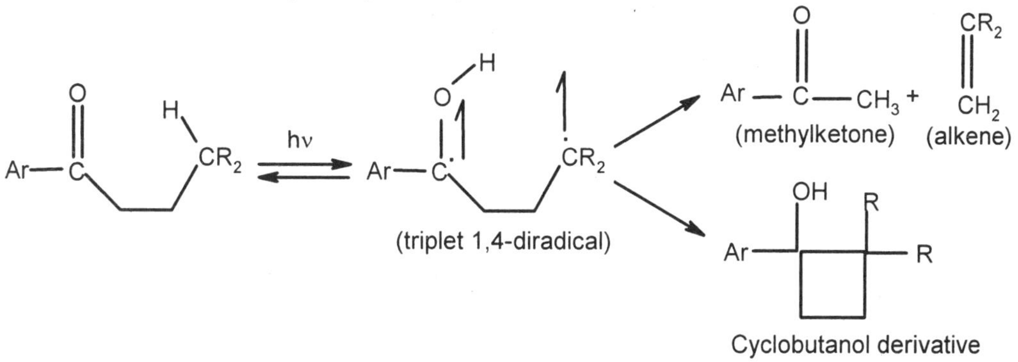

The electron-releasing substituents (like methyl or methoxy) at the *para*-position in the aromatic nucleus decrease the rate-constant and quantum yield for the Type II cleavage. It is very much interesting to note that *p*-hydroxy, *p*-amino and *p*-phenyl substituents inhibit the reaction completely. The phenomenon is supposed to be a consequence of an increase in π-electron availability for which π - π * triplet excitation appears more significant in comparison to n- π * triplet excitation, as the former process becomes lower in energy than the latter.

iii) α-Substitution pattern largely controls the proportion of cyclization product (cyclobutanol from radical recombination) and elimination product (from α,β-bond cleavage) in *Type II reaction*[19]. As the crowding at the α-position increases, rate of radical recombination enhances.

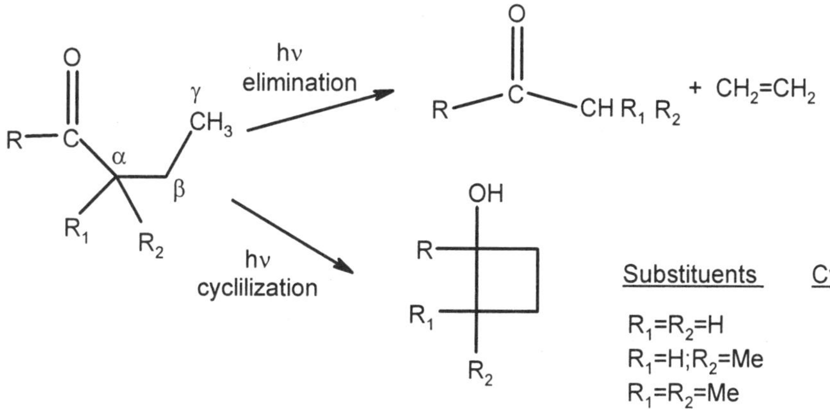

Substituents	Cyclization
$R_1=R_2=H$	10%
$R_1=H; R_2=Me$	29%
$R_1=R_2=Me$	89%

This is due to the effect of eclipsing interactions caused by the substituents at α-position —— may be best understood from the orbital diagram:

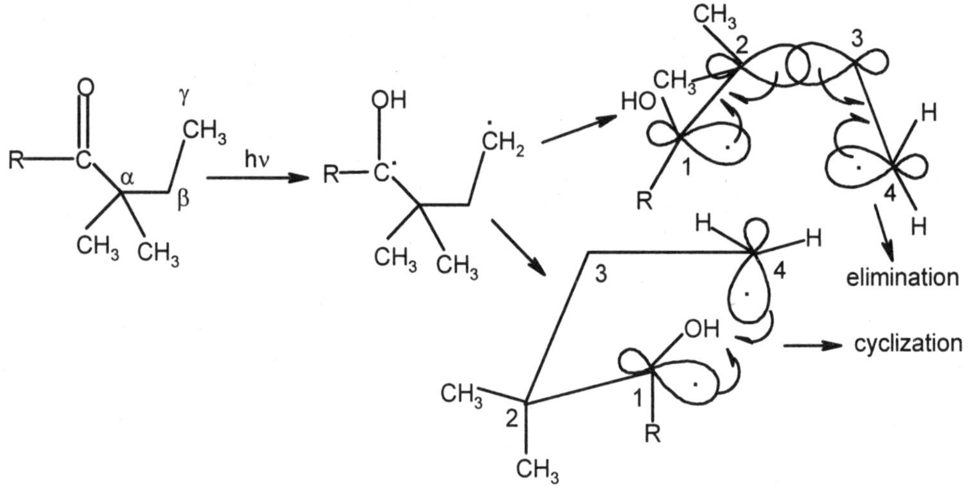

Thus, γ-hydrogen abstraction in the following ketonic compound (**4**) results exclusively in cyclobutanol formation (**5**).

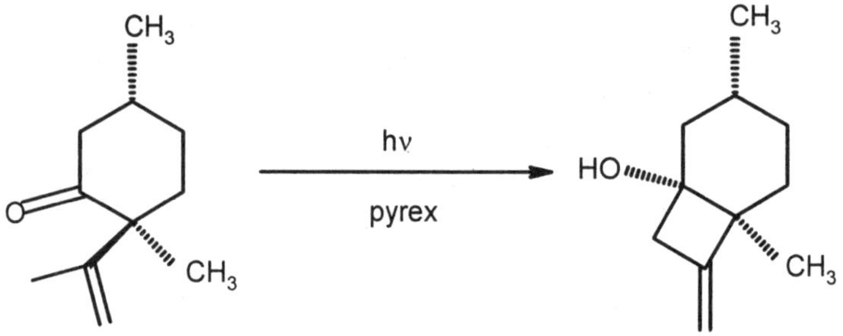

(**4**) → (**5**) [70%]

iv) *Achiral exo*-bicyclohexyl aromatic ketones (*e.g.* **6**) are known to undergo the *Norrish Type II reaction* to produce the following *chiral* ketonic product (**7**) in racemic form[20]; along with this a highly strained tricyclo-compound (**8**) is also formed in about 30% yield.

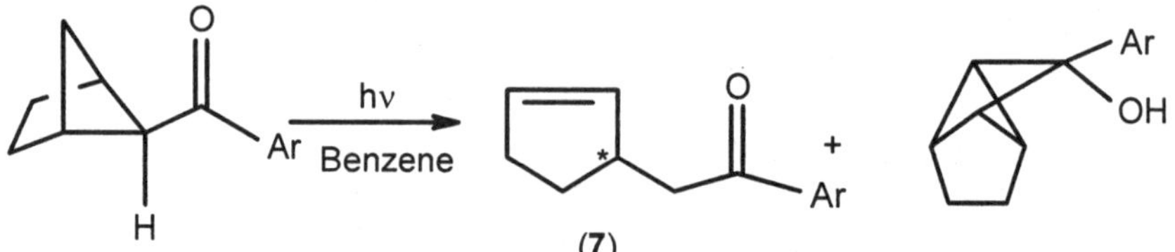

(**6**) → (**7**) (major) + (**8**) (30%)

v) The *Norrish Type II reaction* can also occur for suitably substituted esters[21]. The reaction may take place *via* both singlet and triplet excited states; the singlet state reaction (performed by using excess triplet quencher) is *stereoselective*, but the triplet state reaction is not.

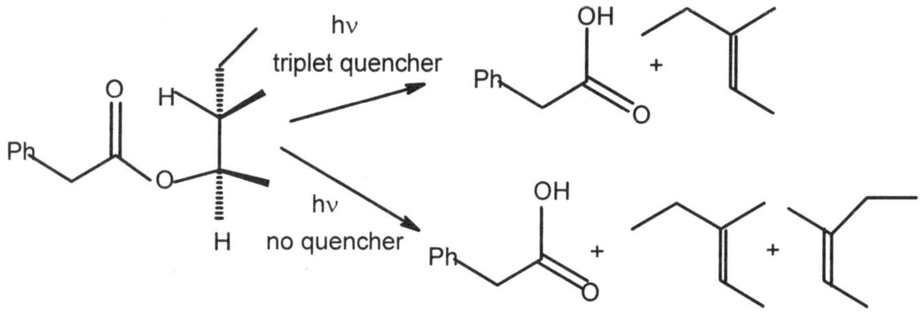

vi) Griesbeck et al.[22] reported that in case of *trans*-conformation of the singlet-excited phthalimide substrate (**9**) the *Norrish Type II cleavage* dominates when R group is methyl, whereas Yang cyclization becomes preferred when 'R' group is *tert*-butyl. The work focuses on the influence of reactive conformation and substitution pattern on the *Norrish Type II* reactivity and selectivity of *singlet-excited* phthalimides.

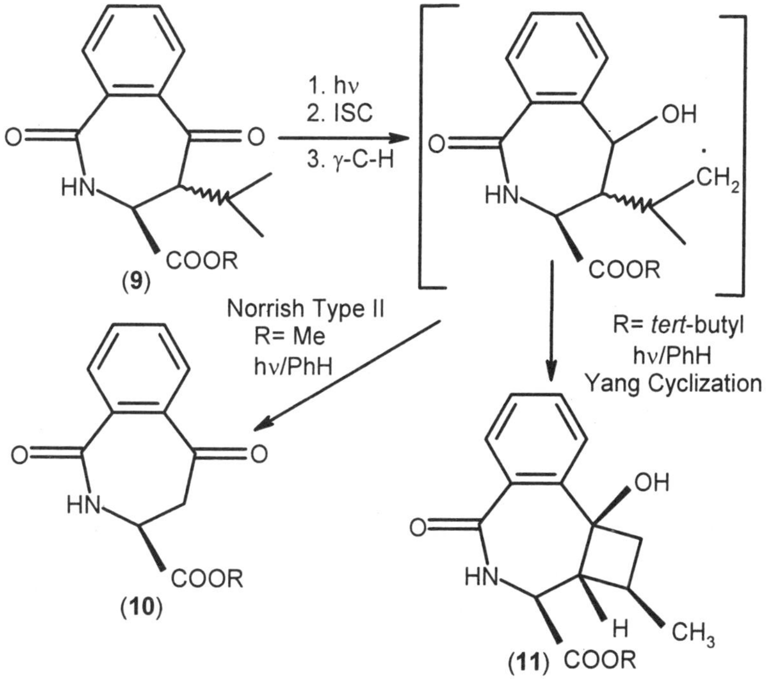

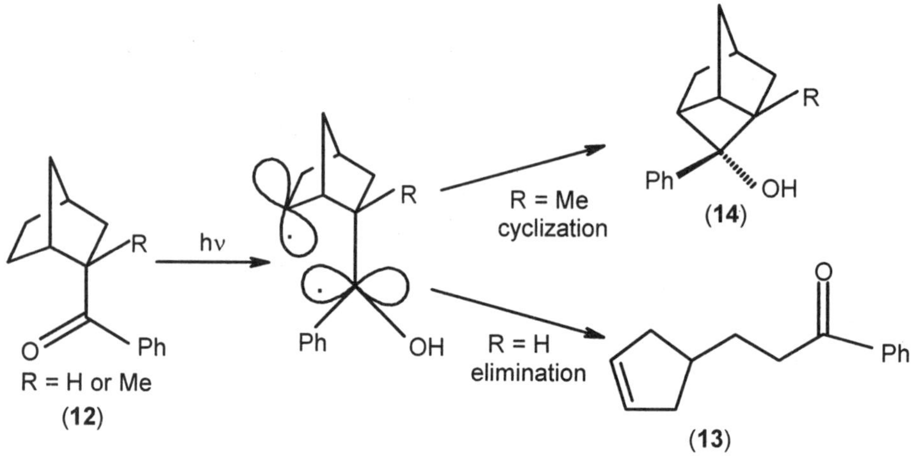

On photo-irradiation *endo*-2-benzoylnorbornane (**12a**, R=H) yields the cleavage product (**13**), but endo-2-benzoyl-2-methylnorbornane (**12b**, R=Me) results in formation of the tricyclo[3,2,1,0]-octane (**14**). The presence of the 2-methyl substituent and the consequent high-energy eclipsing interaction with the phenyl group in the planar transition state required for elimination, results in cyclobutanol formation.

References

1. a) Norrish, R. G. W. and Bamford, C. H. (1937), *Nature*, **140**, 195; b) Norrish, R. G. W. and Bamford, C. H (1936), *Nature*, **138**, 1016.
2. Wagner, P. J. in de Maya, *Rearrangement in Ground and Excited States*, 1980, vol. 3, Academic Press: New York, p. 381.
3. Wayne, R. P., *Principle and Application of Photochemistry*, 1988, Oxford Univ. Press: Oxford, p. 53.
4. a) Wilson, R. M. (1985), *Org. Photochem.*, **7**, 339; b) Cookson, R. C., Hudec, J., Szabo, A. and Usher, G. E. (1968), *Tetrahedron*, **24**, 4313; Wagner, P. J. (1971), *Acc. Chem. Res.*, **4**, 168.
5. a) Wagner, P. J. and Hammend, G. S. (1965), *J. Am. Chem. Soc.*, **87**, 4009; b) Casey, C. P. and Boggs, R. A. (1972), *J. Am. Chem. Soc.*, **94**, 6457.
6. Givens, R. S. and Levi, N. in Patai, *The Chemistry of Acid Derivatives*, 1979, pt. 1, Wiley: New York, p.641.

7   a) Heicklen, J. and Noyes, W. A. (1959), *J. Am. Chem. Soc.*, **81**, 3858; b) Cundall, R. B. and Davies, A. S. (1966), *Proc. Roy. Soc. (London)*, **A 290**, 563; c) Yang, N. C. and Feit, E. D. (1968), *J. Am. Chem. Soc.*, **90**, 505.
8   a) Pitts, J. N. and Blacet, F. E. (1950), *J. Am. Chem. Soc.*, **72**, 2810; b) Martin, G. R. and Sutton, H. C. (1952), *Trans. Faraday Soc.*, **48**, 823.
9   Nicol, C. H. and Calvert, J. G. (1967), *J. Am. Chem. Soc.*, **89**, 1790.
10 Alumbaugh, R. L., Pritchard, G. O. and Rickborn, B. (1965), *J. Phys. Chem.*, **69**, 3225.
11 a) Brunet, V. and Noyes, W. A. (1958), *Bull. Soc. Chim, France*, 121; b) Srinivasan, R. (1962), *J. Am. Chem. Soc.*, **84**, 2475.
12 a) Ausloos, P. and Rebbert, R. E. (1961), *J. Am. Chem. Soc.*, **83**, 4897; b) Borkswaski, R. P. and Ausloos, P. (1961), *J. Phys. Chem.*, **65**, 2257.
13 Rebbert, R. E. and Ausloos, P. (1964), *J. Am. Chem. Soc.*, **86**, 4803.
14 Dougherty, J. (1965), *J. Am. Chem. Soc.*, **87**, 4011.
15 Wagner, P. J. (1971), *Acc. Chem. Soc.*, **4**, 168.
16 Turro, N. J. and Wan, J. K. S. (1968), *J. Am. Chem. Soc.*, **90**, 2186.
17 Kim, T. Y., Ko, E. S., Park, B. S., Yoon, H. and Chae, W. K. (1997), *Bull. Korean Chem. Soc.*, **18**(4), 439.
18 Arnold, J. K., Stringle, L. B. and Worketin, M. S. (2000), *Org. Lett.*, **2**(21), 3381.
19 Lewis, F. D. and Hilliard, T. A. (1970), *J. Am. Chem. Soc.*, **92**, 6672.
20 Padwa, A. and Eisenberg, W. (1970), *J. Am. Chem. Soc.*, **92**, 2590.
21 de Costa, D. P., Bennett, A. K. and Pincock, J. A. (1999), *J. Am. Chem. Soc.*, **121**, 3785.
22 Griesbeck, A. G., Henz, A., Kramer, W. and Wamser, P. (1998), *Tetrahedron Lett.*, **39**, 1549.

- Wagner, P.J. and Hammond, G.S. (1966), *J. Am. Chem. Soc,.* **88**, 1245.
- Wagner, P.J. and Zepp, R.G. (1972), *J. Am. Chem. Soc.*, **94**, 287.
- Wagner, P.J., Kelso, P.A. and Zepp, R.G. (1972), *J. Am. Chem. Soc.*, **94**, 7480.
- Adam, W., Grabowski, S. and Wilson, R.M. (1989), *Chem. Ber.*, **122**, 561.
- Kell, A. J., Stringle, D. L. and Workentin, M. S. (2000), *Org Lett.*, **19**, 3381.
- De Feyter S., Diau, E. W. and Zewail, A. H. (2000), *Angew. Chem. Int. Ed. Engl.*, **39**(1), 260.
- Saphier, S., Sinha, S. C. and Keinan, E. (2003), *Angew. Chem. Int. Ed. Engl.*, **42**(12), 42.

9

Paterno – Buchi Reaction

Introduction

The reaction involving [2+2] photochemical cycloaddition between an electronically excited carbonyl group and a ground state olefin yielding an oxetane is widely known as *Paterno-Buchi reaction*. Paterno and Chieffi[1] discovered the photochemical formation of oxetanes and Buchi, Inman and Lipinsky[2] carried out the initial investigations into its mechanism. For their contributions, the reaction is named as *Paterno-Buchi reaction*.

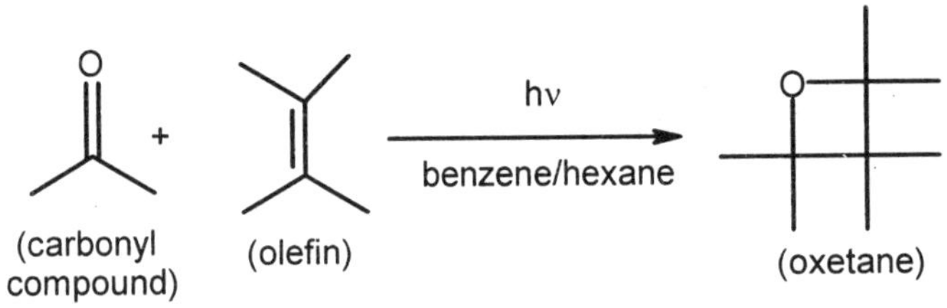

General scheme

The carbonyl functionality may be of aliphatic/ aromatic aldehydes or ketones or also of quinones. The pioneering example of this type of reaction is the photocycloaddition of benzaldehyde with 2-methylbut-2-ene, carried out by Paterno[1] in 1909 using UV light — the less substituted end of the C=C adds to the carbonyl oxygen due to steric factors.

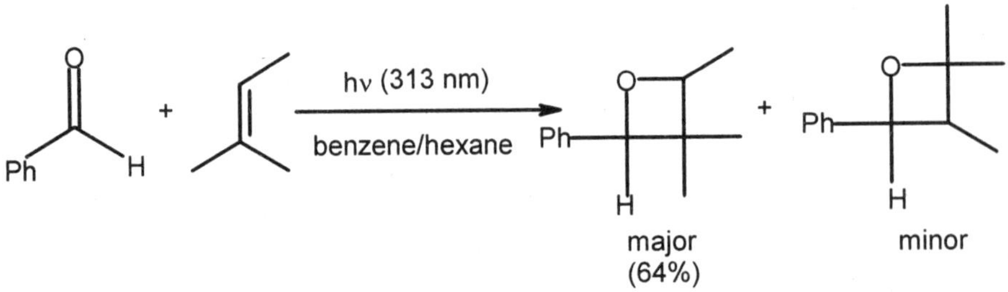

However, those carbonyl compounds that undergo rapid *Norrish type-I* (*e.g.* cyclobutanone) or *type-II* (*e.g.* 5-methylhex-2-one) reaction would relatively be unsuitable counterparts in *Paterno - Buchi reaction*. Yields in the *Paterno - Buchi*

reaction are variable, ranging from very low to fairly good (90%). There may occur several simultaneous side reactions also.

Discussion on mechanism

Mechanistic studies of this classical reaction have revealed that the reaction course differs with varying nature of the carbonyls and olefins involved. But there is no doubt that the light absorbing species in these reactions is the carbonyl component. In general, the mechanism involves the addition of an excited state of the carbonyl compound to the ground state of the olefin. Both singlet and triplet excited states have been shown to add to olefins to yield oxetanes. A good number of chemists studied the mechanistic course of *Paterno-Buchi reaction* from various angles. The essence of their findings would be presented here in brief.

Benzaldehyde on irradiation in presence of 2-methylbut-2-ene yields the four possible stereoisomeric oxetanes[2] (**Scheme 1**). The reaction is supposed to involve n-π* excitation of the carbonyl functionality; the excited singlet species then undergoes a rapid *intersystem crossing* (ISC) to a triplet one, which ultimately adds to the singlet ground state olefin. Loss of configuration about the alkene occurs and mixtures of isomeric oxetanes having all possible orientations are formed.

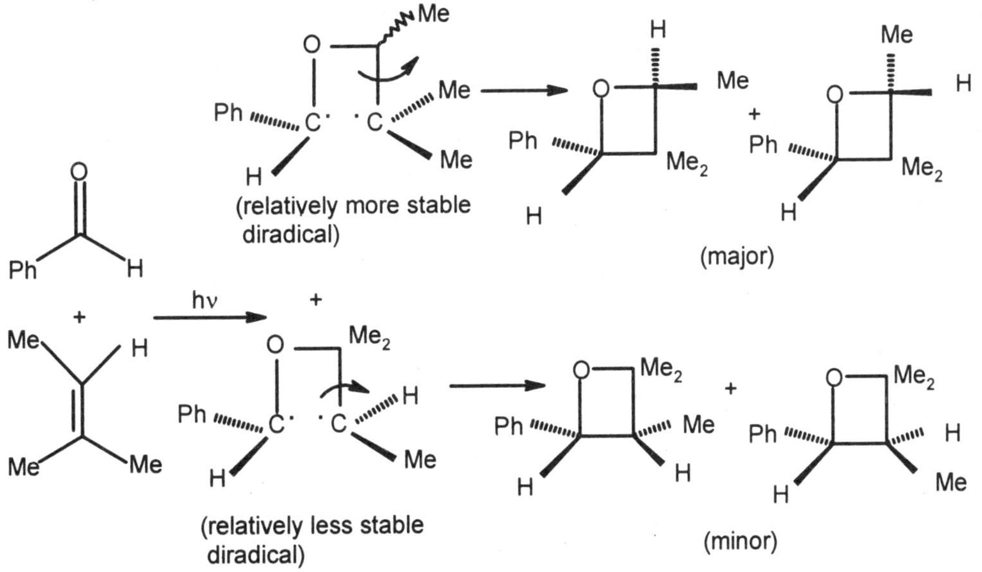

Scheme 1

The products have been considered to arise from a 1,4-diradical intermediate and hence the oxetane expected from the most stable diradical intermediate is supposed to be predominated. The reaction must be *non-concerted* since the addition of a triplet carbonyl to the ground state olefin results in the formation of triplet 1,4-diradical, which must undergo spin inversion before ring closure.

Similarly, photochemical addition of an aldehyde to 2-butene may yield four isomeric oxetanes (A - D). *Cis*-2-butene would give rise to oxetanes A & B, while *trans*-2-butene would give raise to C&D if the additions proceed *stereospecifically*.

$$RCHO + 2\text{-butene} \longrightarrow \text{oxetane}$$
(1)

1a: R= C_6H_5-; 1b: R= *p*-MeO- C_6H_5-; 1c: R= *m*-MeO-C_6H_5-

But the experimental results as reported by Yang *et al.*[3] revealed that the oxetane fractions isolated from the irradiations of 1a,b or c in pure *cis*- and *trans*-2-butenes were indistinguishable from each other. The results clearly indicate that the reaction must proceed *via* a long-lived 1,4-diradical intermediate, which would attain all four conformations (A–D) in equilibrated proportions before cyclizations to oxetanes. The formations of A & D were found to become favoured over those of B & C approximately by a ratio 3:1; this preference may be attributed to the favourable conformations of the concerned diradicals in the transition state during cyclization as they contain both the R and the 4-methyl groups in the pseudoequatorial conformations (e′).

It is an established fact that ketones which are not reduced upon irradiation in presence of alcohol (e.g. isopropanol) cannot be expected to form oxetanes; this has been found to be effective[4-6] in case of the ketones —— 4-aminobenzophenone, 4,4′-bis-(dimethylamino)-benzophenone, xanthone, 1-naphthylphenylketone and 2-naphthylphenylketone. These ketones are neither reduced on irradiation in isopropanol, nor being effected to give oxetanes on irradiation with olefins (isobutylene)[5,6]. It is known that photo-induced reduction involves hydrogen abstraction from the alcohol by the n-π* state of the ketone[5]. The n-π* singlet and triplet states both have radical character (an unpaired electron localized on oxygen)[7].

Moore et al.[8] showed that reduction of benzophenone involves the n-π* triplet, however, in some cases reaction may involve either of the states or the both[9]. Hence, it can be inferred that photochemical addition resulting oxetane formation also requires this n-π* state of the carbonyl counterpart. On this basis, Arnold et al.[6] outlined a very simple scheme (**Scheme 2**) for *Paterno-Buchi reaction* as —

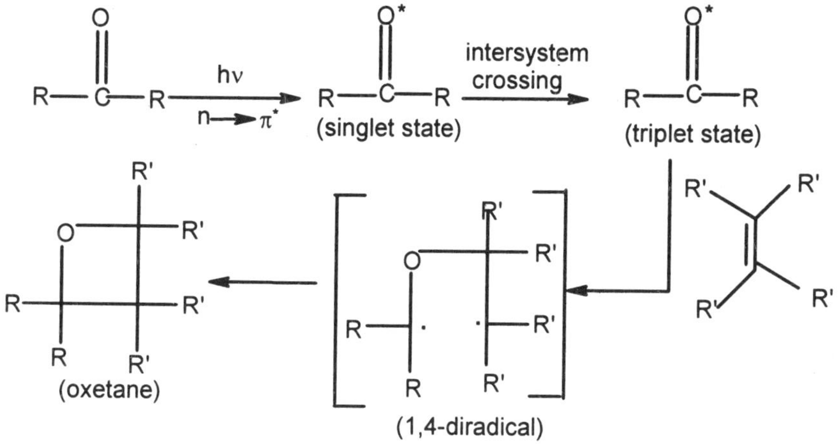

Scheme 2

It is also an observed fact that benzophenone reacts with both cis- & *trans*-2-butene to give the same ratio of *cis*- & *trans*-oxetanes (**Scheme 3**) and causes extensive isomerization of the starting olefin. Any intermediate in this reaction must, therefore, be incapable of retaining its stereochemical integrity and the results are not in disagreement with a 1,4-diradical intermediate[10,11], provided that rotation about the C – C bond is rapid enough relative to ring closure. Diradical intermediate has also recently been detected by using spectroscopic methods[12].

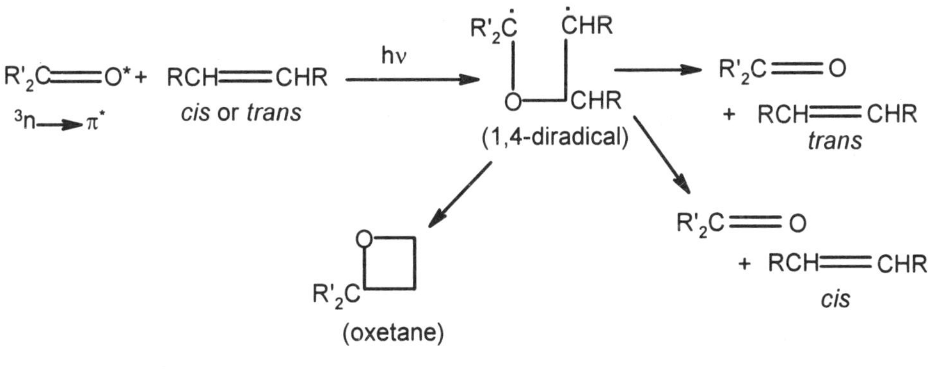

Scheme 3

A marked difference between aliphatic and aromatic carbonyl compounds is very often noticed during the course of oxetane formation. Aliphatic carbonyl compounds are usually found to follow stereoslective path in a higher degree than their aromatic analogues; hence the photoaddition reaction is thought to involve the n-π* singlet state of the aliphatic carbonyl compoud, that proceeds with a high degree of stereoselectivity and little isomerization. For example, benzophenone reacts with 2-butene with a non-stereoselective manner, while acetaldehyde with a high degree of stereoselective fashion (**Scheme 4**). This can be rationalized by assuming the fact that initial singlet n-π* excited state of benzophenone undergoes intersystem crossing at an extremely rapid rate (which is several times faster than that of aliphatic carbonyl compounds[13]), while photo-addition reaction of acetaldehyde with 2-butene takes place mainly from its ^{1}n-π* (singlet) excited state.

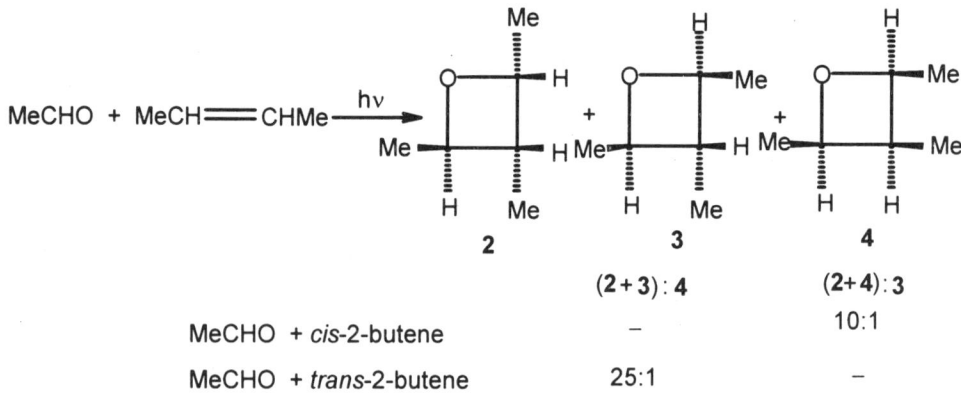

Scheme 4

The acetaldehyde reaction is highly stereoselective, but not stereospecific suggesting that at least part of the reaction is not *concerted*. A singlet diradical intermediate will undergo ring closure appreciably faster than bond inversion. The photochemical addition of acetone to *cis*- & *trans*-1-methoxybut-1-ene results in the formation of four iosomeric oxetanes. This addition reaction is thought to involve both the singlet and triplet excited states of acetone; the stereochemistry of the initial olefin being almost completely retained when it is attacked by the singlet state, but when attacked by the triplet state the total stereochemical integrity is lost.

Based upon the contributions from various laboratories[14-17], an alternate reaction course for the *Paterno-Buchi reaction* has been suggested (**Scheme 5**); this course satisfactorily explains both olefin isomerization as well as oxetane formation. The proposed course of reaction involves an electron-transfer complex, called as *exciplex* that could subsequently collapse to a diradical intermediate. The n-π* singlet excited state of the carbonyl compound, being electron deficient, complexes with the double bond of the olefin forming the *exciplex*. *Exciplex* formation is favoured as the olefin is more highly alkylated. The lack of isomerization of olefins on irradiation with aldehydes results an inefficient intersystem crossing from the singlet n-π* excited state or the singlet *exciplex*. The stereochemistry of oxetane formation will depend on the relative rate of ^{1}[exciplex] formation and intersystem crossing (ISC).

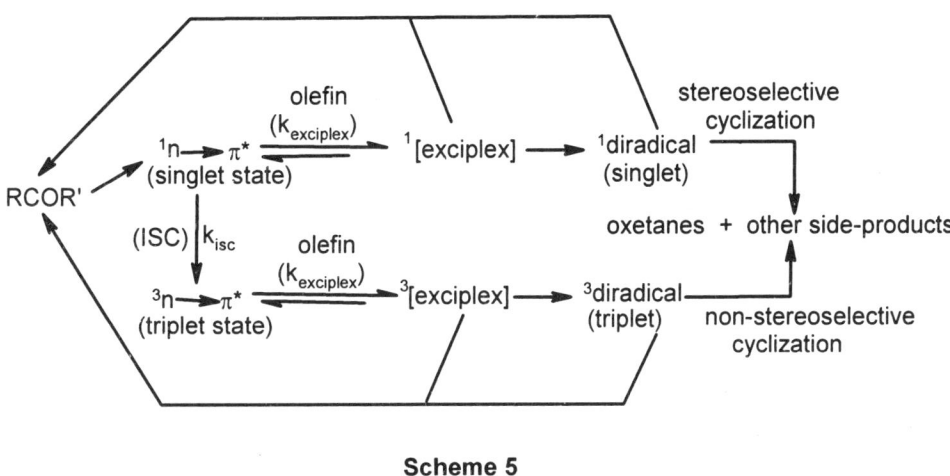

Scheme 5

Competing energy transfer reaction

A major competing reaction is the energy transfer to the olefin. Since an *exciplex* may be involved in some oxetane-forming reactions, it is anticipated that net energy transfer to produce an excited olefin may compete with oxetane formation.

It is a general rule[18] that if an olefin possesses an electronically excited state of lower energy and of the same multiplicity as the interacting n-π* state, energy transfer to the olefin becomes a competitive alternative to oxetane formation. It is reported[19] that benzophenone (E_T= 70 Kcal/mole)[20] on irradiation in presence of dienes (having triplet energy, E_T of the order of 60 Kcal/mole)[20] produces no oxetane, but the products derived from the n-π* triplet of the dienes. At the same time, Arnold et al.[6] also were unable to obtain oxetanes when benzophenone was irradiated in the presence of the olefins such as 1,1-diphenylethylene, stilbene, methylacrylate and 1,2-dichloroethylene. This also accounts for the fact that irradiation of acetone with norbornene yields dimers of norbornene rather than oxetane formation[21], while in case of benzophenone (with relatively lower E_T) and norbornene a fair yield of oxetane is obtained without a trace of norbornene dimer[6] (**Scheme 6**).

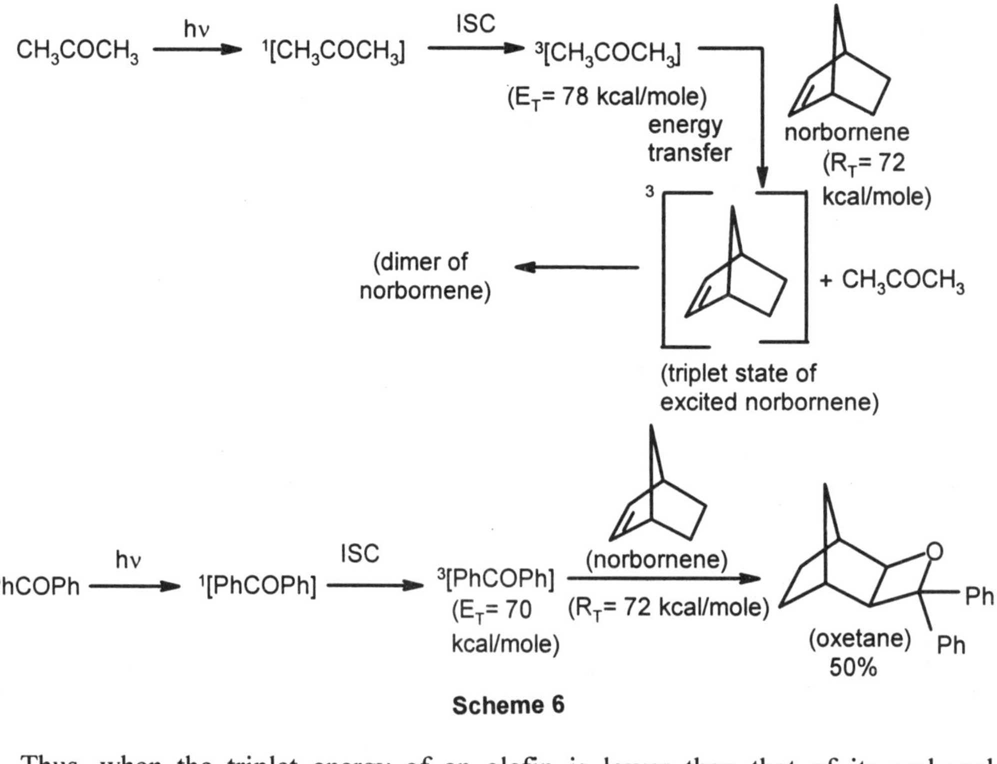

Scheme 6

Thus, when the triplet energy of an olefin is lower than that of its carbonyl counterpart, the triplet-triplet energy transfer may take place effecting only the dimerization of the olefin instead of oxetane formation.

Examples and applications relating to Paterno-Buchi reaction

Paterno-Buchi reaction is very much useful in the synthesis of four membered rings bearing oxygen heteroatoms (product known as oxetane derivatives); the reaction requires simple experimental set-up. *Paterno-Buchi reaction* finds tremendous applications in synthetic organic chemistry. A few examples are cited here:

1) 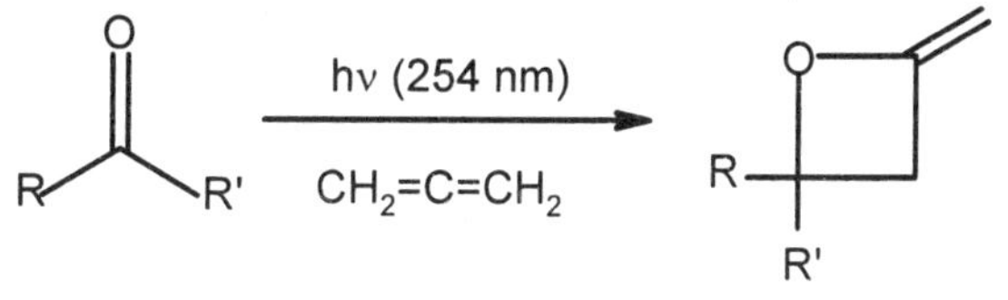 [Reaction of R-CO-R' with CH$_2$=C=CH$_2$ under hv (254 nm) to give oxetane product]

2) [Reaction under hv with Me$_2$C=CMe$_2$] 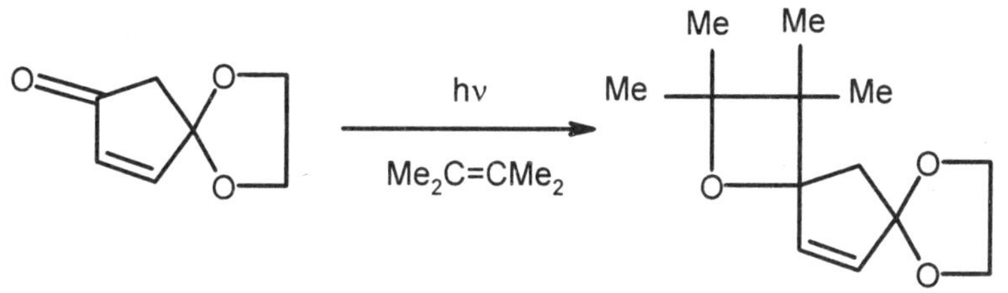 [Ref. 22]

3) 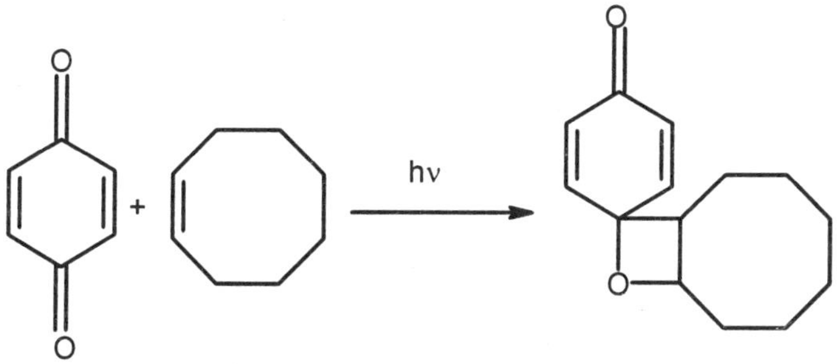 [Reaction of benzoquinone + cyclooctene under hv]

4) 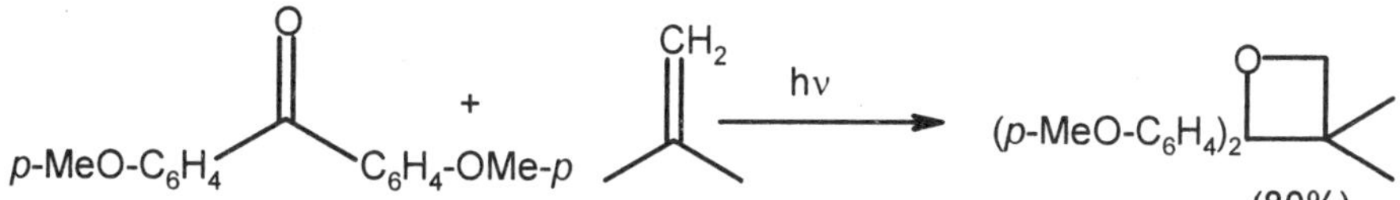 [Reaction of (p-MeO-C$_6$H$_4$)$_2$C=O + CH$_2$=C(Me)$_2$ under hv to give oxetane (80%)]

5) Cruciani et al.[23] reported the formation of an oxetane compound, 1-oxaspiro (5,3) non-5-ene (**5**) on photocyclo-addition reaction of cyclohex-2-ene-1-one with tetramethoxyethylene in benzene solution.

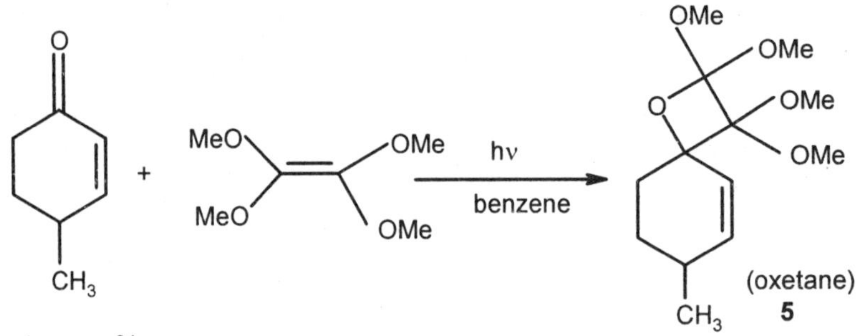

[Scheme showing 4-methylcyclohex-2-enone + 1,1,2,2-tetramethoxyethylene → oxetane **5**, hν, benzene]

6) Aoyama et al.[24] extended the use of *Paterno-Buchi reaction* in the field of heterocyclic chemistry also:

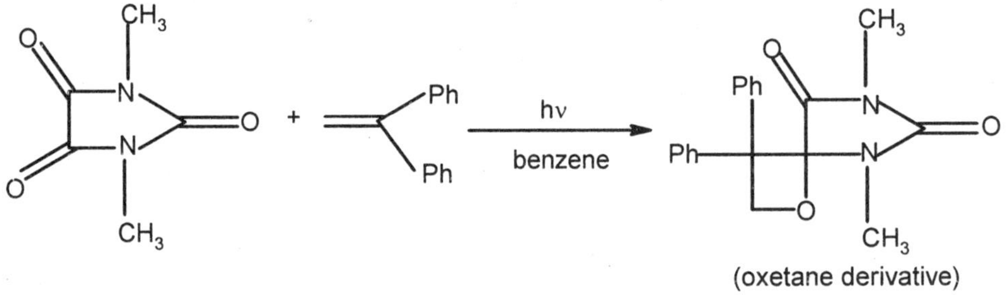

[Scheme showing 1,3-dimethylparabanic acid derivative + 1,1-diphenylethylene → oxetane derivative, hν, benzene]

7) Photochemocal reactions of aromatic carbonyl compounds with silyl ketene acetals have been explored by Yoon et al.[25]

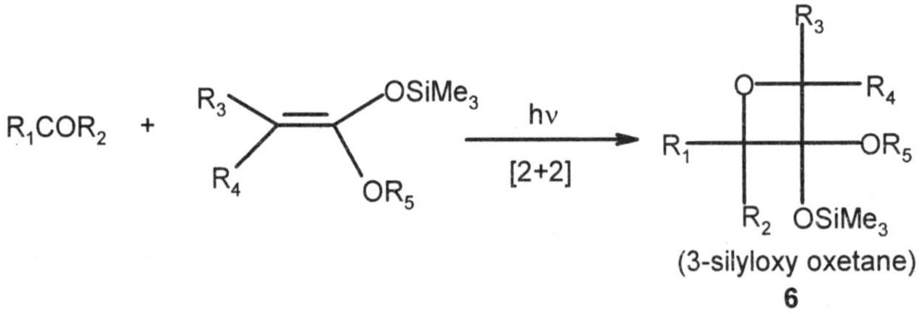

$$R_1COR_2 + \text{silyl ketene acetal} \xrightarrow[\text{[2+2]}]{h\nu} \text{(3-silyloxy oxetane)}\ \mathbf{6}$$

6a: R_1=Ph, R_2=H, R_3=R_4=R_5=Me
6b: R_1=R_2=Ph, R_3=R_4=R_5=Me

8) An unexpected *Paterno-Buchi reaction* taken place in the crystalline state has recently been reported by Kang et al.[26]. The solid-state crystals of aryl-1-phenylcyclopentyl ketones (**7**) on irradiation afforded the novel oxetanes (**8**) and (**9**) *regiospecifically*. The formation of the oxetanes is believed to occur through *Norrish type-II* and hydrogen abstraction producing an alkene and an aldehyde, followed by a *Paterno-Buchi reaction* within the lattice cage.

(7) (8) (9)

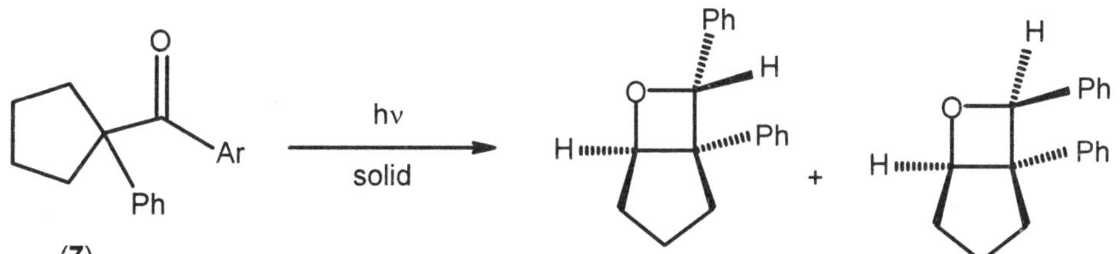

9) Shreiber *et al.*[27] used the *Paterno-Buchi reaction* as one of the key step in their synthesis of an important antifungal metabolite, (+)-avenaciolide (**10**).

(10)

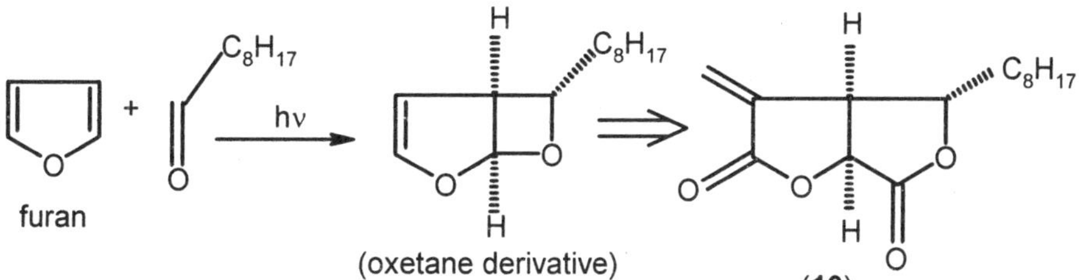

10) The reaction can also occur in an *intramolecular* fashion, provided the functional groups are suitably oriented within the same molecule. Thus, 2-allylcycloheptanone on photolysis furnishes the oxetane (**11**), which can be converted to a synthetically useful compound, azulene on dehydrogenation.

2-allylcycloheptanone (**11**) azulene (25%)

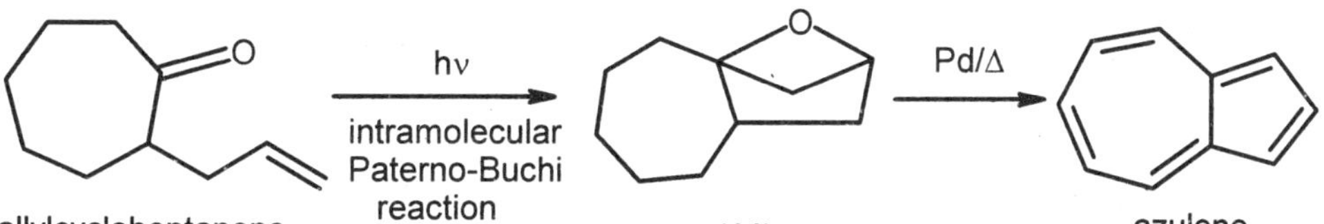

11) Photocycloaddition of aliphatic ketones to olefins substituted with electron-withdrawing groups (particularly cyano group) involves the addition of singlet n-π* state of the ketones to yield oxetanes where the stereochemistry of the olefins is retained.

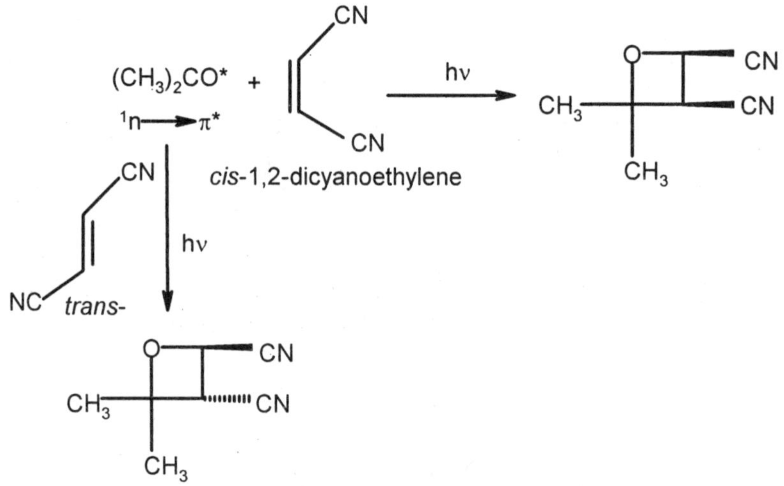

12) Gorman et al.[28] reported an interesting photocycloaddition of n-π* state of ketones to a σ-bond. They carried out the photocycloaddition reaction of benzophenone with quadricyclene (12). The major products (13) & (14) are developed due to the attack of n-π* excited state from the *exo* face to form a diradical, which cyclizes to the observed products.

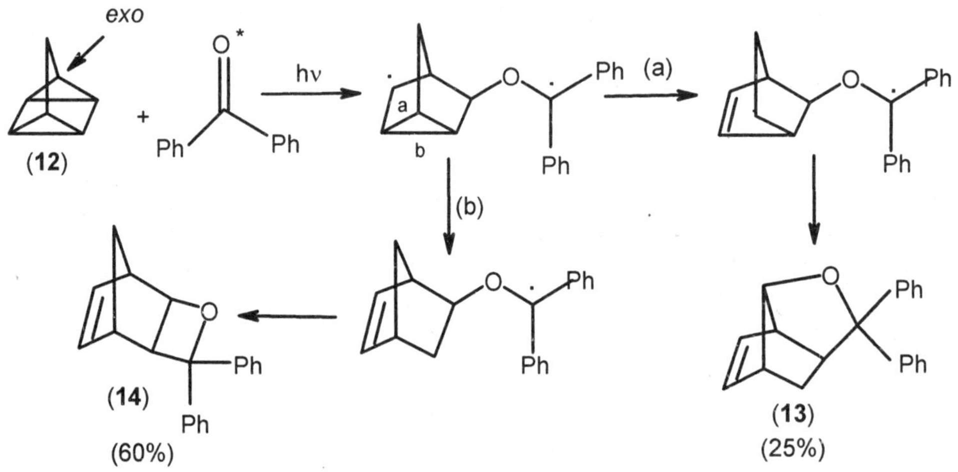

13) Acetylenes also participate in the *Paterno–Buchi reaction* expecting to produce oxetenes. These products are highly susceptible to undergo electrocyclic ring opening to give unsaturated carbonyl compounds[29].

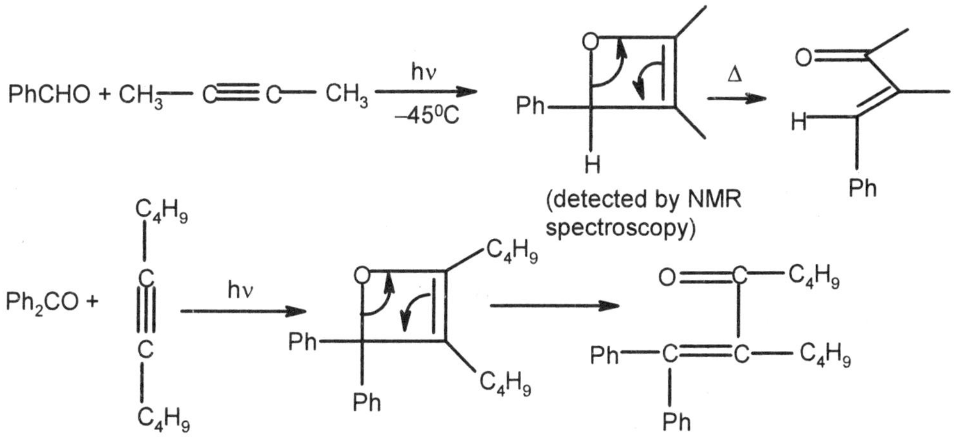

14. The *Paterno–Buchi reaction* provides a unique way to synthesize functionalized oxetanes. Park et al.[30] have described diastereoselective synthesis of highly functionalized oxetanes by means of photocycloaddition of 2,2-diethoxy-3,4-dihydro-2H-pyran (**15**) with achiral aldehydes in a very general way which promises to be widely applicable.

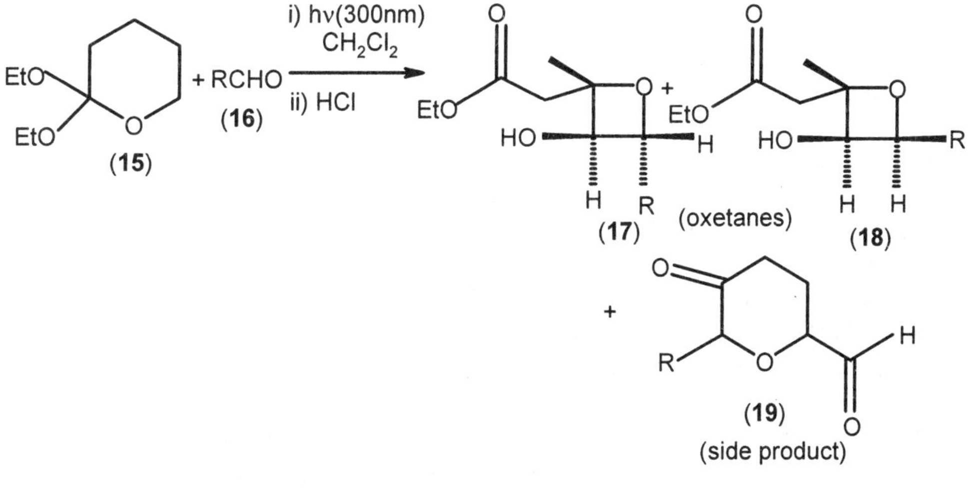

Carbonyl substrate (**16**)	Oxetanes (**17:18**)	Side product (**19**)
17a : R= Ph	57% (92:8)	8%
17b: R= - CH_2CH_2Ph	38% (88:13)	5%
17c: R= - C_6H_{13}	31% (91:9)	<5%

References

1. Paterno, E. and Chieffi, G. (1909), *Gazz. Chim. Ital.*, **39**, 341.
2. Buchi, G., Inman, C.G. and Lipinsky, E.S. (1954), *J. Am. Chem. Soc.*, **76**, 4327.
3. Yang, N.C., Kimura, M. and Eisenhardt, W. (1973), *J. Am. Chem. Soc.*, **95**(15), 5058.
4. Pitts (Jr.), J.N., Johnson (Jr.), H.N. and Kuwana, T. (1962), *J. Phys. Chem.*, **66**, 2456.
5. Beckett, A. and Porter, G. (1963), *Trans. Far. Soc.*, **59**, 2038.
6. Arnold, D.R., Hinman, R.L. and Glick, A.H. (1964), *Tetrahedron Lett.*, No. 22, 1425.
7. Kasha, M., "*Ultaviolet Radiation Effects: Molecular Photochemistry in 'Comparative Effects of Radiation'*, Ed. M. Barton, J.S. Kirby-Smith and J.L. Magee, 1960, Wiley and Sons, New York, p.72.
8. Moore, W.M., Hammond, G.S. and Foss, R.P. (1961), *J. Am. Chem. Soc.*, **83**, 2789.
9. Bridge, N.K. and Porter, G. (1958), *Proc. Roy. Soc.*, **244A**, 276.
10. Saltiel, J., Neuberger, K.R. and Wrighton, M. (1969), *J. Am. Chem. Soc.*, **91**, 3658.
11. For other evidence for these diradical intermediates, see references cited in Griesbeck, A.G., and Stadmuller, S. (1990), *J. Am. Chem. Soc.*, **112**, 1281.
12. Freilich, S.C. and Peters, K.S. (1981), *J. Am. Chem. Soc.*, **103**, 6255; for a review, see Griesbeck, A.G., Mauder, H. and Stadmuller, S. (1994), *Acc. Chem. Rev.*, **27**, 70.
13. Rentzepis, P.M. and Mitschele, C.J. (1970), *Anal. Chem.*, **42**(14), 20A.
14. Arnold, D.R. (1968), *Advan. Photochem.*, **6**, 301.
15. Yang, N.C. and Eisenhardt, W. (1971), *J. Am. Chem. Soc.*, **93**, 1277.
16. Caldwell, R.A. and James, S.P. (1969), *J. Am. Chem. Soc.*, **91**, 5184.
17. Kochevar, I.H. and Nagner, P.J. (1970), *J. Am. Chem. Soc.*, **92**, 5742.
18. Arnold, D.R., Hinman, R.L. and Glick, A.H. (1964), *Tetrahedron Lett.*, 1724; Arnold, D.R., Trecker, D.J. and Whipple, E. B. (1965), *J. Am. Chem. Soc.*, **87**, 2596.
19. Hammond, G.S., Turro, N.J. and Fischer, A. (1961), *J. Am. Chem. Soc.*, **83**, 4674; Hammond, G.S. and Liu, R.S.H. (1963), *ibid*, **85**, 477; Hammond, G.S., Turro, N.J. and Leermakers, P.A. (1962), *J. Phys. Chem.*, **66**, 144; Trecker, D.J., Bardon, R.L. and Henry, J.P. (1963), *Chem. and Ind.*, 652.
20. Lewis, G.N. and Kasha, M. (1944), *J. Am. Chem. Soc.*, **66**, 2100.
21. Scharf, D. and Korte, E. (1963), *Tetrahedron Lett.*, 821.

22 Yoshida Z., Kimura, M. and Yoneda, S. (1975), *Tetrahedron Lett.* **16**(12), 1001.
23 Cruciani, G., Rathjen, H. and Margaretha, P. (1990), *Helv. Chim. Acta*, **73**, 856.
24 Aoyama, H., Hatori, H. and Omote, Y. (1989), *J. Org. Chem.*, **54**, 2359.
25 Yoon, U.C., Kim, M.J., Moon, J.J., Oh, S.W., Kim, H.J. and Mariano, P.S. (2002), *Bull. Korean Chem. Soc.*, **23**(19), 1218.
26 Kang, T. and Schefer, J.R. (2001), *Org. Lett.*, **3**(21), 3361.
27 Screifer, S.L. and Hoveyda, A.H. (1984), *J. Am. Chem. Soc.*, **106**, 7200.
28 Gorman, A.A. and Leyland, P.L. (1972), *Tetrahedron Lett.*, 5345; *ibid*, 5085 (1973).
29 Buchi, G., Kofron, J. T., Koller, E. and Rosenthal, D. (1956), *J. Am. Chem. Soc.*, **78**, 876; Bryce-Smith, D., Fray, G.I. and Gilbert, A. (1964), *Tetrahedron Lett.*, 2137; Zimmerman, H.E. and Craft, L. (1964), *ibid*, 2131.
30 Park, S-K, Lee, S-J, Back, K. and Yu, C-M. (1998), *Bull. Korean Chem. Soc.*, **19**(1), 35.

- Fleming, S. A. and Gao, J. J. (1997), *Tetrahedron Lett.*, **38**, 5407.
- Hubig, S. M., Sun, D. and Kochi, J. K. (1999), *J. Chem. Soc., Perkin Trans. 2*, 781.
- D'Auria, M., Racioppi, R. and Romaniello, G. (2000), *Eur. J. Org. Chem.*, 3265.
- Bach, T. (2000), *Synlett*, 1699.
- Abe, M., Tachibana, K., Fujimoto, K. and Nojima, M. (2001), *Synthesis*, 1243.
- D'Auria, M., Emanuele, L., Poggi, G., Racioppi, R. and Romaniello, G. (2002), *Tetrahedron*, **58**, 5045.
- Griesbeck, A. G. (2003), *Synlett*, 451.

10

Stork Enamine Reaction

Introduction

Enamines usually refer to α,β-unsaturated amines; Wittig and Blumenthal[1] introduced the term 'enamine'. An enamine having a hydrogen atom on nitrogen generally becomes unstable and rearranges to the corresponding *imine*. This behaviour is analogous to the rearrangement of vinyl alcohol (*'enols'*) to carbonyl compounds.

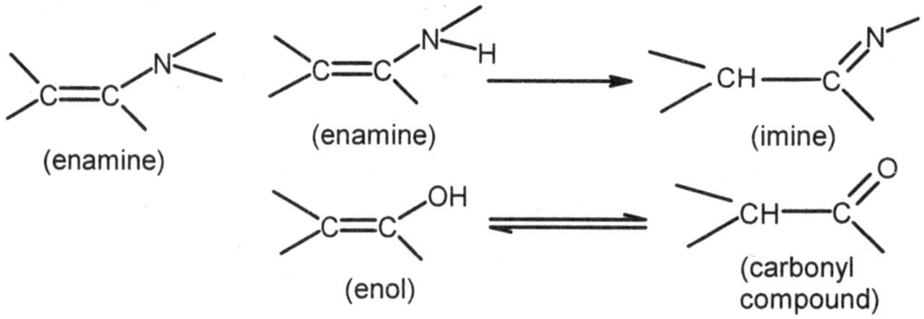

Conversely, enamines with no hydrogen on nitrogen atom are stable towards rearrangement and that is why they are usually prepared by the reaction of a secondary amine with a carbonyl compound in presence of acid catalyst. Enamines constitute an important group of intermediates in organic synthesis. Although enamine reactivity has been known since 1883,[2-4] there is indeed no doubt that the generality and wide ranging applicability of the reaction of enamines with electrophiles had not been understood until the pioneering work of Stork *et al.*[5] in 1954. They investigated widely the versatility of enamines in the cause of C – C bond formation. When an enamine (which is essentially a 'nitrogen enolate") is treated with an alkyl halide or acyl halide followed by hydrolysis, the net result is the alkylation or acylation of carbonyl compound at the α-position. Such protocol of C – alkylation and acylation of a carbonyl compound *via* an enamine intermediate has subsequently been known as *Stork enamine reaction*[6].

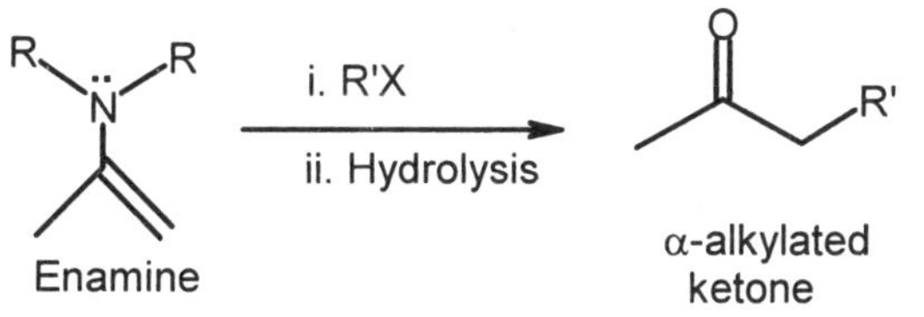

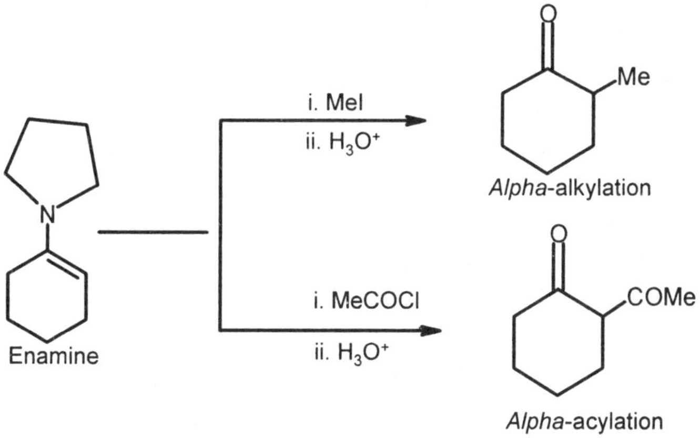

Scheme

Enamine + i. R'X / ii. Hydrolysis → α-alkylated ketone

1-(cyclohex-1-en-1-yl)pyrrolidine (Enamine):
- i. MeI / ii. H_3O^+ → Alpha-alkylation (2-methylcyclohexanone)
- i. MeCOCl / ii. H_3O^+ → Alpha-acylation (2-acetylcyclohexanone)

Thus, *Stork method* provides an alternative route to ketone alkylation. This method has emerged more advantageous over many others in providing preferential monoalkylaion of ketones; alkylation usually takes place on the less substituted side of the starting ketonic compound[7]. However, the product of the *Stork reaction* is largely dependent on the experimental conditions. As it has been observed that the same electrophilic reagent may furnish completely different types of product if relatively trivial changes are made in the solvent, reaction temperature, amine moiety, molar proportion of reagents, etc. Such diversity, of course, provokes the researchers to work in this field. A voluminous work on the *Stork enamine reaction* has already been done, on the basis of which, the definition of *Stork reaction* may be extended so as to include the conversion of an aldehyde or a ketone into a *C*-alkylated, acylated, and carbocyclic or heterocyclic derivative by reaction of an electrophile with an enamine intermediate.

Preparation of Enamines

Enamines can be synthesised by the following methods:

a) Reaction of carbonyl compounds with amines

Enamines are best formed by the reaction of an aldehyde or a ketone bearing at least one α-hydrogen atom and a secondary amine. They are usually prepared by refluxing the carbonyl compound and amine in a suitable solvent such as benzene, toluene or xylene under acid catalysis[8-10]. The water eliminated may be removed from the condensate by means of a Dean and Stark head or using "molecular sieves"[11] (e.g. zeolites). Besides, a chemically inert drying agent (i.e. anhydrous K_2CO_3 or $MgSO_4$) may also be used in the reaction mixture. However in case of aldehydic substances, another general method is also available; this method involves the reaction of an aldehyde with 2 equivalents of secondary amine in the cold in presence of anhydrous K_2CO_3[12a] or Na_2CO_3[12b] giving rise to 1,1-diamine (aminal), which ultimately affords the enamine on destructive distillation. The rate of formation of an enamine depends on the basicity of the amine used as well as on the steric requirements of both the amine and the carbonyl compound. It has been observed that increased substitutions alpha (α) to the carbonyl group of a ketone reduces the rate of enamine formation. Cyclic amines usually react faster than open-chain amines[8]. The most commonly used cyclic amines are piperidine, morpholine and pyrrolidine. Conditions for optimal synthesis of morpholine enamines of acyclic ketones have also been studied using molecular sieves (5A) and $TiCl_4$ as water scavenger[13]. Application of the $TiCl_4$ route[14] to the preparation of enamines of alkylmethylketones has been found to occur under kinetic control.

b) Horner–Wittig reaction

Horner–Wittig reaction is a very successful method for the synthesis of both aldehyde and ketone enamines. A few examples are cited:

Ph CH=O $\xrightarrow{i}$ Ph$_2$P(O)-CH(NR$_2$)-CH(OH)Ph $\xrightarrow{ii}$ Ph-CH=CH-NR$_2$

Ph$_2$CO $\xrightarrow{i, ii}$ Ph$_2$C=CH-NR$_2$ [NR$_2$ = morpholino moiety]

Ph$_2$CO $\xrightarrow{iii, iv}$ Ph$_2$C=C(Ph)-NR$_2$

Reagents: i) Ph$_2$P(O)CH$_2$NR$_2$, BuLi, THF, 0^0; ii) KH, THF, R.T., 3hrs   [Ref. 15-17]
iii) (Et$_2$O)P(O)CH(Ph)NR$_2$; iv) NaH, dioxan, 40^0

Norbornanone $\xrightarrow[\text{BuLi, THF, }-78^0]{(EtO)_2P(O)CH_2NR_2}$ norbornylidene-NR$_2$ [Ref. 18]

[NR$_2$ = morpholino moiety]

c) Base catalysed rearrangement

Allylamines undergo base catalysed rearrangement to yield enamines; the process corresponds to a gain in stability attributed to pπ-conjugation in the enamine[19]. Base catalysed rearrangement of **1** yields the enamine **2** exclusively as *cis*-isomer[20,21].

(1) → *Cis*-isomer (2) ⤫ *Trans*-isomer

Ollis *et al.*[22] reported an interesting series of rearrangements where the diallylammonium bromide (**3**) is induced to undergo a base-catalysed [3,2]-sigmatropic rearrangement, followed further by a thermal [3,3]-sigmatropic rearrangement to furnish ultimately the enamine (**5**). On hydrolysis the aldehyde (**6**) is obtained.

(3) $\xrightarrow[\text{R.T., 8 hrs}]{[H_2O, Na_2S]}$ (4) $\xrightarrow{170^0}$ (5) $\xrightarrow{H_3O^+}$ (6)

d) Reactions of alkenes and alkynes

Enamines may be obtained by an aminomercuration–demercuration procedure involving addition of aromatic amines to the inner carbon of non-activated terminal acetylenes[23]. It has also been observed that aliphatic amines add to the terminal carbon of perfluoroalkylacetylenes in absence of catalyst[24]. Such as —

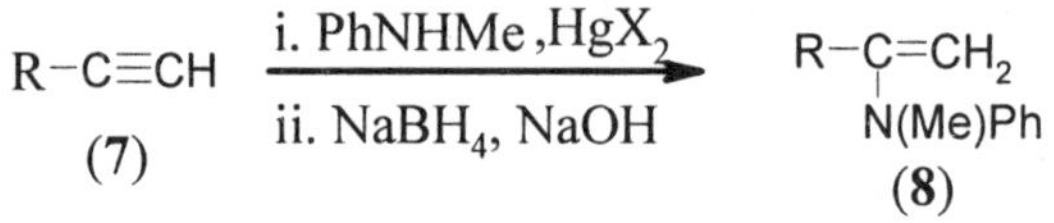

$$R-C\equiv CH \xrightarrow[\text{ii. NaBH}_4,\text{ NaOH}]{\text{i. PhNHMe, HgX}_2} R-\underset{N(Me)Ph}{C}=CH_2$$
(7) → (8)

[R= n-Bu, n-C_6H_{13}, Ph]

$$R'F_2C-C\equiv CH \xrightarrow{R_2N} R'F_2C-CH=CH-NR_2$$
(9) → (10)

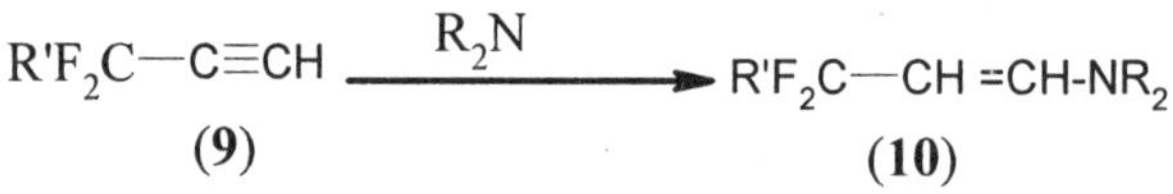

[R_2N = Et_2N, i-Bu_2N, piperidino, pyrrolidino, morpholino; R' = n-C_7H_{15}]

A bicyclic acetylene (12) is supposed as the reactive intermediate in the conversion of 2-or 3-chlorobicyclo [3.2.1] oct-2-enes (11) to enamines (13 & 14) in presence of base[25].

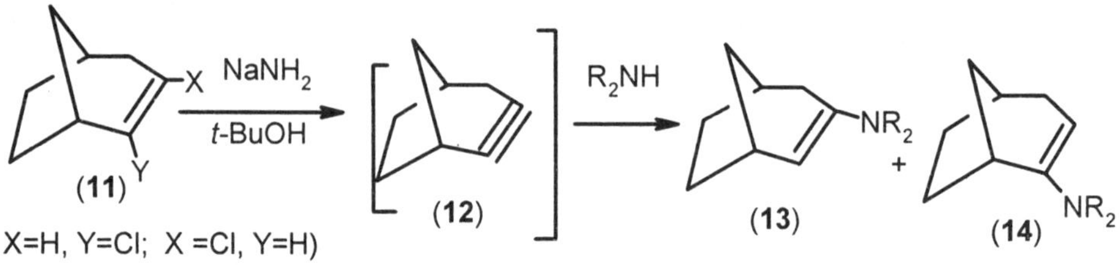

(11) (X=H, Y=Cl; X =Cl, Y=H) $\xrightarrow[t\text{-BuOH}]{\text{NaNH}_2}$ [(12)] $\xrightarrow{R_2NH}$ (13) + (14)

e) Cyclopropane enamines

Cyclopropane enamines (16) may be prepared from cyclopropyl ketones (15) using TiCl₄ as catalyst —

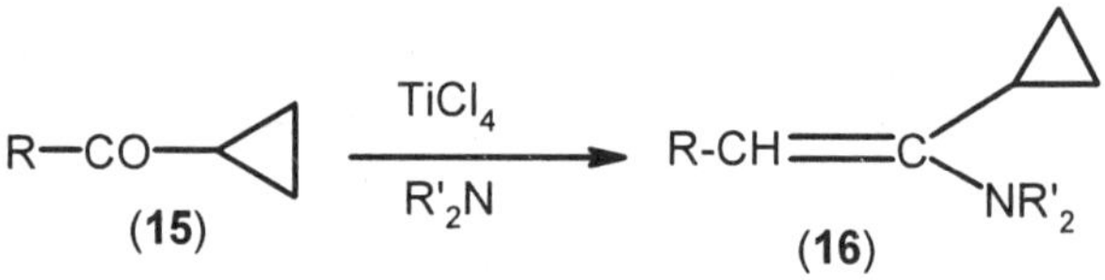

$$R-CO-\triangleleft \xrightarrow[R'_2N]{TiCl_4} R-CH=C\underset{NR'_2}{\overset{\triangle}{}}$$
(15) → (16)

R = Me, Et, cyclopentyl
R'₂N = Me₂N, Et₂N, pyrrolidine, morpholine

But, it has been observed that ring opening of cyclopropane occurs in the compounds (**15**) where 'R' group is cyclopropyl or aryl; in those cases enamines of the type (**17**) are obtained.

$$R-\underset{\underset{NR'_2}{|}}{C}=CH-CH_2-\underset{\underset{NR'_2}{|}}{C}H_2$$

(**17**)

Structure and reactivity

In case of unsymmetrical ketones, mixtures of structurally isomeric enamines are usually obtained. Enamine stabilization is due to interaction of the alkene π-system with the unshared electron pair in a p-orbital on nitrogen; for greater conjugation coplanarity between the bonds of unsaturated carbon atoms and those to nitrogen is necessary. Thus, this stability factor determines the predominant isomer. Johnson et al.[26] showed that these isomers undergo rapid acid catalyzed equilibration —

Scheme 1: a= (quasi) axial; e= (quasi) equatorial; t= tetrasubstituted

The proportion of isomers varies with the amine used. The pyrrolidine enamines of 2-methylcyclohexanones (**18**) exit as <10% in the tetra-substituted form (**19t**) whereas in the case of morpholine, it may constitute 30-65% of the enamine mixture[26]. Optically active (+)-methylpiperidine furnished only the more substituted enamine (**19t**). Similar differences exit between pyrrolidine and morpholine enamines of 2-alkoxycyclo-hexanones also[27]. These differences can be attributed to different conjugating ability and steric requirements of the amine moiety.

The isomer having the tetra-substituted double bond (**19t**) would be destabilized by severe steric interactions ($A^{1,3}$ starin)[28] between the methyl and amine α-methylene groups, if these lie in same plane. In the ground state these steric interactions can be reduced by rotation about the N – C (sp^2) bond. However, this will reduce the nitrogen lone pair interaction with the double bond; hence a balance between these two opposing requirements must clearly exist in order to minimize the energy of **19t**. The less substituted double bond isomer exists in two conformations in which the 'Me' substituent is quasi-axial (**19a**) or quasi-equatorial (**19e**). Conformation **19e** is to some extent destabilized due to the presence of allylic interactions (**Scheme 1**; $A^{1,2}$, less severe than $A^{1,3}$ strain)[28] and that is why the quasi-axial isomer (**19a**) is the most stable one, when other factors remain equal.

Enamine stability has immense impact on its reactivity. An important consequence is that α,α-dialkylation of ketones *via* their enamines is rarely observed because of the fact that for maximum orbital interaction between nitrogen lone pair and the π-electrons of the double bond; the starred groups (as shown in **19t**) must become coplanar. A product-like transition state would then be destabilized by increasing $A^{1,3}$-interactions. Further alkylation or acylation, therefore, takes place at the less substituted α'-position (of the ketone), but at a reduced rate owing to developing 1,3-diaxial interactions (as shown in **19a**) or the developing steric interactions associated with a twist or boat conformations if reaction occurs from the other side of the bond (*i.e.* equatorial approach).

A further consequence of allylic strain is that when an equatorial 2-substituent cannot be converted into the axial conformer by ring-inversion, as in the case of *cis*-4-*t*-butyl-2-methylcyclohexanone, then epimerisation to the *trans*-isomer occurs. This offers a method for the conversion of a more stable *cis*-diequatorial 2,4-disubstituted cyclohexanone to the less stable *trans*-isomer[26]. Allylic strain also expians why enamines of 3-methylcyclohexanones exist mainly as isomer (**21**)[29]. Alkylation[30], acylation[31] and halogenation[32] of the enamine mixture (**20 & 21**) have been shown to yield the 2-substituted-5-methylcyclohexanone (**22**) as the major product. At least in case of product-like transition states, formation of 2-substituted-3-methylcyclohexanone will be inhibited by developing steric interactions as shown in **Scheme 2**:

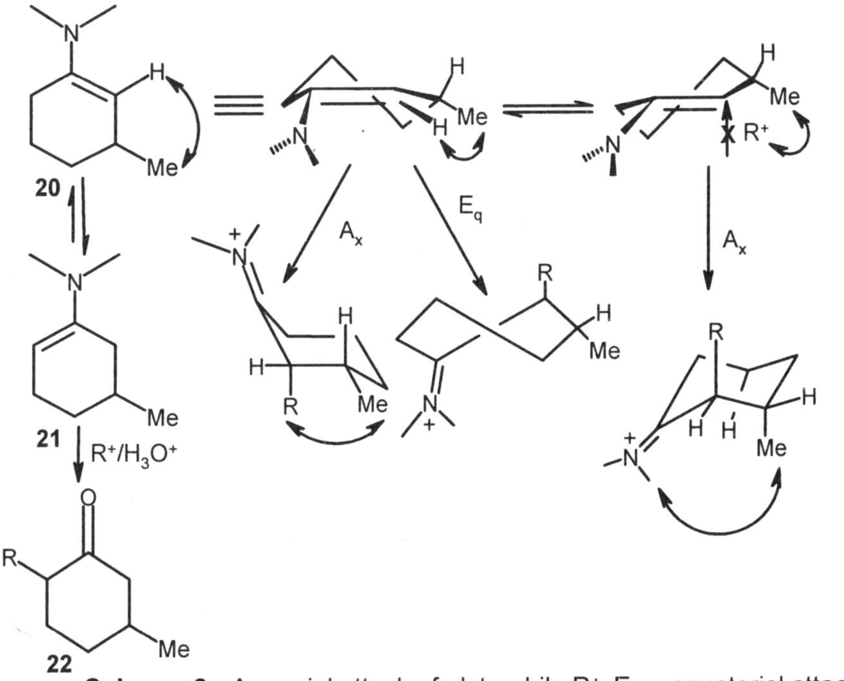

Scheme 2: A_x = axial attack of elctrophile R^+; E_q = equatorial attack

Regarding acyclic enamines, an important investigation by Pocar *et al.* has demonstrated that (i) the less substituted enamine is the more reactive isomer and (ii) interconversion of enamine isomers is largely dependent on the reagent and experimental conditions used. As for example —— the dimethyl amine enamine of methyl isopropyl ketone exist as a 50:50 mixture of (**23**) & (**24**). However reaction with phenylisocyanate gives only 3-dimethylamino-4-methyl-2-pentenoic acid anilide (**25**) in 100% yield[33]. Under the reaction conditions, the more substituted enamine (**24**), that suffers from $A^{1,3}$-strain, rearranges to the more reactive less substituted form (**23**). Conversely, if the equilibrium mixture (**23** & **24**) is treated with 4-nitrophenylazide at room temperature, only triazoline (**26**) is obtained and the other unreacted isomer (**24**) can be isolated by rapid distillation[34].

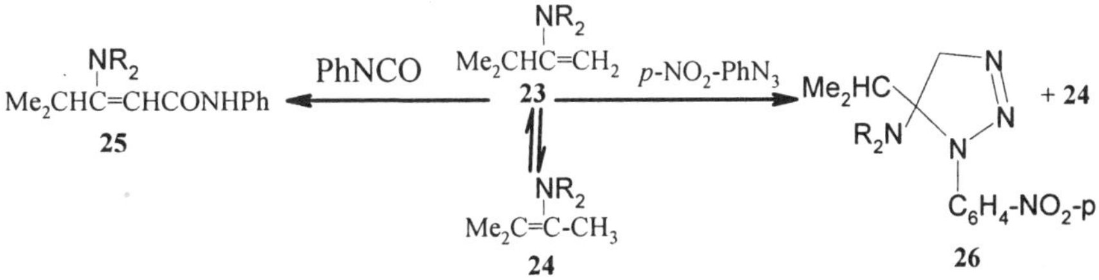

Scheme 3

Acylated enamines also exist as an equilibrium mixture of more (**27**) and less (**28**) substituted forms and the isomeric distribution again varies with the nature of the amine used. It has been observed that pyrrolidine enamine appears to exist mainly in the more substituted (*i.e.* conjugated form, **27**), whereas morpholine and piperidine enamines usually remain in the less substituted form (**28**)[35,36].

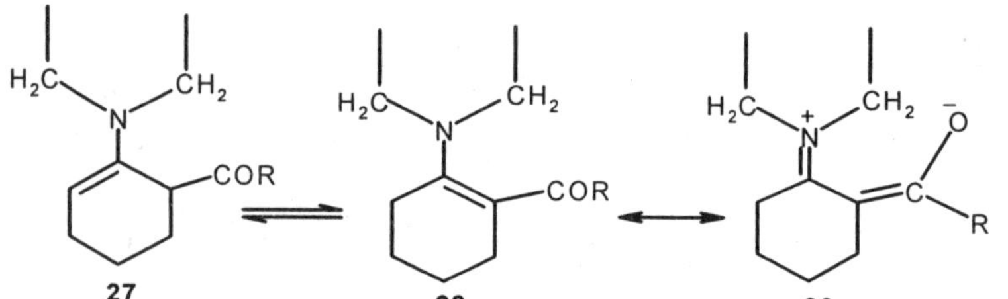

The other important difference is that the more substituted acyl enamine (**27**) may now capable to react in an alternate fashion[37,38] not at the β–carbon but at the oxygen (as shown in canonical structure **29**).

The reactivity of an enamine depends on the amine moiety and on the degree of substitutions at the α- & β-positions. Alkyl substitution at C-α of an enamine enhances the enamine reactivity, while substitution at C-β decreases the reactivity. The order of reactivity is, therefore, normally $R_2NC(R)=CH_2$ > $R_2NC(R)=CHR$ > $R_2NCH=CHR$ > $R_2NCH=CR_2$. Thus cyclic and acyclic ketones enamines are more readily *C*-alkylated than aldehyde enamines. In case of enamines from cyclic ketones, spectroscopic evidences suggest that reactivity may vary with the ring-size[39] of the ketone as in the order 5>12>8>6>7.

General Mechanism

A general mechanistic pathway of enamine alkylation, followed by hydrolysis may be shown as:

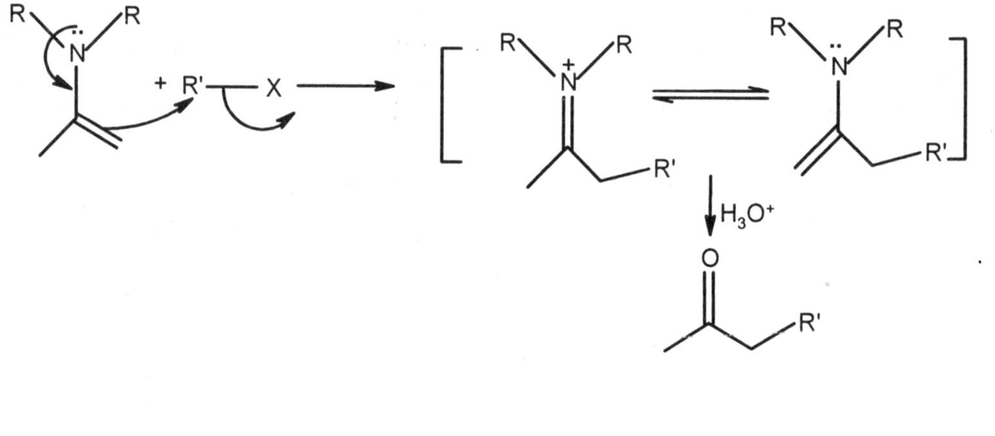

Stork Enamine Reaction

Nucleophilic displacement of a halide by the β-carbon of enamine (**30**) gives an α-alkyl iminium salt (**31**) via an S_N2 process. This iminium intermediate (**31**) can isomerize under the reaction conditions to the less substituted enamine (**32**). This is an equilibrium process and hydrolysis yields the corresponding α-alkyl carbonyl compound (**33**). The alkylation occurs most effectively with reactive halides, since it is essentially an S_N2 reaction.

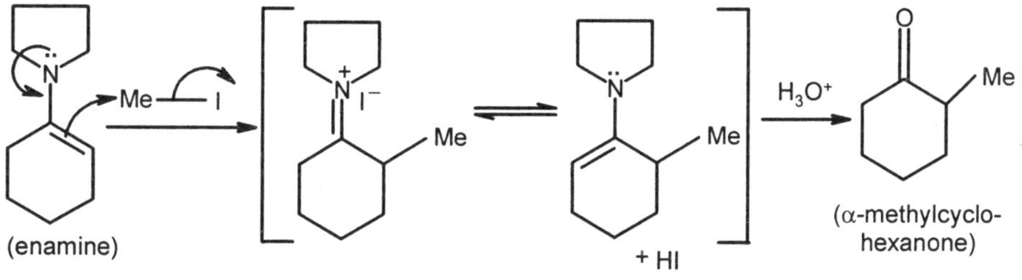

Examples

Few typical alkylating and acylating reactions are cited below:

Alkylation

i) Reaction with allyl bromide —

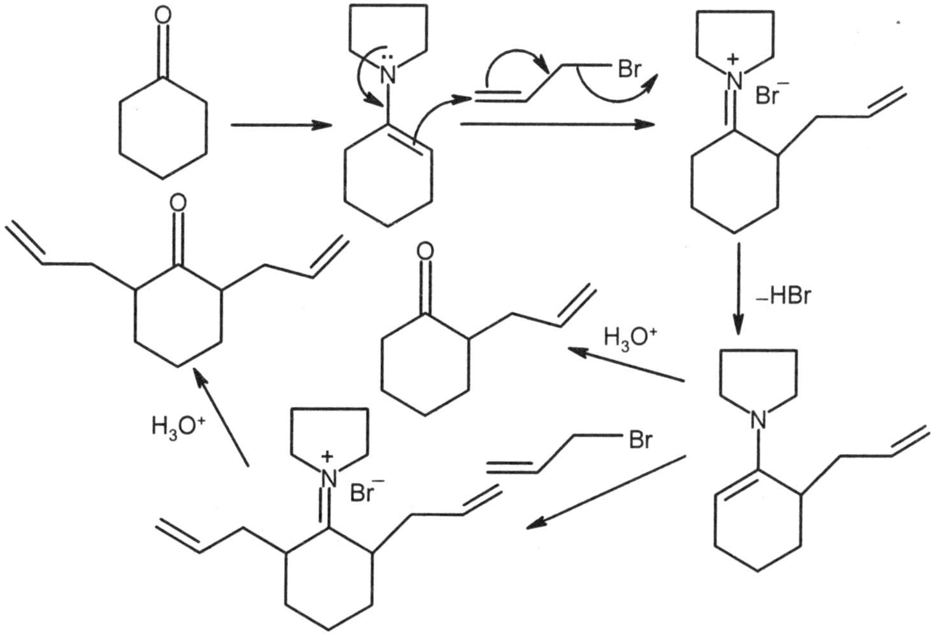

ii) Enamines undergo reaction with perfluoroalkyl iodides through free radical path[40].

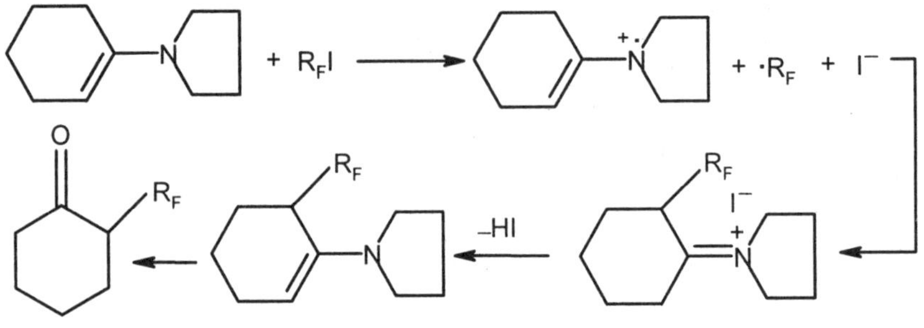

$(R_F = CF_3, C_2F_5, C_6F_{13})$

iii) 1,4-Dicarbonyl compounds may also be prepared by propenylation of enamines followed by ozonolysis[41] or propenylation followed by hydration[42].

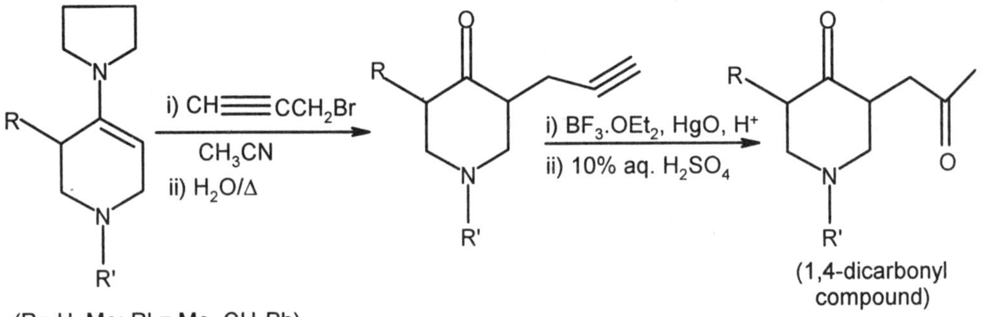

(R= H, Me; R' = Me, CH$_2$Ph)

Acylation
i) The morpholine enamine of acetone (34) reacts with acetyl chloride to give keto iminium salt (35), which on hydrolysis results in the formation of keto aldehyde (36)[43].

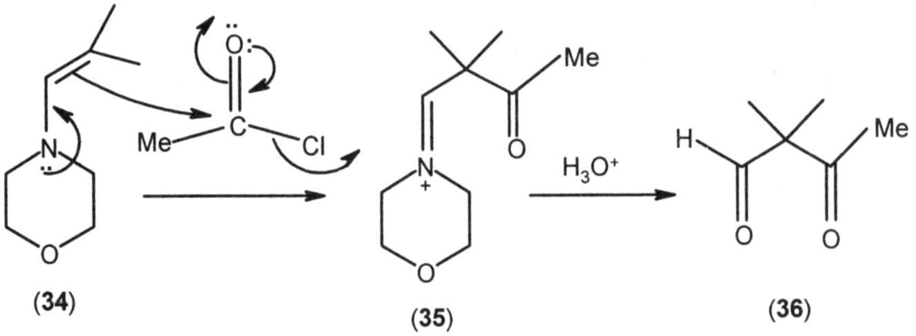

(34)   (35)   (36)

ii) Acylation of pure regio-isomer **37b** or the enamine mixture **37a,b** gives enaminoketone (**38**) which on hydrolysis forms 1,3-dicarbonyl compound (**39**) in good yield (40-90%)[44].

$$R-C(R)=C(NR_2)-CH_3 \quad (37a) \rightleftharpoons R-CH(R)-C(NR_2)=CH_2 \quad (37b) \xrightarrow{\text{i) RCOCl} \atop \text{ii) Et}_3\text{N}} R-CH(R)-C(NR_2)=CHCOR \quad (38) \xrightarrow{H_3O^+} R_2CH-CO-CH_2-CO-R \quad (39)$$

References

1. Wittig, G. and Blumenthal, H. (1927), *Ber.*, **60**, 1085.
2. Collie, J. N. (1884), *Liebigs Ann.*, **226**, 316.
3. Benary, E. (1909), *Chem. Ber.*, **42**, 3912.
4. Robinson, R. (1916), *J. Chem. Soc.*, **109**, 1038.
5. Stork, G., Terrell, R. and Szmuszkovicz, J. (1954), *J. Am. Chem. Soc.*, **76**, 2029; Stork, G. and Landesman, H. (1956), *J. Am. Chem. Soc.*, **78**, 5128.
6. Surrey, A. R., *Name Reactions in Organic Chemistry*, Academic Press, New York, 1961, p.231.
7. Stork, G. and Dowd, S. R. (1963), *J. Am. Chem. Soc.*, **85**, 2178.
8. Stork, G., Brizzolara, A., Landesman, H., Szmuszkovicz, J. and Terrell, R (1963)., *J. Am. Chem. Soc.*, **85**, 207.
9. Herr, M. E. and Heyl, F. W., *J. Am. Chem. Soc.*, **74**, 3627(1952); **75**, 1918, 5927(1953); **77**, 488(1955).
10. Benzing, E. (1959), *Angew. Chem.*, **71**, 521.
11. Breck, D. W. (1964), *J. Chem. Edu.*, **41**, 678.
12. Newman, H. and Fields, T. L. (1972), *Tetrahedron*, **28**, 4051; Mannich, C. and Devidsen, H. (1936), *Ber.*, **69**, 2106.
13. Carlson, R., Phan-Tan-Luu, R., Mathieu, D., Ahouande, F. S., Babadjamian, A., Metzger, J. (1978), *Acta Chem. Scand. B*, **32**, 335.
14. White, W. A. and Weingarten, H. (1977), *J. Org. Chem.*, **42**, 1663.
15. Broekhof, N. L. J. M., Jonkers, F. L., van der Gen, A. (1979), *Tetrahedron Lett.*, 2433.
16. Bhome, H., Haake, M., Auterhoff, G. (1972), *Arch. Pharmaz.*, **305**, 88.

17 Broekhof, N. L. J. M., Jonkers, F. L., van der Gen, A. (1980), *Tetrahedron Lett.*, 2671.
18 Martin, S. F. and Gompper, R. (1974), *J. Org. Chem.*, **39**, 2814.
19 Martinez, S. J. and Joule, J. A. (1978), *Tetrahedron*, **34**, 3027.
20 Sauer, J. and Prahl, H. (1966), *Tetrahedron. Lett..*, 2863; *Chem. Ber.*, **102**, 1917(1969).
21 Ahmed, M. G., Ahmed, S. A. and Hickmott, P. W. (1980), *J. Chem. Soc., Perkin 1*, 2383.
22 Jemison, R. W., Ollis, W. D., Sutherland, I. O. and Tannock, J. (1980), *J. Chem. Soc. Perkin 1*, 1462.
23 Barluenga, J. and Zanar, F. (1975), *Synthesis*, 704.
24 Le Blanc, M., Santini, G. and Riess, J. G. (1975), *Tetrahedron Lett.*, 4151; Le Blanc, M., Santini, G., Gallucci, J. and Riess, J. G. (1977), *Tetrahedron*, **33**, 1453.
25 Brunet, J. J., Fixari, B. and Caubere, P. (1974), *Tetrahedron*, **30**, 2931.
26 Johnson, F., Duquette, L. G., Whitehead, A. and Dorman, L. C. (1979), *Tetrahedron*, **35**, 1675.
27 Forchiassin, M., Risaliti, A. and Russo, C. (1979), *Gazz. Chim. Ital.*, **109**, 33.
28 Johnson, F. (1968), *Chem. Rev.*, **68**, 375.
29 Malhotra, S. K., Moakley, D. F. and Johnson, F. (1967), *Chem. Commun.*, 448.
30 Valentin, E., Pitacco, G. and Colonna, F. P. (1972), *Tetrahedron Lett.*, 2837; Colonna, I. P., Valentin, E., Pitacco, G. and Risaliti, A. (1973), *Tetrahedron*, **29**, 3011.
31 Descotes, G. and Querou, Y. (1916), *Comt. Rend.*, **263C**, 1231.
32 Laskovics, F. M. and Schulman, E. M. (1977), *J. Am. Chem. Soc.*, **99**, 6672.
33 Pocar, D., Stradi, R. and Gioia, B. (1968), *Gazz. Chim. Ital.*, **98**, 958.
34 Pocar, D., Stradi, R. and Bianchetti, G. (1970), *Gazz. Chim. Ital.*, **100**, 1135.
35 Colonna, F. P., Pitacco, G. and Valentin, E. (1971), *Tetrahedron*, **27**, 5481.
36 Opitz, G. and Tempel, E. (1966), *Analen*, **699**, 74.
37 Helmers, R. (1965), *Acta Chem. Scand.*, **19**, 2139.
38 Jacquier, R. and Maurey, G. (1967), *Bull. Soc. Chim. Fr.*, 320.
39 Nagarajan, K. and Rajappa, S. (1969), *Tetrahedron. Lett.*, 2298.
40 Cantacuzene, D. and Dome, R. (1975), *Tetrahedron Lett.*, 2031.
41 Marshall, J. A. and Flynn, G. A. (1977), *Synth. Commun.*, **7**, 417.
42 Alam, M., Baty, J. D., Jones, G. and Moore, C. (1969), *J. Chem. Soc.(C)*, 1520; Herion, G. F. and Quinn, F. X. (1970), *J. Org. Chem.*, **35**, 3054.
43 Alt, G. H. in *Enamines: Synthesis, Structure and Reactions*, Cook, A. G. (Ed.), 1969, Marcel Dekker, New York; Inukai, T. and Yoshizawa, R. J. (1967), *J. Org. Chem.*, **32**, 404.

44 Nilsson, L. (1979), *Acta Chem. Scand.*, **B33**, 203; Nilsson, L. (1979), *Acta Chem. Scand.*, **B33**, 710.

- Szablewski, M. (1994), *J. Org. Chem.*, **59**, 954.
- Hammadi, M. and Villemin, D. (1996), *Synth. Commun.*, **26**, 2901.
- Li, J. J., Trivedi, B. K., Rubin, J. R. and Roth, B. D. (1998), *Tetrahedron Lett.*, **39**, 6111.
- Yehia, N. A. M., Polborn, K. and Muller, T. J. (2002), *Tetrahedron Lett.*, **43**, 6907.
- Kesel, A. (2003), *Biochem. Biophys. Res. Commun.*, **300**, 793.

11

Arndt – Eistert Homologation

Introduction

Arndt-Eistert homologation reaction[1,2] involves the conversion of a carboxylic acid to its higher homologue through enhancing the length of the carbon chain by one methylene group.

$$RCOOH \longrightarrow RCH_2COOH$$

The reaction occurs in the following steps:

$$RCOOH \xrightarrow{SOCl_2} RCOCl \xrightarrow[-CH_3Cl, -N_2]{2CH_2N_2} RCOCHN_2 \text{ (}\alpha\text{-diazoketone)}$$

$$RCH_2COOH \xleftarrow{H_2O} RCH=C=O \xleftarrow{Ag_2O}$$
(ketene intermediate)

The conversion of α-diazoketone into the corresponding ketene intermediate proceeds through a rearrangement path, known as *Wolff rearrangement*[3]. The rearrangement can be brought about by photolysis, thermolysis or with silver oxide (other catalysts, *e.g.* colloidal platinum and copper, may also be used). It has also been observed that in *situ*-generated silver nanoclusters (Ag_n) during the reduction of either silver (I) oxide or other salts presumably catalyze the rearrangement[4]. The highly reactive ketene, as formed, will then react readily with any nucleophiles present in the system, *e.g.* water, alcohol, ammonia, amine to yield the corresponding carboxylic acid, ester, amide or substituted amide, respectively. The 'R' group may be alkyl or aryl, and at the same time may bear a number of functionalities including unsaturation, but not the groups that are acidic enough to react with diazomethane or diazoketone formed.

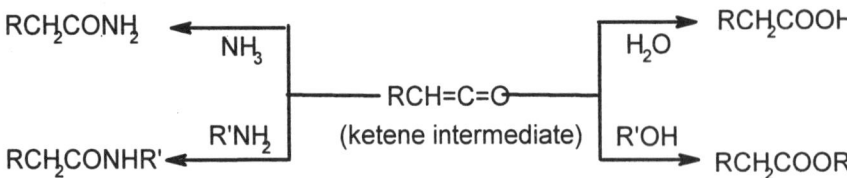

Mechanism

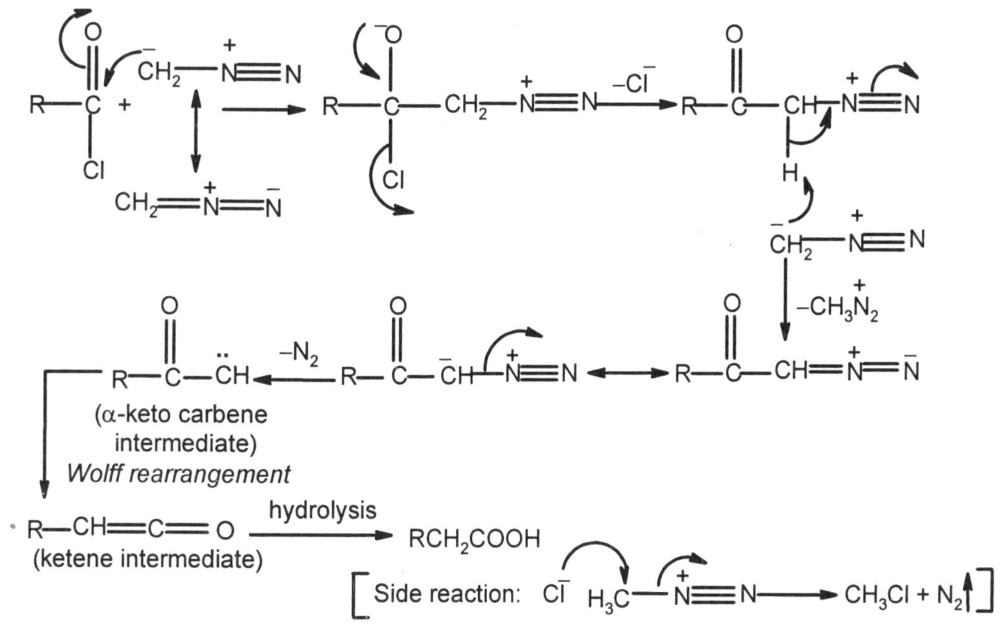

[Mechanism scheme showing: RCOCl reacting with CH₂=N⁺=N⁻ (diazomethane), via tetrahedral intermediate with CH₂–N⁺≡N and Cl⁻, loss of Cl⁻ to give R–CO–CH₂–N⁺≡N, deprotonation to R–CO–CH–N⁺≡N ↔ R–CO–CH=N⁺=N⁻ (diazoketone), with loss of CH₃N₂⁺ (side reaction forming CH₂–N⁺≡N), then loss of N₂ to give R–CO–CH: (α-keto carbene intermediate), Wolff rearrangement to R–CH=C=O (ketene intermediate), hydrolysis to RCH₂COOH.]

[Side reaction: Cl⁻ + H₃C–N⁺≡N → CH₃Cl + N₂↑]

Critical Views

(a) Excess of diazomethane (2 eqv.) is to be used to consume the liberated hydrochloric acid.

$$RCOCl + CH_2N_2 \longrightarrow RCOCHN_2 + HCl$$
$$CH_2N_2 + HCl \longrightarrow CH_3Cl + N_2$$

Otherwise, diazoketone as formed would be lost by reacting with HCl to form chloromethyl ketone.

$$RCOCHN_2 + HCl \longrightarrow RCOCH_2Cl + N_2$$

(b) Carbonyl carbon of diazoketone becomes the carbonyl carbon of the resulting functionality:

$$R^{13}COCHN_2 \xrightarrow[H_2O]{Ag_2O} R^{13}CH_2COOH$$

(c) The mechanistic pathway of *Arndt-Eistert homologation* generally shows a free carbene intermediate that rearranges to a corresponding ketene. Stable ketenes (*e.g.* Ph₂C=C=O) have been isolated and many others have been trapped[5]. Though there is much evidence in support of the presence of free carbene intermediate, it is believed that at least in some cases the two steps — splitting of nitrogen and migration of 'R' group — are *concerted* and a free carbene is absent[6].

(d) The migrating group 'R' retains its configuration in the corresponding product of *Arndt-Eistert homologation*; this fact has been experimentally established by the following transformations involving an optically active acid (**I**) —

[Scheme showing cyclic transformation:
(I) n-C_4H_9, Ph, COOH, CH_3 (chiral center *)
→ Arndt-Eistert homologation →
(II) n-C_4H_9, Ph, CH_2COOH, CH_3 (chiral center *)
↓ 1. CH_2N_2 2. PhMgBr
n-C_4H_9, Ph, $CH_2C(Ph)_2OH$, CH_3
↓ Δ, $-H_2O$
n-C_4H_9, Ph, $CH=C(Ph)_2$, CH_3
↑ CrO_3, AcOH (chromic acid oxidation)
back to (I)]

The acid (**II**), obtained by *Arndt-Eistert homologation* of (**I**), on *Barbier-Weiland degradation* produced further the original one (**I**) having the same configuration as previously was.

(e) When the *Wolff rearrangement* is effected photochemically, one more pathway involving oxirene intermediate can also take place in addition to the normal one. In this bypass way, the initially formed ketocarbene may undergo a carbene-carbene rearrangement through an oxirene intermediate[7]. Evidence has been received from ^{14}C-labelling experiments[8].

[Scheme:
$R-\overset{14}{C}(=O)-C(R')-\overset{+}{N}\equiv N \xrightarrow{h\nu, -N_2} R-\overset{14}{C}(=O)-\ddot{C}-R'$ (ketocarbene-I)
$\rightleftharpoons R-\overset{14}{C}=\!=\!C-R'$ with O bridge (oxirene intermediate, detected by laser spectroscopy)
⇌ $R-\ddot{C}-\overset{14}{C}(=O)-R'$ (ketocarbene-II)

normal pathway from ketocarbene-I → $RR'\overset{14}{C}=C=O$ ketene → $RR'\overset{14}{CH}COOH$
rearrangement → $R\overset{14}{R'C}=C=O$ ketene → $R\overset{14}{R'}CHCOOH$
(both are obtained)]

It has been found that oxirene involving path is less pronounced for diazoketones where R' = H. An intermediate believed to be an oxirene has been detected by laser spectroscopy[9]. This path is not observed in thermal *Wolff rearrangement*. It is also believed that the singlet-excited state of the carbene species intervenes to the oxirene pathway[10], but the triplet-excited state always follows the normal rearrangement pathway[11].

Applications

1. α-Naphthoic acid →(1. SOCl$_2$; 2. CH$_2$N$_2$)→ Diazo-α-acetonaphthone →(Ag$_2$O, H$_2$O, Δ)→ α-Naphthylacetic acid

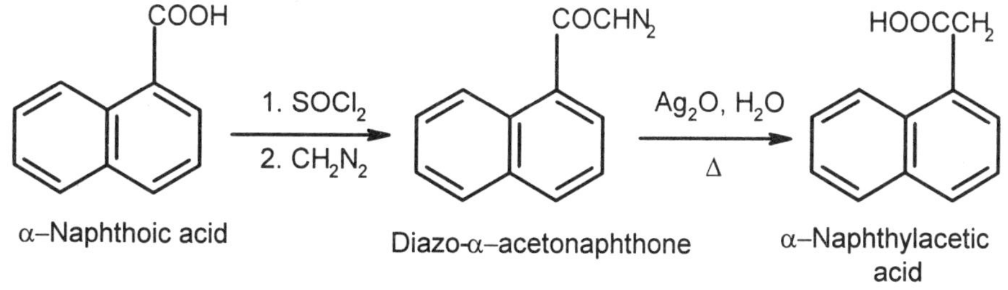

2. 3-NO$_2$-C$_6$H$_4$-CH=CHCH$_2$COOH →(1. SOCl$_2$; 2. CH$_2$N$_2$; 3. Ag$_2$O, H$_2$O)→ 3-NO$_2$-C$_6$H$_4$-CH=CH(CH$_2$)$_2$COOH

3. 3,4,6-Trimethoxybenzoic acid →(1. SOCl$_2$; 2. AgNO$_3$ - NH$_3$)→ (CH$_2$CONH$_2$ derivative) →(reduction)→ Mescaline (produces euphoria)

4. (cyclohexenyl)-CH$_2$COOH →(1. SOCl$_2$; 2. CH$_2$N$_2$; 3. hν/MeOH)→ (cyclohexenyl)-CH$_2$CH$_2$COOMe

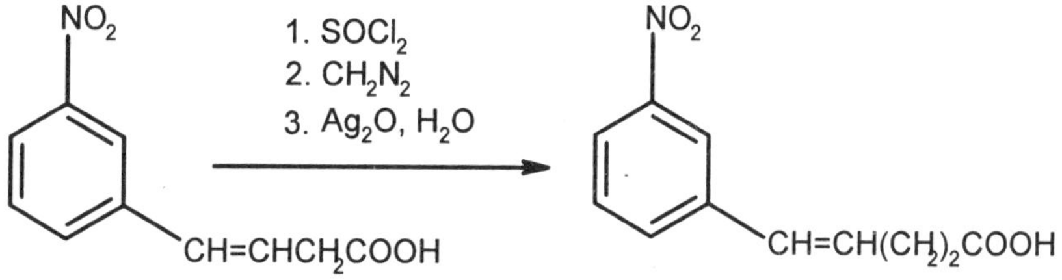

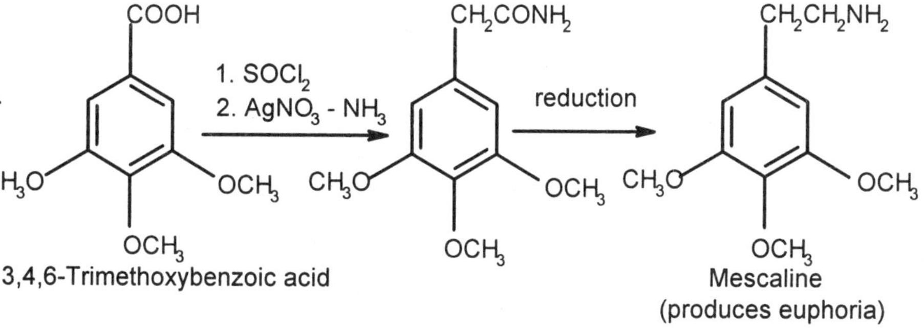

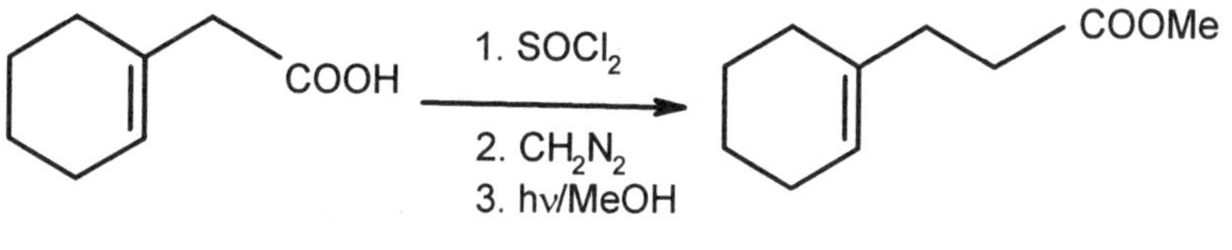

5. Furan-CH₂CH₂-COOH → (1. SOCl₂; 2. CH₂N₂; 3. hν/MeOH) → Furan-CH₂CH₂CH₂-COOMe

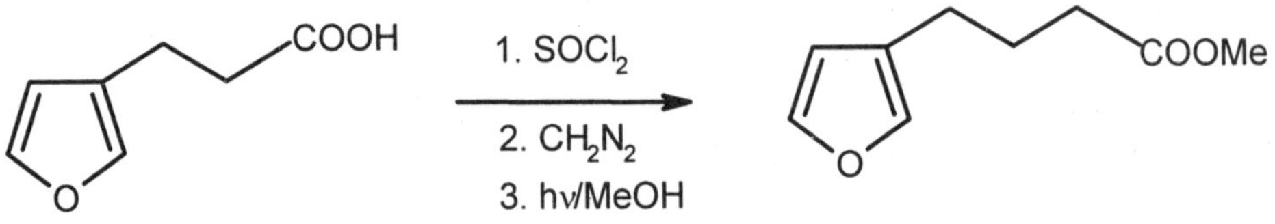

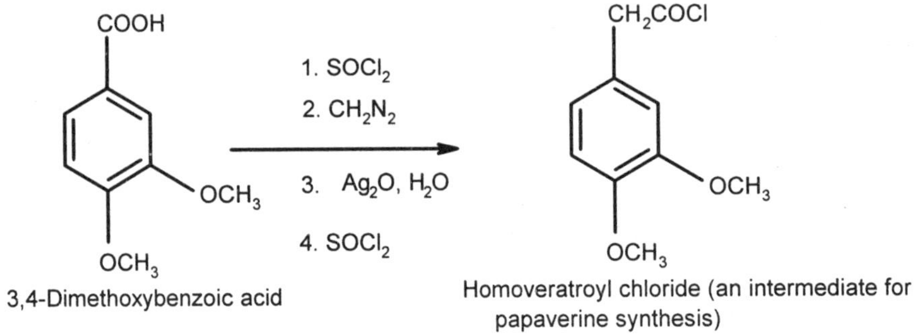

6. 3,4-Dimethoxybenzoic acid → (1. SOCl₂; 2. CH₂N₂; 3. Ag₂O, H₂O; 4. SOCl₂) → Homoveratroyl chloride (an intermediate for papaverine synthesis)

7. [7-Nitro-4-methyl-benzoxazin-3-one-2-yl with α-methyl-CH-COOH substituent] → (EtOCOCl/Et₃N or DCC, Me₃SiCHN₂ (trimethyl silyl diazomethane)) → [corresponding diazoketone] → (Ag⁺ ⁻OOCPh/Et₃N, MeOH/ultrasound) → [homologated carboxylic acid]

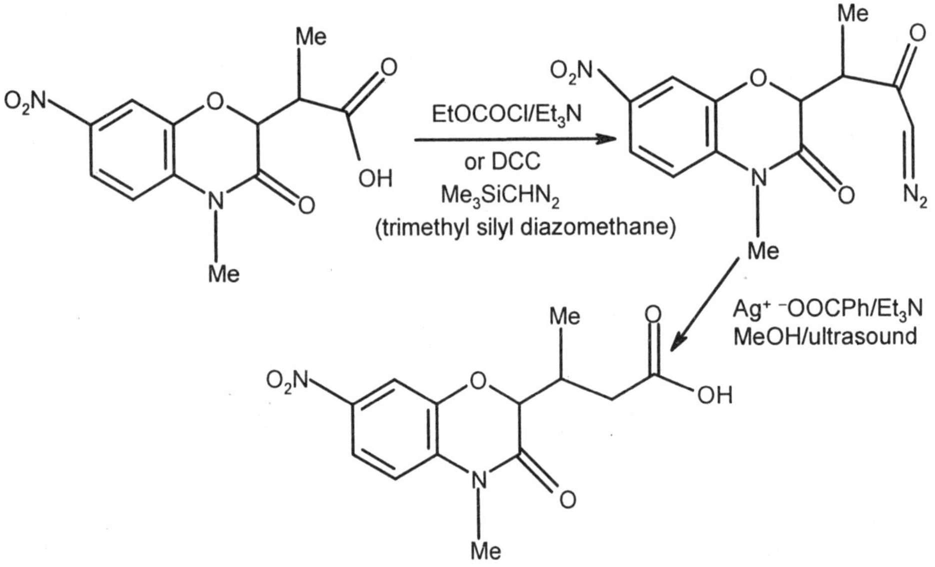

8. [bicyclic dilactone with CH₂COOH substituent] → (i) SOCl₂, pyridine; ii) CH₂N₂; iii) Ag₂O, MeOH) → [bicyclic dilactone with CH₂CH₂COOMe substituent]

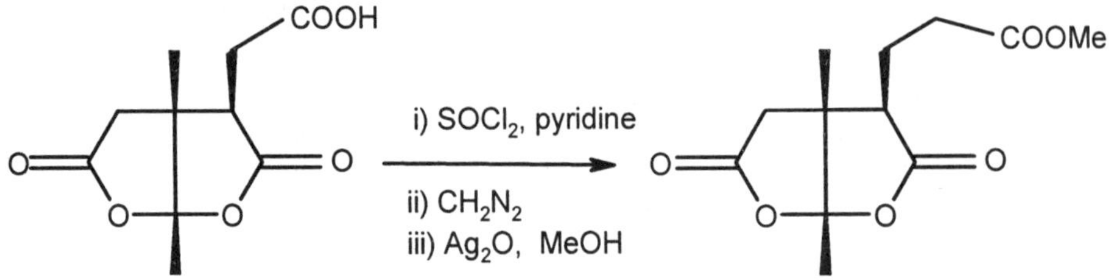

9.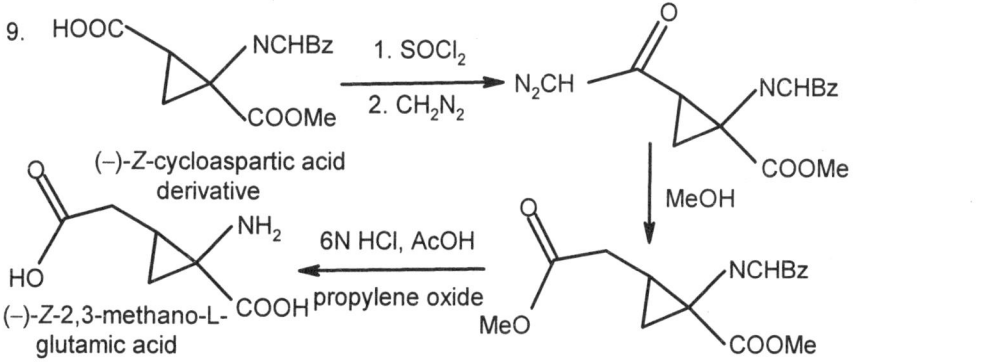

References

1. Arndt, F. and Eistert, B. (1935), *Ber. Dtsch. Chem. Ges.*, **68**, 200.
2. Meier, H. and Zeller, K.(1975), *Angew. Chem. Int. Ed. Engl.*, **14**, 32. [Review]
3. McAlonan, H., Stevenson, P. J., Thompson, N. and Treacy, A. B. (1997), *Synlett*, 1359.
4. Sudrik, S. G., Maddanimath, T., Chaki, N. K., Chavan, S. P., Sonawane, H. R. and Vijayamohanan, K. (2003), *Organic Letters*, **5**(13), 2355.
5. Kirmse, W. and Horner, L. (1956), *Chem. Ber.*, **89**, 2759; Horner, L. and Spietschka, E. (1956), *Chem. Ber.*, **89**, 2765.
6. Kirmse, W., *Carbene Chemistry*, 2nd ed., Academic Press, NY, 1971, p. 476; Torres, M., Ribo, J., Clement, A. and Strausz, O. P. (1983), *Can. J. Chem.*, **61**, 996; Tomoika, H., Hayashi, N., Asano, T. and Izawa, Y. (1983), *Bull. Chem. Soc. Japan*, **56**, 758.
7. Lewars, Y. (1983), *Chem. Rev.*, **83**, 519. [Review]
8. Fenwick, J., Frater, G., Ogi, K. and Strausz, O. P. (1973), *J. Am. Chem. Soc.*, **95**, 124; Zeller, K. (1978), *Chem. Ber.*, **112**, 678.
9. Tanigaki, K., Ebbensen, T. W. (1987), *J. Am. Chem. Soc.*, **109**, 5883.
10. Csizmadia, I. G., Gunning, H. E., Gosavi, R. K. and Strausz, O. P. (1973), *J. Am. Chem. Soc.*, **95**, 133.
11. McMahan, R. J., Chapman, O. L., Hayes, R. A., Hess, T. C. and Krimmer, H. (1985), *J. Am. Chem. Soc.*, **107**, 7597.

- Katrizky, A. R., Zhang, S. and Fang, Y. (2000), *Org. Lett.*, **2**, 3789.
- Katrizky, A. R., Zhang, S., Mostafa Hussain, A. H. and Fang, Y. (2001), *J. Org. Chem*, **66**, 5606.
- Cesar, J. and Sollner Dolenc, M. (2001), *Tetrahedron Lett.*, **42**, 7099.
- Vasanthakumar, G.-R. and Babu, V. V. S. (2002), *Synth. Commun.*, **32**, 651.
- Chakravarty, P. K. *et al.* (2003), *Bioorg. Med. Chem. Lett.*, **13**, 147.

12

Benzilic acid Rearrangement

Introduction

Benzil undergoes an interesting base-catalyzed rearrangement reaction yielding salt of benzilic acid. Probably this is the first intramolecular rearrangement in organic reactions, observed by Liebig in 1838.[1]

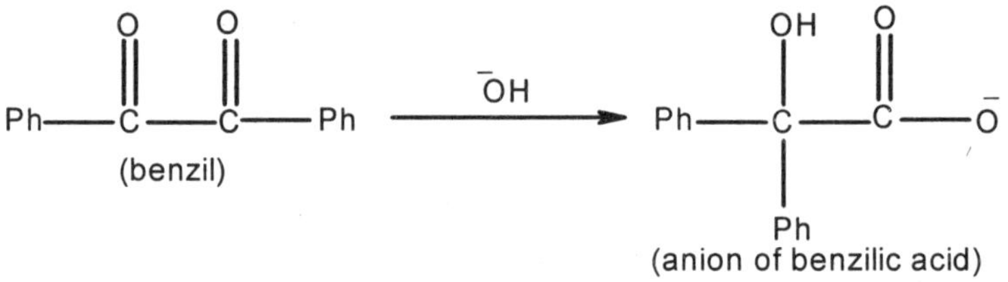

$$Ph-\underset{\underset{(benzil)}{}}{\overset{O}{\underset{\|}{C}}-\overset{O}{\underset{\|}{C}}-Ph} \xrightarrow{^-OH} Ph-\underset{\underset{Ph}{|}}{\overset{OH}{\underset{|}{C}}}-\overset{O}{\underset{\|}{C}}-\bar{O}$$

(anion of benzilic acid)

Any reaction of this kind, where α–diketones rearrange to give the salts of α–hydroxy acids on treatment with base, is known as *benzil-benzilic acid rearrangement* or simply *benzilic acid rearrangement*.

Mechanism

Ingold[2] proposed the common experimentally evidenced mechanism for *benzilic acid rearrangement* as shown below. This mechanism is being well-accepted still to-day.

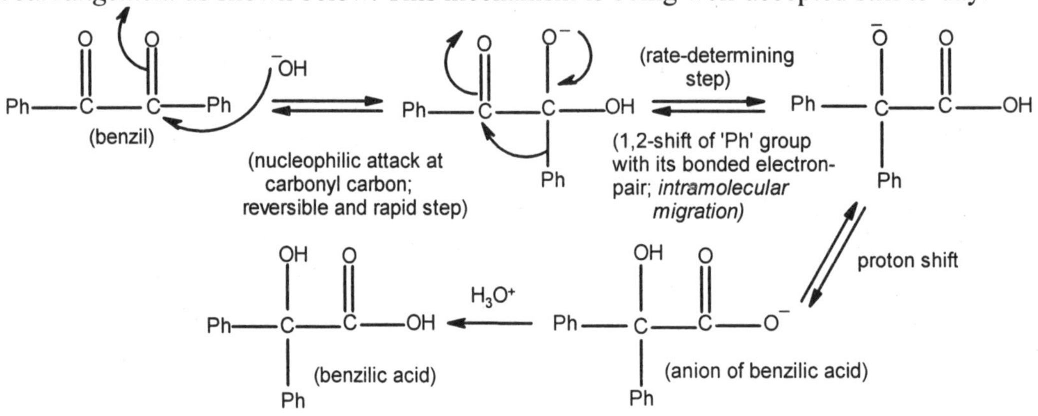

[Mechanism scheme:]

Ph—C(=O)—C(=O)—Ph (benzil) + ⁻OH ⇌ Ph—C(=O)—C(O⁻)(OH)—Ph (nucleophilic attack at carbonyl carbon; reversible and rapid step)

⇌ (rate-determining step) Ph—C(O⁻)(Ph)—C(=O)—OH (1,2-shift of 'Ph' group with its bonded electron-pair; *intramolecular migration*)

⇌ proton shift Ph—C(OH)(Ph)—C(=O)—O⁻ (anion of benzilic acid)

→ H₃O⁺ → Ph—C(OH)(Ph)—C(=O)—OH (benzilic acid)

Critical Views

a) If alkoxide ion (such as, methoxide ion) rather than hydroxide ion is used, corresponding ester will be formed directly —

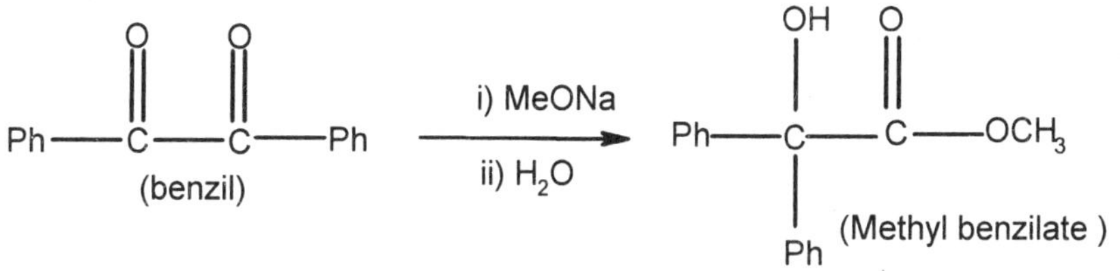

Ph—CO—CO—Ph (benzil) →[i) MeONa][ii) H$_2$O] Ph—C(OH)(Ph)—CO—OCH$_3$ (Methyl benzilate)

But regarding choice of alkoxide ion, one should have to be careful — since alkoxide ions that get oxidized readily (such as, EtO$^-$, Me$_2$CHO$^-$) are not suitable as they reduce the benzil to benzoin.

b) That the first step of the reaction is reversible and rapid is evidenced from the fact that benzil exchanges ^{18}O at a faster rate than it undergoes rearrangement, observed when the reaction is carried in presence of H$_2$O^{18}.

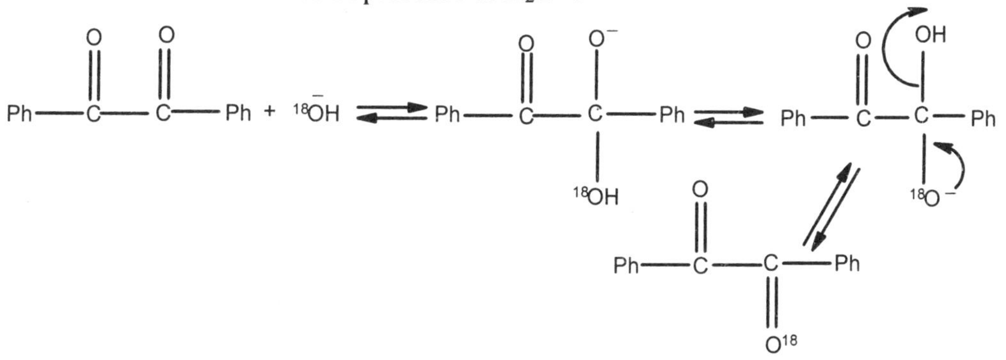

Ph—CO—CO—Ph + ^{18}OH$^-$ ⇌ Ph—CO—C(O$^-$)(^{18}OH)—Ph ⇌ Ph—CO—C(O)(OH)—Ph
⇌ Ph—C(^{18}O)(O$^-$)—CO—Ph

c) It has been observed that rate of the reaction in D$_2$O-dioxane is almost twice as fast as in H$_2$O-dioxane. The observed increase in the reaction rate is attributed to the greater basicity of deuteroxide ions in comparison to hydroxide ions.

d) Besides aromatic 1,2-diketones, the reactions may also be applicable to aliphatic diketones,[3] and to α–keto aldehydes.

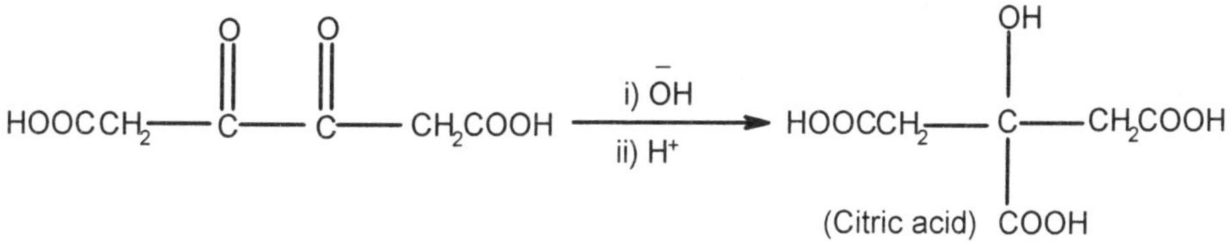

HOOCCH$_2$—CO—CO—CH$_2$COOH →[i) OH$^-$][ii) H$^+$] HOOCCH$_2$—C(OH)(COOH)—CH$_2$COOH (Citric acid)

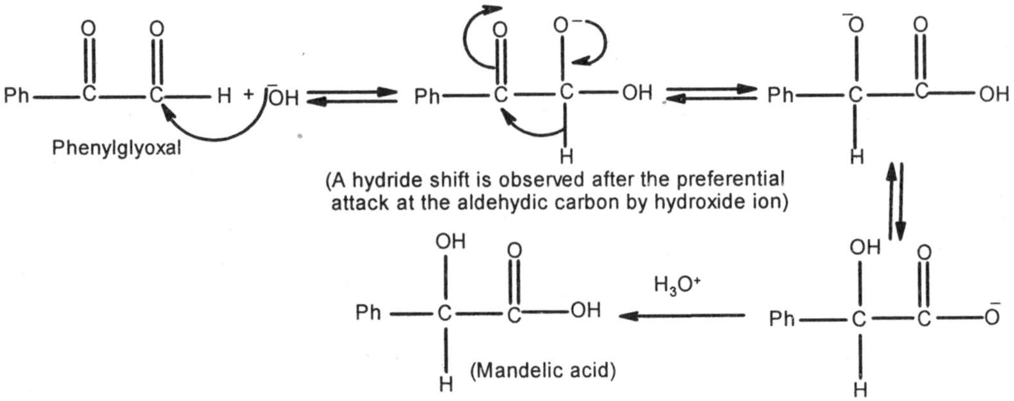

e) A special feature of the *benzilic acid rearrangement* is the migratory aptitude of aryl moiety when the aryl groups are different. It may be expected that aryl group with electron-releasing substituents (*e.g. p*-Me, *p*-MeO, *p*-Cl etc.) will migrate preferentially, but the reverse is found true. The aryl groups substituted with weaker electron-donating functionalities are observed to migrate with more ease than those substituted with stronger electron-donating groups. Thus —

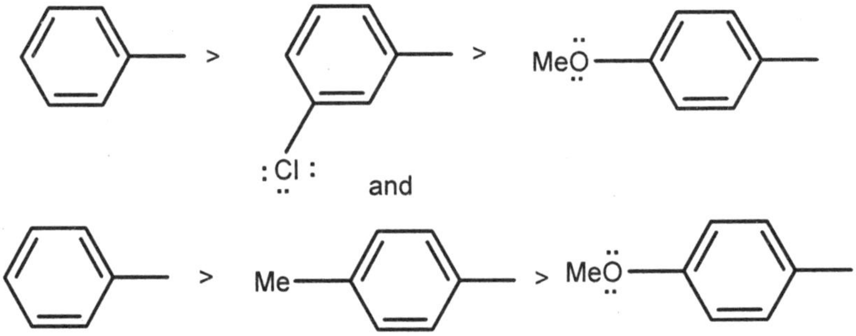

The phenomenon may be argued in this manner — out of the two aryl moieties, the one that is more electron-releasing has better tendency to neutralize (by +R, +I and/or hyperconjugative effect) the positive charge of the carbonyl carbon to which it is attached when the >C=O is polarized. Thus it is the other >C=O group that will link with the incoming nucleophile (base), and consequently the aryl group attached to this 'other' >C=O group migrates preferentially.

f) Toda *et al.*[4] already developed a method for carrying out *benzil-benzilic acid rearrangement* in the solid-state at 80°C, and some reactions were found to proceed quickly in comparison to those in solution. Recently, Yu *et al.*[5] have carried out the solid-state *benzil – benzilic acid rearrangement* using microwave – irradiation and reported that use of microwave-irradiation not only reduces the reaction time, but also enhances the yields. Besides, the authors have claimed that under microwave-irradiation, there is not much difference between electron-donating and electron-withdrawing groups on benzil — both reactions give satisfactory yields in a very short time (<1 minute in most of the cases as they have studied). This reaction protocol provides an advantage over the traditional heating method in organic solvents.

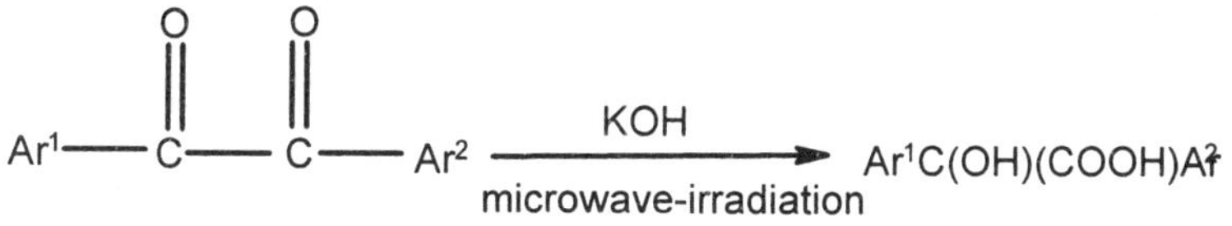

$$Ar^1-\underset{O}{\underset{\|}{C}}-\underset{O}{\underset{\|}{C}}-Ar^2 \xrightarrow[\text{microwave-irradiation}]{KOH} Ar^1C(OH)(COOH)Ar^2$$

Ar^1	Ar^2	%Yield
C_6H_5	C_6H_5	86
p-MeO- C_6H_4	p-MeO- C_6H_4	93
o-Cl- C_6H_4	o-Cl- C_6H_4	98

Applications

Benzilic acid rearrangement finds enormous applications in organic synthesis. A few examples are cited:

1) 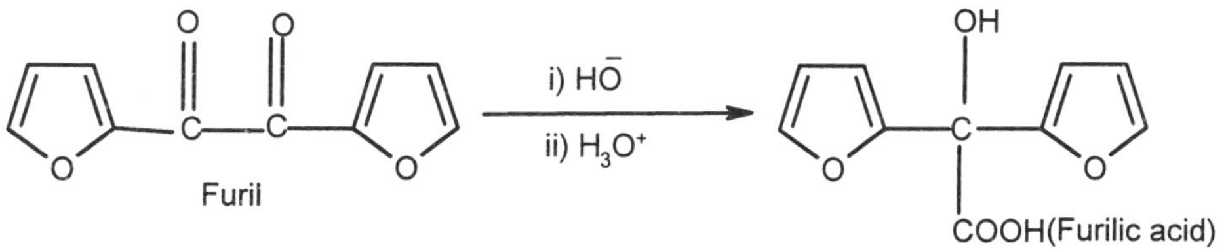 Furil → i) HO⁻, ii) H_3O^+ → Furilic acid [COOH]

2) 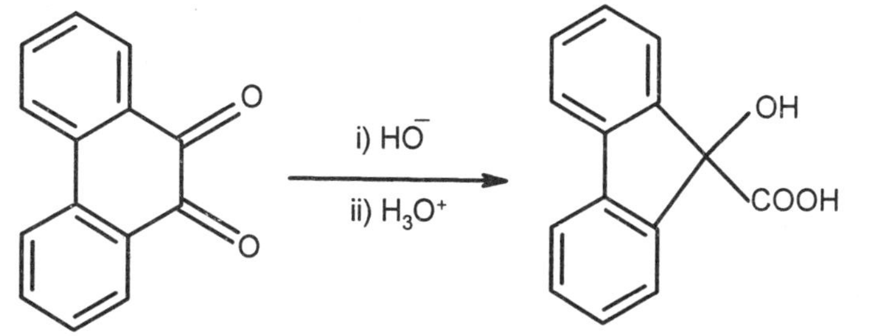 Phenanthroquinone → i) HO⁻, ii) H_3O^+ → 9-Hydroxyfluorene-9-carboxylic acid

3) Cyclohexane-1,2-dione as well as cyclobutane-1,2-dione undergoes *benzilic acid rearrangement* leading to the formation of 1-hydroxy-cyclopentane-1-carboxylic acid

and 1-hydroxy-cyclopropane-1-carboxylic acid, respectively leading to ring contraction.

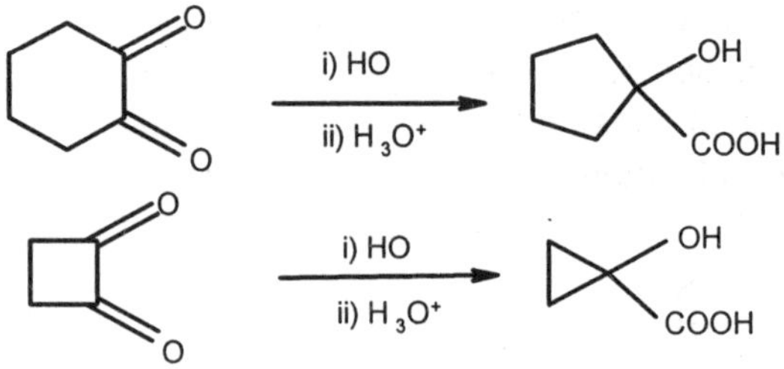

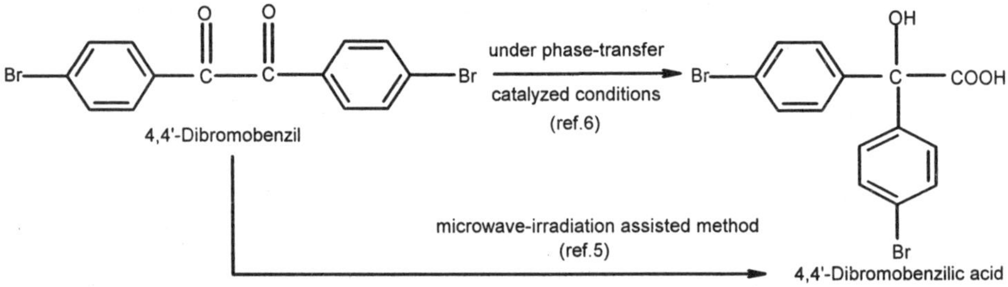

References

1. Liebig, J. (1838), *Justus Liebigs Ann. Chem.*, **25**; Jens, A. (1870), *Justus Liebigs Ann. Chem.*, **155**, 77.
2. Ingold, C. K. (1928), *Ann. Rept. Prog. Chem. (Chem. Soc. London)*, **25**, 124.
3. Schaltegger, A. and Bigler, P. (1986), *Helv. Chim. Acta*, **69**, 2666.
4. Toda, F. (1995), *Acc. Chem. Res.*, **28**, 480; Toda, F. *et al.* (1990), *Chem. Lett.*, 373.
5. Yu. H.-M. Chen, S.-T., Tseng, M.-J., Chen, S.-T. and Wang, K.-T. (1999), *J. Chem. Res(S).*, 621.
6. Maeorg, U., Soomets, U., Perkson, A., Linask, L. and Raidaru, G. (1994), *Mendeleev Commun.*, **4**(3), 99.

- Selman, S. and Eastham, J. F. (1960), *Rev. Chem. Soc.*, **14**, 221. [Review]
- Schaltegger, A. and Bigler, P. (1986), *Helv. Chim. Acta*, **69**, 1666.
- Rajyaguru, I. and Rzepa, H. S. (1987), *J. Chem. Soc., Perkin Trans. 2*, 1819.
- Hatsui, T., Wang, J.-J., Ikeda, S.-Y. and Takeshita, H. (1995), *Synlett*, 35.
- Zhang, K., Corrie, J. E. T., Munasinghe, V. R. N. and Wan, P. (1999), *J. Am. Chem. Soc.*, **121**, 5625.
- Fohlisch, B., Radl, A., Schwetzler–Raschke, R. and Henkel, S. (2001), *Eur. J. Org. Chem.*, 4357.

13

Benzoin Condensation

Introduction

Benzoin condensation involves condensation between two molecules of aromatic aldehydes (identical or different) under the catalytic influence of a cyanide ion, forming α-hydroxy ketone (a dimer, known as *benzoin*).

$$2\text{Ar-CHO} \xrightarrow[\Delta]{\overset{-}{\text{CN}}} \text{Ar-CH(OH)COAr}$$
(aromatic aldehyde)

Boiling alcoholic solution of the aldehyde containing small amount of KCN or NaCN carries out the condensation. Two molecules of the aldehyde behave differently. One of them donates the aldehydic hydrogen to the oxygen (of aldehydic function) of the other; hence the former is known as the 'donor' and the latter as the 'acceptor'. The reaction is reversible in nature. The reaction is successful only for non-enolisable aldehydes.

$$\text{Ph-CHO} + \text{Ph-CHO} \xrightleftharpoons{\overset{-}{\text{CN}}} \text{Ph-CO-CH(OH)-Ph}$$ 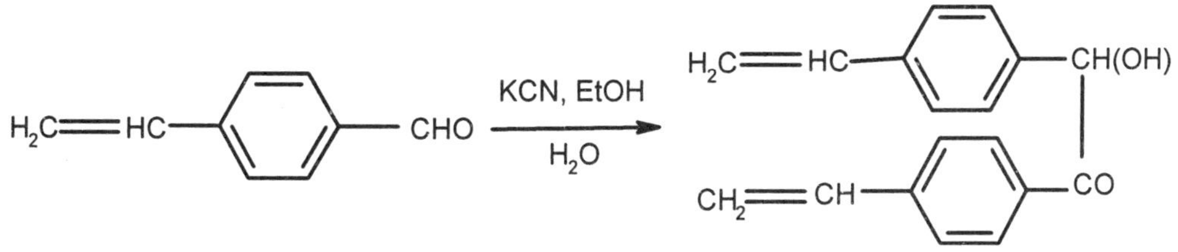
(benzaldehyde) (benzoin)

$$H_2C{=}HC{-}C_6H_4{-}CHO \xrightarrow[H_2O]{\text{KCN, EtOH}} \text{product (62%)}$$

Mechanism

The generally accepted mechanism[1] for *benzoin condensation* is given below:

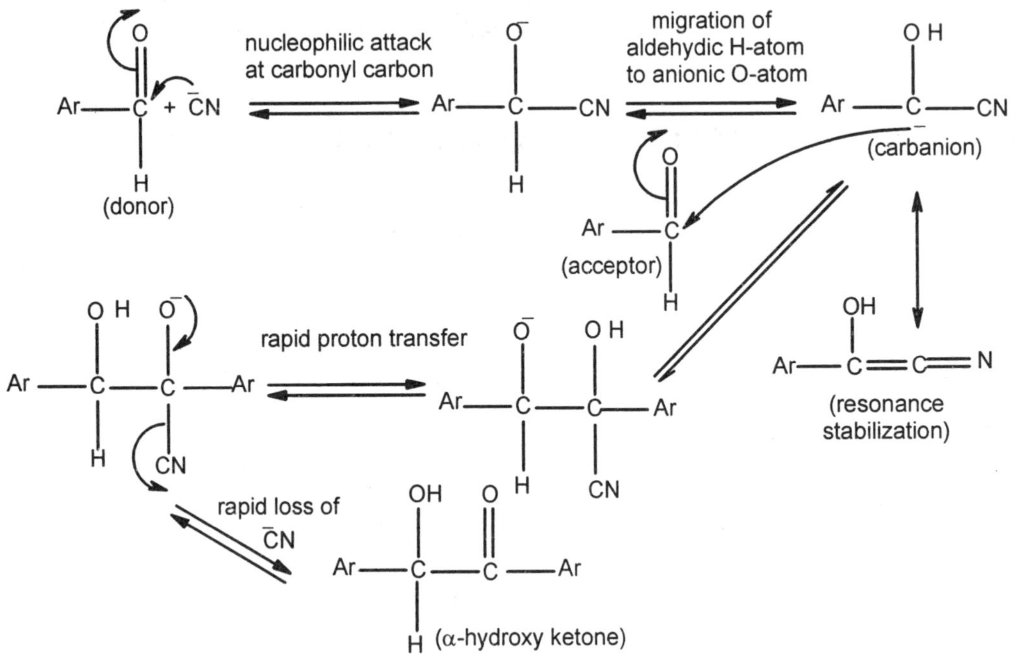

Critical Views

a) Cyanide — a unique catalyst:

Benzoin condensation is specifically catalyzed by *cyanide ions*. This unique catalytic property of cyanide arises out of the fact that the single species at the same time can perform the following functions —

i) it can act as a good nucleophile and attacks the electron-deficient carbonyl carbon of 'donor' aldehyde molecule.

ii) the key step of this reaction is the formation of carbanion developed due to transfer of aldehydic proton. This is possible only when the acidity of this aldehydic proton is appreciable and such acidity is achieved by strong electron-withdrawing power of cyanide group.

iii) at the same time, cyanide group takes part in stabilizing the carbanion as formed through resonance, and

iv) it also acts as a good leaving group in the last step.

b) Besides cyanides, few other catalysts have been reported to be in use. Thus the ylides, formed in the solutions of certain *N*-alkylthiazolium salts, have been found to catalyze the *benzoin condensation* effectively.[2]

Benzoin Condensation

$$R-\overset{+}{N}\underset{Y^-}{\underset{|}{}}\!\!\!\!\!\!\!\!\!\!\overset{H}{\overset{|}{C}}\!\!\!\!\!\!\!\!\!\!S \quad \underset{pH\sim 7}{\rightleftharpoons} \quad R-\overset{+}{N}=\!\!\!\!\!\!\!\!\overset{\bar{}}{C}\!\!\!\!\!\!\!\!S \text{ (Ylide)}$$

$$\text{PhCHO} \xrightarrow[\text{MeOH}]{\text{Ph—CH}_2\text{—N}^+(Cl^-)\text{thiazole-Me, Et}_3\text{N}} \text{PhCH(OH)COPh} \quad 79\%$$

Thiamine has recently been reported to act as a catalyst. It is heat-sensitive and may decompose if heated too vigorously. Instead of running this reaction at elevated temperatures for a few hours, thiamine-catalyzed reaction is generally carried out closer to room temperature for 24 hours or more. Too much water will force aldehyde out of solution preventing an efficient reaction. Too little water also prevents the thiamine hydrochloride from dissolving. Some amount of the base reacts with the thiamine hydrochloride to produce thiamine, which is the active catalyst.

$$2 \text{ PhCHO} \underset{\text{EtOH/NaOH}}{\overset{\text{Thiamine hydrochloride}}{\rightleftharpoons}} \text{PhCH(OH)COPh}$$

N, N'-Disubstituted o-phenylenediamines may also act as effective catalysts in the benzoin condensation of heterocyclic aldehydes.[3]

c) When a mixture of two different aromatic aldehydes is treated with aqueous ethanolic KCN, the mixture undergoes the condensation reaction, called *crossed benzoin condensation*. The products would be '*mixed*' benzoins as well as the '*single*' benzoins.

$$\text{Ar}^1\text{-CHO} + \text{Ar}^2\text{-CHO} \xrightarrow[\Delta]{\bar{C}N} \text{Ar}^1\text{-CH(OH)COAr}^2 + \text{Ar}^2\text{-CH(OH)COAr}^1$$
$$+ \text{Ar}^1\text{-CH(OH)COAr}^1 + \text{Ar}^2\text{-CH(OH)COAr}^2$$

Such situation will develop only when both the aldehyde molecules can act as donor as well as acceptor. However, among them if one can act only as donor, then the nature of benzoin product as obtained becomes simplified — as in the case of:

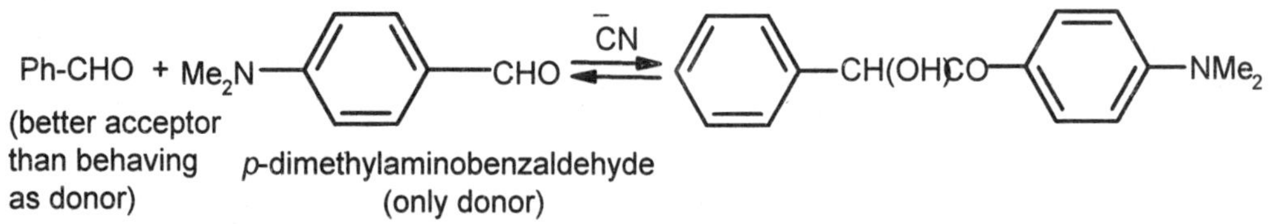

$Ph\text{-}CHO$ + Me_2N—⟨⟩—CHO ⇌ (CN⁻) ⟨⟩—CH(OH)CO—⟨⟩—NMe_2

(better acceptor than behaving as donor)   p-dimethylaminobenzaldehyde (only donor)

d) The reaction is completely a reversible process; this may be established by the fact that when benzoin of the type $Ar^1CH(OH)COAr^1$ is heated with the aldehyde Ar^2CHO in the presence of KCN, a mixed benzoin is obtained.

$$Ar^1CH(OH)COAr^1 + 2Ar^2CHO \underset{}{\overset{CN^-}{\rightleftharpoons}} 2Ar^1CH(OH)COAr^2$$
(mixed benzoin)

e) The utility of the traditional *benzoin condensation* is well documented, but its principal limitation arises in the coupling of two different aldehydes — where a mixture of four different benzoins is formed. The product distribution of the *crossed benzoin condensation* is generally determined by the relative stability of the four possible products. To overcome this limitation, recently Linghu and Johnson[5] have offered a new catalyzed silyl benzoin addition reaction of acylsilanes and aldehydes that control in directing α-silyloxy ketones synthesis as the only adduct.

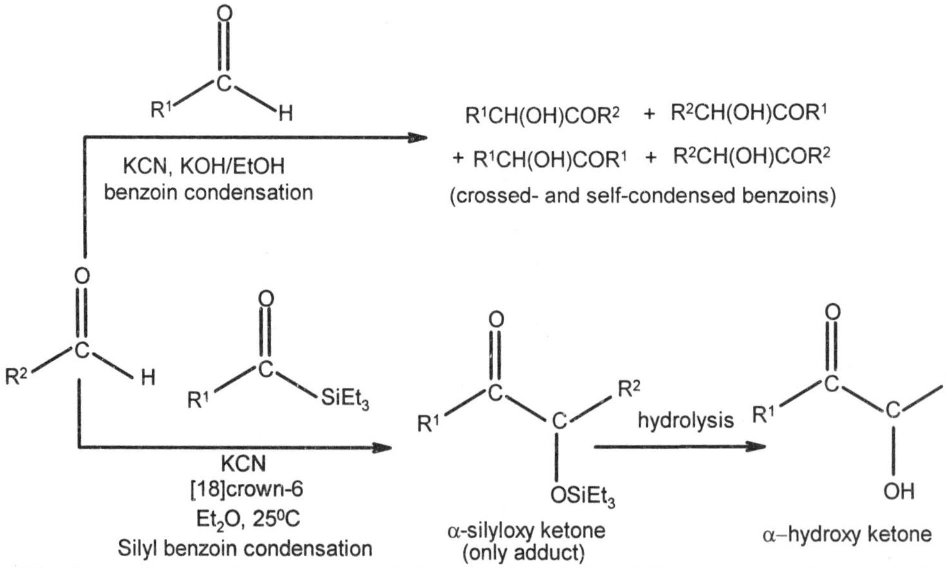

R^1-CHO —[KCN, KOH/EtOH benzoin condensation]→ $R^1CH(OH)COR^2$ + $R^2CH(OH)COR^1$ + $R^1CH(OH)COR^1$ + $R^2CH(OH)COR^2$
(crossed- and self-condensed benzoins)

R^2-CHO + R^1-CO-$SiEt_3$ —[KCN, [18]crown-6, Et_2O, 25°C, Silyl benzoin condensation]→ R^1-CO-C(R^2)(H)(OSiEt₃) α-silyloxy ketone (only adduct) —[hydrolysis]→ R^1-CO-C(R^2)(H)(OH) α-hydroxy ketone

The investigators have reported the results of the following reactions relating to silyl benzoin condensation (shown in the following table):

R^1	R^2	%yield of α-silyloxy ketones
Ph	Ph	90
Ph	4-ClPh	82
4-ClPh	Ph	86
4-MeOPh	Ph	85
4-MeOPh	4-ClPh	80
Ph	Isopropyl	66
4-ClPh	*n*-hexyl	75

The proposed mechanism for the cross silyl benzoin condensation is given as:

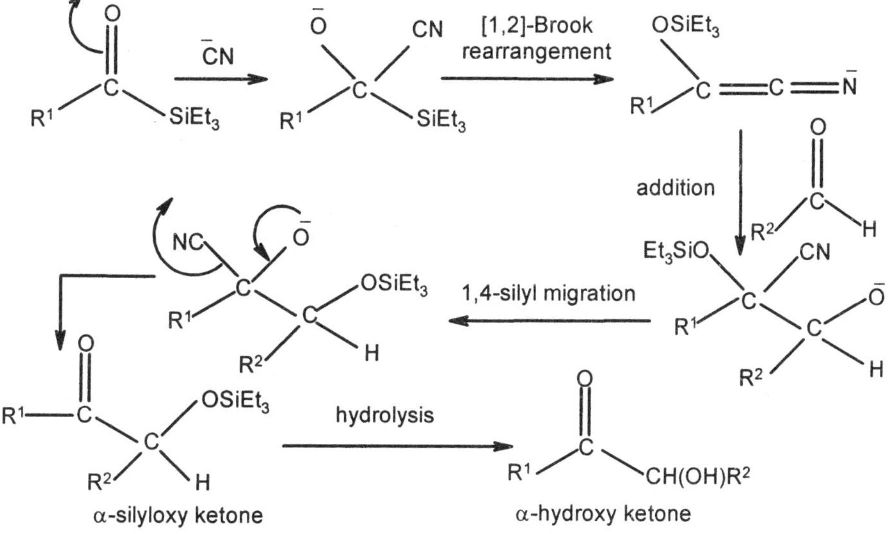

CN
[1,2]-Brook rearrangement
addition
1,4-silyl migration
hydrolysis
α-silyloxy ketone
α-hydroxy ketone

Applications

1) Benzoin condensation can be accomplished for glyoxals (RCOCHO) also:

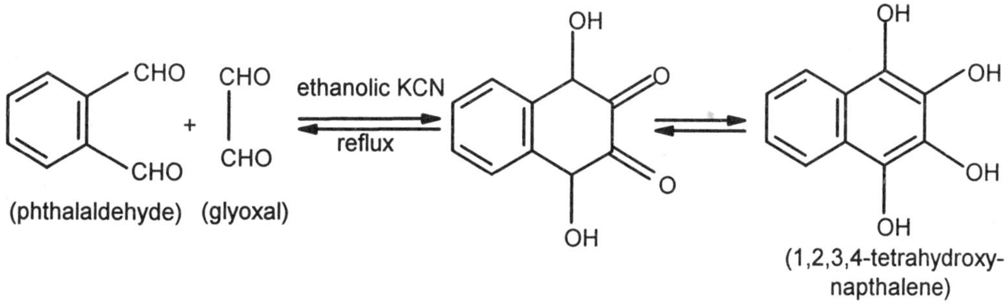

ethanolic KCN
reflux
(phthalaldehyde)
(glyoxal)
(1,2,3,4-tetrahydroxy-napthalene)

The crossed benzoin condensation using glyoxals may be applied for synthesis of polynuclear hydrocarbons.

2) Benzaldehyde on bezoin condensation forms benzoin that can easily be converted into a variety of compounds.

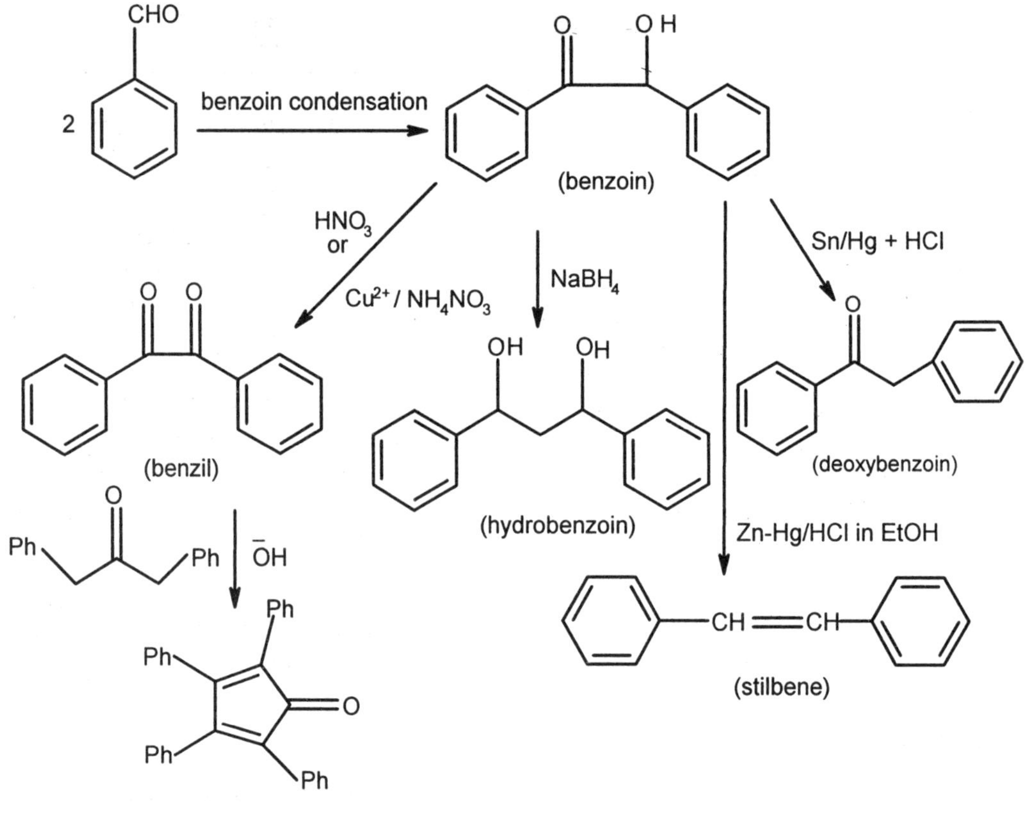

3) Heterocyclic aldehydes also undergo benzoin condensation reaction:

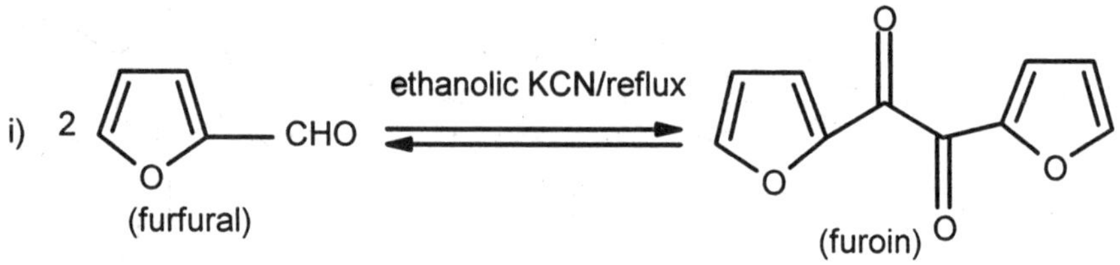

i) 2 (furfural) —CHO $\xrightleftharpoons{\text{ethanolic KCN/reflux}}$ (furoin)

ii) 2 (thiophene-2-aldehyde) —CHO ⇌ (ethanolic KCN/reflux) → 1,2-di(thiophen-2-yl)ethane-1,2-dione

References

1. Lapworth, A. (1903), *J. Chem. Soc.*, **83**, 995; **85**, 1206(1904); Kuebrich, J. P., Schowen, R. L. (1971), Wang, M. and Lupes, M. E., *J. Am. Chem. Soc.*, **93**, 1214.
2. Yano, Y., Tamura, Y. and Tagaki, W. (1980), *Bull. Chem. Soc. Japan*, **53**, 740; Stetter, H., Ramsch, Y. and Kuhlmann, H. (1976), *Synthesis*, 733.
3. Morkovnik, A. S., Khrustalev, V. N., Lindeman, S. V., Struchkov, Y. T. and Morkovnik, Z. S. (1995), *Mendeleev Commun.*, **5** (No. 1), 11.
4. Linghu, X. and Johnson, F. S. (2003), *Angew. Chem. Int. Ed.*, **42**, 2534.

- Macaione, D. P. and Wentworth, S. E. (1974), *Synthesis*, 716.
- Ide, W. S. and Buck, J. S. (1948), *Org. React.*, **4**, 269. [Review].
- Stetter, H. and Kuhlmann, H. (1991), *Org. React.*, **40**, 407. [Review].
- Kluger, R. (1997), *Pure Appl. Chem.*, **69**, 1957.
- Demir, A. S., Dunnwald, T., Iding, H., Pohl, M. and Muller, M. (1999), *Tetrahedron: Asymmetry*, **10**, 4769.
- White, M. J. and Leeper, F. J. (2001), *J. Org. Chem.*, **66**, 5124.
- Enders, D. and Kallfass, U. (2002), *Angew. Chem. Int. Ed.*, **41**, 1743.
- Duenkelmann, P., Kolter–Jung, D., Nitsche, A., Demir, A. S., Siegert, P., Lingen, B. and Baumann, M. (2002), *J. Am. Chem. Soc.*, **124**, 12084.
- Linghu, X. and Johnson, S. (2003), *Angew. Chem. Int. Ed. Engl.*, **42**, 2534.

14

Birch Reduction

Introduction

The reaction where aromatic rings are partially reduced to nonconjugated 1,4-cyclohexadienes by treatment with metallic sodium, potassium or lithium in a mixture of liquid ammonia and an alcohol is known as *Birch reduction*.[1] Such reductions are known as *dissolving metal reductions*. The alcohol (oftenly used alcohols are ethanol, methanol, isopropanol or *t*-butylalcohol) acts as a proton source. Presence of iron salts in commercial ammonia lowers the yields, and thus use of distilled ammonia is suggested. The reaction is usually carried out at the boiling temperature of ammonia (-33^0C). *Birch reduction* provides a convenient method for preparing a wide variety of interesting and useful cyclic dienes. The simplest example of this kind of reaction is the reduction of benzene to 1,4-cyclohexadiene —

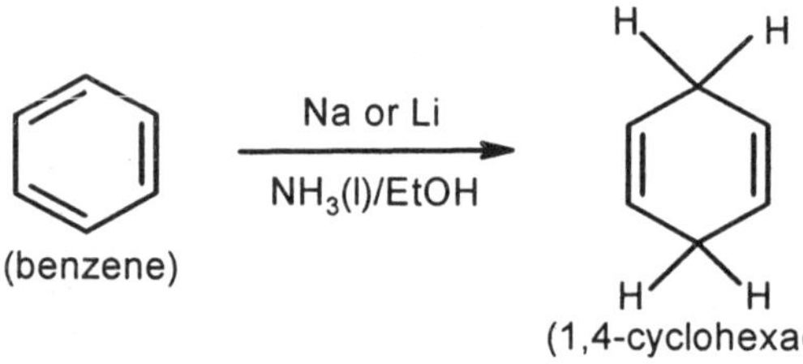

$$\text{(benzene)} \xrightarrow[\text{NH}_3(l)/\text{EtOH}]{\text{Na or Li}} \text{(1,4-cyclohexadiene)}$$

Mechanism

The generally accepted mechanism for *Birch reduction* is depicted below, taking the example of unsubstitued benzene as substrate. The mechanism involves solvated electrons, which are transferred from the metal to the solvent, and hence to the ring.[2]

The mechanism of reduction involves formation of radical anion (**1**) followed by protonation by an alcohol to give free radical (**2**), which is reduced by receiving further an electron transferred from the metal to form a resonance-stabilized carbanion (**3**). Finally, cyclohexadienyl carbanion (**3**) accepts proton from another molecule of alcohol to yield cyclohexadiene (**4**). Protonation to the radical anion is supposed to be the *rate-limiting step*. Such protonation eliminates the possibilities of formation of undesirable side products arising out of dimerization and/or polymerization of the radical anion.

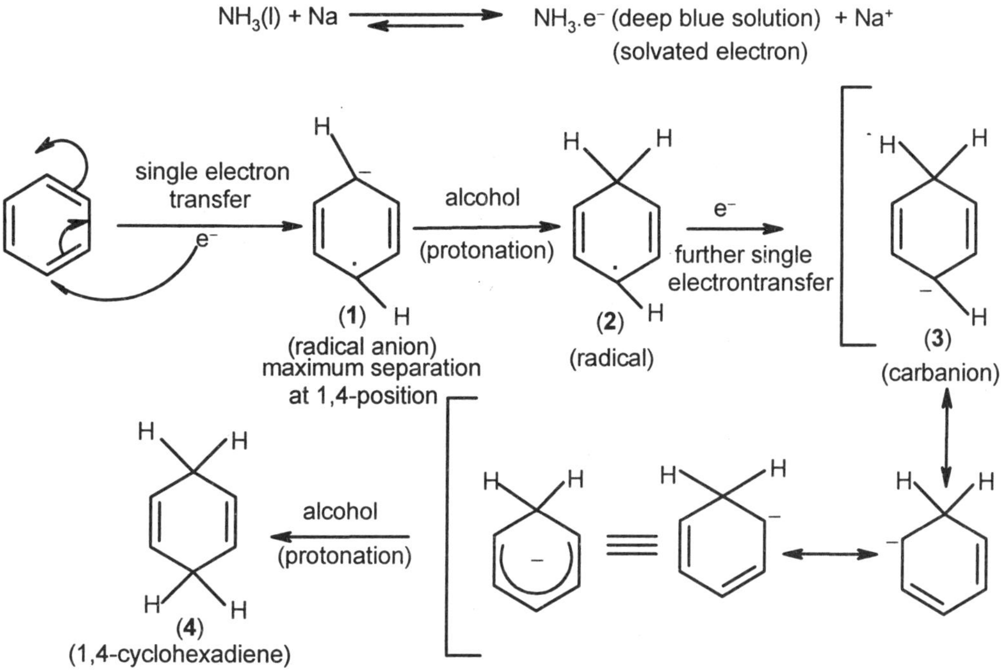

$$NH_3(l) + Na \rightleftharpoons NH_3.e^- \text{ (deep blue solution)} + Na^+$$
$$\text{(solvated electron)}$$

[Mechanism scheme showing:
- Benzene + single electron transfer (e⁻) → (1) radical anion (maximum separation at 1,4-position)
- (1) + alcohol (protonation) → (2) radical
- (2) + e⁻ (further single electron transfer) → (3) carbanion
- (3) resonance structures
- → alcohol (protonation) → (4) 1,4-cyclohexadiene]

Critical Views

a) The relative rate and regioselectivity of *Birch reaction* is largely dependent upon the nature of substituents in the aromatic nucleus.[3] The substituents have profound impact on the stability of the radical anion. As such, electron-donating and electron-withdrawing substituents behave differently towards *Birch reduction*.

(i) Effect of electron-donating substituents
The electron-donating groups (such as alkyl or alkoxyl) retard the rate of the reaction, and usually remain at the nonreduced positions of the product.

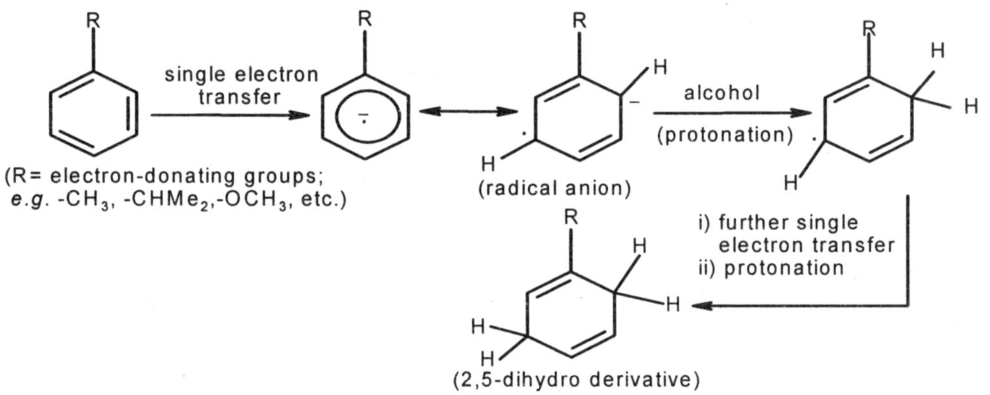

[Scheme: Birch reduction mechanism with electron-donating group R]

(R= electron-donating groups; e.g. -CH$_3$, -CHMe$_2$, -OCH$_3$, etc.)

(radical anion)

(2,5-dihydro derivative)

In fused aromatic ring systems (like naphthalene), an electron-donating substituent directs the reduction in the unsubstituted ring.[4]

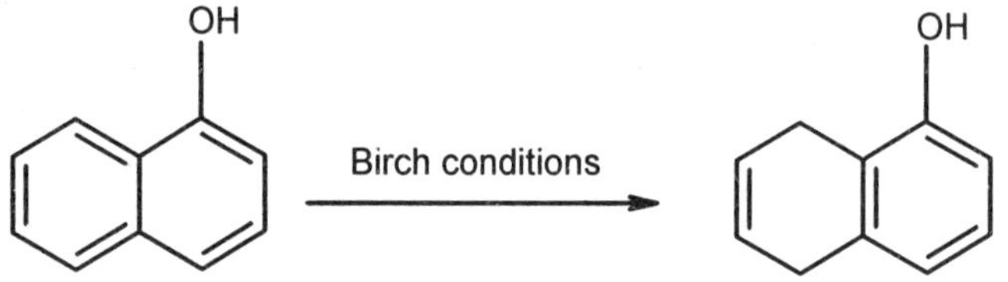

[Scheme: 1-naphthol → 5,8-dihydro-1-naphthol under Birch conditions]

(ii) Effect of electron-withdrawing substituents

The electron-withdrawing groups (such as –COOH, –COOR, –CONH$_2$, –Ph, –SiMe$_3$ etc.) enhance the rate of the reaction, and generally remain at the reduced positions of the product.

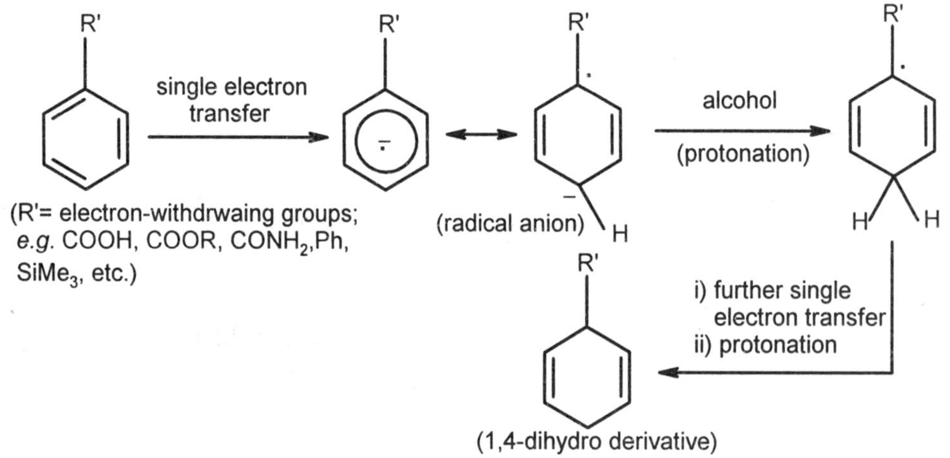

[Scheme: Birch reduction mechanism with electron-withdrawing group R']

(R'= electron-withdrwaing groups; e.g. COOH, COOR, CONH$_2$, Ph, SiMe$_3$, etc.)

(radical anion)

(1,4-dihydro derivative)

An electron-withdrawing substituent in a fused aromatic system, like naphthalene, promotes reduction of that ring to which it is attached.[4]

b) Though an isolated olefinic bond is not usually reduced under *Birch conditions* (exceptions as found in an end methylene or in strained systems), but olefinic bonds conjugated with aromatic systems or with carbonyl groups are readily reduced by $Na/NH_3(l)$. Under *Birch conditions* halogen, nitro, aldehyde, and keto groups may also be reduced. Besides, alkynes are also reduced with $Na/NH_3(l)$ or $Li/NH_3(l)$ affording *trans*-alkenes.

c) As already mentioned that strongly electron-releasing groups (for example, -OCH₃) deactivate the aromatic ring towards *Birch reduction*, and in case of such deactivated systems lithium is often used together with a co-solvent (very often THF) and a weaker proton source (*t*-butylalcohol). The stronger reducing agent, combined with a weaker proton source, promotes the reduction. It is noticed that the rate of the *Birch reduction* increases with metals in the order of K > Na > Li. It is also reported that Na^+ ions retard the fast *Li-Birch reduction*, and while Li^+ ions accelerate the slow *Na-Birch reduction*.

d) Asymmetric *Birch reductions* created a lot of interest in the past few years.[5] Recently, Donohoe *et al.*[6] reported that higher enantioselectivity in the reduction of furans could be achieved by attaching a silyl group to the nucleus.

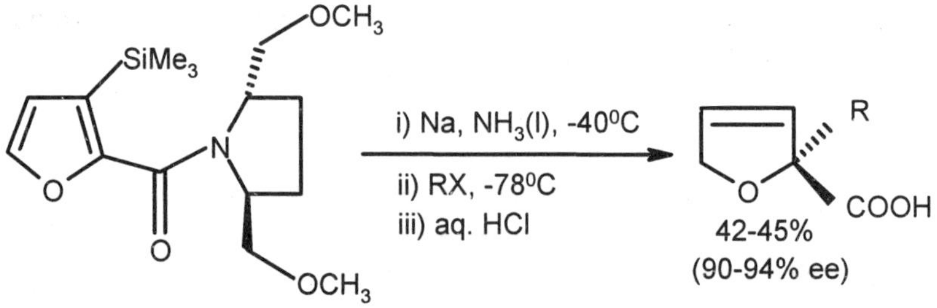

Applications

Applications of *Birch reduction* in organic syntheses are multifold, and this beautiful reaction offers a simple technique for carrying out a variety of useful conversions; this reaction find immense utility as a key step in the syntheses of a number of organic compounds of interest. A few examples of reaction undergoing *Birch reduction* are cited:

1) Birch reduction offers a convenient method for preparing 1,4-dihydroderivatives of aromatic compounds —

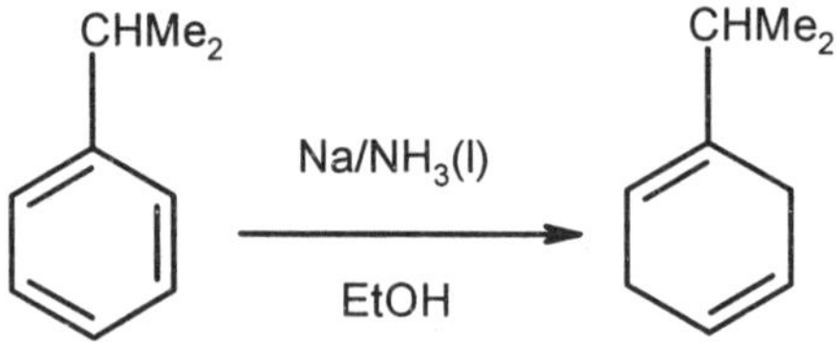

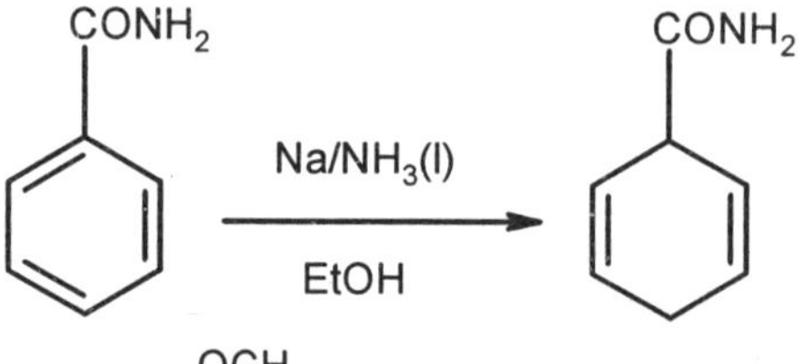

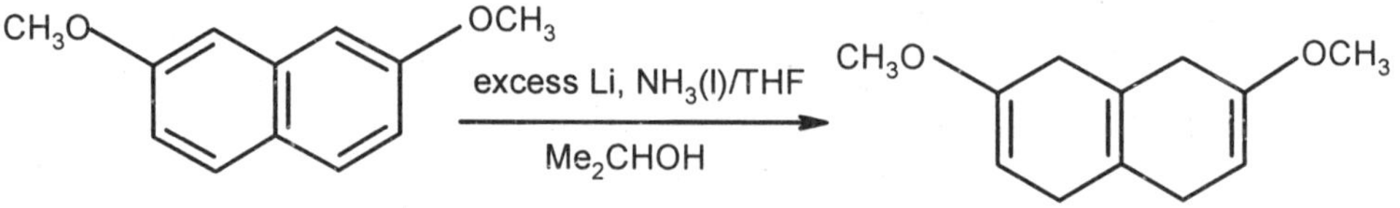

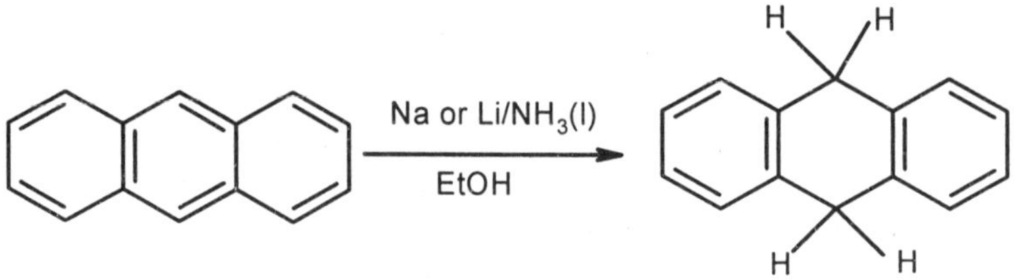

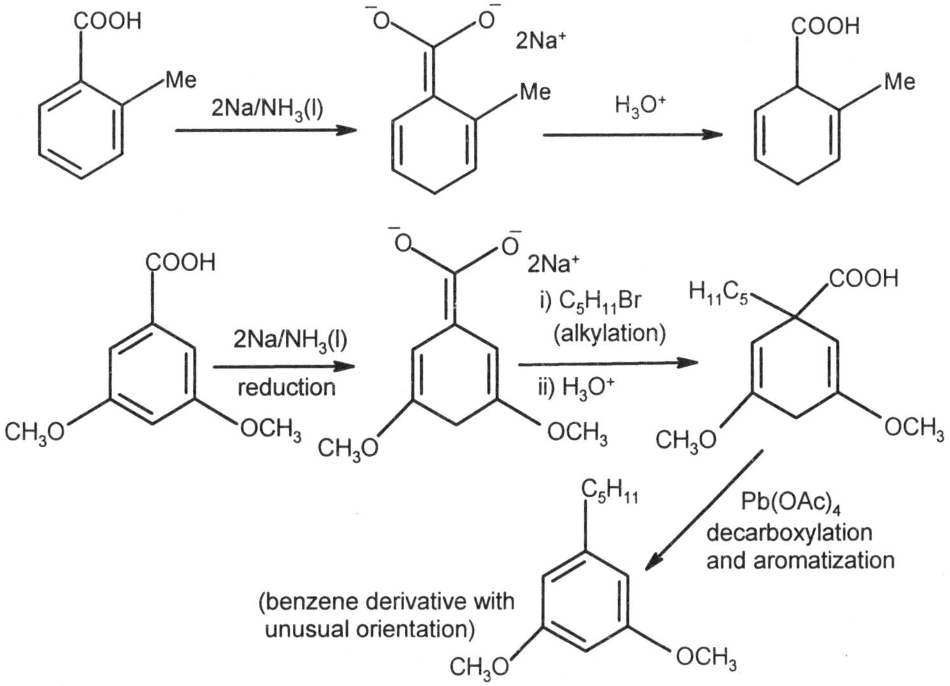

Birch Reduction

2) Selective reduction of a benzene ring in the presence of another reducible group can be achieved by applying *protective-technique* —

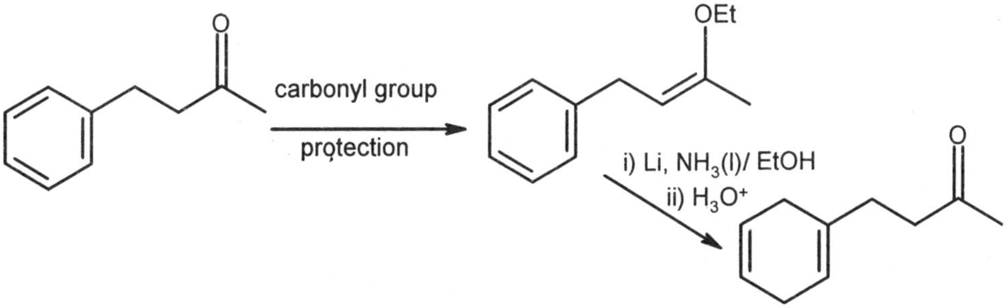

3) Substituted cyclohexenones can be prepared by using *Birch reductive technique* —

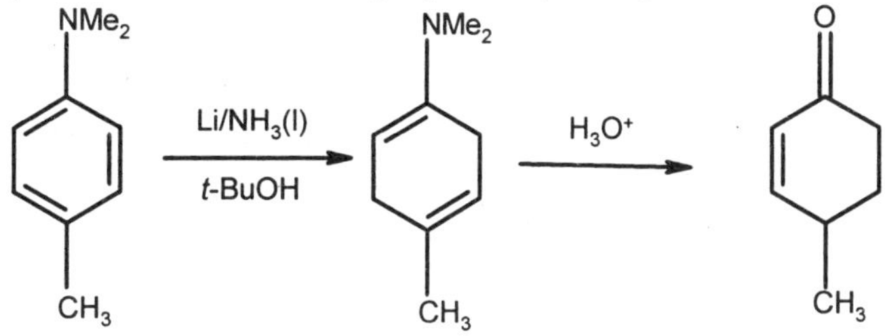

References

1. Birch, A. J. (1944), *J. Chem. Soc.*, 430; for reviews see Birch, A. J. (1950), *Quart. Rev.(London)*, **4**, 69; Birch, A. J. and Smith, H. (1958), *Quart. Rev. (London)*, **12**, 17; Rabidean, P. W. and Marcinow, Z. (1992), *Org. React.*, **42**, 1; Birch, A. J. (1996), *Pure Appl. Chem.*, **68**, 553.
2. Birch, A. J. and Nasipuri, D. (1959), *Tetrahedron*, **6**, 148.
3. Zimmerman, H. E. and Wang, P. A. (1993), *J. Am. Chem. Soc.*, **115**, 2205; Rabidean, P. W. (1989), *Tetrahedron*, **45**, 1579; Zimmerman, H. E. and Wang, P. A. (1990), *J. Am. Chem. Soc.*, **112**, 1280.
4. Rabidean, P. W. and Burkholder, E. G. (1978), *J. Org. Chem.*, **43**, 4283.
5. Schultz A. G. (1999), *Chem Commun*, 1263. [Review]
6. Donohoe T. J. *et al.* (2001), *Tetrahedron Lett.*, **42**, 5841.

- Rabidean, P. W. and Marcinow, Z. (1992), *Org. React.*, **42**, 1. [Review]
- Schultz, A. G. and Pettus, L. (1997), *J. Org. Chem,*, **62**, 6855.
- Pross, A. and Radom, L. (1978), *J. Am. Chem. Soc.*, **100**, 6572.
- Birch, A. J., Hinde, A. L. and Radom, L. (1980), *J. Am. Chem. Soc.*, **102**, 3370.
- Ohta, Y., Doe, M., Morimoto, Y. and Kinoshita, T. (2000), *J. Heterocycl. Chem.*, **37**, 751.
- Guo, Z. and Schultz, A. G. (2001), *J. Org. Chem.*, **66**, 2154.
- Yamaguchi, S., Hamada, E., Yokoyama, H., Hirai, Y. and Shithani, S. (2002), *J. Heterocycl. Chem.*, **39**, 335.
- Jiang, J. and Lai, Y.-H. (2003), *Tetrahedron Lett.*, **44**, 1271.

15

Bischler - Napieralski Reaction

Introduction

Bischler-Napieralski synthesis[1] involves cyclodehydration of β-arylethylamides to 3,4-dihydroisoquinoline derivatives in the presence of phosphorus oxychloride. Phosphorus pentachloride, phosphorus pentaoxide, polyphosphoric acid and zinc chloride are also the common cyclization agents used. The cyclized product is then readily dehydrogenated to the corresponding isoquinoline using Pd/C, sulphur or diphenyl sulphide.

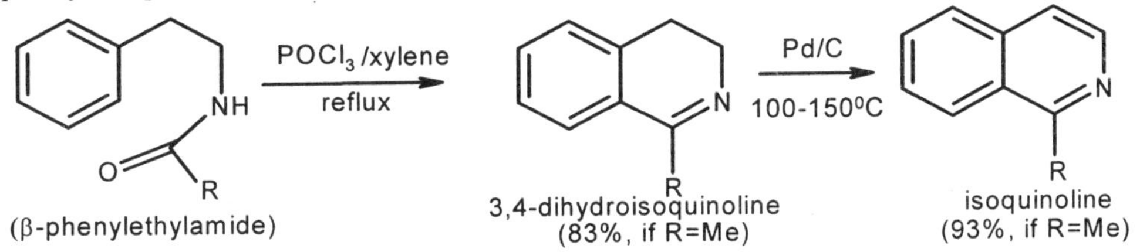

Mechanism

The general mechanistic steps for the reaction are given as:

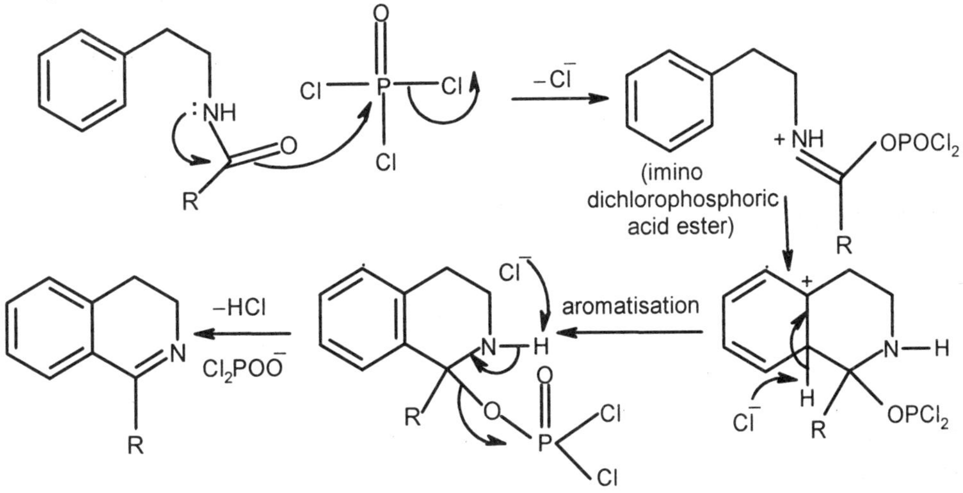

During the reaction sequence, the electrophilic intermediate is very probably an imino chloride or imino phosphate.[2] The former (*i.e.* ArCH$_2$CH$_2$N=CRCl) has been isolated when the amide is treated with PCl$_5$, and it cyclizes efficiently to 3,4-dihydroisoquinoline on heating with a Lewis acid *via* isonitrilium salts (*i.e.* ArCH$_2$CH$_2$N$\overset{+}{\equiv}$CR).[3]

Critical Views

(a) Since the cyclising step involves electrophilic attack on the aromatic ring, feasibility of the method enhances in case of activated rings. Thus, the presence of electron-donating groups on the aromatic ring makes the reaction an efficient one, while electron-withdrawing groups on the aromatic nucleus make it inefficient. For example, cyclisation of *N*-[2-(4-nitrophenyl)-ethyl]-benzamide results in a lower yield of product:

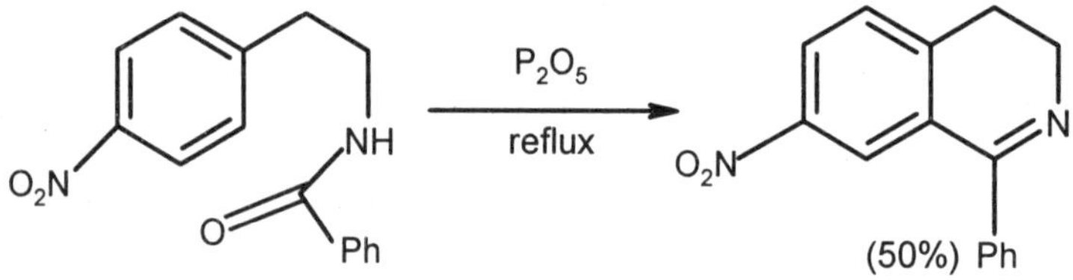

In case of *meta*-substituted substrates, 6-substituted isoquinoline derivatives are obtained exclusively.

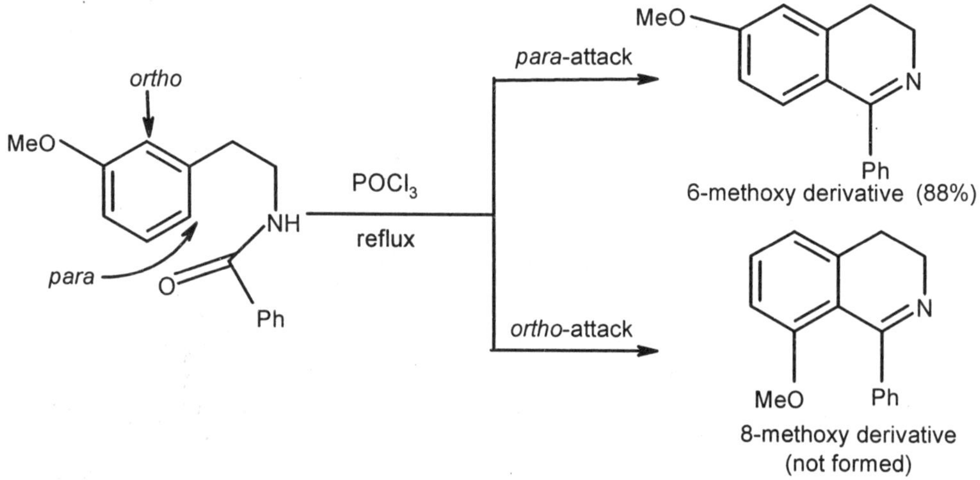

The 6-methoxy derivative is the exclusive product, because the ring closure occurs at *para* to the methoxy group, an electron-donating function.[4]

(b) *Pictet – Gams modification*
To obtain a fully aromatic isoquinoline, *Bischler – Napieralski reaction* sequence had been slightly modified — in this case the amide of a β-methoxy- or β-hydroxy-β-arylethanamine is heated with the usual types of cyclization catalyst. It is uncertain whether dehydration to an unsaturated amide or to an oxazolidine is an initial stage in the overall sequence.[5]

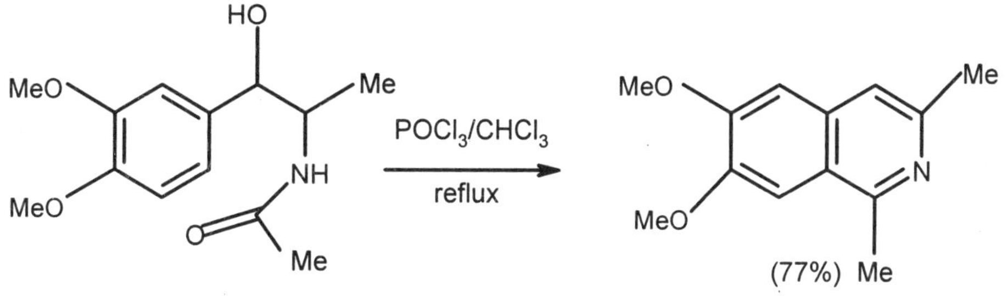

(c) *Abnormal product formation*
Reaction of *N*-[2-(4-methoxyphenyl)-ethyl]-benzamide (**I**) with P_2O_5 (and $POCl_3$) gives 7-methoxy-1-phenyl-3,4-dihydroisoquinoline (**II**) [a normal *Bischler-Napieralski reaction* product] and 6-methoxy-1-phenyl-3,4-dihydroisoquinoline (**III**) [an abnormal product].

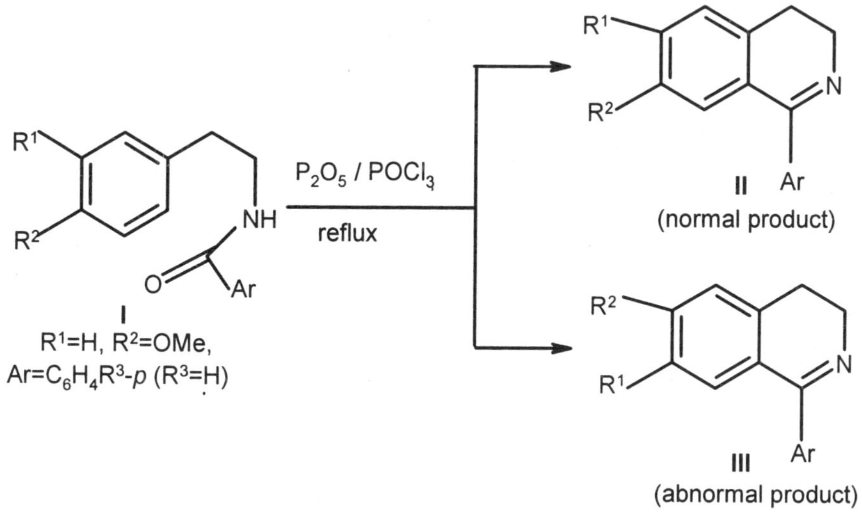

Doi *et al.*[6] examined the effects of reaction conditions and of substituents on the reaction leading to abnormal product. More interestingly, use of cyclization reagents has a profound effect on the reaction mechanism. From their experimental observation:

Entry No.	Substrate (I)	Reagents	Product ratio II : III		% Yield
1	R^1=H, R^2=OMe, R^3=OMe	$POCl_3$/xylene	100	0	28
2	R^1=H, R^2=OMe, R^3=OMe	P_2O_5/ xylene	37	63	18
3	R^1=H, R^2=OMe, R^3=OMe	$POCl_3/P_2O_5$(2:1)/ xylene	16	84	64
4	R^1=H, R^2=OMe, R^3=OMe	$POCl_3/P_2O_5$(9:1)/ xylene	<5	>95	77
5	R^1=H, R^2=OMe, R^3=H	$POCl_3/P_2O_5$(9:1)	67	33	57
6	R^1=H, R^2=Me, R^3=OMe	$POCl_3/P_2O_5$(9:1)	100	0	61
7	R^1=H, R^2=Cl, R^3=OMe	$POCl_3/P_2O_5$(9:1)	100	0	19

The investigator suggested a detailed mechanistic scheme (**Scheme-A**) to explain their observation. In the reaction of **I** with $POCl_3$, cyclization to **II** may proceed *via* dichlorophosphoric acid ester **IV** [**I** → **IV** → **V** → **II**].[7] However, in the presence of P_2O_5, formation of a nitrilium ion intermediate (**VI**) may become the main route.[2] Electrophilic attack by the nitrilium cation at C-2 of the phenyl group (path-b) yields the intermediate (**VII**), which is subsequently aromatized to **II** [**I** → **VI** → **VII** → **II**]; whereas the attack at C-1 gives the spiro compound (**VIII**) which is isomerized to **III** *via* **IX** (path-a; **I** → **VI** → **VIII** → **IX** → **III**]. It is observed that when R^2 is a methoxy group and R^1 is hydrogen, path 'a' becomes the main route because of the electron-donating effect of the methoxy group.

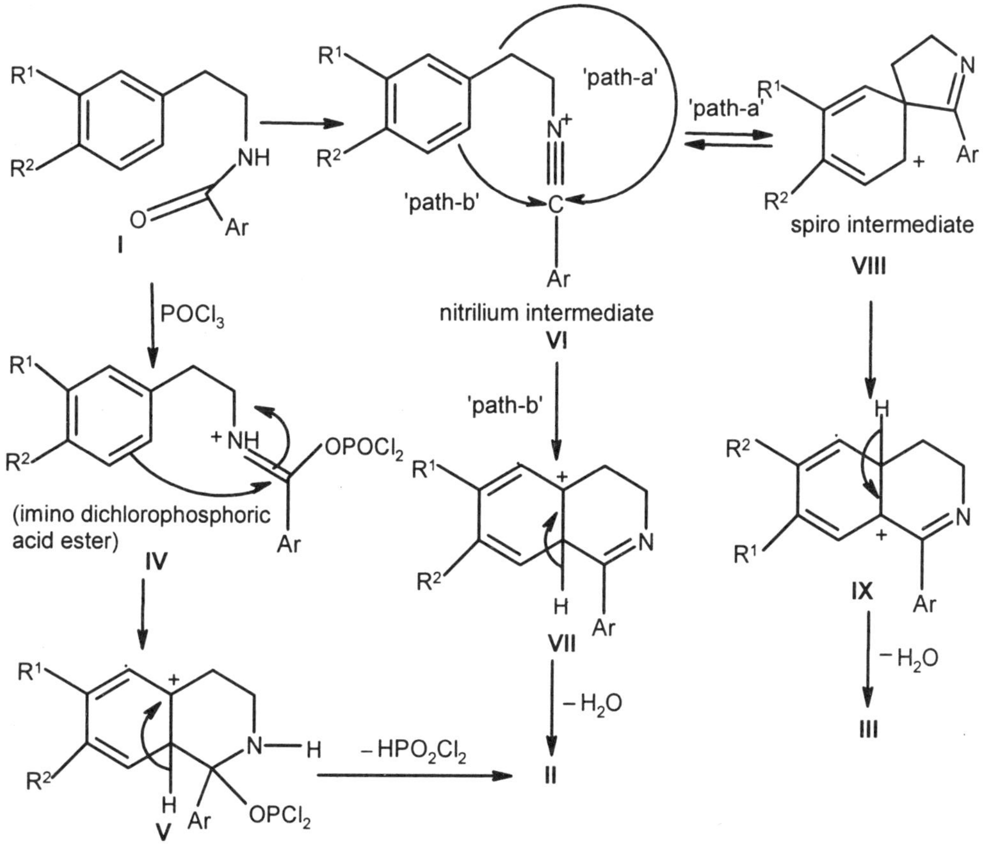

Bischler - Napieralski Reaction

(Scheme-A)

Another example of abnormal product formation:

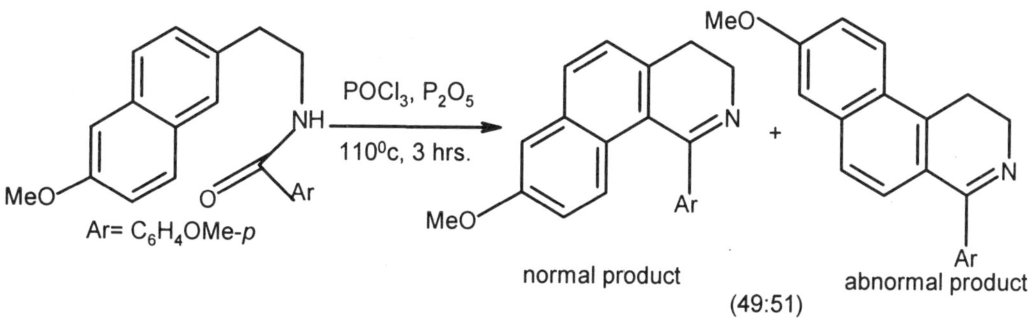

normal product + abnormal product
(49:51)

Ar = C_6H_4OMe-p

Applications

This classical method finds immense applications in synthetic organic chemistry. A few examples are cited:

1. 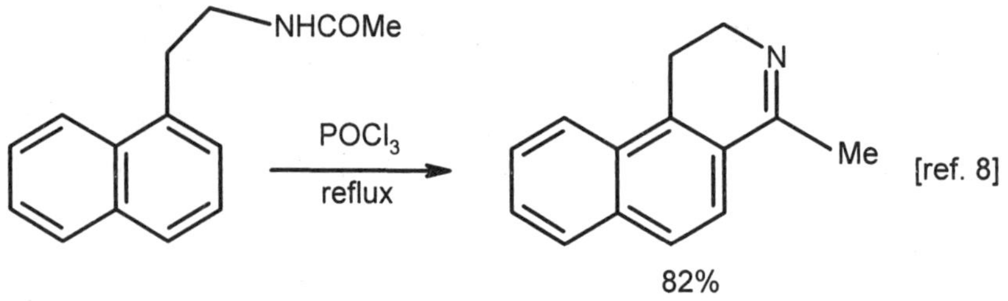 [Naphthalene-CH₂CH₂-NHCOMe] → (POCl₃, reflux) → dihydrobenzo[h]isoquinoline with Me substituent, 82% [ref. 8]

2. 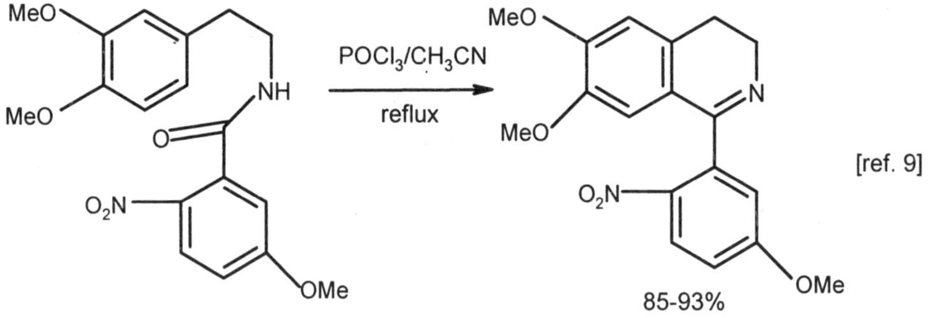 [3,4-dimethoxyphenethylamine acylated with 2-nitro-5-methoxybenzoyl] → (POCl₃/CH₃CN, reflux) → 6,7-dimethoxy-1-(2-nitro-5-methoxyphenyl)-3,4-dihydroisoquinoline, 85–93% [ref. 9]

3. [Pyrazole-indole amide substrate with Ar-CO-NH] → (POCl₃/CH₃CN, reflux) (Ar = Ph; 2-thienyl) → fused pyrazolo-azepino-indole product [ref. 10]

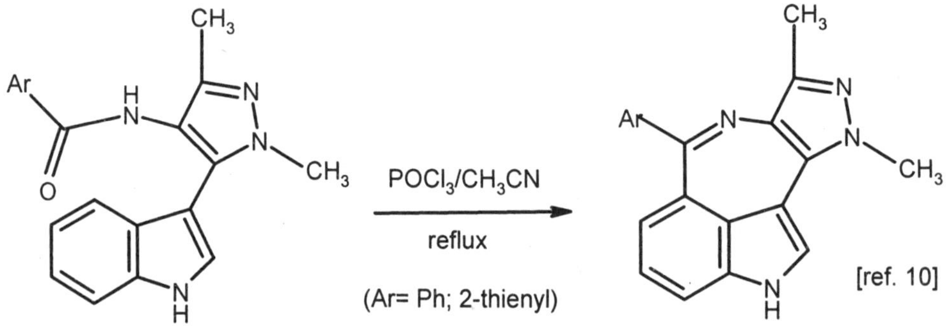

References

1. Bischler, A. and Napieralski, B. (1893), *Ber. Dtsch. Chem. Ges.*, **26**, 1903.
2. Fodor, G. and Nagubandi, S. (1980), *Tetrahedron*, **36**, 1279.

3   Fodor, G., Gall, G. and Phillips, B. A. (1972), *Angew. Chem. Int. Ed. Engl.*, **11**, 919; Nagubandi, S. and Fodor, G. (1980), *J. Heterocycl. Chem.*, **17**, 1457; Ban, Y., Wakamatsu, T. and Mori, M. (1977), *Hereocycles*, **6**, 1711.
4   Reynolds, G. A. and Nauser, C. R. (1955), *Org. Syn. Coll.*, **vol. 3**, 593.
5   Fitton, A. O., Frost, J. R., Zakaria, M. M. and Andrew, G. (1973), *J. Chem. Soc., Chem. Commun.*, 889.
6   Doi, S., Shirai, N. and Sato, Y. (1997), *J. Chem. Soc., Perkin Trans. 1*, 2217.
7   Mundy, B. P. and Ellerd, M. G., in *Name Reactions and Reagents in Organic Synthesis*, John Wiley & Sons, New York, 1988, p.32.
8   Kessar, S., Jit, P., Mundra, K. and Lumb, A. (1971), *J. Chem. Soc. Part C*, 266.
9   Hilger, C. S., Fugmann, B. and Steglich, W. (1985), *Tetrahedron Lett.*, 5975.
10 Abu Safieh, K. A., El – Abadelah, M. M., Abu Zarga, M. H., Sabri, S. S., Voelter, W. and Mossmer, C. M. (2001), *J. Heterocycl. Chem.* **38**, 623.

- Larsen, R. D., Reamer, R. A., Corley, E. G., Davis, P., Grabowski, E. J. J., Reider, P. J., Shinkai, I. (1991), *J. Org. Chem.*, **56**(21), 6034.
- Ishikawa, T., Saitop, T., Noguchi, S., Ishii, H., Ito, S. and Hata, T. (1995), *Tetrahedron Lett.*, **36**(16), 2795.
- Sanchez-Sancho, Francisco, Mann, Enrique, Herradon and Bernardo (2000), *Synlett*, 04.
- Miyatani, K., Ohno, M., Tatsumi, K., Ohishi, Y., Kunitomo, J.-I., Kawasaki, I., Yamashita, M. and Ohta, S. (2001), *Heterocycles*, **55**, 589.
- Ishikawa, T., Shimooka, K., Narioka, T., Noguchi, S., Saito, T., Ishikawa, A.; Yamazaki, E., Harayama, T., Seki, H. and Yamaguchi, K. (2000), *J. Org. Chem*, **65**(26); 9143.
- Abu Safieh, K. A., El – Abadelah, M. M, Sabri, S. S, Voelter, W., Mossmer, C. M. and Stroebele, M. (2002), *Z. Naturforsch.*, **57b**, 1327.
- Nicoletti, M., O'Hagan, D., Slawin, A. M. Z. (2002), *J. Chem. Soc., Perkin Trans. 1*, 116.

16

Bouveault - Blanc Reaction

Introduction

Reduction of carboxylic esters to the corresponding alcohols using sodium in ethanol is known as *Bouveault-Blanc reaction*[1].

$$\underset{R}{\overset{O}{\underset{\|}{C}}}\!-\!OR' \xrightarrow[\text{reflux}]{\text{Na/EtOH}} R\!-\!CH_2OH + R'OH$$

Aldehydes and ketones are also reduced by this procedure.

$$RCHO \xrightarrow{\text{Na/EtOH}} RCH_2OH$$

$$RCOR \xrightarrow{\text{Na/EtOH}} RCH(OH)R$$

Bouveault-Blanc procedure was very much popular for the reduction of carboxylic esters before the discovery of LAH — 'in practice, prior to the discovery of complex metal hydrides, this was the only method for the reduction of carboxylic esters. *Bouveault-Blanc procedure* is still in use when selective reduction of ester group in presence of carboxylic acid functionality is required, and as an inexpensive substitute for LAH reductions of esters in industrial production. In absence of proton donors, dimerization will take place leading to acyloin condensation product.

Mechanism

Sodium serves as single-electron reducing agent, and ethanol as proton donor. It is very interesting to note that this reaction occurs in presence of sodium in ethanol, not in sodium ethoxide, which is the basic product that forms once sodium is dissolved into the alcohol. Thus, it is important that the sodium is dissolving as the reaction takes place, since only then is the free electrons available.
The following mechanistic path has been suggested for the reaction[2].

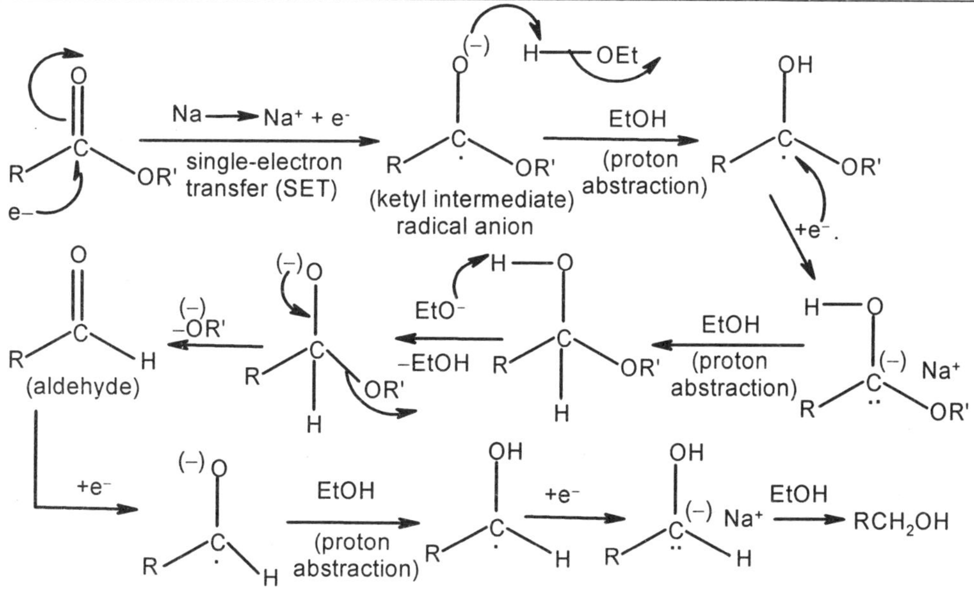

The ketyl intermediate has been isoleted[3].

Critical Views

(a) During reduction of an aldehyde or ketone, formation of thermodynamically more stable alcohol usually predominates, but in cases of strained or sterically hindered systems, the less-stable isomer may also predominate[4].

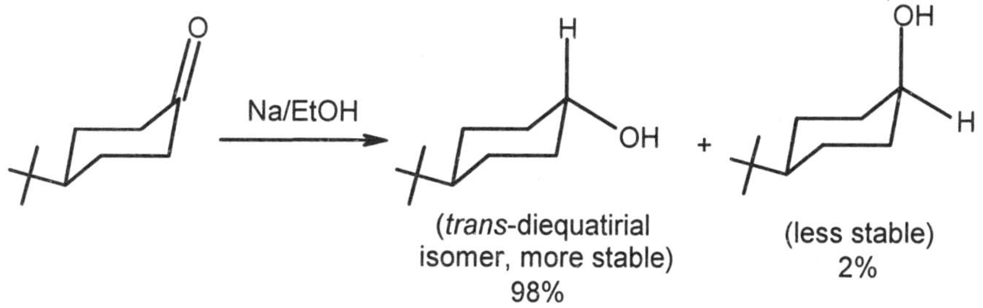

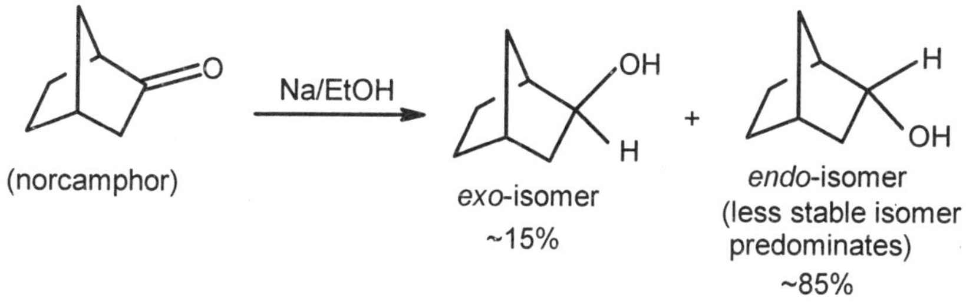

(b) One of the major advantages of *Bouveault-Blanc procedure* is its ability to reduce selectively the ester group in presence of non-conjugated C-C double bond, and of carboxylic acid functionality. Thus, the C-C double bond in compound (**1**) remains intact in the reduced product (**2**).

$$CH_3(CH_2)_7CH=CH(CH_2)_7COOC_2H_5 \xrightarrow{Na/EtOH} CH_3(CH_2)_7CH=CH(CH_2)_7CH_2OH$$

cis-isomer (**1**) → *cis*-isomer (**2**) [Ref. 5]

Similarly, only the ester group in compound (**3**) is reduced by Na/EtOH and during work-up the resulting product (**4**) yields the lactone (**5**).

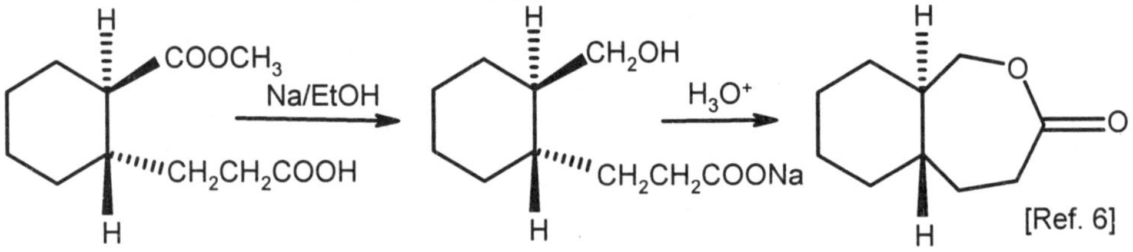

(3) → Na/EtOH → (4) → H₃O⁺ → (5) [Ref. 6]

(c) It is observed that reductive elimination of a methoxy group under *Bouveault-Blanc* conditions takes place when it is flanked on both sides by other two methoxyls[7]. Sharda and Krishnamurthy[8] carried out *Bouveault-Blanc reduction* of some 1,2,3-trimethoxybenzene derivatives leading to the loss of central methoxy group, and used this technique to prepare 3,5-dimethoxybenzoic acid from 3,4,5-trimethoxybenzoic acid. The reaction schemes are shown below:

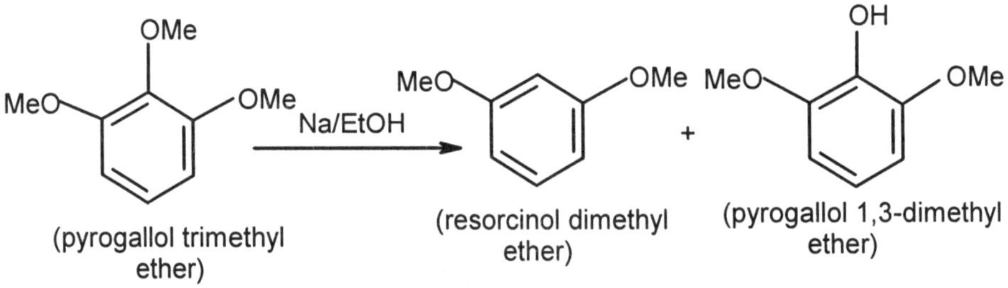

(pyrogallol trimethyl ether) → Na/EtOH → (resorcinol dimethyl ether) + (pyrogallol 1,3-dimethyl ether)

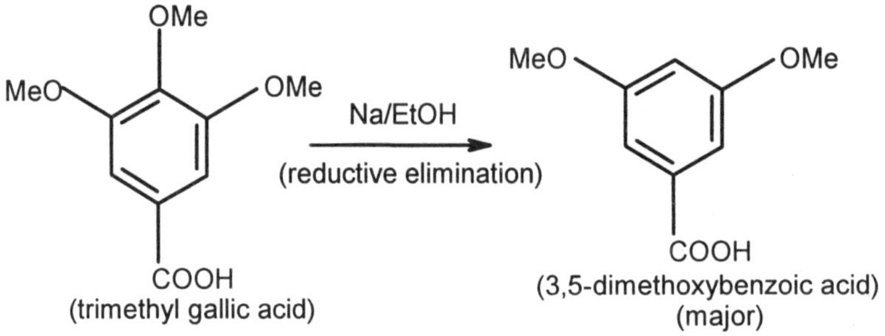

(trimethyl gallic acid) → Na/EtOH (reductive elimination) → (3,5-dimethoxybenzoic acid) (major)

The phenomenon can be rationalized in this way: in 1,2,3-trimethoxy benzene derivatives the middle methoxy group is of different character from the other two methoxyls (as evident from PMR study). The central methoxy group is out of the plane of benzene ring — buttressing effects[9] are also responsible for the above phenomenon. The mechanism of the reactions can also be explained by SR_N1 pathway:

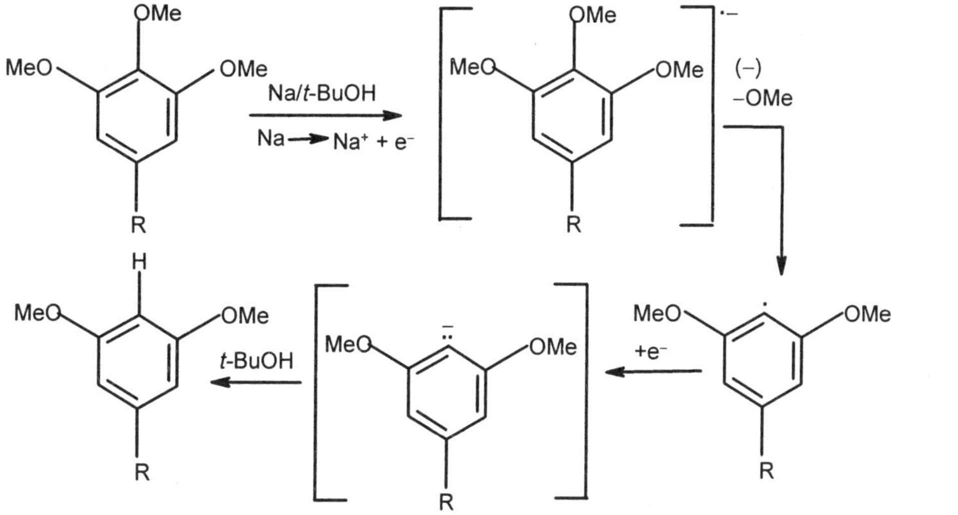

Applications

(i) Carbon-carbon double bond conjugated with an ester group is reduced along with the ester functionality under *Bouveault-Blanc procedure*.

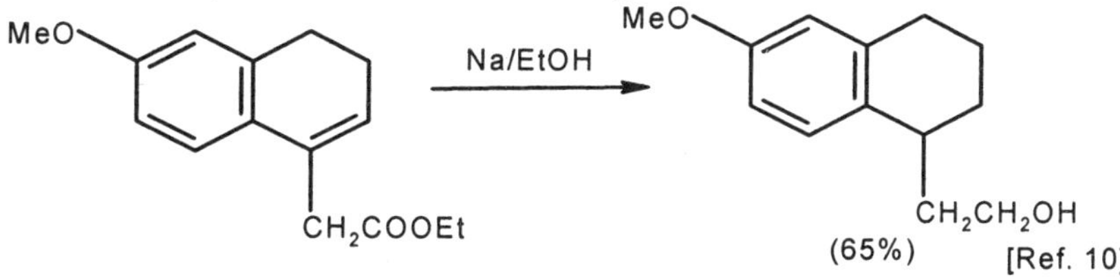

(ii) $n\text{-}C_{11}H_{23}CO_2Et \xrightarrow{\text{Na/EtOH}} n\text{-}C_{11}H_{23}CH_2OH$
(65-75%)   [Ref. 11]

(iii) 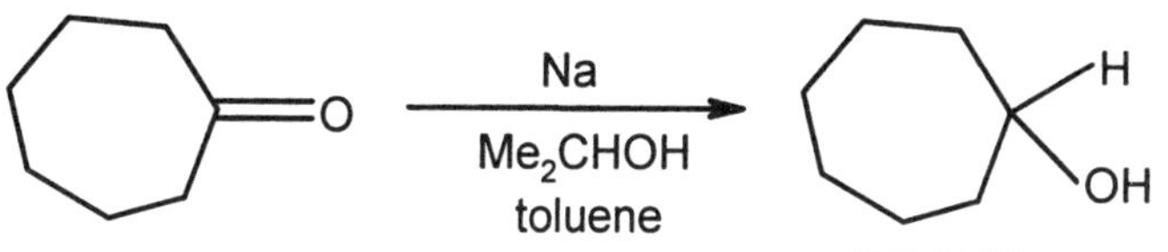 [cycloheptanone] $\xrightarrow[\text{toluene}]{\underset{Me_2CHOH}{Na}}$ [cycloheptanol]
(85-90%)   [Ref. 12]

(iv) The diester (**6**) produces a diol (**7**) through this process.

$$\text{(6)} \xrightarrow{\text{Na/EtOH}} \text{(7)}$$

References

1. Bouveault, L. and Blanc, G. (1903), *Comp. Rend.*, **136**, 1676; *Bull. Soc. Chim. France*, **31**, 666 (1904).
2. Pradhan, S. K. (1986), *Tetrahedron*, **42**, 6351; Huffman, J. W. (1983), *Acc. Chem. Res.*, **16**, 399; Giordano, C., Perdoncin, G. and Castaldi, G. (1985), *Angew. Chem. Edn. Engl.*, **24**, 499.
3. Rautenstrauch, V. and Geoffroy, M. (1976), *J. Am. Chem. Soc.*, **98**, 5035; *ibid*, **99**, 6280(1977).
4. Huffman, J. W. and Charles, J. T. (1968), *J. Am. Chem. Soc.*, **90**, 6486; Coulombeau, A. and Rassat, A. (1968), *Chem. Commun.*, No. 24, 1587.
5. Adkins, H. and Gillespie, R. H. (1955), *Org. Synth.*, **Coll. Vol. 3**, 671.
6. Paqultte, L. A. and Nelson, N. A. (1962), *J. Org. Chem.*, **27**, 2272.
7. Asahina, Y., Hiyasaka, H. and Sekisawa, T. (1936), *Ber. dtsch. Chem. Ges.*, **69B**, 1643.
8. Sharda, R. and Krishnamurthy, H. G. (1980), *Indian J. Chem.*, **19B**, 405.
9. Horton, W. J. and Spence, J. T. (1958), *J. Am. Chem. Soc.*, **80**, 2453; Burling, E. D., Jefferson, A. and Scheinmann, T. (1965), *Tetrahedron*, **21**, 2653.
10. Haberland, G. (1986), *Ber.*, **69**, 1380.
11. Ford, S. G. and Marvel, C. S. (1943), *Org. Synth.*, **Coll. Vol. 2**, 372.
12. Dev, S. (1956), *J. Indian Chem. Soc.*, **33**, 769.
- Ruehlmann, K., Seefluth, H., Kiriakidis, T., Michael, G., Jancke, H. and Kriegsmann, H. (1971), *J. Organomet. Chem.*, **27**, 327.
- Banerji, J., Bose, P., Chakrabarti, R. and Das, B. (1993), *Indian J. Chem.*, **32B**, 709.
- Seo, B. I., Wall, L. K., Li, H., Buttrum, J. W. and Lewis, D. E. (1993), *Syn. Commun.*, **23**, 15.
- Zhang, Y. and Ding, C. (1997), *Huaxue Tongbao*, 36.

17

Cannizzaro Reaction

Introduction

The *Cannizzaro reaction*[1] involves self oxidation-reduction between two molecules of aromatic or aliphatic aldehydes having no α–hydrogens when treated with a strong base. In this reaction, one molecule of aldehyde reduces the other to a primary alcohol and is itself reduced to the corresponding carboxylic acid salt normally in equimolar amounts.

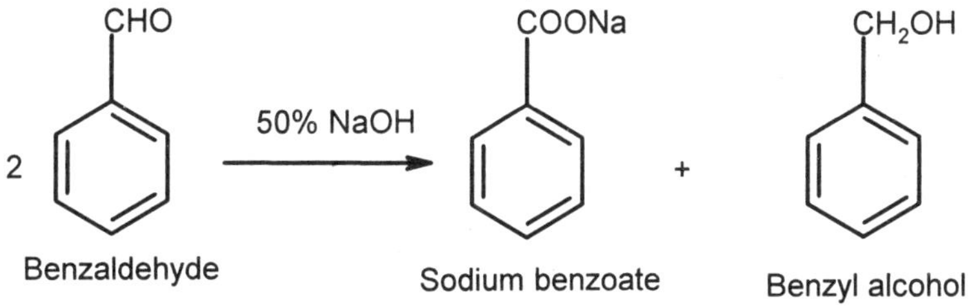

$$2\ PhCHO \xrightarrow{50\%\ NaOH} PhCOONa + PhCH_2OH$$

Benzaldehyde → Sodium benzoate + Benzyl alcohol

$$2\ Me_3CCHO \xrightarrow{50\%\ NaOH} Me_3CCOONa + Me_3CCH_2OH$$

Mechanism

The *Cannizzaro reaction* proceeds through hydride ion transfer mechanism — aldehydic hydrogen of one molecule of the aldehyde is transferred along with its bonded electron pair to the carbonyl carbon of the other one. The kinetic study of the reaction revealed that though the reaction in general follows third order kinetics (Rate α [aldehyde]^2[$^-$OH]), there are evidences of maintaining fourth order kinetics (Rate α [aldehyde]^2[$^-$OH])^2) particularly at higher base concentrations.[2]

Case-I:

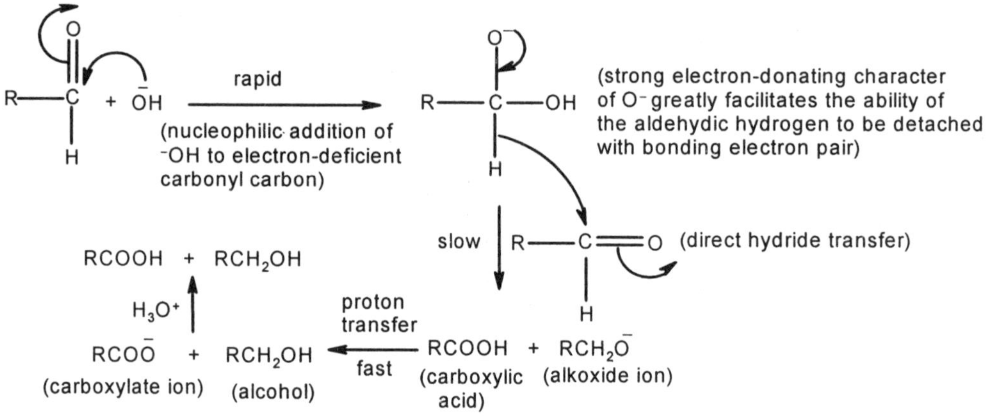

[Mechanism scheme: RCHO + ⁻OH → rapid nucleophilic addition of ⁻OH to electron-deficient carbonyl carbon gives R-C(O⁻)(OH)H; the strong electron-donating character of O⁻ greatly facilitates the ability of the aldehydic hydrogen to be detached with bonding electron pair; slow direct hydride transfer to another RCHO gives RCOOH + RCH₂O⁻ (carboxylic acid + alkoxide ion); fast proton transfer gives RCOO⁻ + RCH₂OH (carboxylate ion + alcohol); H₃O⁺ gives RCOOH + RCH₂OH.]

Case-II:

[Mechanism scheme: RCHO + ⁻OH → rapid nucleophilic addition of ⁻OH to electron-deficient carbonyl carbon gives R-C(O⁻)(OH)H; loss of proton in basic solution gives R-C(O⁻)(O⁻)H (di-ion); slow hydride transfer to another RCHO gives RCOO⁻ + RCH₂O⁻ (carboxylate ion + alkoxide ion); H₃O⁺ gives RCOOH + RCH₂OH (carboxylic acid + alcohol).]

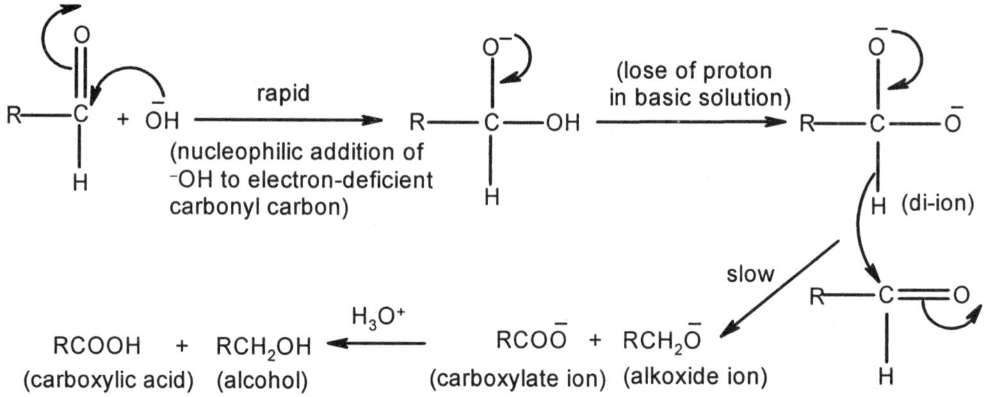

That the hydride ion is directly transferred from one molecule of the aldehyde to the other one and not from the medium was established from the experimental observation that the recovered alcohol contained no deuterium when the reaction was run in D_2O.[2]

Critical Views

a) Aldehydes with an α-hydrogen do not give the reaction, because when these compounds are treated with base the aldol reaction takes place much faster. An exception is observed in the case of cyclopentanecarboxyaldehyde.[3]

b) Crossed Cannizzaro reaction

Cannizzaro reaction taking place between two dissimilar aldehydes is termed as *crossed Cannizzaro reaction*. In practice, the reductant aldehyde (is oxidized to carboxylate ion) used is formaldehyde having high reductive reactivity.

$$Me_3CCHO + HCHO \xrightarrow[ii)\ H_3O^+]{i)\ \bar{O}H} Me_3CCH_2OH + HCOOH$$

$$C_6H_5CHO + HCHO \xrightarrow[ii)\ H_3O^+]{i)\ \bar{O}H} C_6H_5CH_2OH + HCOOH$$

furan-2-CHO + HCHO $\xrightarrow[ii)\ H_3O^+]{i)\ \bar{O}H}$ furan-2-CH$_2$OH + HCOOH

c) Intramolecular (internal) Cannizzaro reaction

1,2-Dialdehydes and α-ketoaldehydes on treatment with a strong base undergo *intramolecular* or *internal Cannizzaro reaction*.

$$\underset{(\alpha\text{-ketoaldehyde})}{R-\underset{\|}{\underset{O}{C}}-\underset{\underset{H}{|}}{\underset{\|}{\underset{O}{C}}}} + \bar{O}H \xrightarrow{rapid} R-\underset{\|}{\underset{O}{C}}-\underset{\underset{H}{|}}{\underset{|}{C}}-OH \xrightarrow[\text{(slow)}]{\text{intramolecular hydride ion transfer}}$$

$$\underset{(\alpha\text{-hydroxycarboxylic acid})}{RCH(OH)COOH} \xleftarrow{H_3O^+} R-\underset{\underset{H}{|}}{\underset{\underset{OH}{|}}{C}}-\underset{\|}{\underset{O}{C}}-\bar{O} \xleftarrow{\text{intramolecular proton transfer}} R-\underset{\underset{H}{|}}{\underset{\bar{O}}{\underset{|}{C}}}-\underset{\|}{\underset{O}{C}}-OH$$

$$\underset{(glyoxal)}{OHC-CHO} \xrightarrow[ii)\ H_3O^+]{i)\ \bar{O}H} HOH_2C-COOH$$

$$\underset{(phenyl\ glyoxal)}{C_6H_5-CO-CHO} \xrightarrow[ii)\ H_3O^+]{i)\ \bar{O}H} \underset{(mandelic\ acid)}{C_6H_5-CH(OH)COOH}$$

d) In case of substituted benzaldehydes rate of the *Cannizzaro reaction* depends upon the electronic nature of the substituents. It has been observed that electron-withdrawing groups enhance the reactivity of aldehydes — is to be expected since both the aldehyde molecules acquire some negative charge in the transition state, and may be substantiated from the fact that electron-withdrawing groups increase the electron-deficiency on the carbonyl carbon facilitating thereby the nucleophilic attack.

Conversely, electron-releasing substituents decrease the reaction rate and in extreme cases, the reaction may not occur as found in the case of *p*-dimethylaminobenzaldehyde.

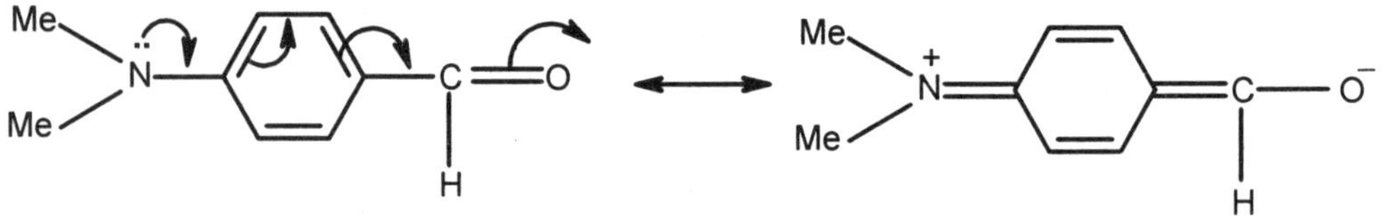

Applications

1) Various types of alcohols and acids can easily be synthesized using the classical, and internal and crossed *Cannizzaro reaction*.

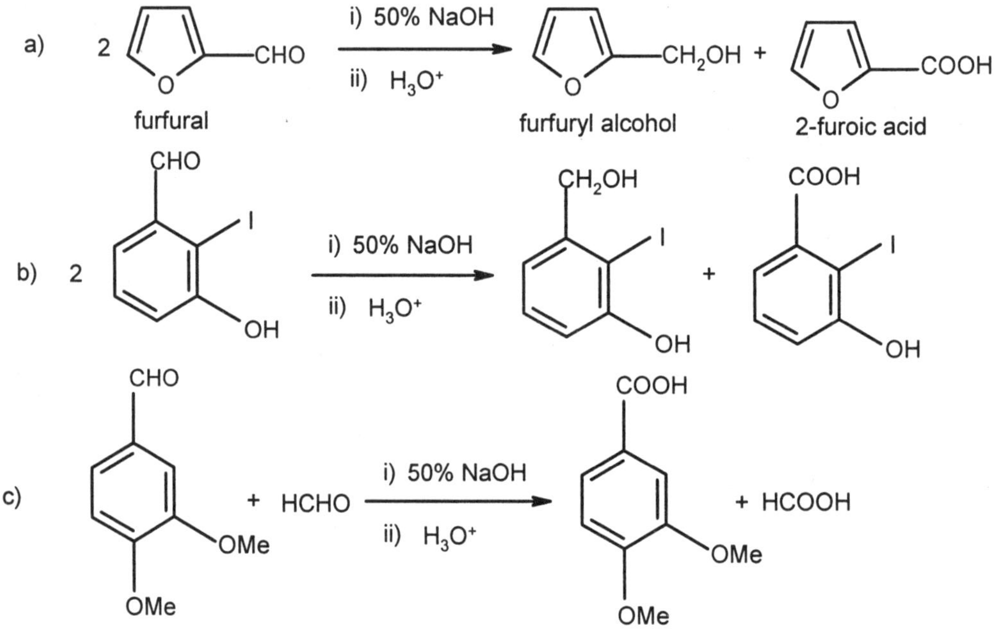

a) 2 furfural + CHO → i) 50% NaOH, ii) H_3O^+ → furfuryl alcohol (CH₂OH) + 2-furoic acid (COOH)

b) 2 (3-iodo-2-hydroxybenzaldehyde) → i) 50% NaOH, ii) H_3O^+ → (CH₂OH product) + (COOH product)

c) (3,4-dimethoxybenzaldehyde) + HCHO → i) 50% NaOH, ii) H_3O^+ → (COOH product with OMe groups) + HCOOH

d) [reaction: phthalaldehyde (benzene with two ortho CHO groups) → i) aq.NaOH/dioxane ii) H₃O⁺ → benzene with ortho CH₂OH and COOH]

e) [reaction: succinaldehyde (OHC–CH₂–CH₂–CHO) → (in presence of rhodium phosphine complex as catalyst) (ring closure) → γ-butyrolactone]   [ref. 4]

lactone

(note that the molecule bears α-hydrogen atoms)

2) *Crossed Cannizzaro reaction* finds application in the industrial manufacture of pentaerythritol. The method follows Tollens' condensation that comprises aldol condensation of acetaldehyde with three molecules of formaldehyde followed by *crossed Cannizzaro reaction* in the last step.

$$CH_3CHO + HCHO \xrightarrow[\text{aldol condensation}]{Ca(OH)_2} HOH_2C-CH_2CHO \xrightarrow[Ca(OH)_2]{2\,HCHO}$$

$$\underset{\text{(pentaerythritol)}}{\underset{HOH_2C}{\overset{HOH_2C}{|}}\!\!\!\!C\!\!\!\!\underset{|}{\overset{|}{}}\!\!\!\!CH_2OH} \xleftarrow[\text{(crosssed Cannizzaro reaction)}]{HCHO \;\; Ca(OH)_2} \underset{HOH_2C}{\underset{|}{\overset{HOH_2C}{\overset{|}{}}}} C-CHO$$

References

1. Cannizzaro, S. (1853), *Justus Liebigs Ann. Chem.*, **88**, 129.
2. Ashby, E. C., Coleman III, D. T. and Gamasa, M. P. (1983), *Tetrahedron Lett.*, 851; Ashby, E. C., Coleman III, D. T. and Gamasa, M. P. (1987), *J. Org. Chem.*, **52**, 4079; Swain, C. G., Powell, A. L., Sheppard, W. A. and Morgan, C. R. (1979), *J. Am. Chem. Soc.*, **101**, 3576; Watt, C. I. F. (1988), *Adv. Phys. Org. Chem.*, **24**, 57.

3 Fredenhagen, H. and Bonhoeffer, K. F. Z. (1938), *Phys. Chem., Abstr. A.*, **181**, 379; Hauser, C. R., Hamrica Jr., P. J. and Stewart, A. T. (1956), *J. Org. Chem.*, **21**, 260.
4 Bergens, S. H., Fairlie, D. P. and Bosnich, B. (1990), *Organometallics*, **9**, 566.
5 van der Maeden, F. P. B., Steinberg, H. and de Boer, T. J. (1972), *Recl. Trav. Chim. Pays-Bas*, **91**, 221.

- Geissman, T. A. (1944), *Org. React.*, **2**, 94.
- Franzen, V. (1955), *Ber*, **88**, 1361; **90**, 623(1957).
- Lachovicz D. R., Gritter R. J. (1963), *J. Org. Chem.*, **28**, 1061.
- Henderickson, J. B., Bogard, T. L. and Fish, M. E. (1970), *J. Am. Chem. Soc.*, **92**, 5538.
- Sen Gupta, A. K. (1968), *Tetrahedron Lett.*, 5205.
- Mehta, G. and Padma, S. (1991), *J. Org. Chem.*, **56**, 1298.
- Sheldon, *et al.* (1997), *J. Org. Chem.*, **62**, 3931.
- Thakuria, J. A., Baruah, M. and Sandhu, J. S. (1999), *Chem. Lett.*, 995.
- Russell, A. E., Miller, S. P. and Morken, J. P. (2000), *J. Org. Chem.*, **65**, 8381.
- Reddy, B. V. S., Srinvas, R., Yadav, J. S. and Ramalingam, T. (2002), *Synth. Commun.*, **32**, 1489.

18

Chichibabin Amination Reaction

Introduction

*C*hichibabin amination reaction[1] offers a method for direct amination of pyridines and related heterocycles using alkali metal amides. Substituted alkali metal amides (*e.g.* RNH^- & R_2N^-) have also been used. The attack takes place preferentially at the α-position unless both such positions are blocked, in which case the 4-position is attached. Aminated product is formed with the evolution of hydrogen.

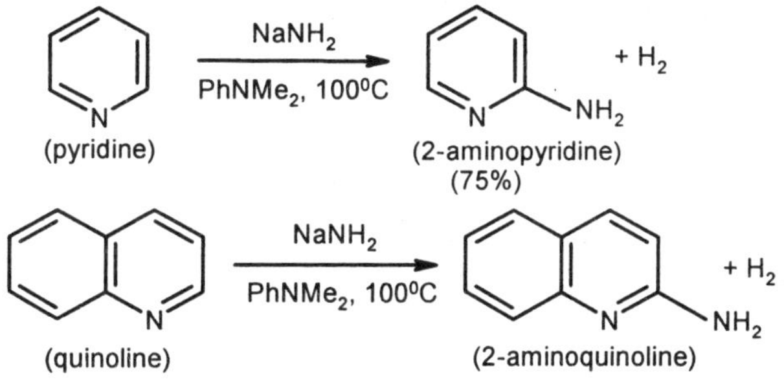

(pyridine) → (2-aminopyridine) (75%) + H_2 [NaNH$_2$, PhNMe$_2$, 100°C]

(quinoline) → (2-aminoquinoline) + H_2 [NaNH$_2$, PhNMe$_2$, 100°C]

Mechanism and Discussion

Regarding the mechanism of *Chichibabin amination reaction*, all the possible evidences are consistent with the following steps:

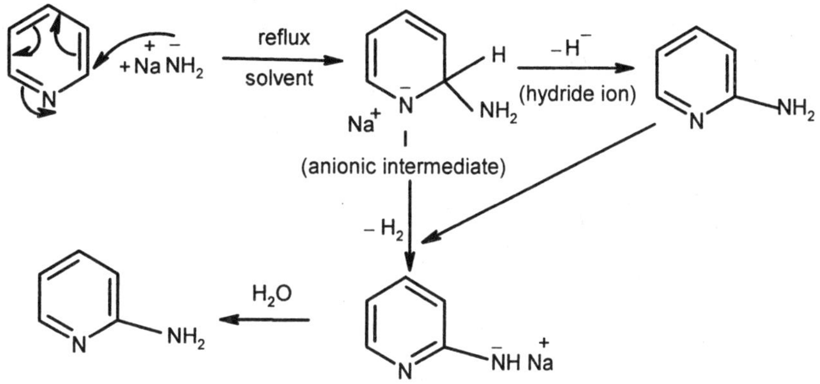

The reaction is initiated by the attack of nucleophile preferentially at C-2. In the second step a hydride ion (H⁻) is eliminated to form aminopyridine, which reacts with anionic intermediate (**I**) to evolve out hydrogen gas.

Evidence for anionic intermediate of type I:
(i) Existence of such type of intermediate (from quinoline) has been established by means of NMR spectroscopy.[2]

(ii) Another probable mechanism involving *pyridyne* type intermediate has been rejected because 3-ethylpyridine yielded only 2-amino-3-ethylpridine[3] and that deuterium is not lost from pyridine-3-d as required for the formation of a *pyridyne*.[4]

It is assumed that preferential attack at C-2 is a consequence of *intramolecular* delivery of the nucleophile, perhaps guided by complexation of ring nitrogen with metal cation. It has been observed that amination of 3-alkylpyridine is *regioselective* for the 2-position.[5]

For amination of 2- or 4-alkylpyridines more vigorous conditions are required, because proton abstraction from the side-chain by the amide occurs first and ring attack must therefore involve a dianionic intermediate.[6] Heteroaromatics having a methyl group bonded at an angular carbon also show diminished reactivity towards this reaction.[7] Nitro-compounds are very much reluctant to undergo *Chichibabin reaction* under the normal conditions of this reaction.[2]

Applications

Synthesis of 2-[3-aminopropyl]-5,6,7,8-tetrahydronaphthyridine has been carried out *via* a one-pot double *Suzuki reaction* and a highly *regioselective intramolecular Chichibabin cyclization*.[8]

References

1. Chichibabin, A. E. and Zeide, O. A. (1914), *J. Russ. Phys. Chem. Soc.*, **46**, 1216.
2. Levitt, L. S. and Levitt, B. W. (1975), *Chem. Ind.* (London), 520.
3. Ban, Y. and Wakamatsu, T. (1964), *Chem. Ind.*, 710.
4. Abramovitch, R. A. *et al.* (1965), *Can. J. Chem.*, **43**, 725; Abramovitch, R. A. and Boutton, G. A. (1967), *Chem. Commun.*, 274.
5. Abramovitch, R. A., Helmer, F. and Saha, J. G. (1964), *Chem. Ind.*, 659.
6. Viscardi, G., Savarino, P., Quagliotta, P., Barni, E. and Bottam, M. (1996), *J. Heterocycl. Chem.*, **33**, 1195.
7. Bergstrom, F. W. (1931), *J. Am Chem. Soc.*, **53**, 3027, 4065; *J. Org. Chem.*, **3**, 233(1938).
8. Palucki, M., Hughes, D. L., Yasuda, N., Yang, C. and Reider, P. J. (2001), *Tetrahedron Lett.*, **42**, 6811(2001).

- Reviews: Vorbruggen, H. (1990), *Adv. Heterocycl. Chem.* **49**, 117; McGill, C. K. and Rappa, A. (1988), *ibid*, **44**,1; Pozharskii, A. F., Simonov, A. M. and doro'kin, V. N. (1978), *Russ. Chem. Rev.*, **47**, 1042; Leffler, M. T. (1942), *Org. Reactions*, **1**, 91.
- Kelly, T. R., Bridger, G. J. and Zhao, C. (1990), *J. Am. Chem. Soc.*, **112**, 8024.
- Tanga, M. J., Bupp, J. E. and Tochimoto, T. K. (1994), *J. Heterocycl. Chem.*, **31**, 1641.
- Seko, S. and Miyake, K. (1998), *Chem. Commun.*, 1519.
- Katritzky, A. R., Qiu, G., Long, Q.-H., He, H.-Y. and Steel, P. J. (2000), *J. Org. Chem.*, **65**, 9201.

19

Claisen Condensation

Introduction

When carboxylic esters containing an α–hydrogen are treated with a strong base (such as sodium ethoxide or sodium triphenylmethide), two molecules of the substrate undergo self-condensation to give β-keto ester and an alcohol. This aldol-type condensation reaction is known as *Claisen condensation* or *Claisen ester condensation*.[1]

$$2RCH_2COOR' \xrightarrow[\text{2. }H_3O^+]{\text{1. NaOEt /EtOH}} RCH_2COCH(R)COOR' + R'OH$$

β-keto ester
(one ester molecule is acylated at the carbon alpha to its carbonyl group)

The reaction is usually carried out by means of sodium ethoxide in ethanol solution. In the reaction course, an enolate is first formed from an ester molecule, which subsequently undergoes nucleophilic addition to the carbonyl group of another ester molecule to yield enolate of β-keto ester. Acidification in the final step furnishes the isolable neutral form of the product. Thus, one molecule of the ester gives rise to an enolate, while the second one acts as an acylating agent. The classic example of *Claisen condensation* is the condensation of two molecules of ethyl acetate to give synthetically useful ethyl acetoacetate (or known as acetoacetic ester).

$$2CH_3COOC_2H_5 \xrightarrow[(-C_2H_5OH)]{\text{NaOEt /EtOH}} CH_3-\underset{\underset{Na^+}{\bar{O}}}{C}=CH-\underset{\underset{}{\overset{O}{\|}}}{C}-OEt \xrightarrow{\substack{\text{AcOH}\\\text{(acidification)}}} CH_3COCH_2COOEt$$

(ethylacetoacetate)

Mechanism

The well-accepted mechanism for *Claisen Condensation* is given as:

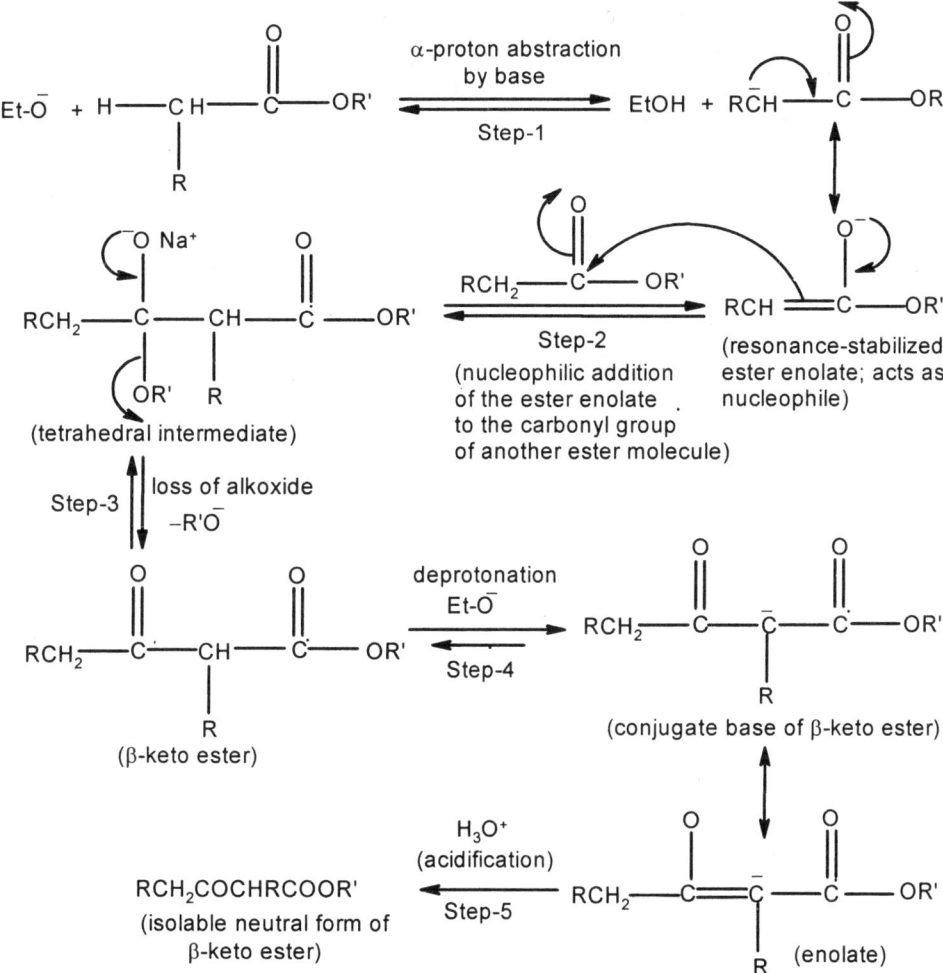

Critical Views

a) From the scrutiny of the reaction mechanistic path it is revealed that to get a good yield of the β-keto ester, the product must have to be converted to its enolate in the basic reaction mixture (Step-4). If the β-keto ester is not converted completely to its enolate, it is attacked at the ketonic functional group by an alkoxide leading to its decomposition. Accordingly, for the reaction to take place presence of at least one α-hydrogen is necessary in the product. At the equilibrium, condition of the last stage of the reaction (*i.e.* Step 4) the enolate anion of β-keto ester predominates and it is converted to the desired product (Step 5) by adding an organic acid, such as acetic acid. Hence, at least two protons must be present at the α-carbon of an ester in order for the equilibrium to favour product formation — *i.e.* it may be concluded that *Claisen condensation* is possible for esters of the type RCH_2COOR', but not for $R_2CHCOOR'$ under the usual conditions of the reaction. Thus, ethyl 2-methylpropanoate cannot undergo *Claisen condensation* under the customary conditions of the reaction —

$$2Me_2CH-\overset{O}{\underset{}{C}}-OEt \xrightleftharpoons{EtONa} Me_2CH-\overset{O}{\underset{}{C}}-\underset{Me}{\overset{Me}{C}}-\overset{O}{\underset{}{C}}-OEt + EtOH$$

(ethyl 2-methylpropanoate)

(ethyl 2,2,4-trimethyl-3-oxo-pentanoate)
(can't form a stable anion; formed in no more than trace amounts.)

But the reaction in such cases can be made to proceed if an excess of very strong base, such as sodium triphenylmethide [$Ph_3C^-Na^+$], lithium di-isopropylamide [$(Me_2CH)_2N^-Li^+$].[2]

$$2Me_2CH-\overset{O}{\underset{}{C}}-OEt \xrightarrow[\text{(abstraction of acidic α-hydrogen irreversibly)}]{Ph_3\bar{C}Na^+} Me_2C=\overset{\bar{O}}{\underset{}{C}}-OEt$$

(ethyl 2-methylpropanoate) (enolate)

$$\downarrow$$

$$Ph_3CH + Et\bar{O} + Me_2CH-\overset{O}{\underset{}{C}}-\underset{Me}{\overset{Me}{C}}-\overset{O}{\underset{}{C}}-OEt$$

(ethyl 2,2,4-trimethyl-3-oxo-pentanoate)

It is supposed that the strong base (sodium triphenylmethide) removes all the ethanol formed (as ethoxide) during the reaction course, directing the reaction to the right; sodium ethoxide is not able to do so.

b) Crossed-Claisen condensation: When *Claisen condensation* reaction is carried out with two different ester molecules, both having α-hydrogens, a mixture of all four products is generally obtained; the reaction finds less synthetic importance. However, if only one of them bears an α-hydrogen, the *crossed-Claisen reaction* becomes synthetically satisfactory. Among esters lacking α-hydrogen that are commonly used are esters of aromatic acids, formic acid or oxalic acid.

Ph—COOEt + RCH$_2$COOR' $\xrightarrow{\text{1. NaOEt} \quad \text{2. H}_3\text{O}^+}$ Ph—CO—CH(R)COOR'

EtO—CO—OEt + RCH$_2$COOR' $\xrightarrow{\text{1. NaOEt} \quad \text{2. H}_3\text{O}^+}$ HO—CO—CH(R)COOR'
(ethyl carbonate)   (malonic ester derivative)

H—CO—OEt + RCH$_2$COOR' $\xrightarrow{\text{1. NaOEt} \quad \text{2. H}_3\text{O}^+}$ H—CO—CH(R)COOR'
(ethyl formate)

Similarly,

Me$_2$CHCOOCMe$_3$ $\xrightarrow{\text{(Me}_2\text{CH)}_2\bar{\text{N}} \text{Li}^+}$ Me$_2$C=C(OLi$^+$)OCMe$_3$ $\xrightarrow{\text{PhCOCl}}$ PhCOC(Me)$_2$COOMe
(*t*-butyl 2-methyl-propanoate)   (enolate)   (78%)

c) Intramolecular Claisen condensation: Dieckmann condensation
Intramolecular Claisen condensation involves cyclization of an ester molecule (*via* enolate anion) bearing two ester functions within it. The *intramolecular condensation* leads to the formation of cyclic β-keto ester and this special category of *Claisen condensation* is termed as *Dieckmann condensation*.[3]

(CH$_2$)$_n$(CH$_2$COOR)(COOR) $\xrightarrow[\text{(NaOEt or Na or KH)}]{\text{base}}$ (CH$_2$)$_n$(CHCOOR)(C=O) [cyclic]

The *Dieckmann condensation* is most successful for the formation of 5-, 6-, and 7-membered ring.

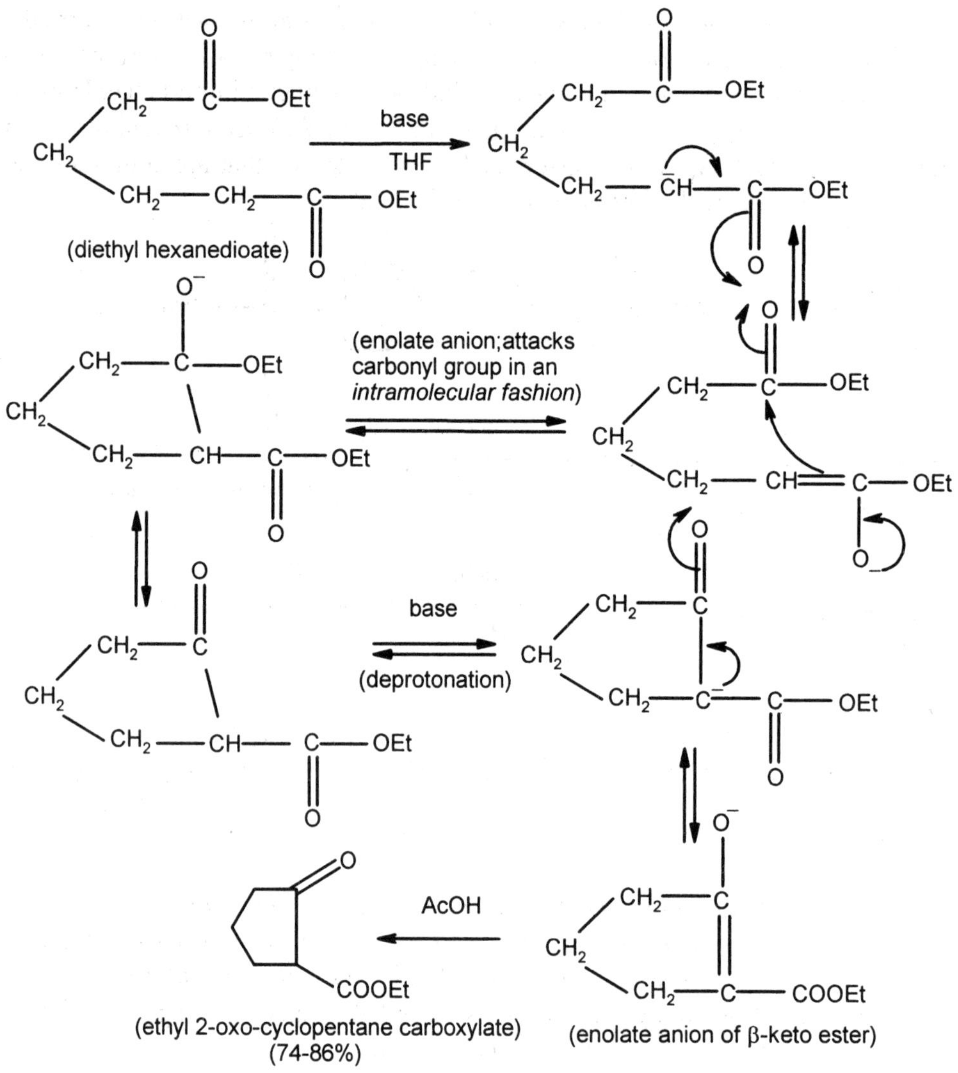

O
CH2 C OEt
CH2
CH2 CH2 C OEt
O
(diethyl hexanedioate)
base
THF
(enolate anion;attacks carbonyl group in an intramolecular fashion)
base
(deprotonation)
AcOH
COOEt
(ethyl 2-oxo-cyclopentane carboxylate)
(74-86%)
(enolate anion of β-keto ester)

Similarly,

COOEt
COOEt
CH2CH2COOEt
i) NaH
ii) H3O+
EtOOC
O
COOEt
(21%)

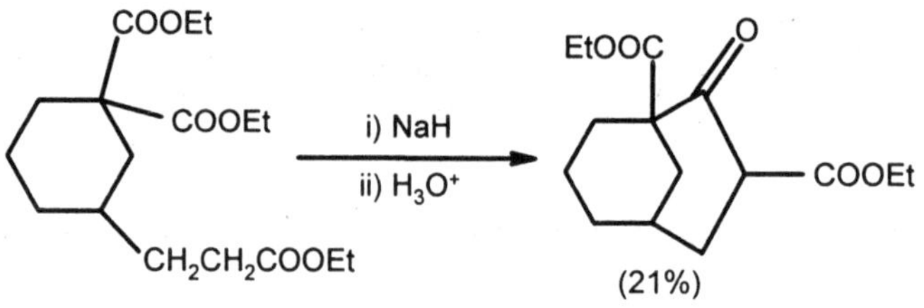

Applications

1) *Claisen condensation* is a very useful reaction and is the key synthetic step of many significant organic molecules. By applying this reaction, synthetically important compounds, such as ethyl acetoacetate (EAA), malonic ester derivatives can be produced, which are the precursor of a variety of organic substances. A few useful applications of *Claisen condensation* are mentioned here:

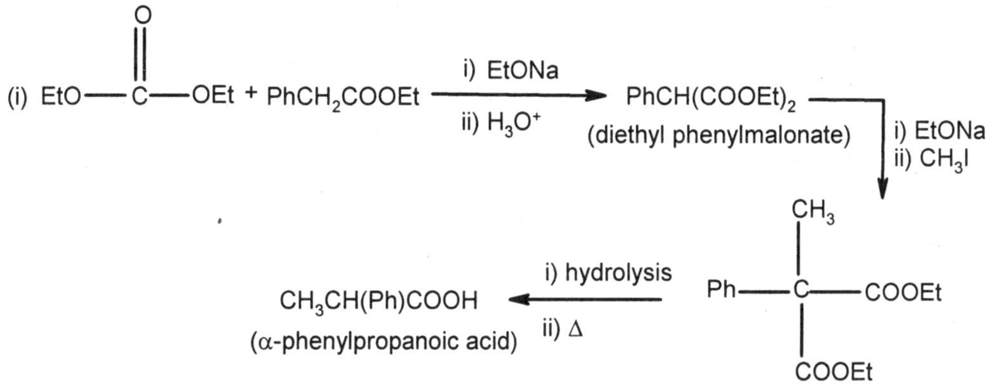

(i) EtO—C(=O)—OEt + PhCH$_2$COOEt $\xrightarrow{\text{i) EtONa} \quad \text{ii) H}_3\text{O}^+}$ PhCH(COOEt)$_2$ (diethyl phenylmalonate) $\xrightarrow{\text{i) EtONa} \quad \text{ii) CH}_3\text{I}}$ Ph—C(CH$_3$)(COOEt)(COOEt) $\xrightarrow{\text{i) hydrolysis} \quad \text{ii) }\Delta}$ CH$_3$CH(Ph)COOH (α-phenylpropanoic acid)

2) Enolate anions of ketones produced in strong bases are made to undergo condensation reaction (with ester molecules) that resembles *Claisen condensation*.

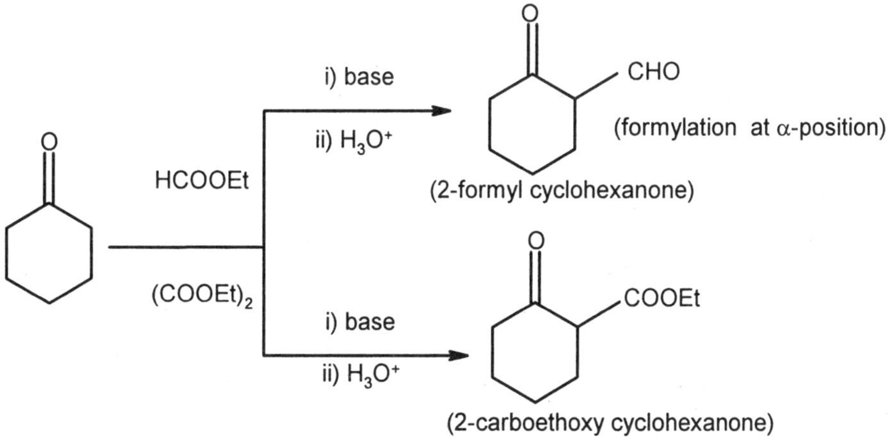

Cyclohexanone + HCOOEt $\xrightarrow{\text{i) base} \quad \text{ii) H}_3\text{O}^+}$ 2-formyl cyclohexanone (formylation at α-position)

Cyclohexanone + (COOEt)$_2$ $\xrightarrow{\text{i) base} \quad \text{ii) H}_3\text{O}^+}$ 2-carboethoxy cyclohexanone

3) *Claisen condensation* is employed as one of the synthetic steps for the preparation of various natural products of interest, such as flavones, isoflavones, thiamine (Vit. B$_1$), nicotine (alkaloid), camphoric acid (terpenoids), *etc.* —

References

1. Claisen, L. and Lowman, O. (1887), *Ber. Dtsch. Chem. Ges.*, **20**, 651; Hauser, C. R. and Hudson, B. E. (1942), *Org. React.*, **1**, 266. [Review]
2. For a discussion, see Garst, J. I. (1979), *J. Chem. Educ.*, **56**, 721.
3. Dieckmann, W. (1894), *Ber. Dtsch. Ges.*, **27**, 102; Schaefer, J. P. and Bloomfield, J. (1967), *J. Organomet. Chem.*, **15**, 1. [Review]
- Ziegler, F. E. (1988), *Chem. Rev.*, **88**, 1423. [Review]
- Tanabe, Y. (1989), *Bull. Chem. Soc. Jpn.*, **62**, 1917.
- Corey, E. J. and Lee, D.-H. (1991), *J. Am. Chem. Soc.*, **113**, 4026.
- Kashima, C., Takahashi, K. and Fukusaka, K. (1995), *J. Heterocycl. Chem.*, **32**, 1775.
- Yoshida, Y., Hayashi, R., Sumihara, H. and Tanabe, Y. (1997), *Tetrahedron Lett.*, **38**, 8727.
- Toda, F., Suzuki, T. and Higa, S. (1998), *J. Chem. Soc., Perkin Trans.*, *1*, 3521.
- Shinoda, M., Sato, Y. and Shishido, K. (1999), *J. Am. Chem. Soc.,* **121**, 6507.
- Tanabe, Y., Hamasaki, R. and Funakoshi, S. (2001), *Chem. Commun.*, 1674.
- Mogilaiah, K. and Kankaiah, G. (2002), *Indian J. Chem, Sect. B*, **41B**, 2194.
- Deville, J. P. and Behar, V. (2002), *Org. Lett.*, **4**, 1403.
- Ho, J. Z., Mohareb, R. M., Ahn, J. H., Sim, T. B. and Rapoport, H. (2003), *J. Org. Chem.*, **68**, 109.
- Mogilaiah, K. and Reddy, N. V. (2003), *Synth. Commun.*, **33**, 73.
- Nubbemeyer, U. (2003), *Synthesis*, 961.

20

Clemmensen Reduction

Introduction

The reduction of carbonyl function of aldehydes and ketones to methylene group with amalgated zinc and concentrated hydrochloric acid is known as *Clemmensen reduction*.[1] This classical method of reduction is still very popular.

$$\underset{R'}{\overset{R}{>}}C=O \xrightarrow[\text{reflux}]{\text{Zn - Hg, HCl}} \underset{R'}{\overset{R}{>}}CH_2$$

The reaction is carried out by refluxing the carbonyl compounds with amalgated zinc and excess of concentrated hydrocholic acid, sometimes in presence of a non-miscible solvent that serves to keep the carbonyl concentration in the aqueous phase low thereby preventing bimolecular condensations at the metal surface. The reduction is useful especially for ketones containing phenolic or carboxylic groups, which remain unaffected. Such reduction is also observed in *Wolff-Kishner reduction*, but *Clemmensen reduction* is easier to perform.

$$CH_3(CH_2)_5CHO \xrightarrow[\text{reflux}]{\text{Zn - Hg, HCl}} CH_3(CH_2)_5CH_3$$
(*n*-heptanal) (*n*-heptane)

$$C_6H_5COCH_3 \xrightarrow[\text{reflux}]{\text{Zn - Hg, HCl}} C_6H_5CH_2CH_3$$
(acetophenone) (ethylbenzene)

Mechanism

The mechanism is still unclear. The actual reduction occurs through a complex mechanism on the surface of zinc. Nakabayashi (1960) proposed that the reaction pathway might involve transfer of electrons from the metal surface to the carbonyl carbon atom.[2]

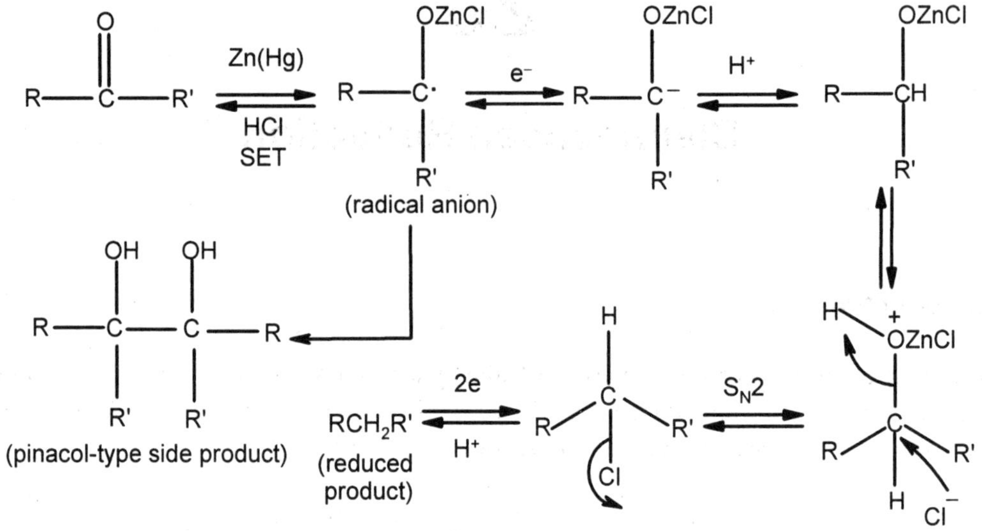

Critical Views

a) Since *Clemmensen reduction* occurs under vigorous conditions, it is not suitable for the reduction of polyfunctional molecules, and at the same time is not recommended for acid-sensitive and also for the substrates of high molecular masses.[3]

However, in such cases a modified method of the reduction has been suggested where a solution of carbonyl compounds in acetic anhydride or ether is treated with powdered zinc and hydrogen chloride gas.[4]

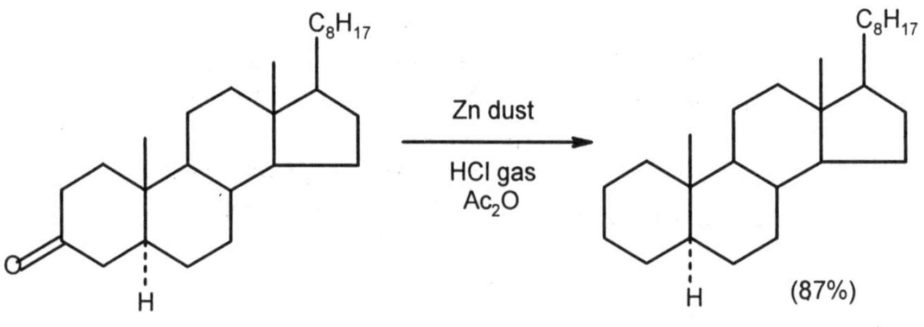

The α,β-unsaturated ketones undergo reduction of both the olefinic and carbonyl groups or if only one group is reduced it is the C=C bond.[5] However, Banerjee et al.[6] reported the reduction of carbonyl group rather than that of conjugated double bond in their experimental observation.

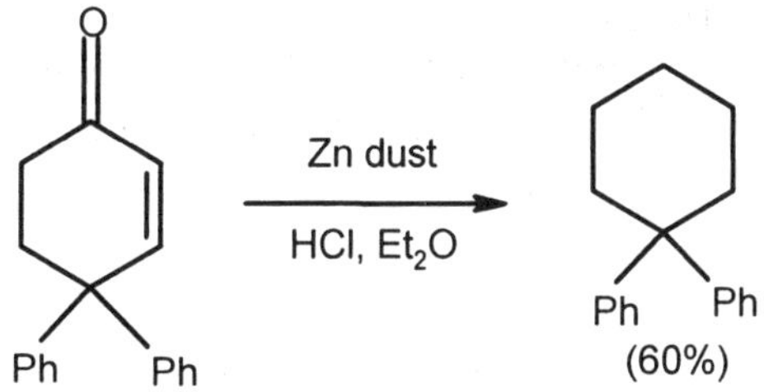

Nevertheless, for stable compounds, it is an effective method of reduction and the reduction is specific for carbonyl groups of aldehydes and ketones containing many other reducible functionalities.

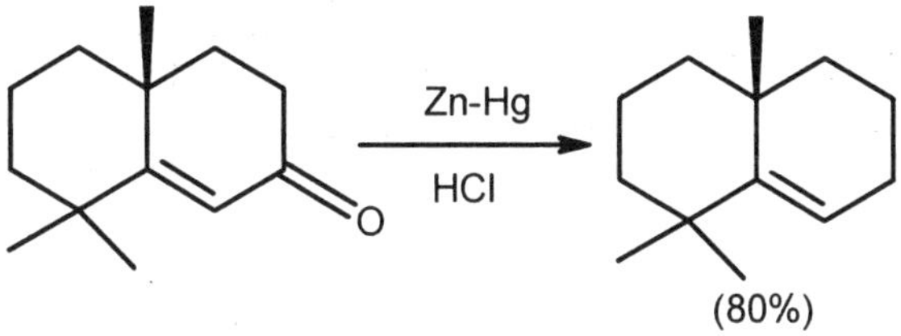

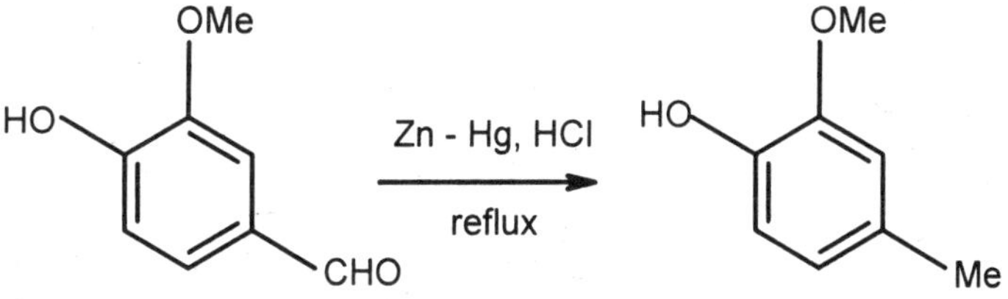

Steric hindrance surrounding the carbonyl function affects the reaction so much as observed in case of the following hindered ketone that would not undergo such reduction under customary condition of the reaction.

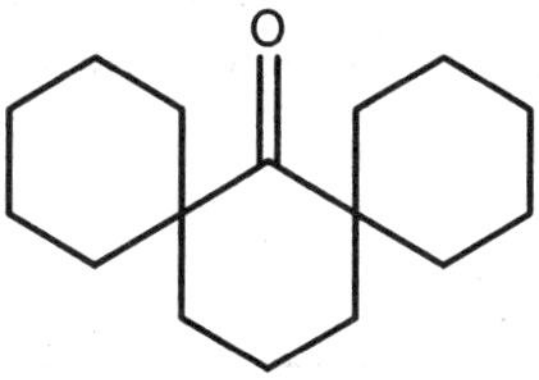

b) α–Hydroxy ketones on *Clemmensen reduction* yield ketones through hydrogenolysis of –OH group or olefins, while 1,3-diketones usually undergo rearrangement forming rearranged mono-ketones.[7]

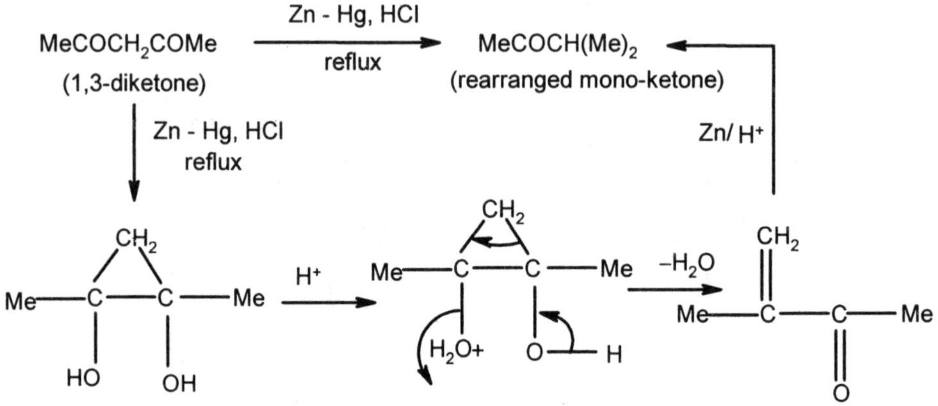

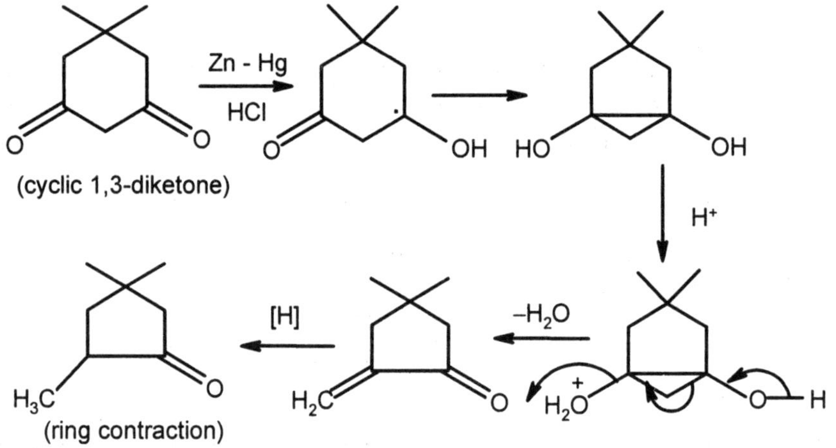

c) Deuterium-labelling of a methylene group is possible through *Clemmensen condition* using DCl / *O*-deuterated solvents (like D$_2$O, CH$_3$COOD, etc.), and this fact demonstrates that both the hydrogen atoms involved in the reductions are derived from the proton donors, rather than hydrogen-atoms donors.[8]

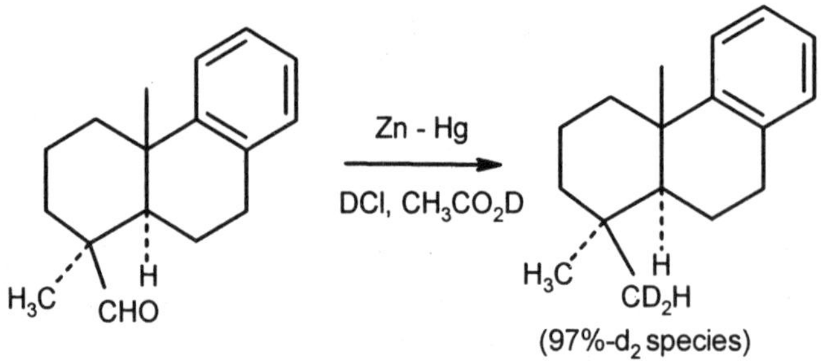

d) The reaction is commonly run as a three-phase mixture of toluene, aqueous hydrochloric acid and zinc amalgam. Since under this condition the aqueous phase contains a low concentration of the ketone and its conjugate acid, as a result of which possibility of bimolecular reduction is minimized. Reduction of the carbonyl compound occurs in the aqueous phase.

Another striking point is that zinc amalgam is used rather than zinc metal in the reaction scheme; this is done to minimize the side reaction leading to the formation of molecular hydrogen from the reaction of zinc with the aqueous acid. Amalgation of zinc raises its hydrogen-overvoltage to the point where it survives as a reducing agent in the acid solution and is not consumed in the reaction with acid to give molecular hydrogen. The choice of acid is confined to the hydrogen halides, which appear to be the only strong acids whose anions are not reduced by zinc amalgam.

Applications

Clemmensen reduction bears much implication in synthetic organic chemistry. Still it is very popular among the chemists for carrying out reductions of a variety of organic substrates possessing other many groups. A few cases are illustrated:

1) cyclohexanone $\xrightarrow{\text{Zn - Hg, HCl, H}_2\text{O}}$ cyclohexane

2) naphthalene-COC_7H_{15} $\xrightarrow{\text{Zn - Hg, HCl, H}_2\text{O}}$ naphthalene-C_8H_{17}

3) pyrole $\xrightarrow{\text{Zn - Hg, HCl, H}_2\text{O}}$ (2,5-dihydro derivative) + (tetrahydro derivative)

4) salicylaldehyde $\xrightarrow{\text{Zn - Hg, HCl, H}_2\text{O}}$ o-cresol (CH_3)

5) Ph-COCH$_2$CH$_2$COOH $\xrightarrow[\text{toluene, reflux} \atop \text{20-30 h}]{\text{Zn - Hg, HCl}}$ Ph-CH$_2$CH$_2$CH$_2$COOH

(4-oxo-4-phenyl butanoic acid) → (4-phenyl butanoic acid) (82 - 89%)

6) Ph-CH$_2$-C(=S)-SCH$_3$ $\xrightarrow{\text{Zn - Hg, HCl, H}_2\text{O}}$ Ph-CH$_2$-CH$_2$-SCH$_3$ (54%)

References

1. Clemensen, E. (1913), *Ber. Dtsch. Chem. Ges.*, **46**, 1837; Martin, E. L. (1942), *Org. Reactions*, **1**, 155.
2. Nakabayashi, T. (1960), *J. Am. Chem. Soc.*, **82**, 3900, 3906, 3909.
3. Buchanan, J. G. St. C. and Woodgate, P. D. (1969), *Quart. Rev. Chem. Soc. London*, **23**, 522.
4. Yamamura, S. and Hirata, Y. (1968), *J. Chem. Soc., C*, 2887; Toda, M., Hirata, Y. and Yamamura, S. (1969), *Chem. Commun.*, No. 16, 919; Toda, M., Hayashi, M., Hirata, Y. and Yamamura, S. (1972), *Bull. Chem. Soc. Jpn.*, **45**, 264.
5. Vedejs, E. (1975), *Org. React.*, **22**, 401.
6. Banerjee, A. K., Alvarez, J., Santan, M. and Carrasco, M. C. (1986), *Tetrahedron*, **42**, 6615.
7. Cusack, N. J. and Davis, B. R. (1965), *J. Org. Chem.*, **30**, 2062.
8. Enzell, C. R. (1966), *Tetrahedron Lett.*, No. 12, 1285.
- Brewster, J. H. (1954), *J. Am. Chem. Soc.*, **76**, 6364.
- Martin, E. L. (1943), *Org. Syn.*, **Coll. Vol. 2**, 499.
- Staschewski, D. (1959), *Angew. Chem.*, **71**, 726.
- Borden, W. T. and Ravindranathan, T. (1975), *J. Org. Chem.*, **36**, 4125.
- Burdon, J. and Price, R. C. (1986), *J. Chem. Soc., Chem. Commun.*, 893.
- Talapatra, S. K., Chakrabati, S., Mallik, A. and Talapatra, B. (1990), *Tetrahedron*, **46**, 6047.
- DiVona, M. L. and Rosnati, V. (1991), *J. Org. Chem.*, **56**, 4269.
- Luchetti, L. and Rosnati, V. (1991), *J. Org. Chem.*, **56**, 6836.

21

Cope Rearrangement

Introduction

Thermal isomerization of 1,5-dienes through [3,3]-sigmatropic changes leading to regioisomeric 1,5-dienes is known as *Cope rearrangement* [1].

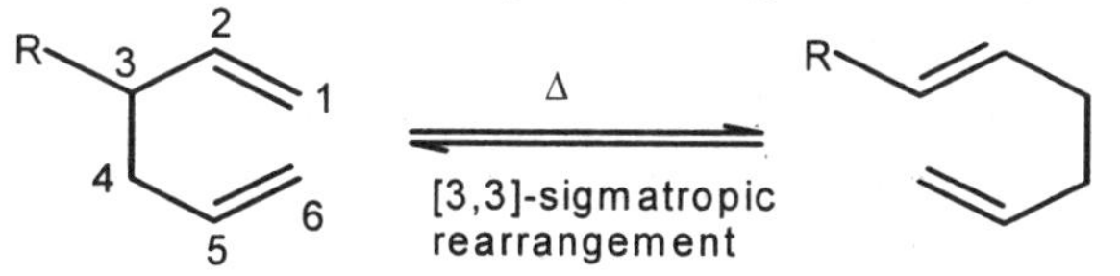

When the diene is symmetrical about the C_3-C_4 bond, the starting material and the product are the same. So, a *Cope rearrangement* can be detected only when the diene becomes unsymmetrical about this bond. The temperature required to bring about the reaction depends on the substituents — a group (if R=acyl, phenyl etc.) on C-3 or C-4 with which the new double bond in the product diene can conjugate lowers the energy of the transition state and the rearrangement occurs at relatively lower temperature (165-85⁰). The reaction is usually reversible and produces an equilibrium mixture of two 1,5-dienes of which the thermodynamically more stable isomer predominates.

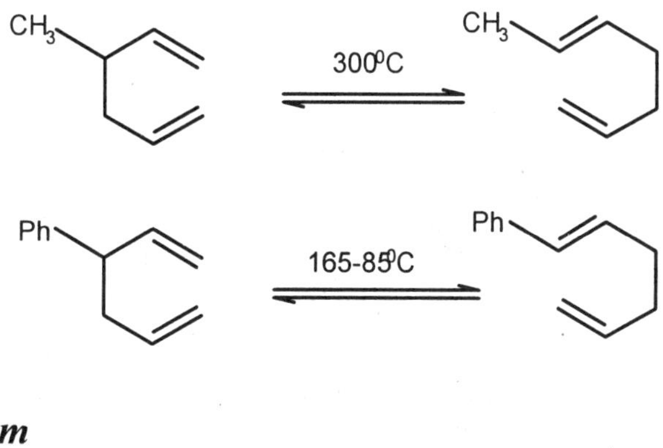

[Ref. 2]

Mechanism

Cope rearrangement is a concerted-step intramolecular process involving a six-membered cyclic transition state (**2**) [3]. This mechanistic approach is supported by the facts that the reaction is of large negative entropy of activation and of high stereoselectivity, and also relatively insensitive toward substituent and solvent effects.

(1) (2) (3)

(six-membered cyclic transition state)

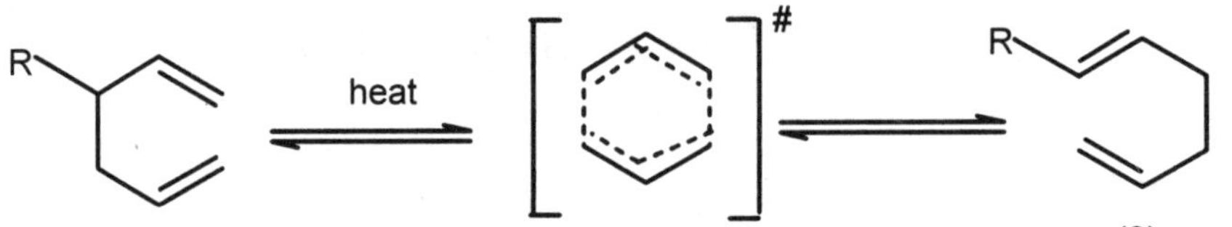

In acyclic 1,5-dienes, the six-membered transition state prefers a chair to a boat conformation[4]. This generalization can be evidenced from the stereospecific rearrangement of *meso*-3,4-dimethyl-1,5-hexadiene (**4**) to *cis* , *trans*- 2,4-octadiene (**5**) that is consistent only with a chair conformation for the transition state, because alternate boat conformation would give rise to the *cis*, *cis* or *trans*, *trans*-2,4-octadiene (which is not the situation)[5].

(**4**) *meso*-isomer

(**5**) H *cis; trans* -isomer

(**4**) *meso*-isomer

(six-membered) (chair conformation)

(**5**) *cis, trans* -isomer

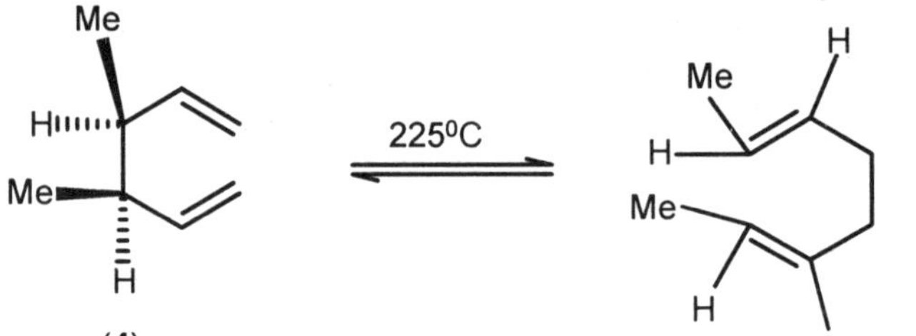

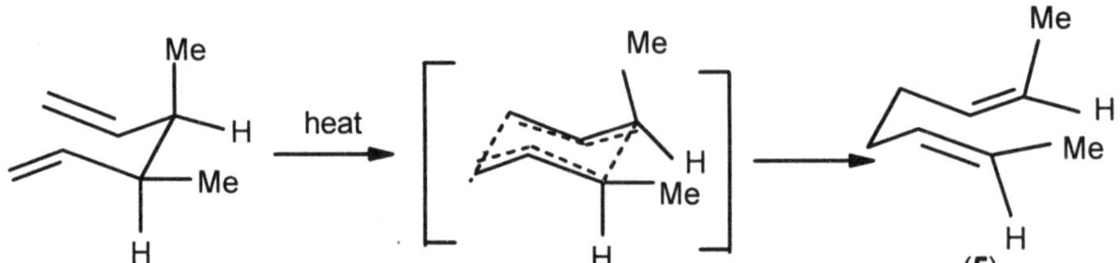

Prefernce for the chair transition state is a consquence of orbital-symmetry relationships[6].

Critical Views

(a) *Oxy-Cope rearrangement*

Rearrangement reaction of 1,5-dienes having a 3-hydroxy substituent (as in **6**) is known as *Oxy-Cope rearrangement*[7]. The reaction is generally not reversible; the rearranged product, an enol (7), rapidly tautomerizes to the corresponding carbonyl compound (**8**). Besides, *oxy-cope rearrangement* is accompanied by a retro-ene reaction (side reaction) so that the products of both reactions are normally isolated.

(**6**) (enol intermediate) (**7**) (carbonyl compound) (**8**)

(retro-ene products)

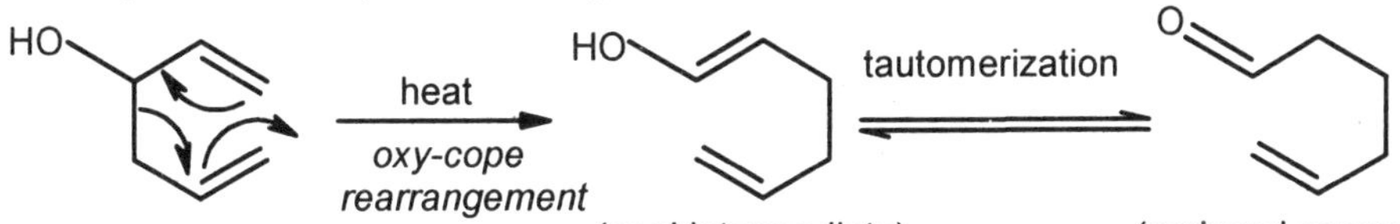

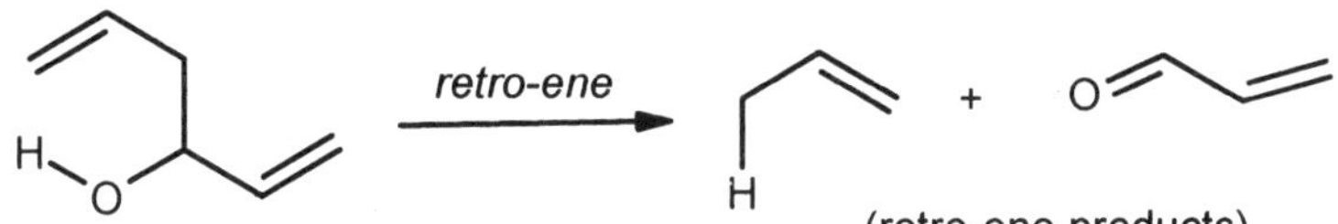

The rate of *oxy-cope rearrangement* is greatly accelerated by factors of 10^{10}-10^{17} when an alkoxide is used instead of the traditional alcohol[8], and to make the alkoxide the starting alcohol is usually treated with base (potassium hydride and potassium hexamethyldisilazide are commonly used bases). The product is then the potassium enolate, which is more stable than the simple potassium alkoxide (starting material), and as the reaction proceeds, conjugatation is growing between $O^{(-)}$ and the new π bond. This may be termed as anionic *oxy-cope rearrangement*[9].

(**6**) (alkoxide)

(product) (**9**)

(enolate) resonance stabilised

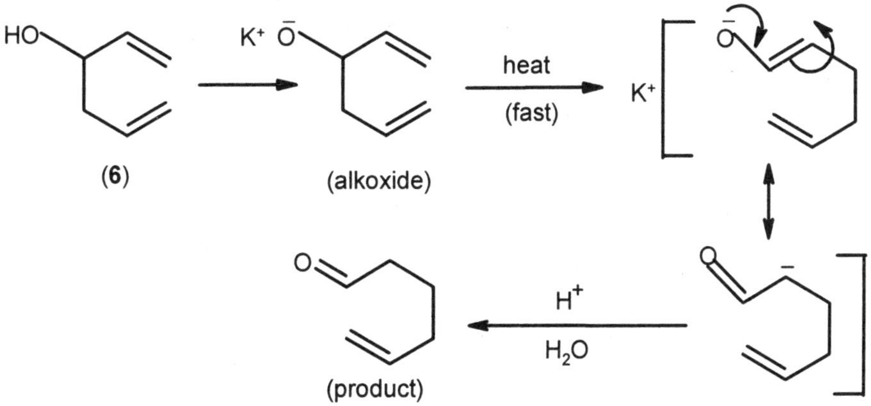

Hence, reactions are being carried out at reduced temperature making the methodology more versatile and minimizing the thermal retro-ene side reaction. Other examples of *anionic oxy-cope rearrangement*[9]:

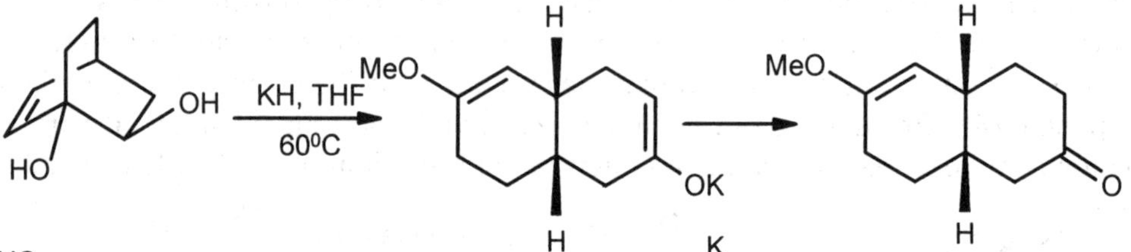

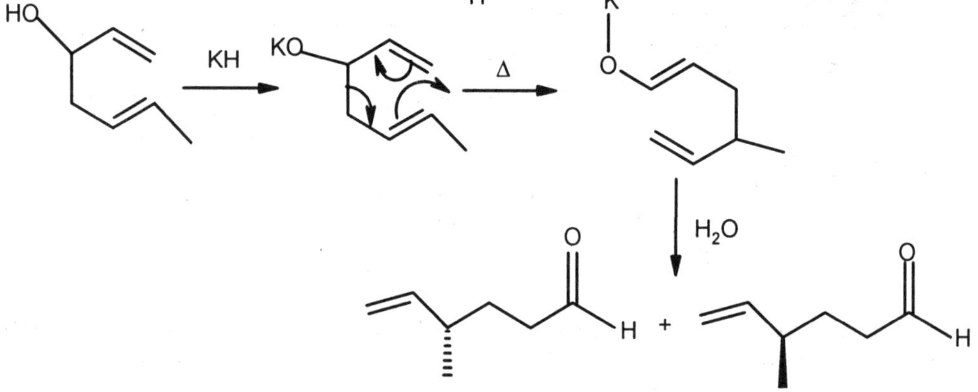

(S/R=64:36)

(b) There is a number of instances where the substrate molecule is so modified by substituents as to make alternate stepwise mechanisms for *Cope rearrangement* possible. The alternatives may be: (i) dissociation into a pair of allylic radicals or into a cation – anion pair, followed by recombination, and (ii) formation of the new σ-bond before the first has started to cleave, leading to a diradical or zwitterionic intermediate. For example — a diradical mechanism for the *Cope rearragnement* of 2,5-diphenyl-1,5-hexadiene (**10**) may be proposed (the phenyl groups can stabilize the radical centres), although there is no evidence for discrete diradical intermediate[10].

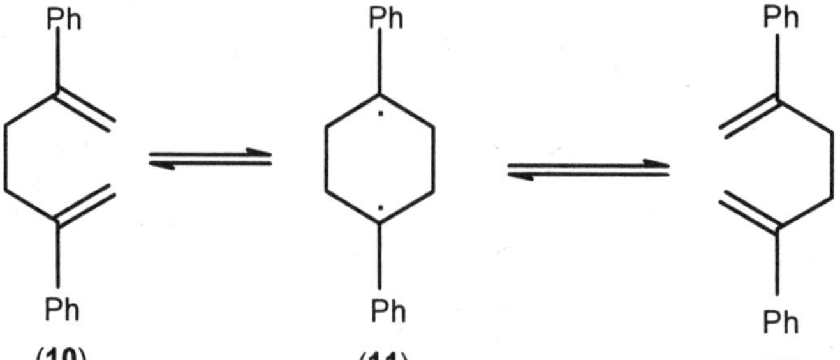

Again, the *Cope rearrangement* of (**13**) in benzene at 80°C involves a zwitterionic intermediate that can be trapped with benzaldehyde[11].

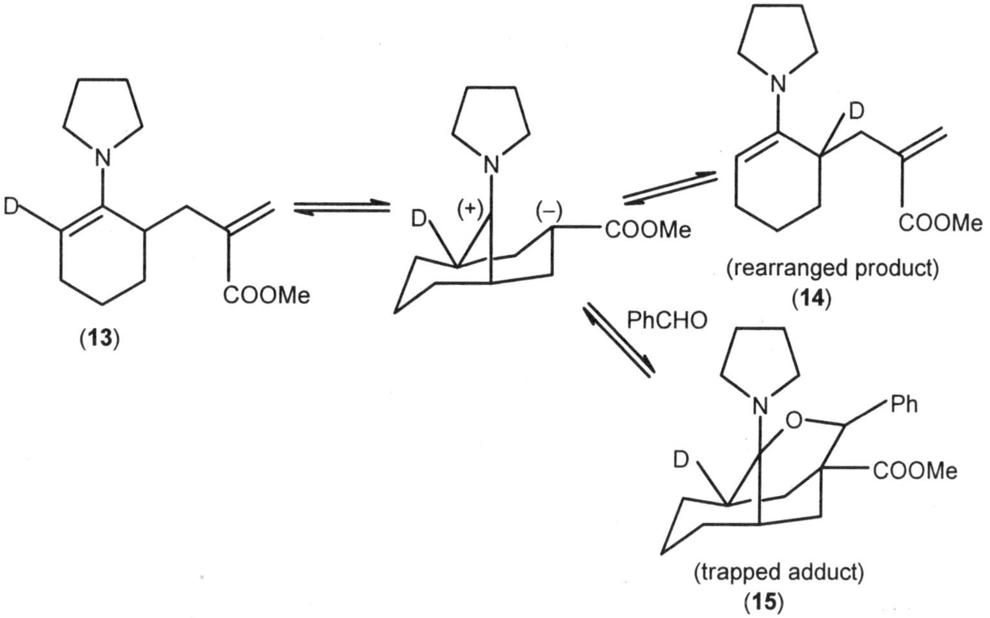

(c) There are several systems that undergo *Cope rearrangement*, but the products bear the same structure as the starting materials; this type of 'degenerate' rearrangement is known as *degenerate Cope rearrangement*[12]. Among the compounds undergoing this rearrangement, bullvalene, barbaralane, semibullvalene, azabullvalene are especially interesting.

Applications

1. 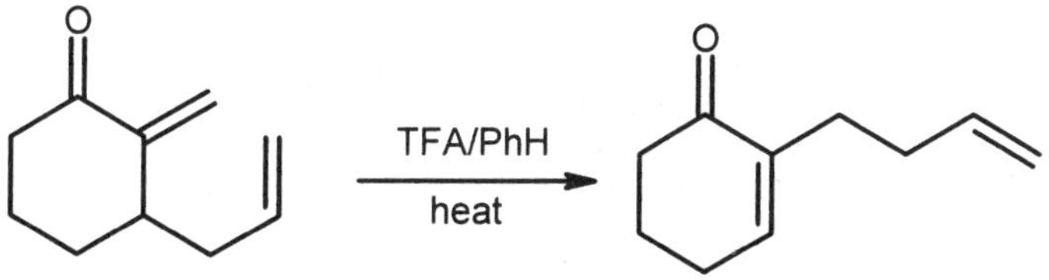 [reaction: cyclohexanone with allyl substituent → TFA/PhH, heat → rearranged cyclohexenone] [Ref. 13]

2. An industrial synthesis of citral (a key intermediate in the synthesis of vitamin A) involves [3,3]-sigmatropic rearrangements — a *Claisen* followed by a *Cope*.

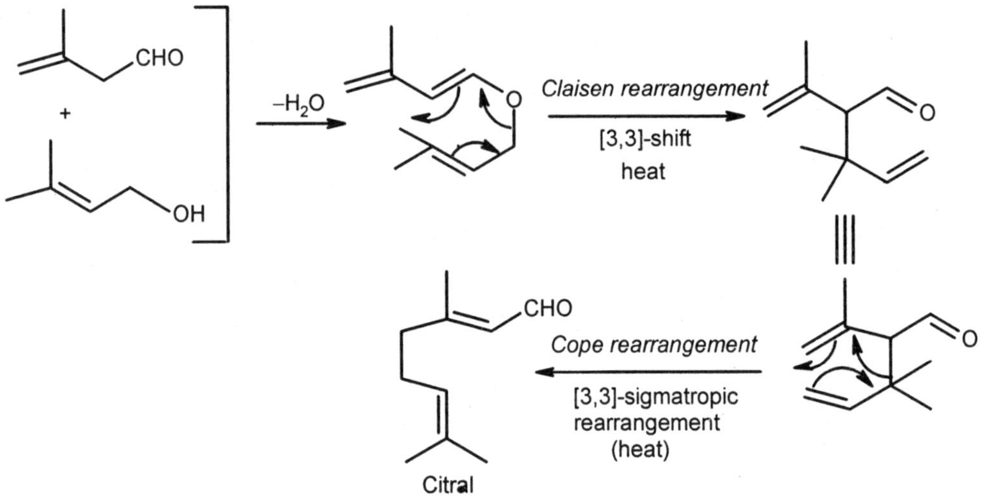

3. Synthesis of a 'bridgehead' alkene is possible by using a potassium alkoxide-accelerated *Cope rearrangement*.

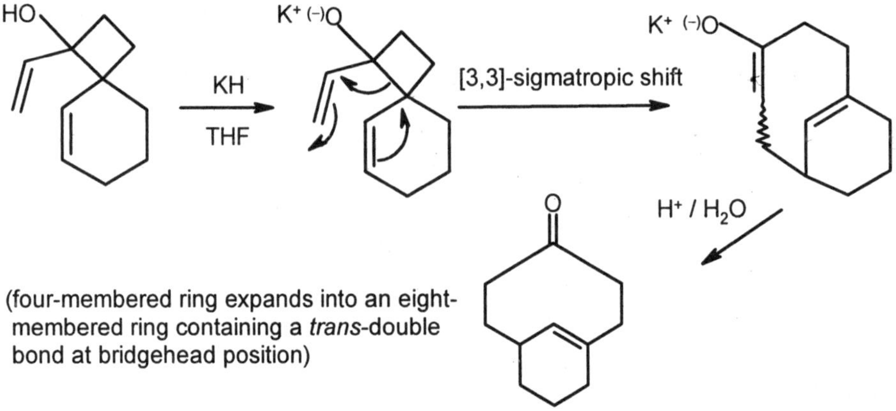

(four-membered ring expands into an eight-membered ring containing a *trans*-double bond at bridgehead position)

4.

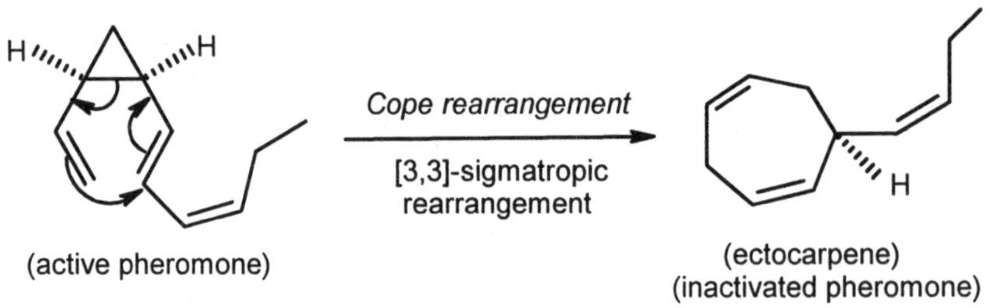

(active pheromone)

(ectocarpene)
(inactivated pheromone)

It is very interesting finding that the cyclopropyl pheromone (active form) inactivates itself, with a half-life of several minutes at ambient temperature by *Cope rearrangement* to the cycloheptadiene; the conversion is driven by release of strain arisen out of three-membered ring.

5. 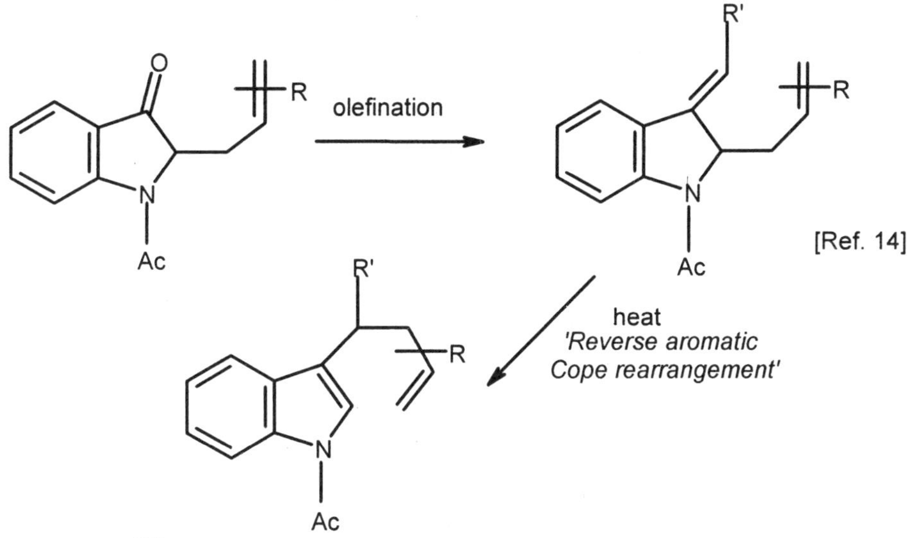 [Scheme: indolin-3-one with N-Ac and 2-CH₂CH₂-alkene-R → olefination → 3-alkylidene-indoline → heat, 'Reverse aromatic Cope rearrangement' → 3-substituted indole with R', R groups] [Ref. 14]

6. 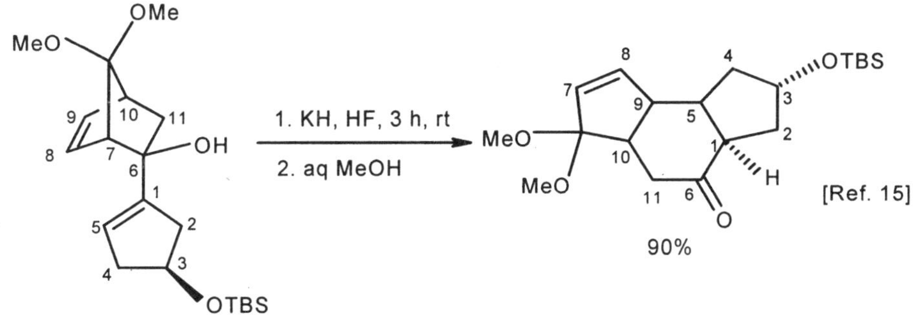 [Scheme: tricyclic substrate with OMe, MeO, OH, cyclopentene-OTBS, numbered 1–11]
   1. KH, HF, 3 h, rt
   2. aq MeOH
   → [bicyclic product with MeO, MeO, OTBS, ketone, numbered 1–11] 90% [Ref. 15]

References

1. Cope, A. C. and Hardy, E. M. (1940), *J. Am. Chem. Soc.*, **62**, 441.
2. Levy, H. and Cope, A.C. (1944), *J. Am. Chem. Soc.*, **66**, 1684.
3. Poupko, R., Zimmermann, H., Muller, K. and Luz, Z. (1966), *J. Am. Chem. Soc*, **118**, 7995.
4. Shea, K. J., Stoddard, G. J., England, W. P. and Hoffner, C. D.(1992), *J. Am. Chem. Soc*, **114**, 2635.
5. Doering, W. and Von, E. and Roth W. R. (1962), *Tetrahedron*, **18**, 67; Gujewski, J. J, Benner, C. W. and Hawkins, C. M. (1987), *J. Org. Chem.*, **52**, 5198; Paquette, L. A., DeRussy, D. T. and Cottrell, C. E. (1988), *J. Am. Chem. Soc.*, **110**, 890.

6. Hoffmann, R. and Woodward, R. B. (1965), *J. Am. Chem. Soc*, **87**, 4389; Fukui, K., and Fujimoto, H. (1966), *Tetrahedron Letts.*, 251.
7. Berson, J. A. and Jones, M. (1964), *J. Am. Chem. Soc*, **86**, 5019; Viola, A. Ioro, E. J, Chen, K. K., Glover, G. M., Nayak, U. and Kocienski, P. J. (1967), *J. Am. Chem. Soc*, **89**, 3462; Berson, J. A. and Walsh. Jr., E. J. (1968), *J. Am. Chem. Soc.*, **90**, 4729; Viola , A., Padilla, A. J. Lennox, D. M., Hecht, A. and Proverb, R. J. (1974), *J. Chem. Soc., Chem Commun*, 491; Paquette, L. A., 1990 *Angew. Chem. Int. Endn. Engl.*, **29**, 609 (review); *ibid, Synlett*, 1990, **67** (review).
8. Evans, D. A. and Golob, A. M.(1995), *J. Am. Chem. Soc*, **97**, 4765; Evans, D. A. and Nelson, J. V. (1980), *J. Am. Chem. Soc*, **102**, 774; Gujewski, J. J. and Gee, K. R. (1991), *J. Am. Chem. Soc*, **113**, 967.
9. Paquette, L. A. (1997), *Tetrahedron*, **53**, 13971 (review).
10. Dewar, M. J. S. and Wade, L. E. (1977), *J. Am. Chem. Soc.*, **99**, 4417; Dewar, M. J. S., Ford, G. P., McKee, M. L., Rzepa, H. S. and Wade, L. E. (1977), *J. Am. Chem. Soc.*, **99**, 5069; Gajewski, J. J. and Conrad, N. D. (1978), *J. Am. Chem. Soc.*, **100**, 6268.
11. Gompper, R. and Ulrich, W. –R. (1976), *Angew. Chem. Int. Edn. Engl.,* **15**, 299.
12. Doering, W. von E. and Roth, W. R. (1963), *Tetrahedron*, **19**, 715; Decock- Le Reverend, B. and Goudmand, P. (1973), *Bull. Soc. Chim. Fr.*, 389; Chang, A. K., Anet, F. A. L., Mioduski, J. and Meinwald, J. (1974), *J. Am. Chem. Soc.*, **96**, 2887; Hoffmann, R. and Stohrer, W. –D. (1971), *J. Am. Chem. Soc.,* **93**, 6941.
13. Dauben, W. G. and Chollet, A. (1981), *Tetrahedron Letts*, 1583.
14. Kawasaki, T., Nonaka, Y., Watanabe, K., Ogawa, A., Higuchi, K., Terashima, R., Masida, K. and Sakamoto, M. (2001), *J. Org. Chem.*, **66**, 1200.
15. Paquette, L. A., Gao, Z., Ni, Z. and Smith, G. F. (1988), *J. Am. Chem. Soc.*, **120**, 2543.
- Schroder, G., Oth, J. F. M. and Mereny, R. (1965), *Angew. Chem. Int. Edn. Engl.*, **4**, 752.
- Corey, E. J. and Kania, R. S. (1998), *Tetrahedron Lett.*, No. 39, 741.
- Davies, H. M. L. (1993), *Tetrahedron*, **49**, 5203 (review).
- Paquette, L. A. (1997), *Tetrahedron*, **53**, 13971(review).
- Schneider, C. (1997), *Synlett*, 815; Miyashi, T., Ikeda, H. and Takahashi, Y. (1999), *Acc. Chem. Res.*, **32**, 815 (review).
- Ogawa, Y., Ueno, T., Karikomi, M., Seki, K., Haga, K. and Uyehara, T. (2002), *Tetrahedron Letts.*, No. 43, 7827.
- Clive, D. L. J. and Ou, L. (2002), *Tetrahedron Letts.*, No. 43, 4559.
- Ogawa, Y., Toyama, M., Karikomi, M., Seki, K., Haga, K. and Uyehara, T. (2003), *Tetrahedron Letts.*, No. 44, 2167

22

Dakin Reaction

Introduction

An aromatic aldehyde or ketone, having an OH or NH_2 group at the *ortho-* or *para-* position, is converted to phenolic derivative on treatment with alkaline hydrogen peroxide followed by hydrolysis; this reaction is known as *Dakin reaction* or *Dakin oxidation*[1].

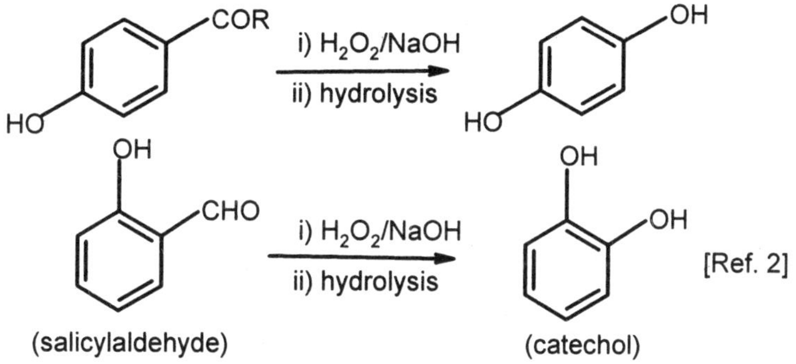

Mechanism

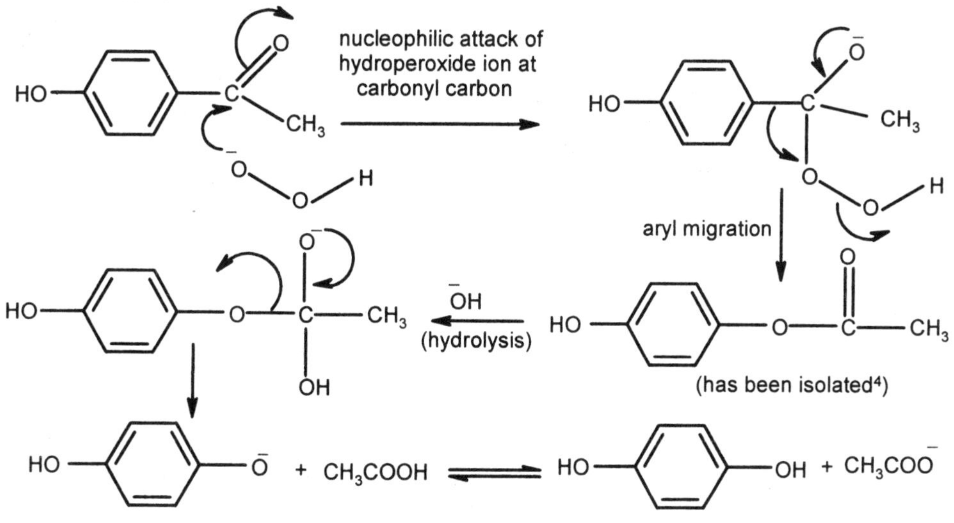

The machanism of *Dakin reaction* is conveniently similar to that of *Baeyer-Villiger reaction*[3].

Critical Views

(a) Matsumoto et al.[5] performed the reaction on aromatic aldehydes with an alkoxy group in the ring, but with no OH or NH_2; in this case acidic H_2O_2 (30-35% aqueous solution in methanol) was used. The oxidation is very much easy to handle.

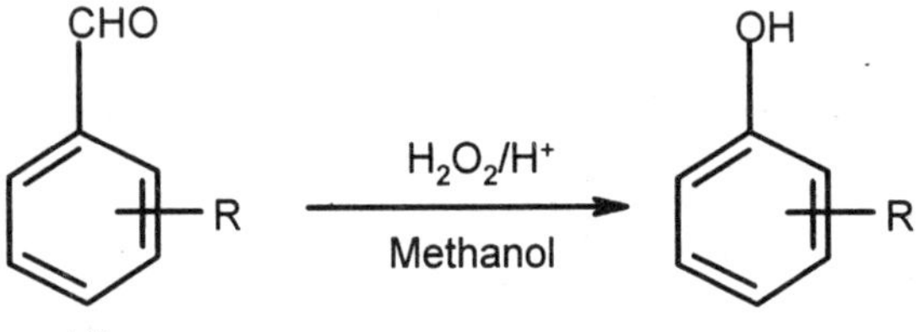

Substrate (1)	% Yield (2)
2-methoxy	94
3-methoxy	31
4-methoxy	90
2,3-dimethoxy	30
2,4-dimethoxy	90
3,4-dimethoxy	60
2,4,6-trimethoxy	89
3,4-methylenedioxy	67

The superiority of this system was strikingly shown in the oxidation of 4-methoxy-2-(3-methyl-2-buten-1-yloxy)benzaldehyde (3a), which on treatment with 31% aqueous H_2O_2 in acidic methanol at room temperature for 4 hours furnished the corresponding phenol (4a) in a 80% yield.

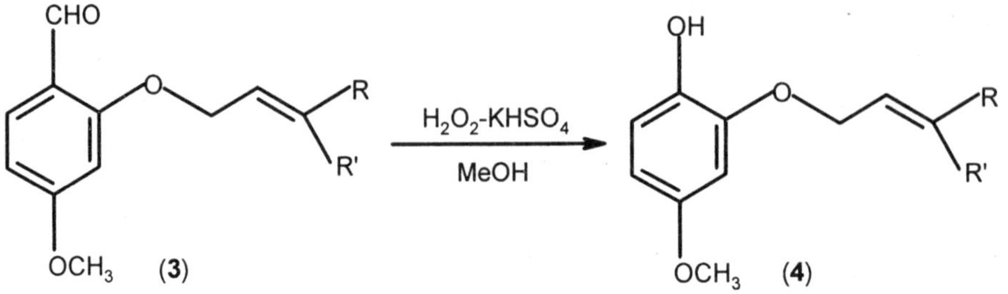

(3a: R=R'=CH_3; b: R=CH_3, R'=H; c: R=R' =H)

(4a: R=R'=CH_3; b: R=CH_3, R'=H; c: R=R' =H)

(b) Urea-hydrogen peroxide adduct is stable, inexpensive and an easily handled reagent. Varma and Naicker[6] have used this reagent in an efficient solid-state oxidation of hydroxylated aldehydes and ketones to the corresponding phenol derivatives.

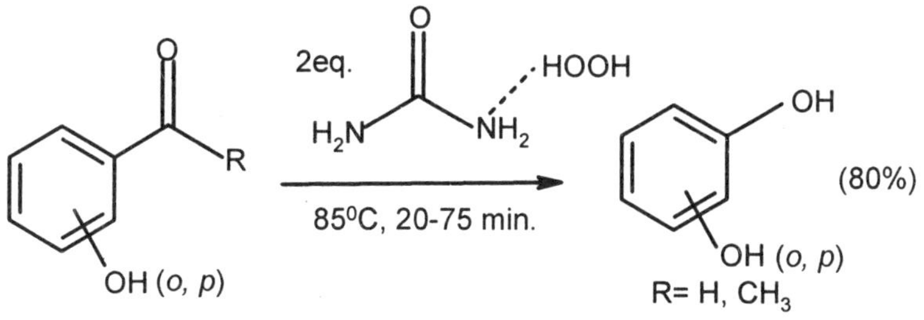

Applications

i) [Ref. 7]

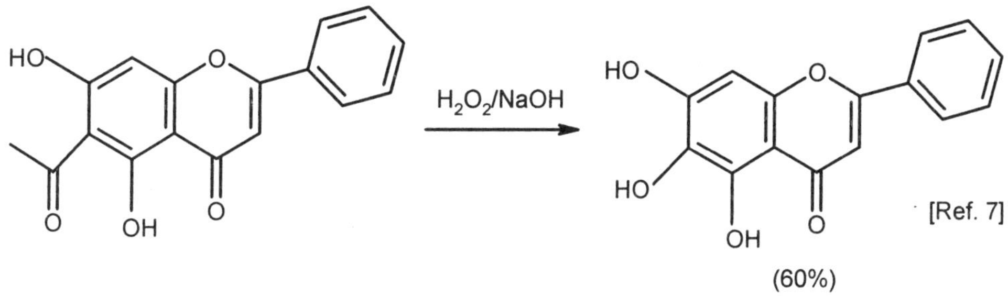

ii) [Ref. 8]

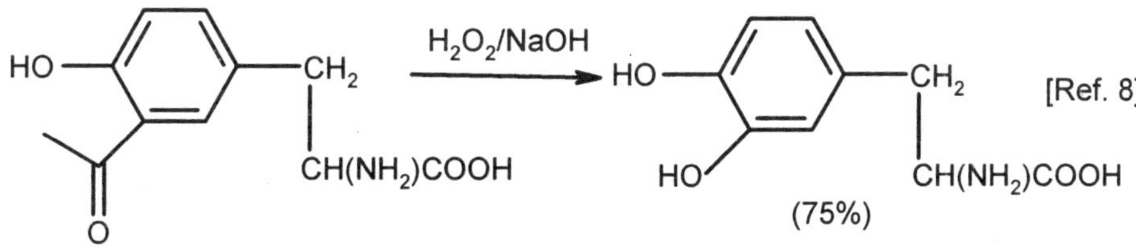

References

1. Dakin, H. D. (1909), *J. Am. Chem. Soc.*, **42**, 477.
2. Bunton, C. A. in *Peroxide Reaction Mechanism*, edited by J. O. Edwards, Interscience.
3. Hocking, M. B., Bhandari, K., Shell, B. and Smyth, T. A. (1982), *J. Org. Chem.*, **47**, 4208.
4. Hocking, M. B., Ko, M. and Smyth, T. A. (1978), *Can. J. Chem.*, **56**, 2646.
5. Matsumoto, M., Kobayashi, H. and Hotta, Y. (1984), *J. Org. Chem.*, **49**, 4740.

6   Varma, R. S. and Naicker, K. P. (1999), *Org. Lett.*, **1**, 189.
7   Agasimundin, Y. and Siddappa, S. (1973), *J. Chem. Soc. Perkin Trans. I*, 503.
8   Bretschneider, H., Hohenlowe-Oehringen, K., Kaiser, A. and Wolcke, U. (1973), *Helvitica Chim. Acta*, **56**, 2857.

- Zhu, J., Beugelmans, R., Bigot, A., Singh, G. P. and Bois-Choussay, M. (1993), *Tetrahedron Lett.*, **34**, 7401.
- Guzman, J. A., Mendoza, V., Garcia, E., Garibay, C. F., Olivares, L. Z. and Maldonado, L. A. (1995), *Synth. Commumn.*, **25**, 2121.
- Jung, M. E. and Lazarova, T. I. (1997), *J. Org. Chem.*, **62**, 1553.
- Roy, A., Reddy, K. R., Mohanta, P. K., Ila, H. and Junjappa, H. (1999), *Synth. Commumn.*, **29**, 3781.

23

Di - π - methane Rearrangement

Introduction

Photochemical rearrangement reaction of 1,4-dienes leading to the formation of vinylcyclopropanes is widely known as *di-π-methane rearrangement*[1].

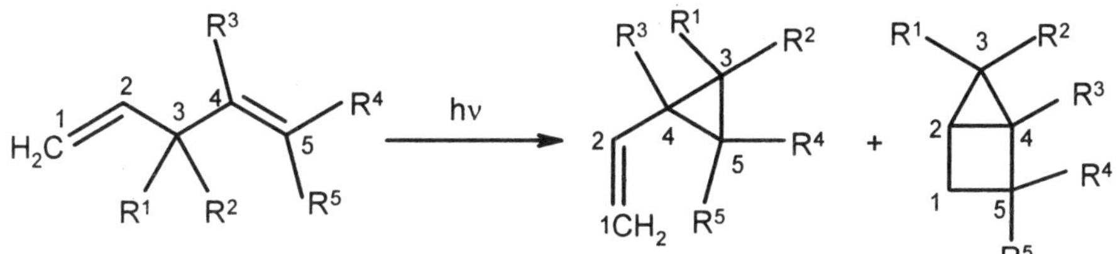

(1,4-diene)   (vinylcyclopropane)   (bicyclic product)
'di-π-methane unit'   'rearranged product'   'side product'

This is the most general photochemical reaction applicable to acyclic, cyclic, bi- and tricyclic 1,4-dienes[2]. Zimmermann and his collaborators had extensively studied the reaction, and that is why, now-a-days, this rearrangement reaction is also called as *Zimmermann reaction*.

Mechanism and Discussion

The reaction is believed to proceed through diradical pathway, though the species is not necessarily intermediate, but may be the transition state[3].

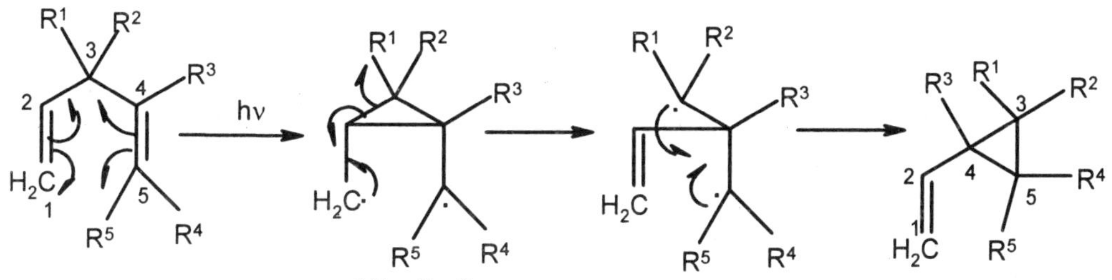

(diradical)

It has been established that for acyclic and monocyclic systems, the rearrangement proceeds through a singlet mechanism (*viz.* photochemical rearrangement of 3,3-dimethyl-1,1,5,5-tertaphenylpenta-1,4-diene to 1,1-dimethyl-2,2-diphenyl-3-(2,2-diphenyl)vinylcyclopropane), whereas for bicyclic or tricyclic systems the reaction follows triplet mechanism (*viz.* acetone-sensitized photochemical rearrangement of barrelene to semibulvalene)[4].

The singlet state-reaction is supposed to be *concerted* and to involve inversion of configuration only at C-3; configurations at C-1 and C-5 remain intact.[5] For example, *cis*-1,1-diphenyl-3,3-dimethyl-1,4-hexadiene (**1**) rearranges to (**2**), in which the side chain is *cis*, but *trans*-1,1-diphenyl-3,3-dimethyl-1,4-hexadiene (**3**) rearranges to (**4**), in which the side chain is *trans*[5]. Similarly, (**5**) and (**6**) yield predominantly (**7**) and (**8**), respectively[6].

Critical Views

(a) The direction of rearrangement is controlled by the ease of electron availability for π-bond formation in the diradical formed; conversely, is directed by the extent of electron delocalization in the diradical formed during the reaction that may be exemplified by the following transformation (**Scheme I**). The substrate (**9**) gives exclusively (**11**), not the (**12**). This can be rationalized from the fact that the benzylic radical of (**10**) can be stabilized by delocalization into the aromatic rings and is not readily available for π-bond formation as required by route 'b'; route 'a' is therefore followed leading to the formation of (**11**) as a sole product.

Scheme I

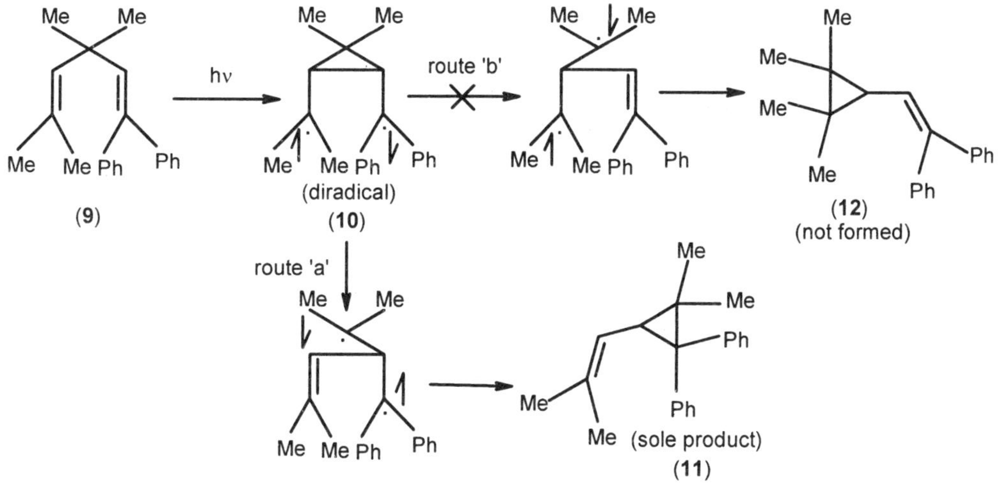
Me
hν
(diradical)
(9)
(10)
route 'b'
route 'a'
(12)
(not formed)
(sole product)
(11)
Me
Ph

(b) The reaction has been extended to allylic benzenes[7], β,γ-unsaturated imines[8] and to triple-bond systems[9]. Asymmetrically sensitized *di-π-methane rearrangements* has also been observed, which leads to the synthesis of optically active products[10].

(c) It has been observed that β,γ-unsaturated ketones undergo photochemical reaction similar to *di-π-methane rearrangement* involving 1,2-acyl migration and formation of a three-membered ring; this particular reaction is known as *oxa-di-π-methane rearrangement*[11]. Here the carbonyl group acts as one of the π-component.

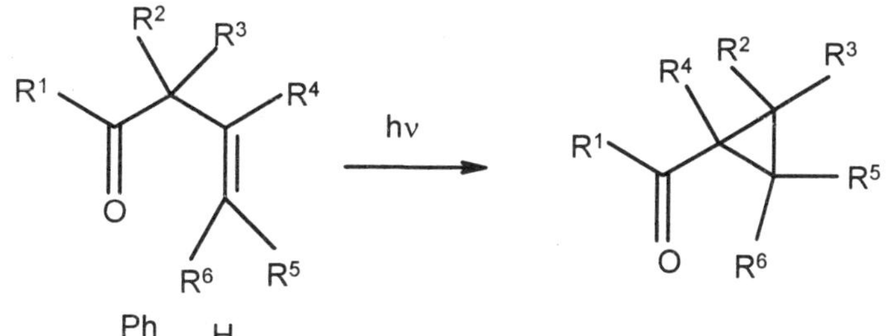
R1
R2
R3
R4
R5
R6
O
hν

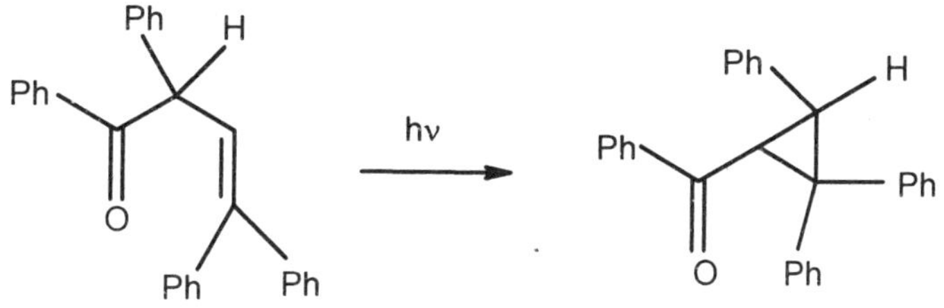
Ph
H
O
hν

Applications

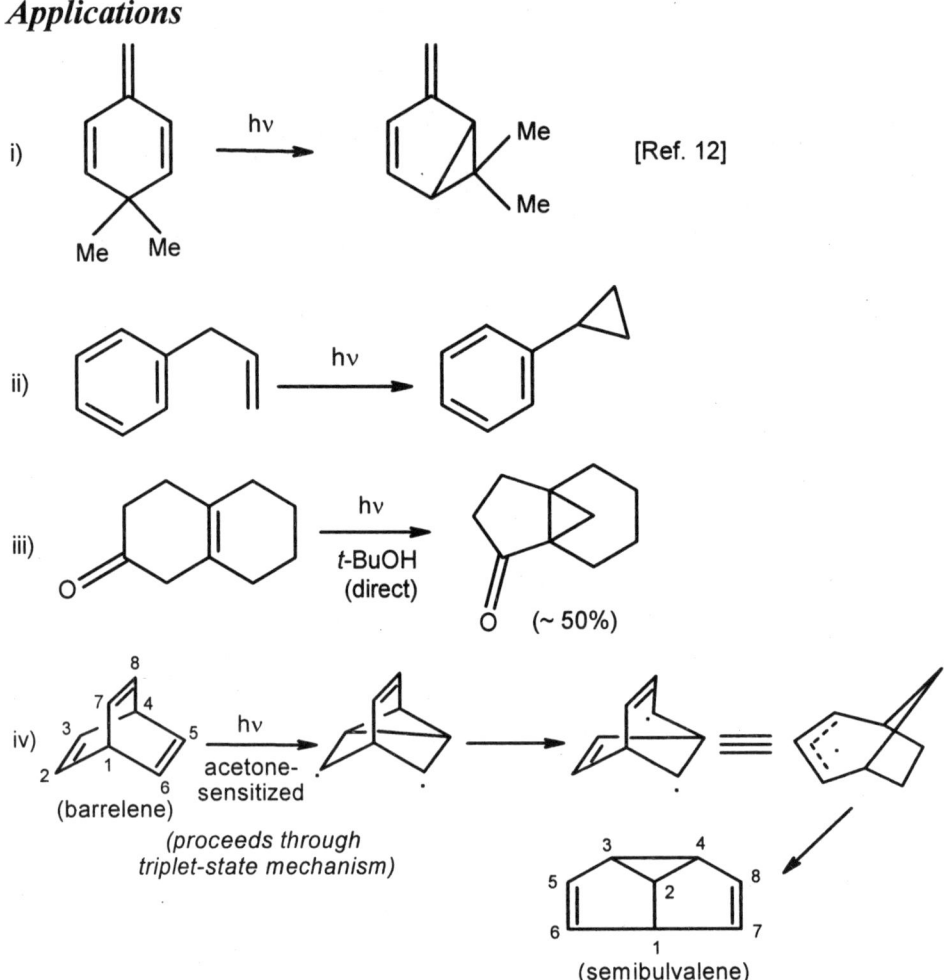

References

1. Zimmermann, H. E. and Grunewald, G. L. (1966), *J. Am. Chem. Soc.*, **88**, 183; Zimmermann, H. E. and Pincocki, J. A. (1973), *J. Am. Chem. Soc.*, **95**, 2957; Hixson, S. S., Mariano, P. S. and Zimmermann, H. E. (1973), *Chem. Rev.*, **73**, 531.
2. Zimmermann, H. E. in *Organic Photochemistry*, A. Pawda (Editor), Marcel Dekker, New York, 1991, **Vol. 11**, p.1.
3. Zimmermann, H. E. and Little, R. D. (1974), *J. Am. Chem. Soc.*, **96**, 5143; Zimmermann, H. E., Boettcher, R. J., Buehler, N. E. and Keck, G. E. (1975), *J. Am. Chem. Soc.*, **97**, 5635.

4   Zimmermann, H. E. and Pratt, A. C. (1970), *J. Am. Chem. Soc.*, **92**, 1407, 1409; Zimmermann, H. E., Binkley, R. W., Givens, R. S., Crundewald, G. L., and Sherwin, M. A. (1969), *J. Am. Chem. Soc.*, **91**, 3316; Zimmermann, H. E., Givens, R. S. and Pagni, R. M. (1968), *J. Am. Chem. Soc.*, **90**, 6096.
5   Zimmermann, H. E. and Pratt, A. C. (1970), *J. Am. Chem. Soc.*, **92**, 6267.
6   Zimmermann, H. E., Baeckstrom, P., Johnson, T. and Kurtz, D. W. (1972), *J. Am. Chem. Soc.*, **94**, 5504; **96**, 1459 (1974).
7   Zimmermann, H. E. and Swafford, R. L. (1984), *J. Org. Chem.*, **49**, 3069; Paqeutte, L. A. and Bay, E. (1984), *J. Am. Chem. Soc.*, **106**, 6693.
8   Armesto, D., Horspool, W. M., Langa, F. and Ramos, A. (1991), *J. Chem. Soc., Perkin Trans. 1*, 223.
9   Griffin, G. W., Chihal, D. M., Perreten, J. and Bhacca, N.S. (1976), *J. Org. Chem.*, **41**, 3931.
10   Hoshi, N., Furukawa, Y., Hagawara, H. and Sato, K. (1980), *Chem. Lett.*, 47.
11   Dauben, W. G., Kellog, M. S., seeman, J. I. and Spitzer, W. A. (1970), *J. Am. Chem. Soc.*, **92**, 1786; Demuth, M. (1991), *Org. Photochem.*, **11**, 37.
12   Zimmermann, H. E., Hackett, P., Juers, D. F., McCall, J. M. and Schroder, B. (1971), *J. Am. Chem. Soc.*, **93**, 3653.
•   Zimmermann, H. E. and Welter, T. R. (1978), *J. Am. Chem. Soc.*, **100**, 4131.
•   Paquette, L. A., Bay, E., Yeh Ku, A., Rondan, N. G., and Houk, K. N. (1982), *J. Org. Chem.*, **47**, 422.
•   Demuth, M. and Mikhail, G. (1989), *Synthesis*, 145.
•   Mehta, G. and Subrahmanyam, D. (1991), *J. Chem. Soc., Perkin Trans. 1*, 395.
•   De Keukeleire, D. and He, S. H. (1993), *Chem. Rev.*, **93**, 359.
•   Janz, K. M. and Scheffer, J. R. (1999), *Tetrahedron Lett.*, **40**, 8725.
•   Zimmermann, H. E. and Cirkva, V. (2000), *Org. Lett.*, **2**, 2365.
•   Jimenez, M. C., Miranda, M. A. and Tormos, R. (2000), *Chem. Commun.*, 2341.
•   Ihmels, H., Mohrschladt, C. J., Grimme, J. W. and Quast, H. (2001), *Synthesis*, 1175.
•   Zimmermann, H. E. and Chen, W. (2002), *Org. Lett.*, **4**, 1155.
•   Tanifuji, N., Huang, H., Shinagawa, Y. and Kobayashi, K. (2003), *Tetrahedron Lett.*, **44**, 751.

24

Etard Reaction

Introduction

The oxidation of an arylmethyl group to aldehydic function with a solution of chromyl chloride in an inert solvent (CCl_4 or CS_2) followed by hydrolysis is known as *Etard reaction*[1].

Initially an insoluble complex (Etard complex) containing two atoms of chromium for each hydrocarbon molecule is formed, which then undergoes rapid hydrolysis.

$$\text{(toluene)} \xrightarrow[CCl_4 \text{ or } CS_2]{2\ CrO_2Cl_2} [\text{PhCH}_3 \cdot (CrO_2Cl_2)_2] \xrightarrow{H_2O} \text{(benzaldehyde)}$$

(insoluble complex)

The yields are high and an added advantage of this oxidation reaction is that the aldehyde group in not oxidized further.

Mechanistic Approach

The mechanism of *Etard reaction* is still not completely clear and is a subject of much discussion.[2] It is believed that an insoluble complex is formed on addition of the reactants, which is subsequently hydrolyzed by water to the corresponding aldehyde. But there is some disagreement regarding the structure of the complex.[3]

The mechanistic approach as shown for the reaction (that bears an intermediate, benzyl ester of chromic acid) appears to be the best fitted one since it serves to explain a number of side reactions observed — such as molecular rearrangements,[3h] the formation of alcohols rather than carbonyl compounds when insoluble intermediates are hydrolyzed in the presence of a reducing agent (SO_2),[3e,f] and the formation of alkyl chlorides in certain cases rather than alcohols or carbonyl compounds.

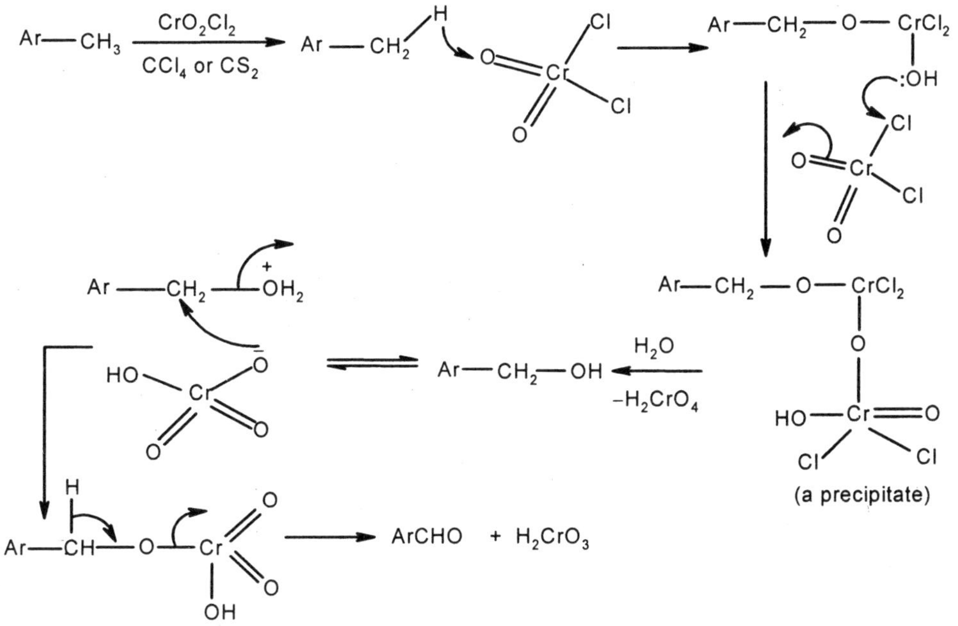

[Mechanism scheme showing Etard reaction pathway with CrO_2Cl_2 in CCl_4 or CS_2, producing ArCHO + H_2CrO_3 via intermediate $Ar-CH_2-O-CrCl_2$ species and precipitate]

However, there are also strong arguments in support of the acylal-type structural proposition for the complex (**A**).[3d]

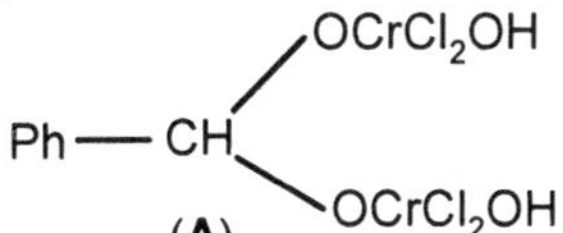

$$Ph-CH\begin{matrix}OCrCl_2OH\\OCrCl_2OH\end{matrix}$$
(**A**)

Still another proposal is that the complex is composed of benzaldehyde coordinated with reduced chromyl chloride.[3g]

Applications

Etard reaction finds much application in synthetic organic chemistry. Formylations of few arylmethyl groups are shown below:

1) 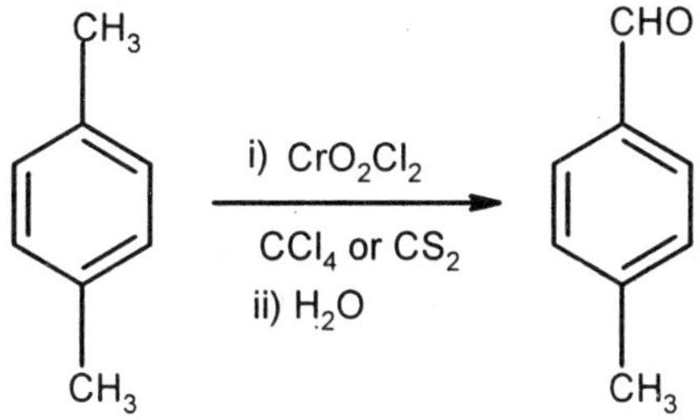 p-xylene $\xrightarrow[\text{ii) }H_2O]{\text{i) }CrO_2Cl_2,\ CCl_4\text{ or }CS_2}$ 4-methylbenzaldehyde [Ref. 4]

2) $p\text{-BrC}_6\text{H}_4\text{CH}_3 \xrightarrow[\text{ii) H}_2\text{O}]{\text{i) CrO}_2\text{Cl}_2 / \text{CCl}_4} p\text{-BrC}_6\text{H}_4\text{CHO}$ (80%)

3) $p\text{-O}_2\text{NC}_6\text{H}_4\text{CH}_3 \xrightarrow[\text{ii) H}_2\text{O}]{\text{i) CrO}_3 / \text{H}_2\text{SO}_4} p\text{-O}_2\text{NC}_6\text{H}_4\text{CHO}$ (89-94%) [Ref. 5]

4) [2-nitro-4-methoxytoluene] $\xrightarrow[\text{ii) H}_2\text{O}]{\text{i) CrO}_2\text{Cl}_2 / \text{CCl}_4}$ [2-nitro-4-methoxybenzaldehyde] (50%) [Ref. 6]

References

1. Etard, A. L. (1980), *Compt. Rend.*, **90**, 524; Hartford, W. H. and Darrin, M. (1958), *Chem. Rev.*, **58**, 1.
2. Nenitzescu, C. D. (1968), *Bull. Soc. Chim. Fr.*, 1349.
3. a) Necsoiu, I., Balaban, A. T., Pascaru, I., Sliam, E., Elion, M. and Nenitzescu, C. D. (1963), *Tetrahedron*, **19**, 1133; b) Wheeler, O. H. (1964), *Can. J. Chem.*, **42**, 706; c) Stairs, R. A. (1964), *Can. J. Chem.*, **42**, 550; d) Wiberg, K. B. and Eisenthal, R. (1964), *Tetrahedron*, **20**, 1151; e) Wheeler, O. H. (1960), *Can. J. Chem.*, **38**, 2137; f) Necsoiu, I., Przemetchi, V., Ghenciulescu, A., Rentea, C. N. and Nenitzesu, C. D. (1966), *Tetrahedron*, **22**, 3037; g) Duffin, H. C. and Tucker, R. B. (1966), *Chem. Ind. (London)*, 12622 & *Tetrahedron*, **24**, 6999 (1968); h) Rentea, C. N., Rentea, M., Necsoiu, I. and Nenitzescu, C. D. (1968), *Tetrahedron*, **24**, 4667.
4. Law, H. D. and Perkin, F. M. (1907), *J. Chem. Soc.*, 258.
5. Nishimura, T. (1963), *Org. Syn.*, **Coll. Vol. 4**, 713; Lieberman, S. V. and Conner, R. (1943), *ibid*, **Coll. Vol. 2**, 441.
6. Boon, W. R. (1940), *J. Chem. Soc.*, S230.

- Rocek, J. (1962), *Tetrahedron Lett.*, 135.
- Schildknecht, H. and Hatzmann, G. (1968), *Angew. Chem. Int. Ed. Engl.*, **7**, 293.
- Luzzio, F. A. and Moore, W. J. (1993), *J. Org. Chem.*, **58**, 512.

25

Fischer – Indole Synthesis

Introduction

F*ischer indole synthesis*[1] involves acid-catalyzed rearrangement of arylhydrazones of aldehydes or ketones leading to the formation of indoles, with elimination of ammonia. The preparation of 2-phenylindole illustrates the process in its simplest form.[2]

Zinc chloride is the catalyst most frequently employed, but many others, including other metal halides, proton (*e.g.* polyphosphoric acid, sulphuric acid) and Lewis acid (*e.g.* BF_3), and certain transition metals, are in use. In many cases the reaction can be carried out simply by heating together the aldehyde or ketone and phenylhydrazine in acetic acid.[3]

Microwave irradiation has been used to facilitate the reaction.[4] The formation of the phenylhydrazone and its subsequent rearrangement take place without the necessity for isolation of the phenylhydrazone; the common practice is the treatment of an aldehyde or a ketone with a mixture of phenylhydrazine and the catalyst. Toluenesulphonic acid, cation exchange resins and phosphorous trichloride have each been recommended for efficient cyclizations, sometime even at or below room temperatures.[5]

Mechanism

The mechanism of the reaction has been widely investigated and the well-accepted mechanism involves following steps. However, the key step of the mechanism is a [3,3]-sigmatropic rearrangement in which a carbon-carbon bond is formed.[6]

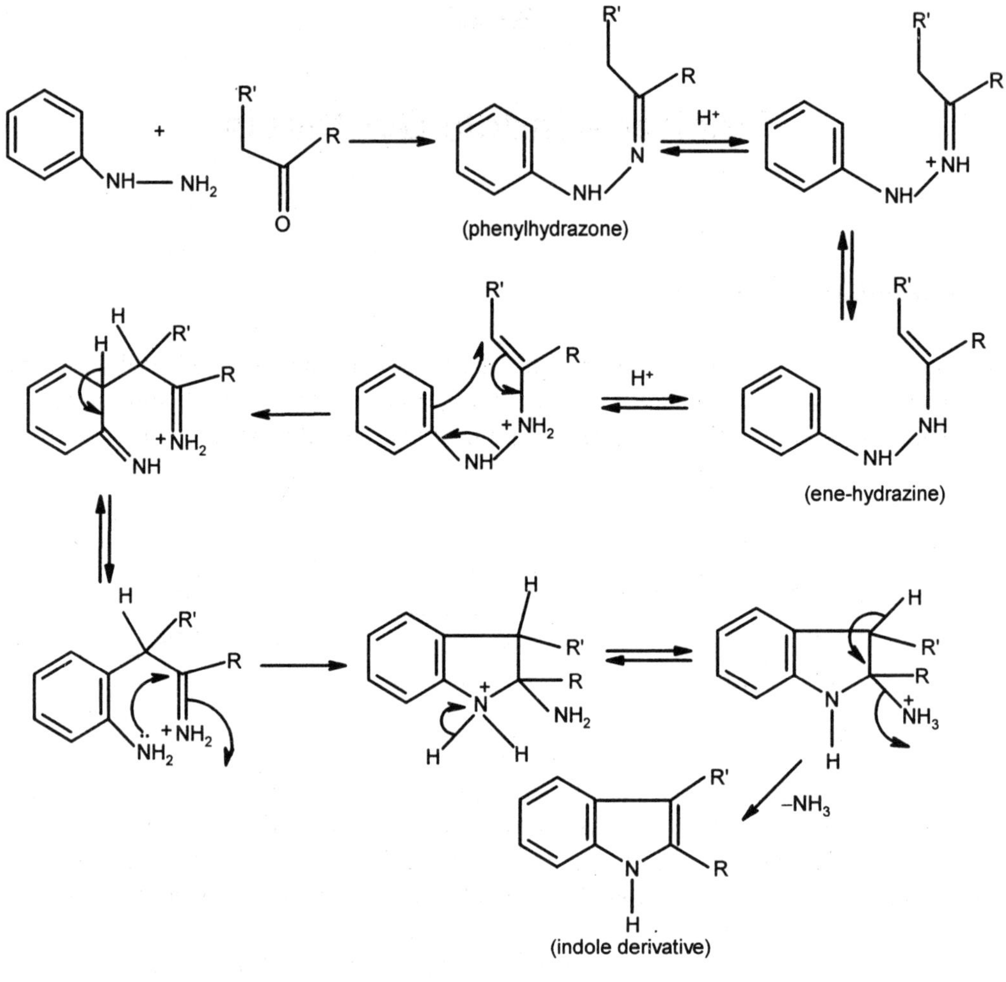

There is considerable evidence in support of the above sequence; for example, labelling studies proved the loss of the β-nitrogen as ammonia, and in some cases intermediates have been detected by ^{13}C and ^{15}N-NMR spectroscopy[7] and have been isolated by different workers.[8]

Critical Views

a) Presence of electron-donating groups (EDG) on the benzene ring of hydrazone derivatives enhances the rate of *Fischer cyclization*, while the rate is retarded by the presence of electron-withdrawing (EWG) substituents.[9] It has been observed that electron-withdrawing substituents at *meta*-position give rise to roughly equal amounts of 4- and 6-substituted indoles, and similarly oriented electron-donating groups produce mainly the 6-substituted products.[10]

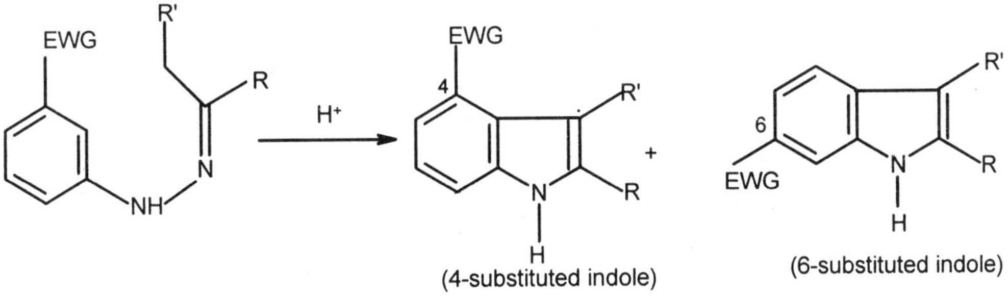

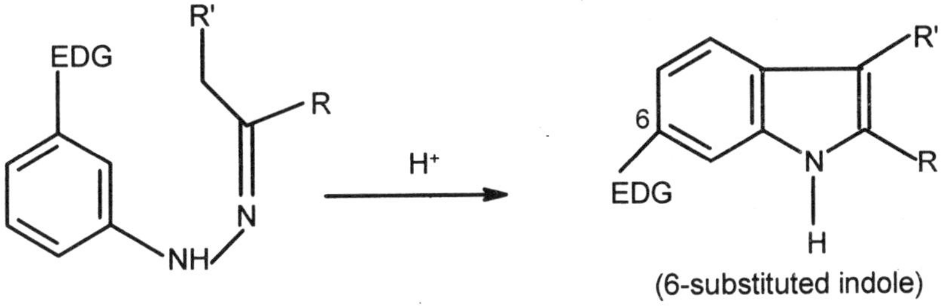

b) In case of phenylhydrazones arising out of unsymmetrical ketones, formation of a mixture of two possible indoles may take place and the ratio of the products is of considerable practical importance.

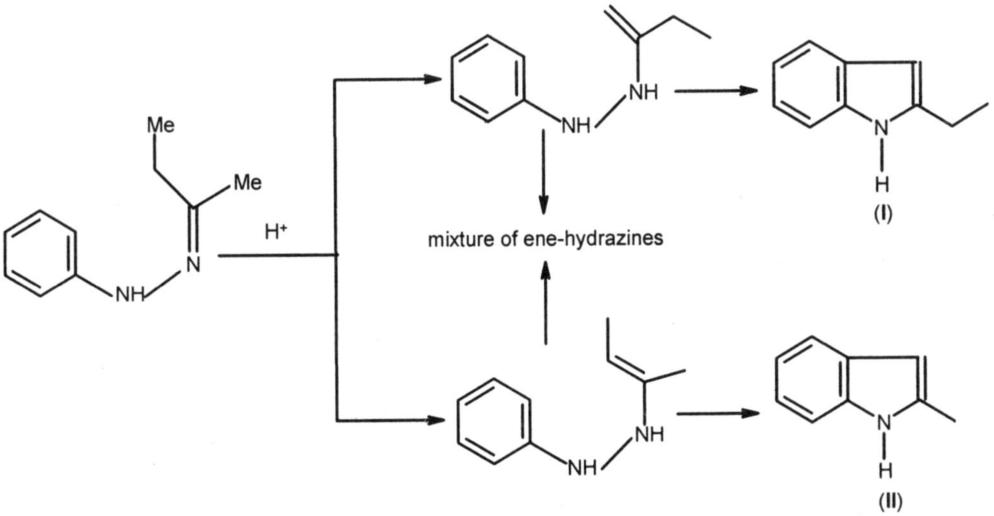

It appears that strongly acidic conditions favour the least substituted ene-hydrazine formation, and hence the product (**I**).[11]

Catalyst	I	II
AcOH	0	100
PPA	50	50
$MeSO_3H$, P_4H_{10}	78	22

c) Another interesting feature of *Fischer reaction* is that when both the *ortho*-positions on the benzene ring of hydrazone derivatives are blocked, there occurs a migration of the group during the reaction — as observed in the case of acetophenone 2,6-dimethylphenylhydrazone when treated with acid.[12]

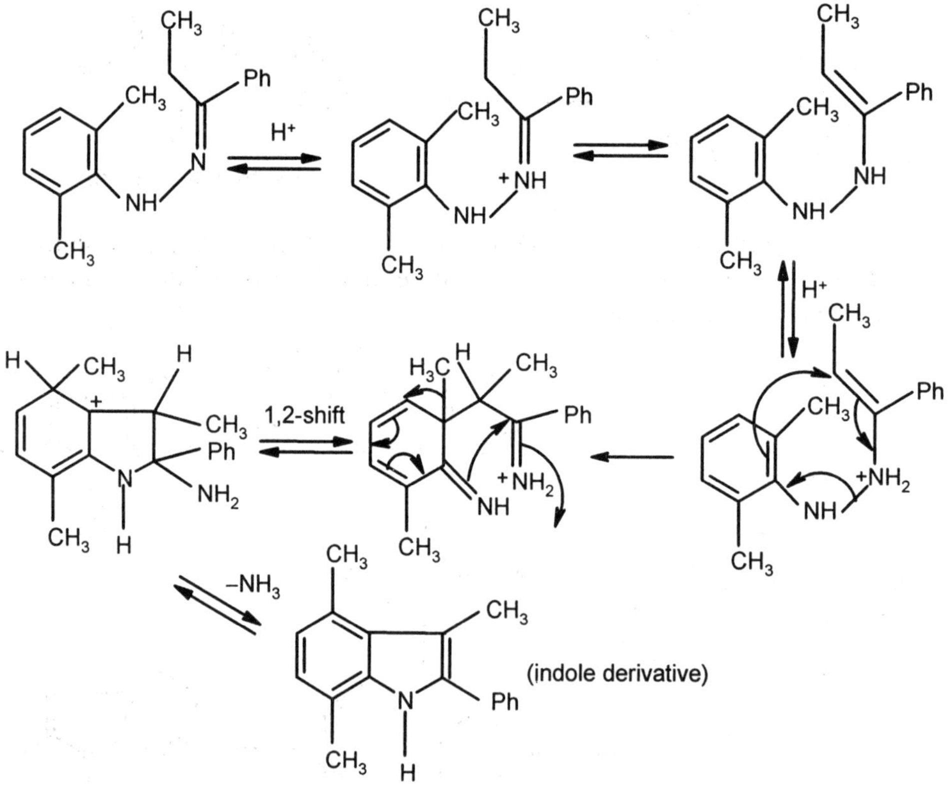

In some cases, a 1,4-shift could also have been observed but the steric effects seem to play an important role to decide between the two modes of shifts.[13]

Applications

1) 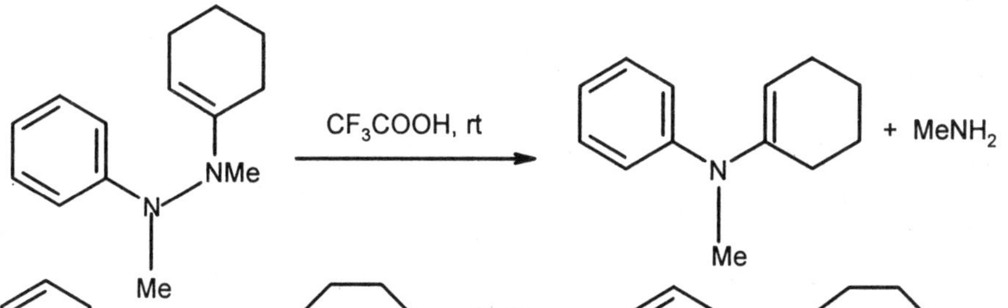 [Scheme: N-methyl-N-phenylhydrazine with cyclohexenyl-NMe group → CF₃COOH, rt → N-methyl-N-phenyl cyclohexenylamine + MeNH₂]

2) 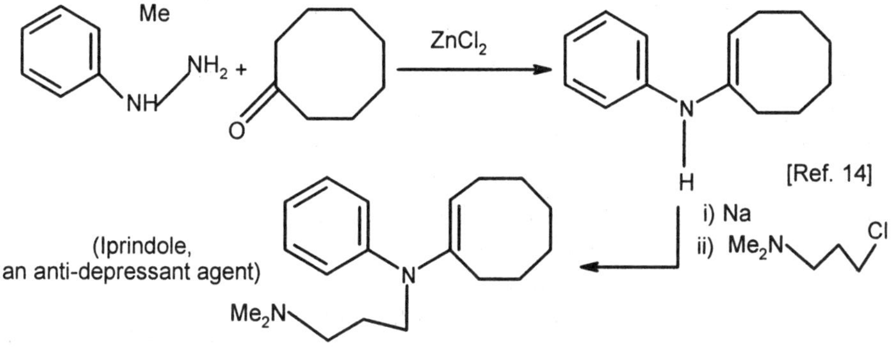 [Scheme: PhNH-NH₂ + cyclooctanone → ZnCl₂ → phenylhydrazone-type intermediate] [Ref. 14]

i) Na
ii) Me₂N~~~Cl

→ (Iprindole, an anti-depressant agent)

3) 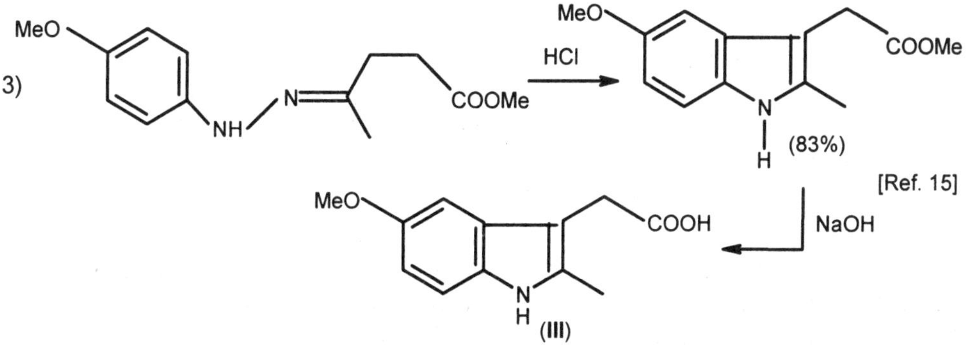 [Scheme: 4-MeO-phenylhydrazone of methyl levulinate → HCl → 5-methoxy-2-methyl-3-(methoxycarbonylmethyl)indole (83%)] [Ref. 15]

→ NaOH → 5-methoxy-2-methyl-3-(carboxymethyl)indole (III)

This type of indole, with a 5-OMe group, is the basis of many pharmaceutical — the acid (**III**) is used to synthesis Indomethacin, a non-steroid anti-inflammatory agent.

4) 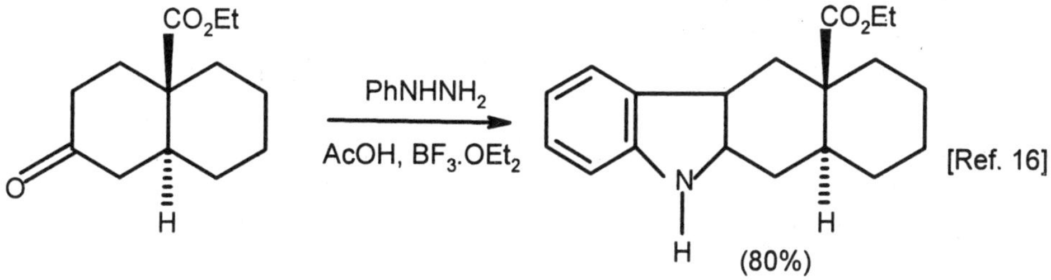 [Scheme: decalone with CO₂Et group + PhNHNH₂, AcOH, BF₃·OEt₂ → fused carbazole product (80%)] [Ref. 16]

References

1. Fischer, E. and Jourdan, F. (1883), *Ber. Dtsch. Chem. Ges.*, **16**, 2241; Fischer, E. and Speier, A. (1895), *Ber. Dtsch. Chem. Ges.*, **28**, 3252; Robinson, B., *The Fischer indole Synthesis*, John Wiley and Sons, Chichester, New York, 1982.
2. Shriner, R. L., Ashley, W. C. and Welch, E. (1955), *Org. Synth.*, **Coll. Vol. III**, 725.
3. Rogers, C. V. and Corson, B. B. (1963), *Org. Synth.*, **Coll. Vol. IV**, 884.
4. Abramovitch, R. A. and Bulman, A. (1992), *Synlett*, 795.
5. Murakami, Y., Yokoyama, Y., Miura, T., Hirasawa, H., Kamimura, Y. and Izaki, M. (1984), *Heterocycles*, **22**, 1211; Baccolini, G., Dalpozzo, R. and Todesco, P. E. (1988), *J. Chem. Soc., Perkin Trans. 1*, 971.
6. Robinson, G. M. and Robinson, R. (1918), *J. Chem. Soc.*, **113**, 639; Hughes, D. L. and Zhao, D. (1993), *J. Org. Chem.*, **58**, 228.
7. Douglas, A. W. (1978), *J. Am. Chem. Soc.*, **100**, 6463.
8. Bajwa, G. S. and Brown, R. K. (1970), *Can. J. Chem.*, **48**, 2293; Allen, C. F. H. and Wilson, C. V. (1943), *J. Am. Chem. Soc.*, **65**, 611.
9. Przheval'skii, N. M., Kostromina, L. Y. and Grandberg, I. I. (1988), *Kim. Geterotsikl. Soeden.*, **24**, 188; *Chem. Abstr.*, **109**, 210837(1988).
10. Benson, S. C., Lee, L. and Snyder, J. K. (1996), *Tetrahedron Lett.*, **37**, 5061.
11. Zhao, D., Hughes, D. L., Bender, D. R., DeMarco, A. M. and Reider, P. J. (1991), *J. Org. Chem.*, **56**, 3001.
12. Carlin, R. B. *et al.* (1964), *J. Am. Chem. Soc.*, **86**, 5300.
13. Miller, B. and Matzeka, E. R. (1980), *J. Am. Chem. Soc.*, **102**, 4722.
14. Rice, L. M., Hertz, E. and Freed, M. E. (1964), *J. Med. Chem.*, **7**, 313.
15. Shaw, E. (1955), *J. Am. Chem. Soc.*, **77**, 4319.
16. Ataraschi, S., Choi, J-K., Ha, D-C., Hart, D.J., Kuzmich, D., Lee, C-S., Ramesh, S. and Wu, S.C. (1997), *J. Am. Chem. Soc.*, **119**, 6226.

- Posvic, H., Dombro, R., Ho, H. and Telinski, T. (1974), *J. Org. Chem.*, **39**, 2575; Schiess, P. and Griedeo, A. (1974), *Helv. Chim. Acta*, **57**, 2643.
- Ishii, H. (1981), *Acc. Chem. Res.*, **14**, 275. [Review]
- Mills, K., Al Khawaja, I. K., Al-Saleh, F. S. and Joule, J. A. (1981), *J. Chem. Soc., Perkin Trans. 1*, 636.
- Miyata, O., Kimura, Y., Muroya, K., Hiramatsu, H. and Naito, T. (1999), *Tetrahedron Lett.*, **40**, 3601.
- Bhattacharya, G., Su, T.-L., Chia, C.-M. and Chen, K.-T. (2001), *J. Org. Chem.*, **66**, 426.
- Kozmin, S. A., Iwama, T., Huang, Y. and Rawal, V. H. (2002), *J. Am. Chem. Soc.*, **124**, 4628.
- Pete, B. and Parlagh, G. (2003), *Tetrahedron Lett.*, **44**, 2537.

26

Friedel – Crafts Reaction

Introduction

Introduction of an alkyl or an acyl group into an aromatic nucleus in the presence of an acid catalyst (usually Lewis acids) is known as *Friedel-Crafts reaction.*[1]

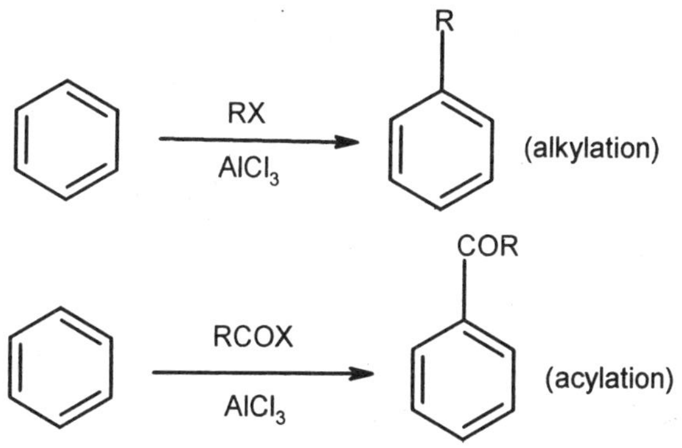

$$C_6H_6 \xrightarrow[AlCl_3]{RX} C_6H_5R \quad \text{(alkylation)}$$

$$C_6H_6 \xrightarrow[AlCl_3]{RCOX} C_6H_5COR \quad \text{(acylation)}$$

Commonly used solvents in this reaction are nitrobenzene, carbon disulphide, petroleum ether, *etc*. The scope and application of *Friedel-Crafts reactions* have proliferated tremendously. Although there is apparently wide variety of types, *Friedel-Crafts reactions* are broadly classified into two general categories — (a) *Friedel-Crafts alkylation* and (b) *Friedel-Crafts acylation*.

(a) Friedel-Crafts Alkylation
Reagents and catalysts:
The generally used alkylating agents are alkyl halides, alkenes and alcohols; but many others such as alkynes, ethers, thiols, sulphate and sulphonates, cycloalkanes (under conditions where they can be converted into carbocations) may also be employed. However, the order of reactivity for all types of reagents is:

allylic ~ benzylic > tertiary > secondary > primary.

In case of alkyl halides, the reactivity order[2] has been observed as: $F > Cl > Br > I$. Thus, benzene on treatment with FCH_2CH_2Cl in the presence of BCl_3 yields $PhCH_2CH_2Cl$.[3]

The most commonly used Lewis acid catalysts are anhydrous $AlCl_3$ and BF_3, but a variety of Lewis acids and also proton acids such as HF, H_2SO_4, H_3PO_4 are in use.[4]

It has been observed that for active halides, a trace amount of a less active catalyst (*e.g.* $ZnCl_2$) is enough — while for less active halides (such as chloromethane), a more active catalyst (*e.g.* $AlCl_3$) is required and also in larger amounts. The order of effectiveness of Lewis acid catalysts has been shown to be[5]: $AlBr_3 > AlCl_3 > GaCl_3 > FeCl_3 > SbCl_5 > ZnCl_4, SnCl_4 > BCl_3, BF_3, SbCl_3$. However, the reactivity order in each case depends on substrates, reagents, and conditions.

Mechanism

The *Friedel-Crafts reaction* follows the mechanistic pathway of aromatic electrophilic substitution involving the formation of *Wheland intermediate*. It is evident that in most cases[6] the electrophile in *Friedel-Crafts alkylation* is a carbocation (in free-state or as a part of an ion-pair). Alcohols and alkenes also furnish carbocation electrophiles on reaction with Lewis or proton acids. In case of tertiary alkyl halides, where formation of stable carbocation is feasible, the alkylating species is the actual carbocation:

$$R-Cl + AlCl_3 \rightleftharpoons [\overset{+}{R}\,\overset{-}{AlCl_4}]_{\text{ion-pair}} \rightleftharpoons [\overset{+}{R} + \overset{-}{AlCl_4}]_{\text{free carbocation}} \xrightarrow{\text{electrophilic attack}}$$

$$\xrightarrow{} [\text{Wheland intermediate}] \xrightarrow{\text{aromatization}} \text{Ar-R} + HCl + AlCl_3$$

In case of primary alkyl halides, alkylation takes place *via* a polarized complex. Secondary alkyl halides are on the boarder line and alkylation may occur through either mechanism:

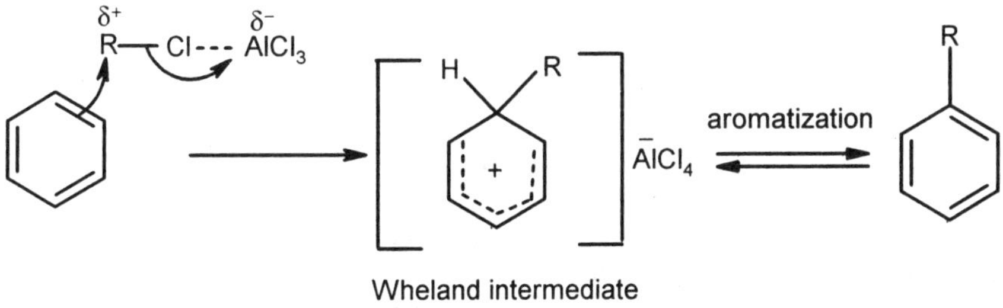

Friedel-Crafts Reaction

[Scheme: R–Cl···AlCl₃ with δ+/δ− attacking benzene → Wheland intermediate [H, R on ring with + and AlCl₄⁻] ⇌ aromatization → C₆H₅R]

Wheland intermediate

(b) Friedel-Crafts Acylation
Reagents and catalysts:

This is a good method for the preparation of aryl ketones. The acylating reagents are usually acyl halides, carboxylic acids, acid anhydrides and ketenes. Among the acyl halides, chlorides are most commonly employed. One of the most common catalysts for intramolecular *Friedel-Crafts acylation* (ring closure) is polyphosphoric acid. The catalysts for usual *Friedel-Crafts acylation* are those of the *Friedel-Crafts alkylation*, but in acylation, a little more than one mole of catalyst is required per mole of acyl halide — because one mole of the catalyst remains complexed with the carbonyl oxygen of the ketone formed. When acid anhydride is the reagent, slightly more than two moles of the catalyst is required — one mole is used up in liberating acyl halide and the other one for complexing with the product.

$$(RCO)_2O + AlCl_3 \longrightarrow RCOCl + RCOOAlCl_2$$

$$C_6H_6 + RCOCl \longrightarrow C_6H_5-\overset{+}{C}(R)=\overset{-}{O}-AlCl_3 \xrightarrow[\text{ice-cold}]{H_2O} C_6H_5COR$$

Mechanism

Depending upon conditions[7] probably two mechanisms may operate for *Friedel-Crafts acylation*. In most cases the attacking species is the acyl cation, either free or as an ion-pair[8]:

$$RCOCl + AlCl_3 \rightleftharpoons \left[R-\overset{+}{C}=\overset{..}{O} \longleftrightarrow R-C\equiv\overset{+}{O} \atop \text{free acyl cation} \atop \text{or} \atop R\overset{+}{C}O\ \overset{-}{A}lCl_4 \atop \text{ion-pair} \right] \overline{A}lCl_4$$

[Mechanism diagram: electrophilic substitution of benzene by acyl cation forming Wheland intermediate, followed by aromatization (−HCl) to aryl ketone (COR), then H₂O ice-cold workup.]

In other mechanism instead of acyl cation, a 1:1 complex attacks the aromatic ring directly[3]:

$$RCOCl + AlCl_3 \rightleftharpoons R-\underset{Cl}{\overset{+}{C}=\overset{}{O}}-\overline{A}lCl_3 \quad \text{(1:1 adduct)}$$

[Mechanism diagram: 1:1 adduct undergoes electrophilic substitution with benzene forming Wheland intermediate with H, R, C, Cl, OAlCl₃ substituents; aromatization −HCl then H₂O ice-cold gives aryl ketone (COR).]

Critical Views

(a) *Friedel – Crafts alkylation* suffers from few limitations. A frequent limitation of the reaction is di- and poly-alkylations of the substrate — this happens since the entering group (alkyl) is the activating one. Another important synthetic limitation is that during the reaction, rearrangement usually takes place in the reactant. For example, benzene treated with *n*-propyl bromide gives mostly isopropyl benzene (cumene) and much less *n*-propyl benzene. It is also interesting to note that the electron-releasing groups like -OH, -OR, -NH$_2$ do not facilitate the reaction, since the catalyst forms complexes with these groups. Moreover, in most cases, *meta*-directing groups deactivate the ring making it too inactive to undergo alkylation — thus nitrobenzene does not respond towards *Friedel-Crafts reaction*.

(b) An extension of *Friedel-Crafts reaction* is the arylation of an aromatic substrate — the reaction is called *Scholl reaction*.[9] The coupling of two aromatic molecules takes place in the presence of both a Lewis and a proton acid.

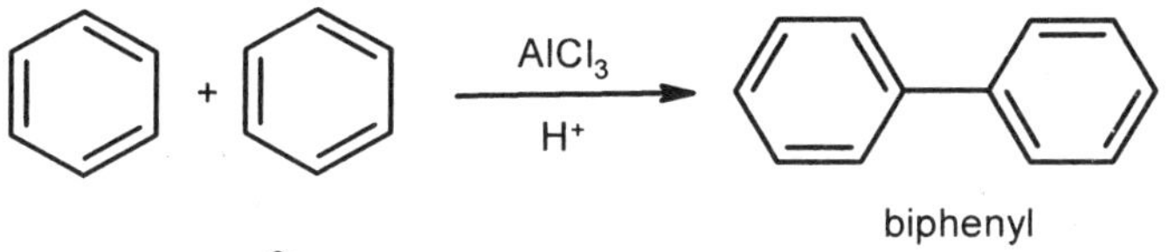

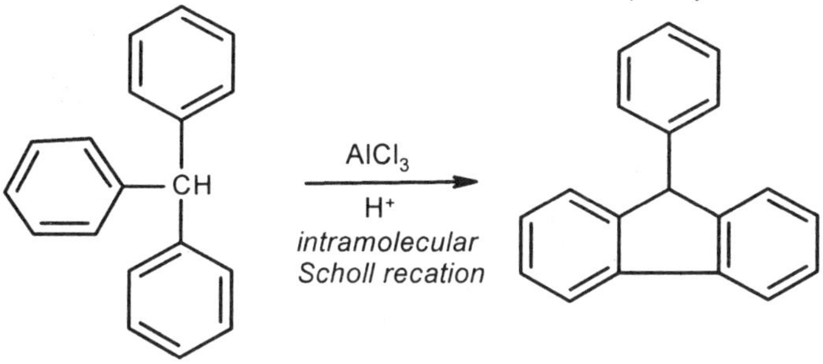

(c) A variety of substrates including heterocyclic compounds (except pyridines and quinolines) undergo *Friedel-Cratfs acylation reaction*. Compounds having *ortho/para*-directing groups such as alkyl, hydroxy, alkoxy, and halogen are easily acylated and furnish exclusively the *para*-products because of the relatively large size of the incoming acyl group. Since the entering acyl group is deactivating unlike the alkyl, a mono-substituted product is usually obtained.

Applications

The contributions of *Friedel-Crafts reaction* to synthetic organic chemistry are remarkable. A variety of compounds can be synthesized with the help of this reaction. Few examples are cited below:

1.

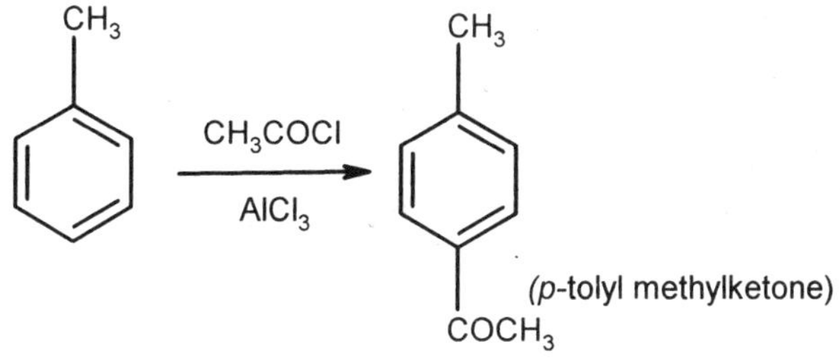

2. PhCH₂CH₂COCl →(AlCl₃) 1-hydrindone (indan-1-one)

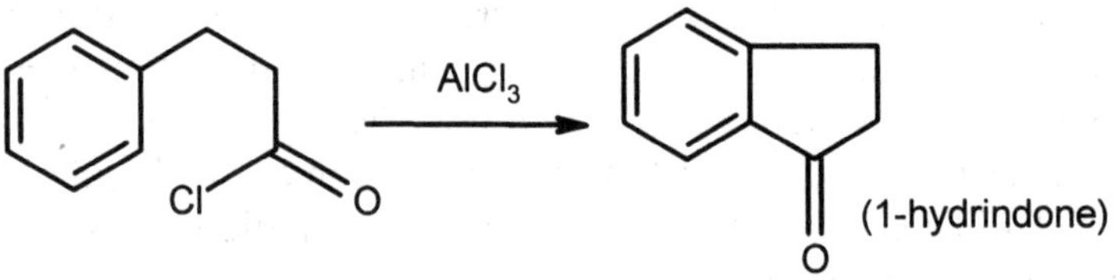

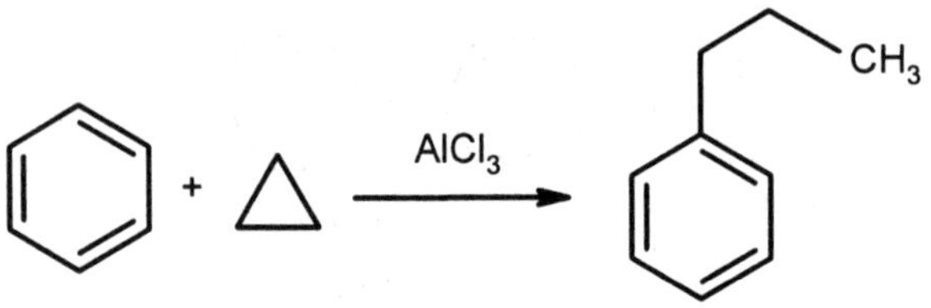

3. C₆H₆ + cyclopropane →(AlCl₃) PhCH₂CH₂CH₃

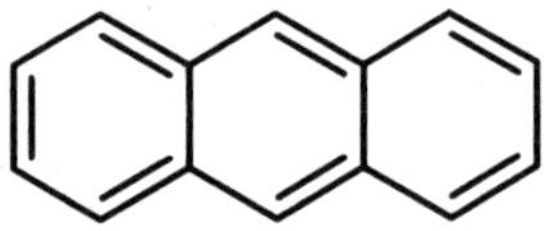

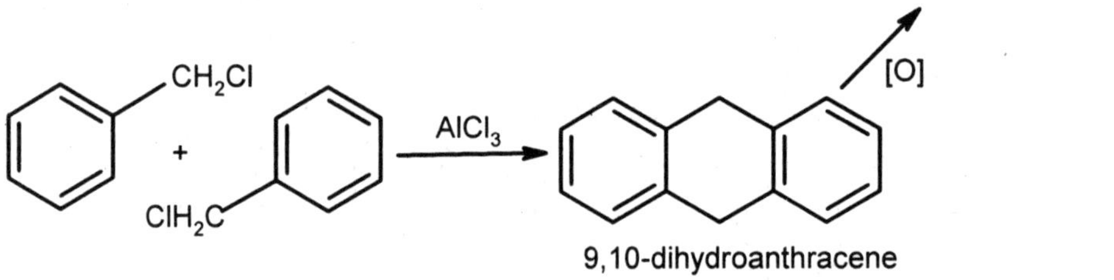

4. PhCH₂Cl + ClCH₂Ph →(AlCl₃) 9,10-dihydroanthracene →[O] anthracene

5. naphthalene + ClCOCOCl (oxalylchloride) →(AlCl₃) (7,8-acenaphthaquinone) →(Na-Hg/HCl) (acenaphthene)

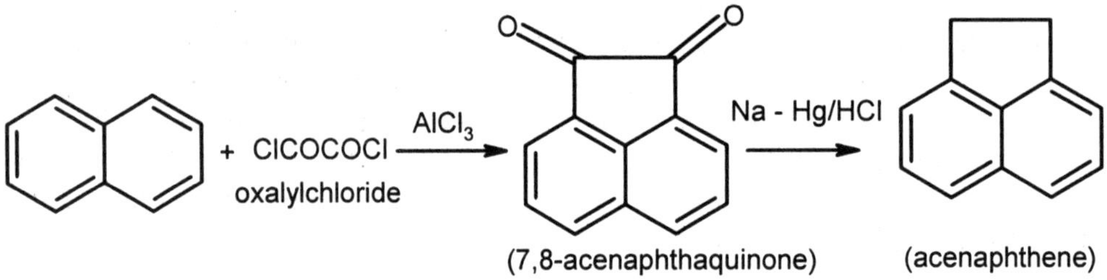

6. C₆H₆ + succinic anhydride →(AlCl₃) PhCO(CH₂)₂COOH →(Na-Hg/HCl) Ph(CH₂)₃COOH →(conc. H₂SO₄) α-tetralone →(i. Na-Hg/HCl, ii. Se) naphthalene

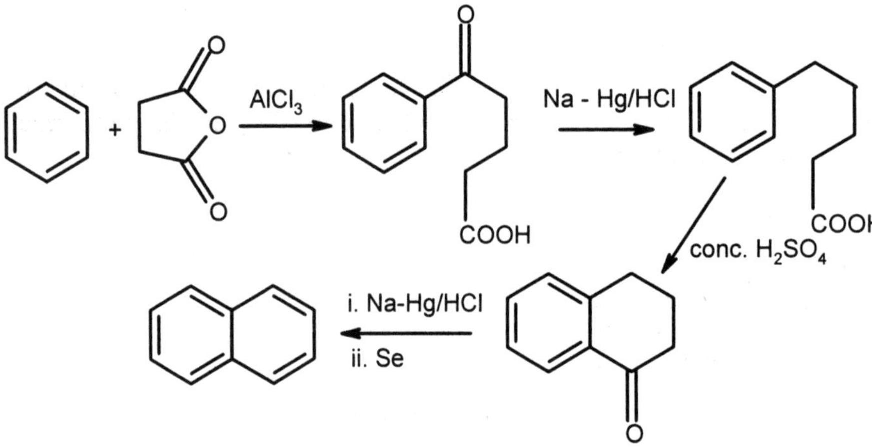

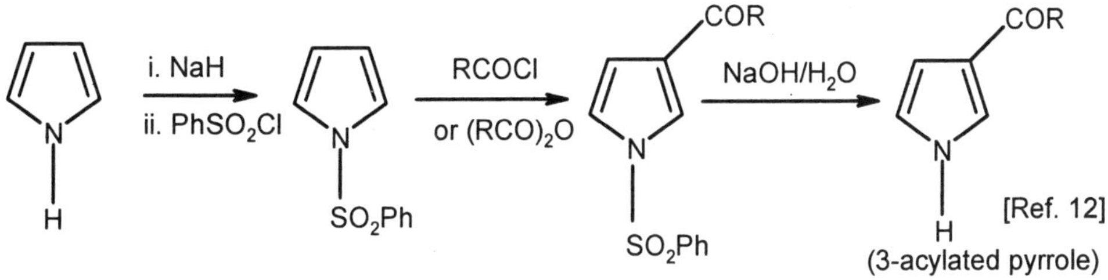

11. Aliphatic and alicyclic compounds also undergo *Friedel-Crafts reactions*. In cases of acylation of olefins, silyl group is usually incorporated into the substrate molecule to avoid strenuous reaction conditions. For example, acylation of trialkyl-

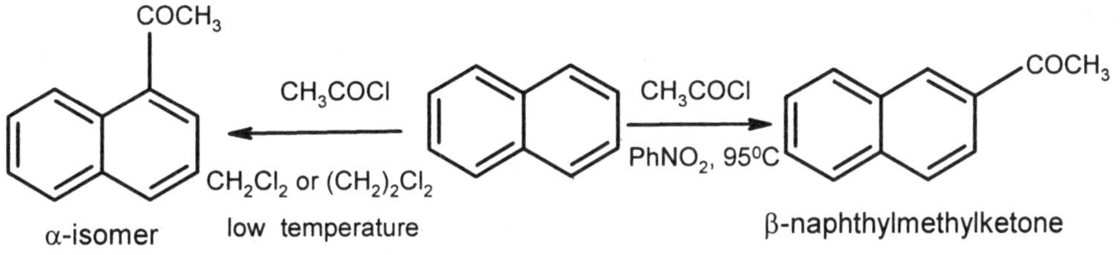

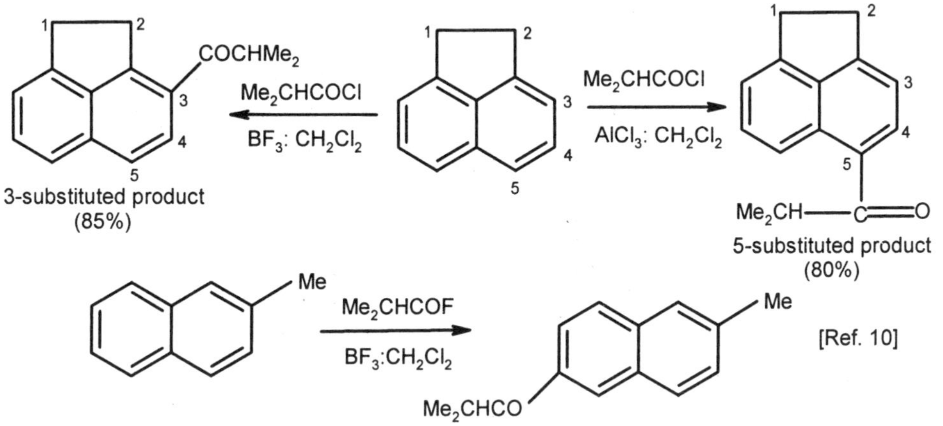

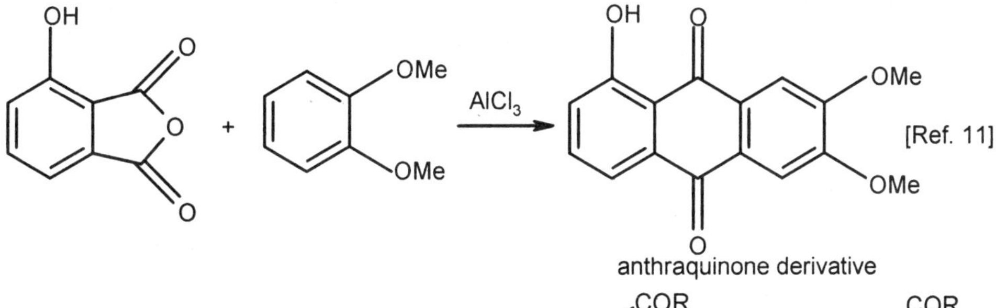

silylolefins under mild Friedel-Crafts conditions produce α,β-unsaturated ketones by replacement of the trialkylsilyl group by the acyl moiety.[13]

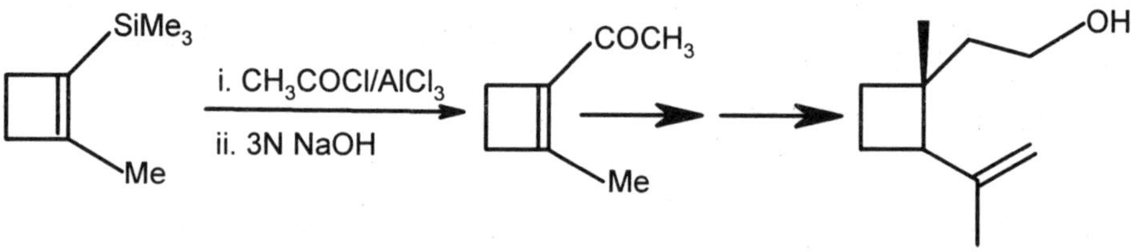

Cyclohexene on reaction with crotonyl chloride yields the cyclopentenone derivative as one of the products, which is an important precursor of terpenoids.[14]

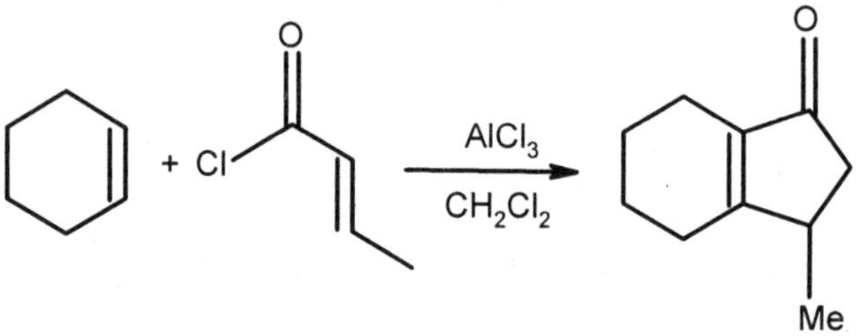

Cyclododecene (*cis* and *trans*) reacts slowly with crotonyl bromide at −78°C to furnish a ketone that is an important intermediate in the synthesis of (+/−)-muscone, a effective perfumery product.[15]

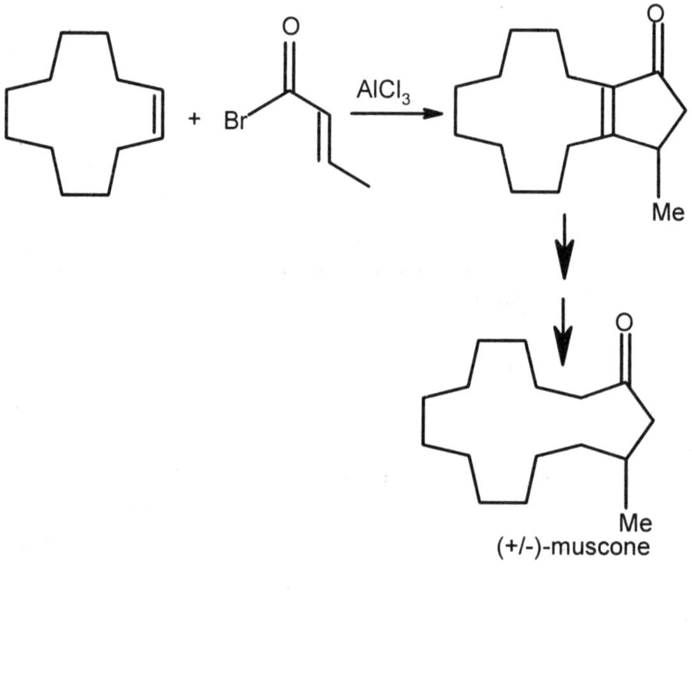

(+/−)-muscone

References

1. Friedel, P. and Crafts, J. M. (1877), *Compt. Rend.,* **84**, 1392.
2. Calloway, N. O. (1937), *J. Am. Chem. Soc.,* **59**, 1474; Brown, H. C. and Jungk, H. (1955), *J. Am. Chem. Soc.,* **77**, 5584.
3. Olah, G. A. and Kuhn, S. J. (1964), *J. Org. Chem.,* **29**, 2317.
4. Olah, G. A., *Friedel-Crafts and Related Reactions,* **vol. 1**, Wiley: NY, 1963 – 1965, pp. 201, 853.
5. Russell, G. A. (1959), *J. Am. Chem. Soc.,* **81**, 4834; Yakobson, G. G. and Furin, G. G. (1980), *Synthesis,* 345. [Review]
6. Brown, H. C. and Jungk, H. (1956), *J. Am. Chem. Soc.,* **78**, 2182; Kalchschmid, F. and Mayer, E. (1976), *Angew. Chem. Int. Ed. Engl.,* **15**, 773.
7. Taylor, R., *Electrophilic Aromatic Substitution,* and Wiley: NY, 1990, p.222.
8. Corriu, R., Dore, M. and Thomassin, R. (1971), *Tetrahedron,* **27**, 6501, 5819; Tan, L. K. and Brownstein, S. (1983), *J. Org. Chem.,* **48**, 302.
9. Kovacic, P. and Jones, M. B. (1987), *Chem. Rev.,* **87**, 357. [Review]
10. Hyatt, J. A. and Raynolds, F. W. (1984), *J. Org. Chem.,* **49**, 384.
11. Danielson, K. (1956), *Acta Chem. Scand.,* **50**, 954.
12. Rokach, J., Hamel, P., Kakushima, M. and Smith, G. M. (1981), *Tetrahedron Lett.,* **22**, 4901.
13. Fleming, I., Dunogues, J. and Smithers, R. H. (1989), *Org. React.,* **37**, 57.
14. Negishi, E., Boardman, L. D., Tour, J. M., Sawada, H. and Rand, C. L. (1983), *J. Am. Chem. Soc.,* **105**, 6344.
15. Hacini, S., Pardo, R. and Santelli, M. (1979), *Tetrahedron Lett.,* 4553.

- Mahato, S. B. (2000), *J. Indian Chem. Soc.,* **77**, 175. [Review]
- Ottoni, O., Neder, A. V. F., Dias, A. K. B., Cruz, R. P. A. and Aquino, L. B. (2000), *Org. Lett.,* **3**, 1005(2000).
- Cheug, Y., Ye, H.-Y., Zhan, Y.-H. and Meth-Cohn, O. (2001), *Synthesis,* 904.
- Fleming, I. (2001), *Chemtracts: Org. Chem.,* **14**, 405. [Review]
- Le Roux, C. and Dubac, J. (2002), *Synlett,* 181.
- Sefkow, M. and Buchs, J. (2003), *Tetrahedron Lett.,* **44**, 193.
- Jorgensen, K. A. (2003), *Synthesis,* 1117.
- Khalaf, A. A. and Albar, H. A. (2004), *J. Indian Chem. Soc.,* **81**, 518.
- Ray, S., Srivastava, N., Sangita and Atual Kumar (2004), *Synth. Commun.,* **34**(13), 2345.

27

Gabriel Synthesis

Introduction

Gabriel synthesis[1] offers a method for the synthesis of primary amines from alkyl halides using phthalimide anion as a protected form of ammonia that cannot alkylate more than once. Hence, the primary amines formed in this reaction will be uncontaminated by secondary or tertiary amines.

Phthalimide has one acidic N–H proton (pKa 8.3) that is abstracted by potassium hydroxide to give the phthalimide anion — which acts as a nucleophile toward primary alkyl halides in a bimolecular nucleophilic substitution (S_N2) process; the resulting product on hydrolysis (either basic or acidic) yields the desired primary amine. The reaction is usually rather slow but can be conveniently speeded by the use of a dipolar aprotic solvent such as N,N-dimethylformamide (DMF)[2] or crown ether[3].

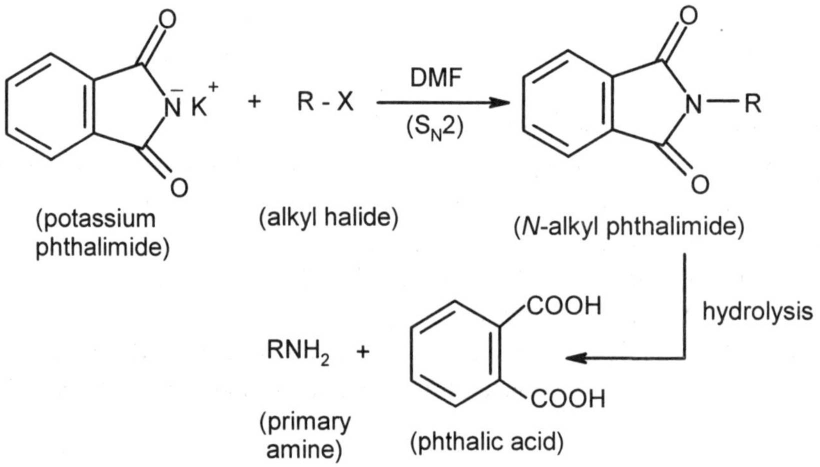

Mechanism

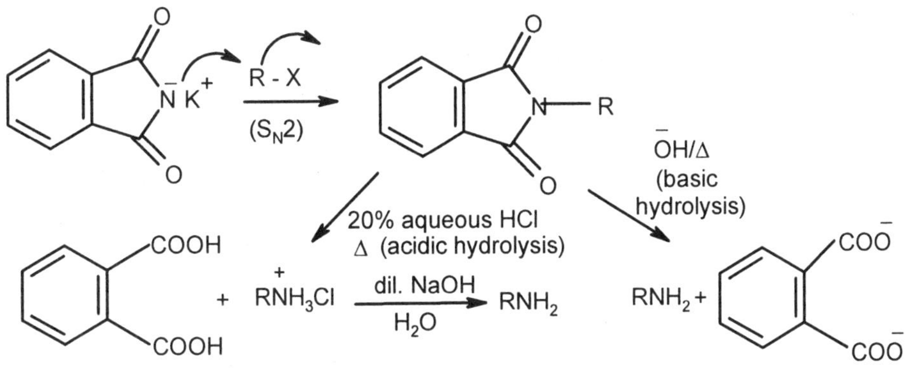

Critical Views

Hydrolysis of *N*-alkylphthalimide, whether acid or base catalyzed, is usually very slow; that's why a more effective method of cleaving the two amide bonds is in common use where the phthalimide derivative is heated with hydrazine so that hydrazine displaces the primary amine, giving the very stable hydrazide of phthalimide. This procedure is the *Ing-Manske modification*[4].

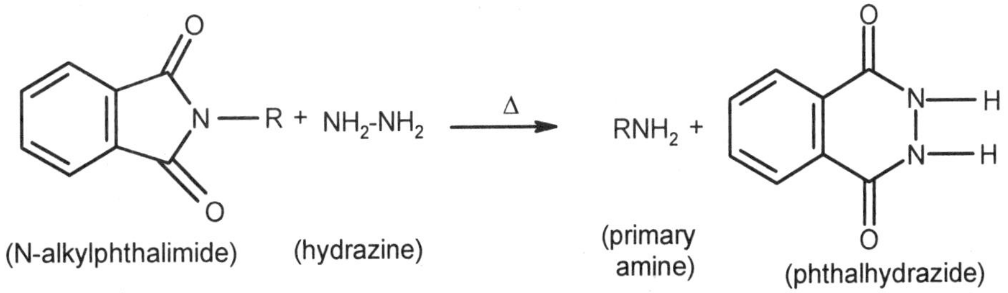

Aryl halides cannot be converted to arylamines by the *Gabriel synthesis*, because they do not undergo nucleophilic substitution with *N*-potassiophthalimide in the first step of the procedure.

Applications

Since phthalimide can undergo only a single alkylation, the formation of secondary and tertiary amines does not occur, and the *Gabriel synthesis* is a valuable procedure for the laboratory preparation of primary amines. Besides simple alkyl halides, α-halo ketones and α-halo esters have been employed as substrates in this method. Alkyl *p*-toluenesulphonate esters have also been used.

i) [Phthalimide potassium salt] + CH$_3$CH$_2$CH$_2$CH$_2$Br $\xrightarrow{\text{DMF}}$ [N-butylphthalimide] —CH$_2$(CH$_2$)$_2$CH$_3$

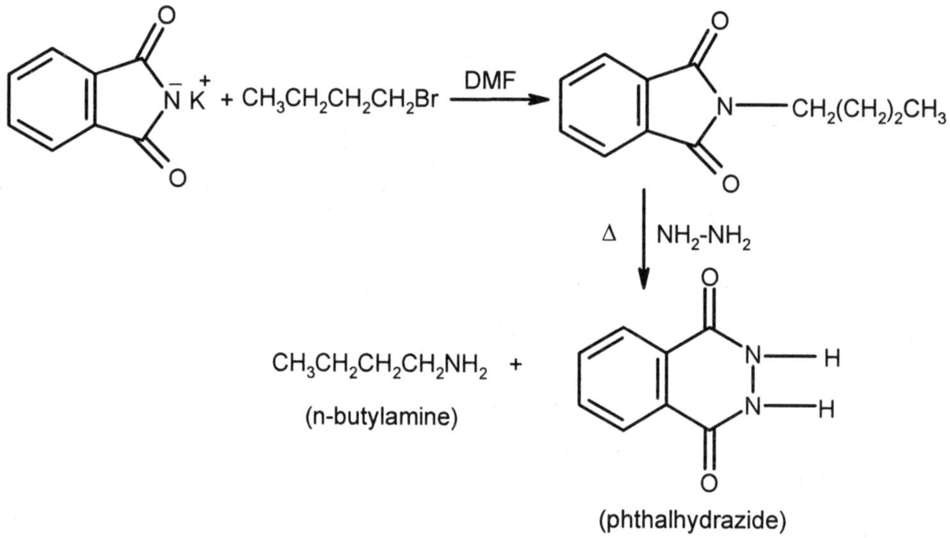

$\xrightarrow{\Delta,\ NH_2-NH_2}$

CH$_3$CH$_2$CH$_2$CH$_2$NH$_2$ + (phthalhydrazide)
(n-butylamine)

ii) [Phthalimide potassium salt] + [benzyl chloride] —CH$_2$Cl $\xrightarrow{\text{DMF}}$ [N-benzylphthalimide] —CH$_2$Ph

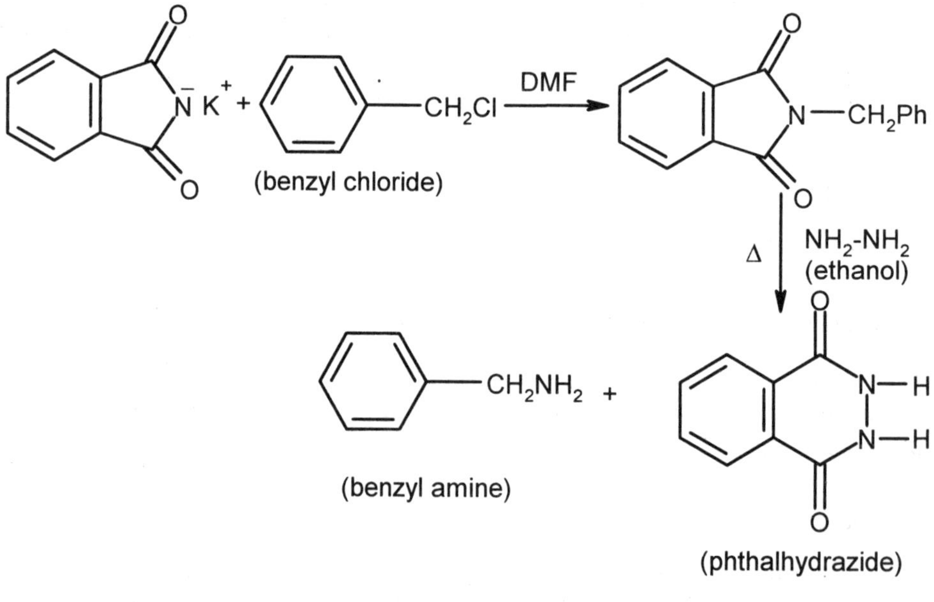

$\xrightarrow[\text{(ethanol)}]{\Delta,\ NH_2-NH_2}$

Ph—CH$_2$NH$_2$ + (phthalhydrazide)
(benzyl amine)

Gabriel Synthesis

iii) 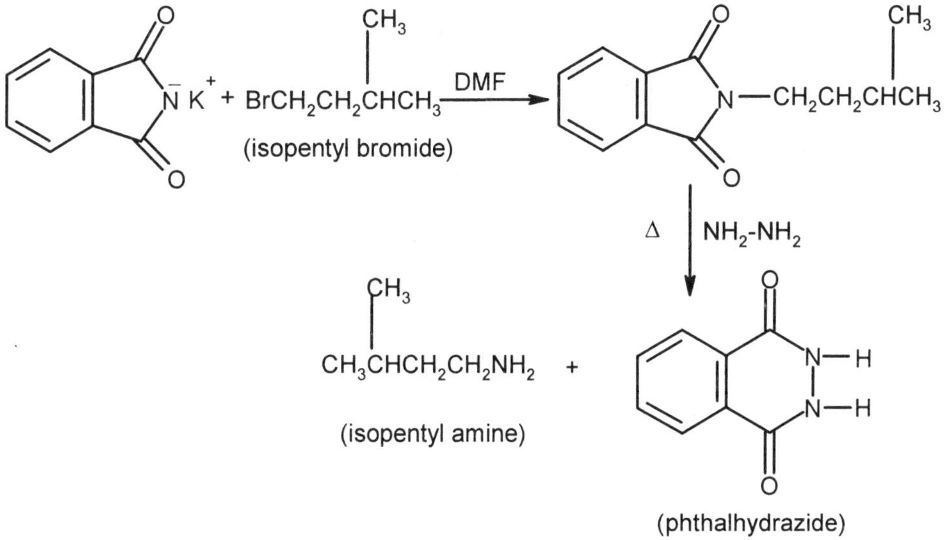 [Reaction: Potassium phthalimide + BrCH₂CH₂CH(CH₃)CH₃ (isopentyl bromide) →(DMF)→ N-substituted phthalimide →(Δ, NH₂-NH₂)→ CH₃CH(CH₃)CH₂CH₂NH₂ (isopentyl amine) + phthalhydrazide]

iv) The phthalimide synthesis is much useful for making amines that also contain other functional groups. 4-Aminobutanoic acid (gamma-aminobutyric acid, commonly known as GABA), which functions as an agent in the transmission of nerve impulses, is synthesized using this method:

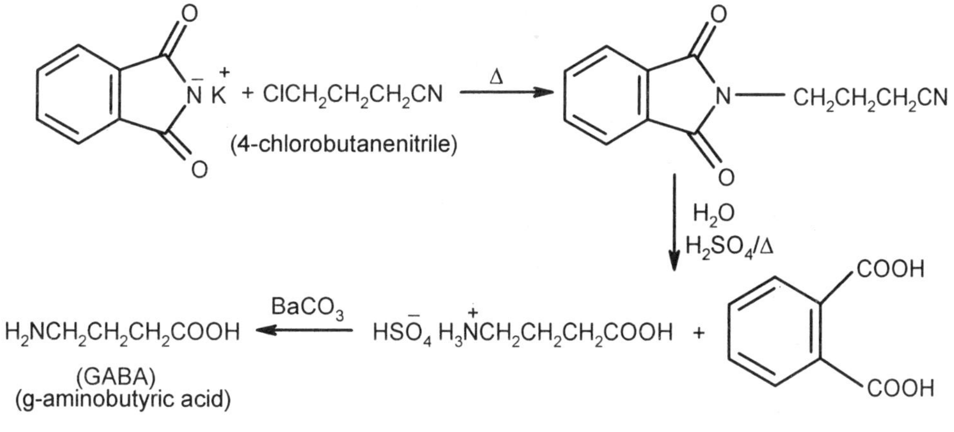

[Reaction: Potassium phthalimide + ClCH₂CH₂CH₂CN (4-chlorobutanenitrile) →(Δ)→ N-CH₂CH₂CH₂CN phthalimide →(H₂O, H₂SO₄/Δ)→ HSO_4^- $H_3N^+CH_2CH_2CH_2COOH$ + phthalic acid (o-C₆H₄(COOH)₂) →(BaCO₃)→ H₂NCH₂CH₂CH₂COOH (GABA, g-aminobutyric acid)]

The carboxylic acid group of the amino acid is protected as the nitrile, which is hydrolyzed in the second step of the process.

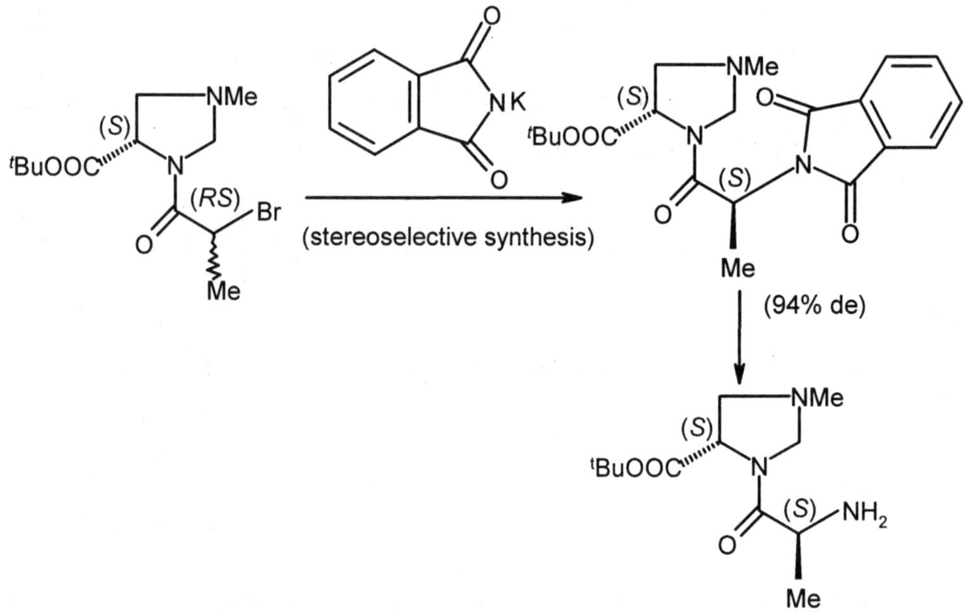

References

1. Gabriel, S. (1887), *Ber. Dtsch. Chem. Res.*, **20**, 2224.
2. Sheehan, J. C. and Bolhofer, W. A. (1950), *J. Am. Chem. Soc.*, **72**, 2786; Landini, D. and Rolla, F. (1976), *Synthesis*, 389.
3. Soai, K., Ookawa, A. and Kato, K. (1982), *Bull. Chem. Soc. Jpn.*, **55**, 1671.
4. Ing, H. R. and Manske, R. H. F. (1926), *J. Chem. Soc.*, 2348; Khan, M. N. (1995), *J. Og. Chem.*, **60**, 4536.

- Press, J. B., Haug, M. F. and Wright, W. B. (1985), *Synth. Commun.*, **15**, 837.
- Slusarska, E. and Zwierzak, A. (1986), *Justus Liebigs Ann. Chem.*, 402.
- Han, Y. and Hu, H. (1990), *Syhthesis*, 122.
- Ragnarsson, U. and Grehn, L. (1991), *Acc. Chem. Res.*, **24**, 285 (Review).
- Toda, F., Soda, S. and Goldberg, I. (1993), *J. Chem. Soc. Perkin Trans. 1*, 2357.
- Mamedov, V. A., Tsuboi, S., Mustakimova, L. V., Hamamoto, H., Gubaidullin, A. T., Litvinov, I. A. and Levin, Y. A. (2001), *Chem. Heterocycl. Compd.*, **36**, 911.

28

Gattermann – Koch Reaction

Introduction

The formylation of benzene and alkylbenzenes using carbon monoxide and hydrogen chloride in the presence of aluminium chloride (catalyst) and a small amount of cuprous chloride (co-catalyst) under high pressure is known as *Gattermann-Koch reaction*[1].

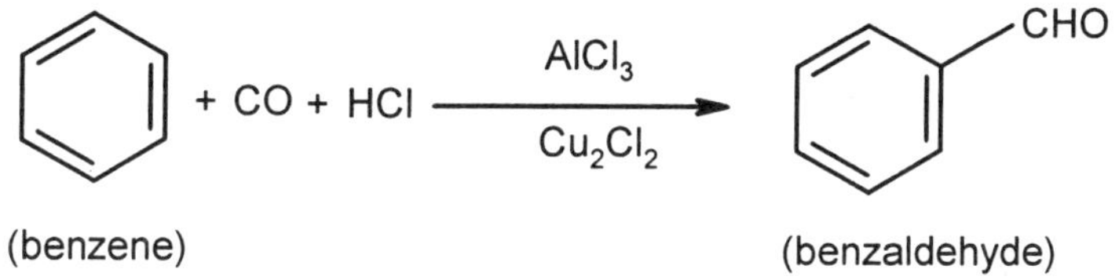

$$\text{benzene} + CO + HCl \xrightarrow{AlCl_3, Cu_2Cl_2} \text{benzaldehyde}$$

(benzene) → (benzaldehyde)

Usually, nitrobenzene or ether is used as solvent. In case of alkylbenzenes, the aldehyde group is introduced into the *para*-position only[2]. The method is used industrially to prepare aryl aldehydes.

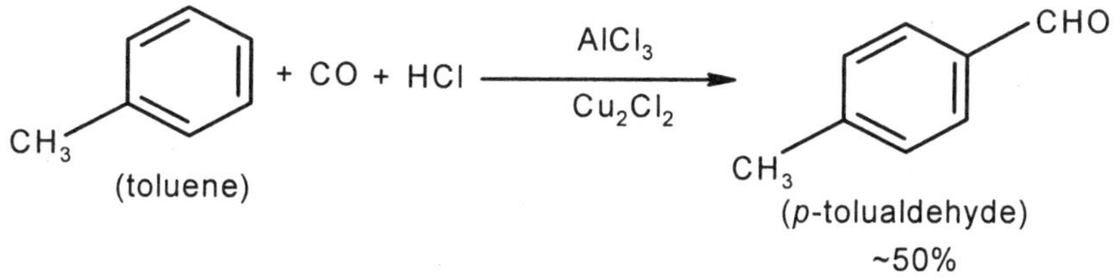

$$\text{toluene} + CO + HCl \xrightarrow{AlCl_3, Cu_2Cl_2} p\text{-tolualdehyde}$$

(toluene) → (*p*-tolualdehyde) ~50%

The *Gattermann-Koch aldehyde synthesis* is not applicable to phenols or their ethers, amino aromatic species and also when the aromatic ring is strongly deactivated (*e.g.* nitrobenzene).

Mechanism

The *Gattermann-Koch formylation* is considered as a typical electrophilic aromatic substitution with high *para* regioselectivity[3]. The most likely electrophile is the acylium ion $[HCO]^+$ in the ion pair, $[HCO]^+[AlCl_4]^-$.

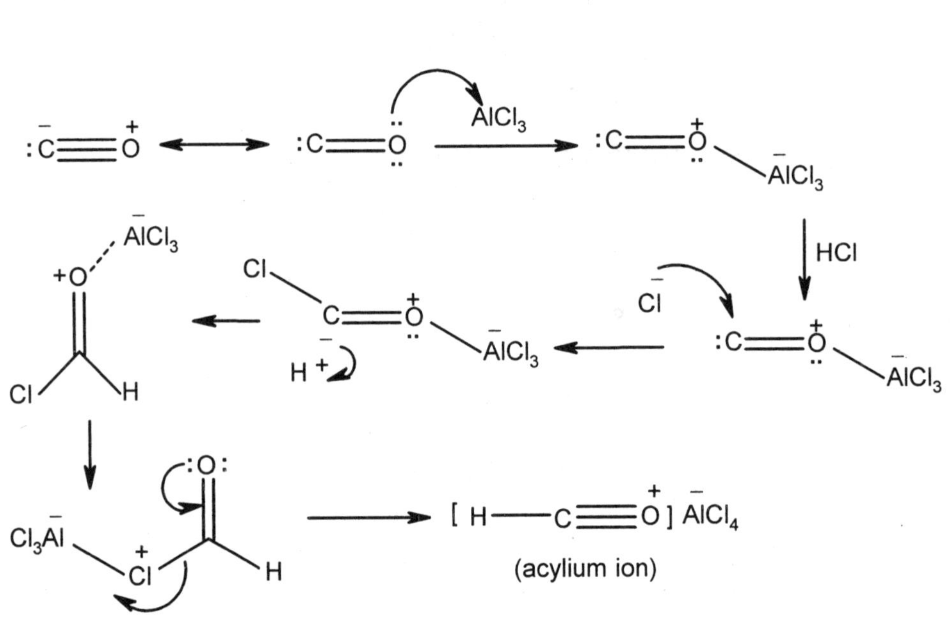

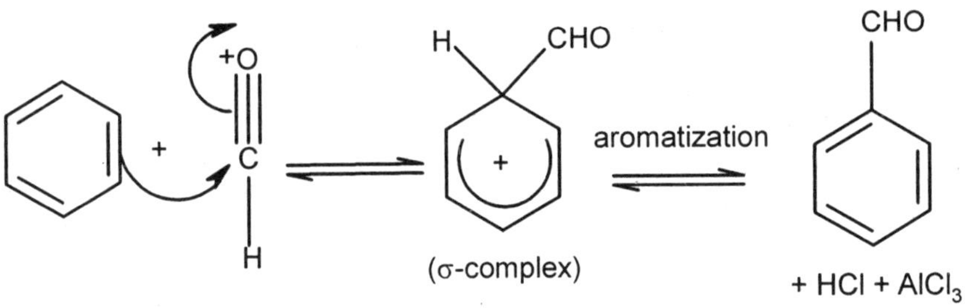

Critical Views

(i) The reaction, in fact, remains in an equilibrium that lies unfavourably for the product formation, but is pulled over to the right by complexing of the aldehyde with the Lewis acid catalyst.

(ii) It is believed that cuprous chloride forms a complex with carbon monoxide, thereby increasing its local concentration and this probably speeds up the formation of acylium ion (attacking electrophile).

(iii) Common factors such as electron density of the aromatic substrate[4], reactivity of electrophiles[5], stability of reaction intermediates[6], and steric factors[7] may influence the regioselectivity.

Applications

(1-methylnaphthalene)

CO + HCl (conventional type) → 4-methyl-1-naphthaldehyde + 1-methyl-2-naphthaldehyde

SbF$_5$ – HF/CO or F$_3$CCOOH – SbF$_5$/CO (modified process) → 4-methyl-1-naphthaldehyde

[Ref. 8]

References

1. Gattermann, L. and Koch, J. (1897), *Chem. Ber.*, **30**, 1622.
2. Moersch, G. W. and Zwiesler, M. L. (1971), *Synthesis*, 647.
3. Olah, G. A., Ohannesian, L. and Arvanaghi, M. (1987), *Chem. Rev.*, **87**, 671; Tanaka, M., Iyoda, J. and Souma, Y. (1992), *J. Org. Chem.*, **57**, 2677.
4. Pederson, E. B., Peterson, T. E., Torssell, K. and Lawesson. (1973), *Tetrahedron*, **29**, 579; Kita, Y., Tohma, H., Hatanaka, K., Tadeka, T., Fujita, S., Mitoh, S., Sakurai, H. and Oka, S. (1994), *J. Am. Chem. Soc.*, **116**, 3684.
5. Olah, G. A., Kobayashi, S. and Nishimura, J. (1973), *J. Am. Chem. Soc.*, **95**, 564.
6. Olah, G. A. and Melby, E. G. (1974), *J. Org. Chem.*, **39**, 1203; Olah, G. A., Hashimoto, I. and Lin, H. C. (1977), *Proc. Natl. Acad. Sci. U.S.A.*, **74**, 4121.
7. Brown, H. C., Bolto, B. A. and Jenson, F. R. (1958), *J. Org. Chem.*, **23**, 414.
8. Tanaka, M., Fujiwara, M., Xu, Q., Souma, Y., Ando, H. and Laali, K. K. (1997), *J. Am. Chem. Soc.*, **119**, 5100; Tanaka, M., Fujiwara, M., Xu, Q., Ando, H. and Raeker, T. J. (1998), *J. Org. Chem*, **63**, 4408.

- Crounse, N. N. (1949), *Org. React.*, **5**, 290 (Review).
- Tanaka, M., Fujiwara, M., Ando, H. and Souma, Y. (1996), *Chem. Commun.*, 159.
- Tanaka, M., Fujiwara, M. and Ando, H. (1995), *J. Org. Chem.*, **60**, 2106.
- de Rege, P. J. F., Gladysz, J. A. and Horvath, I. T. (1997), *Science*, **276**, 776.
- Doana, M. I., Ciuculescu, A., Bruckner, A., Pop,, M. and Filip, P. (2002), *Rev. Roum. Chim.*, **46**, 345.

29

Haller – Bauer Reaction

Introduction

Cleavage of non-enolizable ketones with sodium amide leading to carboxylic amide derivative and a neutral fragment in which the carbonyl group is replaced by hydrogen is called the *Haller-Bauer reaction*[1].

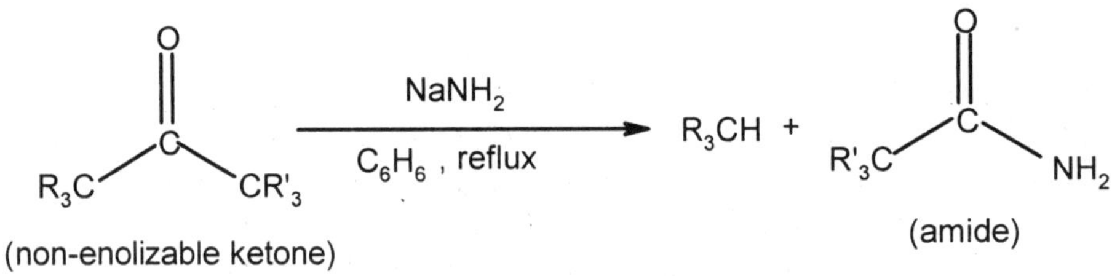

$$R_3C-CO-CR'_3 \xrightarrow[C_6H_6, \text{ reflux}]{NaNH_2} R_3CH + R'_3C-CO-NH_2$$

(non-enolizable ketone)   (amide)

Ketones of the type $ArCOCR_3$ yield R_3CONH_2, which is not easily attainable by other methods:

$$ArCOCR_3 \xrightarrow[C_6H_6, \text{ reflux}]{NaNH_2} Ar-H + R_3CCONH_2$$

The amide derivative can easily be converted to useful trisubstituted acetic acid:

$$R_3CCONH_2 \xrightarrow{HNO_2} R_3CCOOH \text{ (trisubstituted acetic acid)}$$

Mechanism

The NH_2^- nucleophile attacks the carbonyl acrbon followed by C–C σ bond cleavage. The NH_2 loses its proton before the 'CR$_3$' group is cleaved[2].

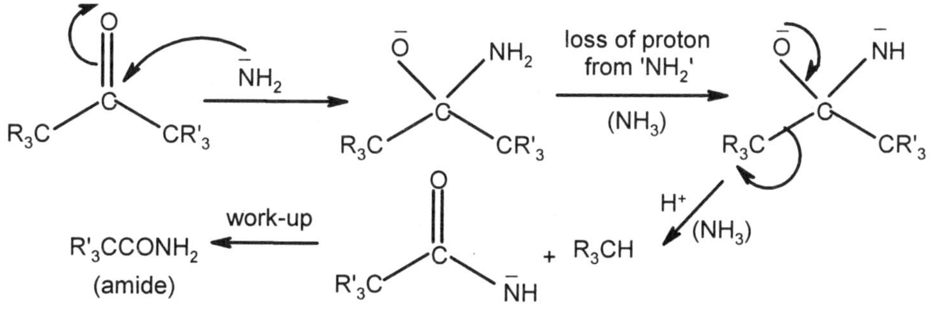

Haller–Bauer Reaction

Crtical Views

(a) Besides sodium amide, base-induced *Haller-Bauer type cleavage* is frequently carried out with aqueous sodium hydroxide in benzene or in 2-methoxyethanol, potassium *tert*-butoxide in dimethylsulphoxide-water, and aqueous sodium hydroxide in tetrahydrofuran.

Mehta and Reddy[3] carried out the *Haller-Bauer reaction* on the substrate (**1**) in presence of 50% aq. NaOH in benzene, and isolated the mixture of products (**2** & **3**) as methyl ester after treatment with diazomethane.

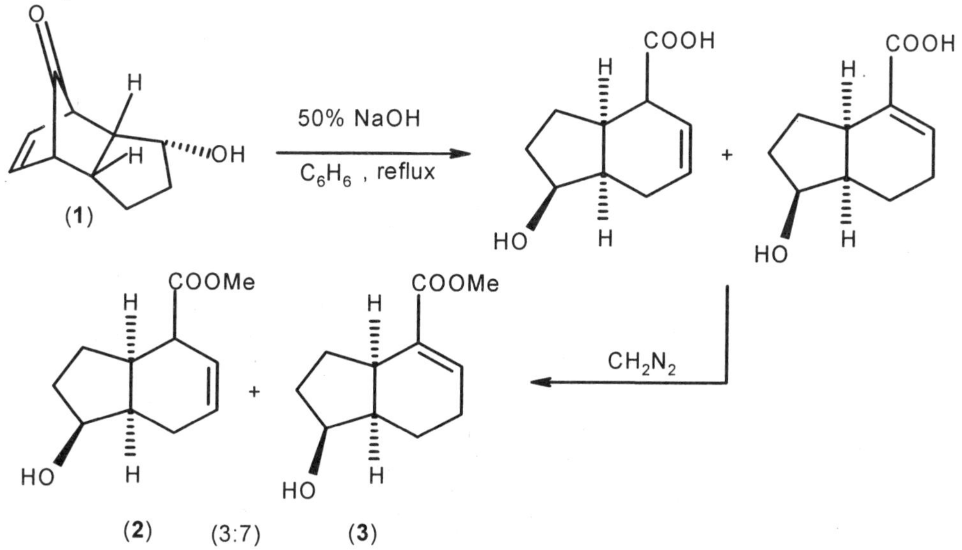

(b) It has been shown that the configuration of optically active alkyl group (of neutral fragment, R_3CH) is retained[4].

Applications

The base-induced *Haller-Bauer type cleavage* of non-enolizable ketones finds immense application in organic synthesis[5]. One or two representative examples are cited:

(i) *Haller-Bauer type cleavage* in norbornane derivatives offers an easy access to *cis*-1,3-disubstituted cyclopentanes, which have been used as intermediate in the synthesis of terpenoids[6] as well as in the synthesis of carbocyclic nucleosides[7].

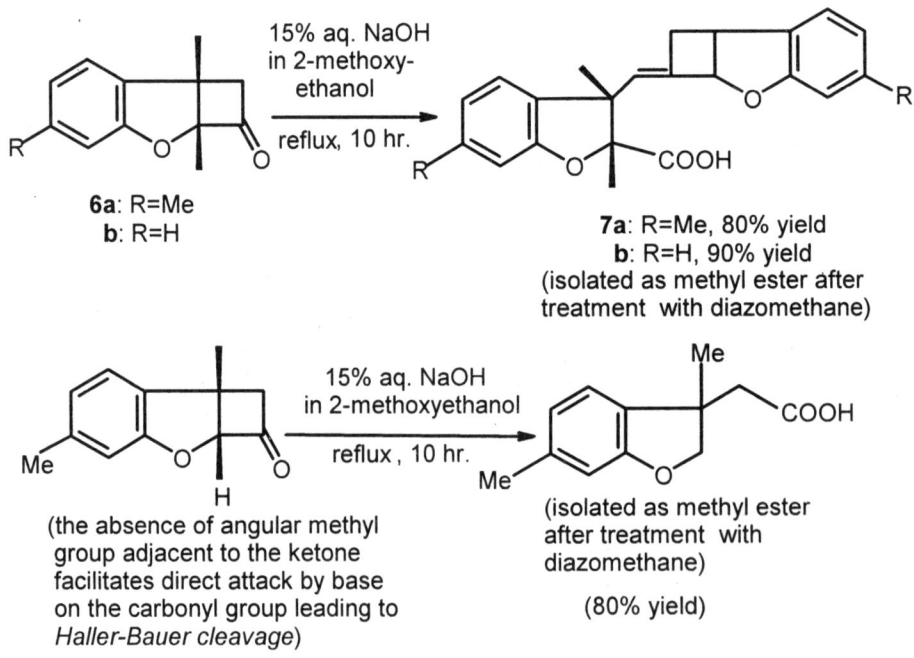

(ii) In 1998 Mitra *et al.*[9] carried out *Haller-Bauer type cleavage* of cyclobutabenzofuranones leading to the formation of useful molecule.

References

1. Haller, A. and Bauer, E. (1908), *Compt. Rend.*, **147**, 824; (1909), *ibid*, **148**, 70, 127; (1909), *ibid*, **149**, 5; (1914), *Ann. Chem.*, **I**, 5; Haller, A. (1922), *Bull. Soc. Chim. France* **31**, 1117.
2. Bunnett, J. E. and Hrutfiord, B. F. (1962), *J. Org. Chem.*, **27**, 4152.
3. Mehta, G. and Reddy, D. S. (1997), *Synlett*, 612.
4. Impasto, F. J. and Walborsky, H. M. (1962), *J. Am. Chem. Soc.*, **84**, 4838; Paquette, L. A. and Gilday, J. P. (1988), *J. Org. Chem.*, **53**, 4972; Paquette, L. A. and Ra, C. S. (1988), *J. Org. Chem.*, **53**, 4978.
5. Mehta, G. and Reddy, D. S. (1996), *SynLett*, **3**, 229; Mehta, G., Reddy, K. S. and Kunwar, A. C. (1996), *Tetrahedron Lett.*, **37**, 2289; Mehta, G. and Praveen, M. (1995), *J. Org. Chem.*, **60**, 279; Paquette, L. A. and Maynard, G. D. (1989), *J. Org. Chem.*, **54**, 5054; Guir, F., Do Khac, D., Benchikh-le-Hocine, M., Fetizon, M., Neuman, A. and Prange, T. (1991), *Acta Cryst.*, **C47**, 2109.
6. Curran, D. P. and Chen, M. (1985), *Tetrahedron Lett.*, **26**, 4991.
7. Crimmins, M. T. (1998), *Tetrahedron*, **54**, 9229.
8. Hamlin, K. E. and Weston, W. A. (1957), *Org. React.*, **9**, 1.
9. Mitra, A., Bhowmik, D. R. and Venkateswaran, R. V. (1998), *J. Org. Chem.*, **63**, 9555.

- Guir, F., Do Khac, D., Benchikh-le-Hocine, M. and Fetizon, M (1993), *Synthesis*, 775.
- Paquette, L. A. and Gilday, J. P. (1990), *Org. Prep. Proc. Int.*, **22**, 167.
- Mehta, G. and Venkateswaran, R. V. (2000), *Tetrahedron*, **56**, 1399.
- Arjona, O., Medel, R. and Plumet, J. (2001), *Tetrahedron Lett.*, **42**, 1287.
- Ishihara, K. and Yano, T. (2004), *Org. Lett.*, **6**, 1983.

30

Hell – Volhard – Zelinsky Reaction

Introduction

The *Hell-Volhard-Zelinsky reaction*[1] offers a method for preparing α-chloro- or bromo carboxylic acids by treating the acid with chlorine or bromine in the presence of a small amount of red phosphorus or phosphorus halide (PBr_3 or PCl_3) as catalyst, followed by water to hydrolyze the intermediate α-chloro/bromo acyl halide.

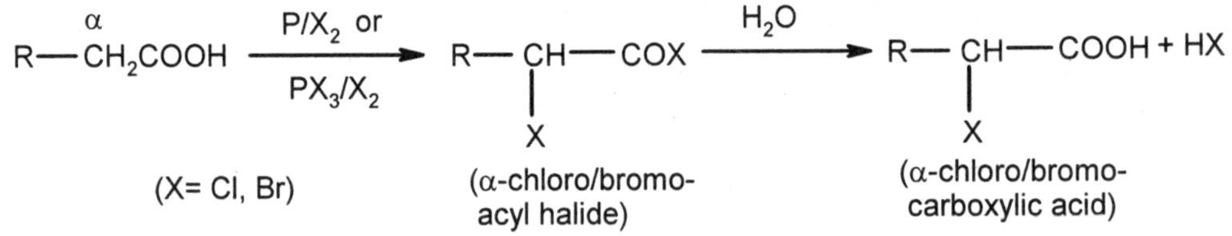

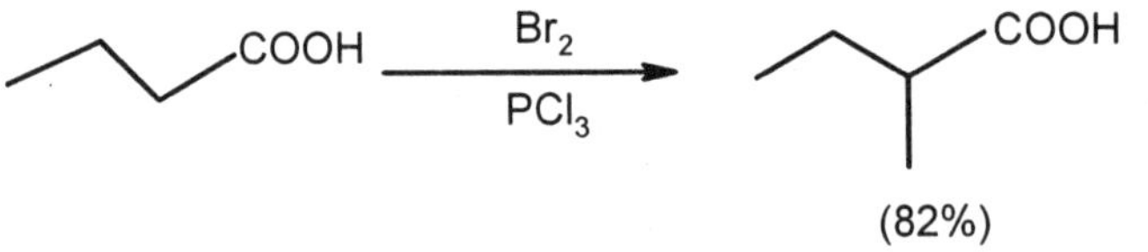

$$R\text{—}CH_2COOH \xrightarrow[PX_3/X_2]{P/X_2 \text{ or}} R\text{—}CH(X)\text{—}COX \xrightarrow{H_2O} R\text{—}CH(X)\text{—}COOH + HX$$

(X = Cl, Br)   (α-chloro/bromo-acyl halide)   (α-chloro/bromo-carboxylic acid)

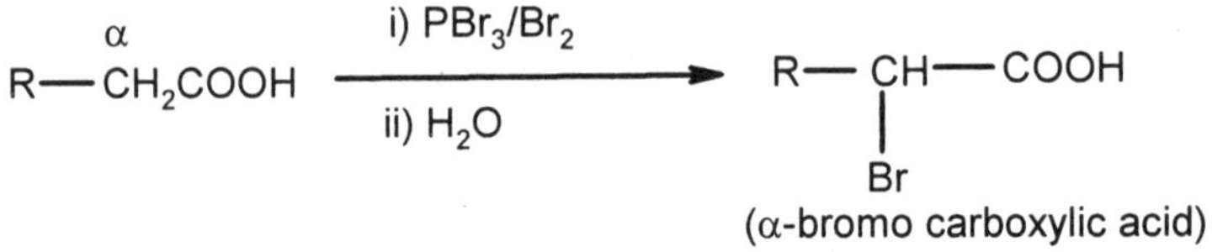

$$R\text{—}CH_2COOH \xrightarrow[ii) H_2O]{i) PBr_3/Br_2} R\text{—}CH(Br)\text{—}COOH$$

(α-bromo carboxylic acid)

$$\text{CH}_3\text{CH}_2\text{CH}_2\text{COOH} \xrightarrow{Br_2 / PCl_3} (CH_3)_2 C(Br) COOH \quad (82\%) \quad [\text{Ref. 2}]$$

The reaction is usually not applicable to iodine or fluorine. The acyl intermediate can undergo bromide exchange with unreacted carboxylic acid that allows the catalytic cycle to continue until the conversion is complete.

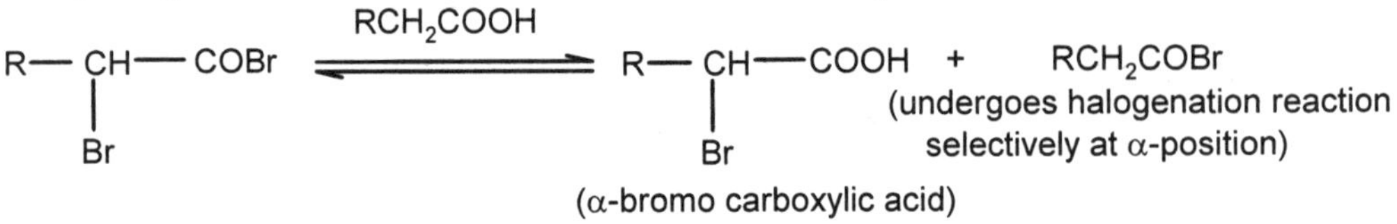

$$R\text{—}CH(Br)\text{—}COBr \; \xrightleftharpoons{RCH_2COOH} \; R\text{—}CH(Br)\text{—}COOH \; + \; RCH_2COBr$$

(α-bromo carboxylic acid)   (undergoes halogenation reaction selectively at α-position)

When there are two α-hydrogens, both of them may be replaced by using excess of the halogen. Though it is often hard to stop the reaction at the monosubstituted stage, it can be achieved by the use of calculated amount of halogen, and due to the fact that enol form of monohalogenated compound becomes less nucleophilic. In case of a di- or polycarboxylic acids, the α-H atoms with respect to each carboxylic group may be replaced by halogen, if sufficient amount of halogen is used. Since the reaction with bromine is specific for α-H atoms, it can be used to detect the presence of α-H in the corresponding acids.

Mechanism

The reaction actually takes place on the acyl halide formed from the carboxylic acid and the catalyst. α-Halogenation of the acid, presumably involves the enol form of this intermediate acyl halide[3]; the enol is nucleophilic, attacking bromine (or chlorine) to give the α-brominated (or chlorinated) acyl halide, which on hydrolysis yields the α-bromo- (or chloro) carboxylic acid.

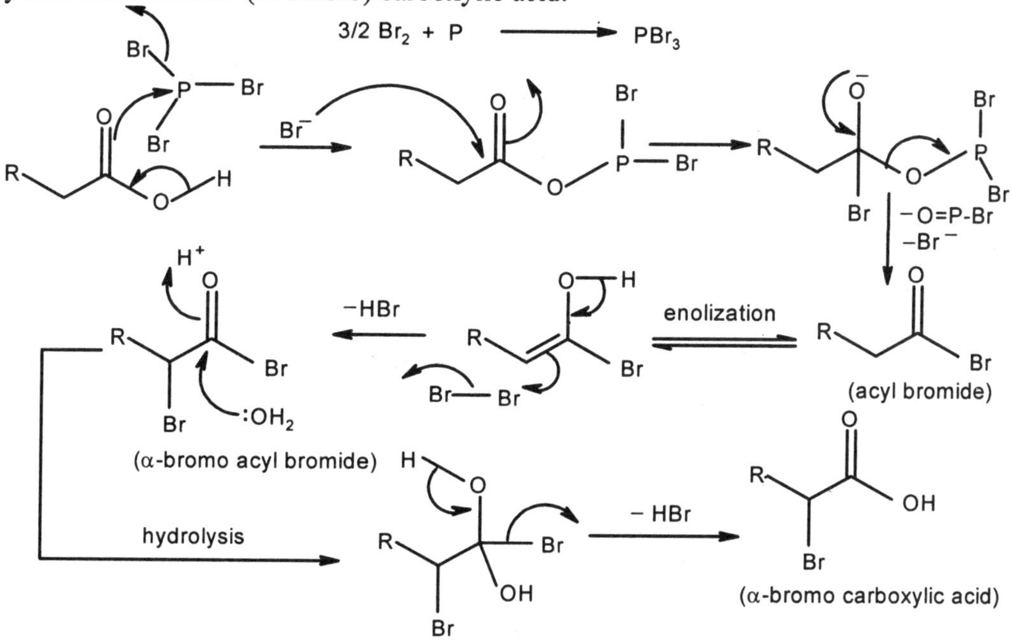

Critical Views

(a) Each molecule of carboxylic acid is α–halogenated while it is in the acyl halide stage. The halogen from the catalyst (PX_3) does not enter the α-position. Thus, the use of Br_2 and PCl_3 results in the α-bromination, not chlorination.

$$\text{C}_6\text{H}_5-\text{CH}_2\text{COOH} \xrightarrow[\text{benzene, 80°C}]{\text{Br}_2,\ \text{PCl}_3} \text{C}_6\text{H}_5-\underset{\underset{\text{Br}}{|}}{\text{CH}}\text{COOH} \quad (60-62\%)$$

(phenylacetic acid) → (α-bromophenylacetic acid)

Therefore, it may be concluded that acyl halides would undergo α-halogenation without catalyst, and this is the fact. Actually, carboxylic acid derivatives that can easily be enolizable (*e.g.* acyl halides, anhydrides) and other many compounds (*e.g.* malonic ester, aliphatic nitro compounds) do not need catalyst.

(b) *N*-Bromosuccinimide in a mixture of sulphuric acid-trifluoroacetic acid can monobrominate simple carboxylic acids[4].

(c) α-Iodination may be achieved in the presence of chlorosulphuric acid ($ClSO_2OH$) as catalyst[5]. α–Iodoacids may also be obtained from the corresponding α-chloro- or α-bromo-acids by warming them with potassium iodide in methanol or acetone.

$$\text{R}-\underset{\underset{\text{Br}}{|}}{\text{CH}}-\text{COOH} + \text{KI} \longrightarrow \text{R}-\underset{\underset{\text{I}}{|}}{\text{CH}}-\text{COOH}$$

Applications

The *Hell–Volhard–Zelinsky reaction* finds useful applications in synthetic organic chemistry. It offers a selective α-chlorination or α-bromination of carboxylic acids. If a derivative of the α-haloacid is desired, the α-halo acyl halide serves as an activated intermediate (similar to acid chloride) for the synthesis of an ester, amide, or other derivatives. If the α-haloacid itself is required, a water hydrolysis completes the synthesis.

$$\text{R}-\underset{\underset{\text{Br}}{|}}{\text{CH}}-\text{COBr} \quad \text{(α-bromo acyl bromide)}$$

$$\xrightarrow{\text{NH}_3} \text{R}-\underset{\underset{\text{Br}}{|}}{\text{CH}}-\text{CONH}_2$$

$$\xrightarrow{\text{R'OH}} \text{R}-\underset{\underset{\text{Br}}{|}}{\text{CH}}-\text{COOR'}$$

$$\xrightarrow{\text{H}_2\text{O}} \text{R}-\underset{\underset{\text{Br}}{|}}{\text{CH}}-\text{COOH}$$

Besides the α-halogen can be displaced by nucleophilic substitution:

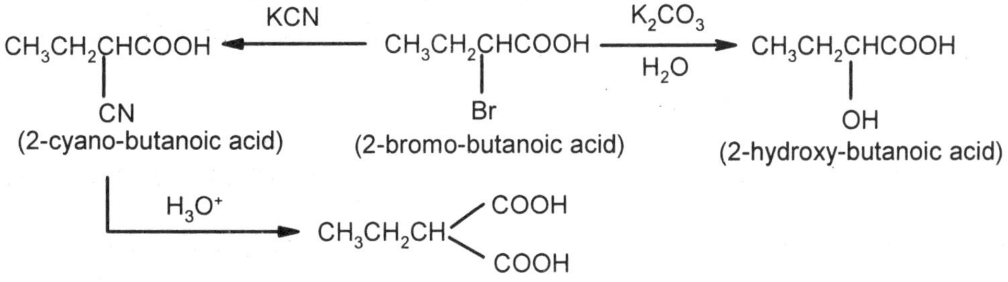

$$CH_3CH_2\underset{CN}{\underset{|}{C}}HCOOH \xleftarrow{KCN} CH_3CH_2\underset{Br}{\underset{|}{C}}HCOOH \xrightarrow[H_2O]{K_2CO_3} CH_3CH_2\underset{OH}{\underset{|}{C}}HCOOH$$

(2-cyano-butanoic acid)   (2-bromo-butanoic acid)   (2-hydroxy-butanoic acid)

$$\xrightarrow{H_3O^+} CH_3CH_2CH{<}\genfrac{}{}{0pt}{}{COOH}{COOH}$$

A standard procedure for the preparation of an α-amino acid involves α-bromo carboxylic acids as the substrate and aqueous ammonia as the nucleophile:

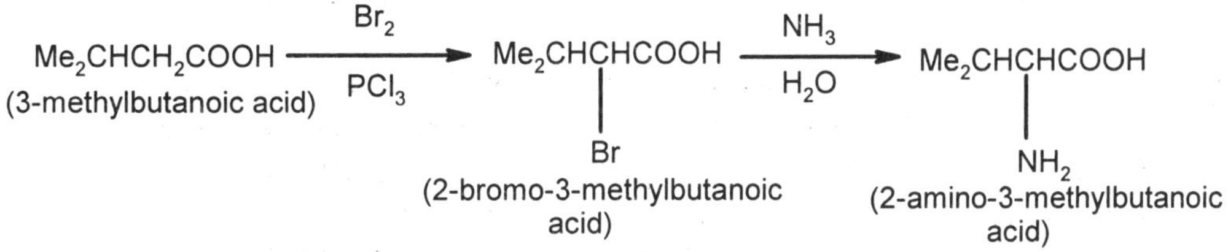

$$Me_2CHCH_2COOH \xrightarrow[PCl_3]{Br_2} Me_2CH\underset{Br}{\underset{|}{C}}HCOOH \xrightarrow[H_2O]{NH_3} Me_2CH\underset{NH_2}{\underset{|}{C}}HCOOH$$

(3-methylbutanoic acid)   (2-bromo-3-methylbutanoic acid)   (2-amino-3-methylbutanoic acid)

Few other examples are:

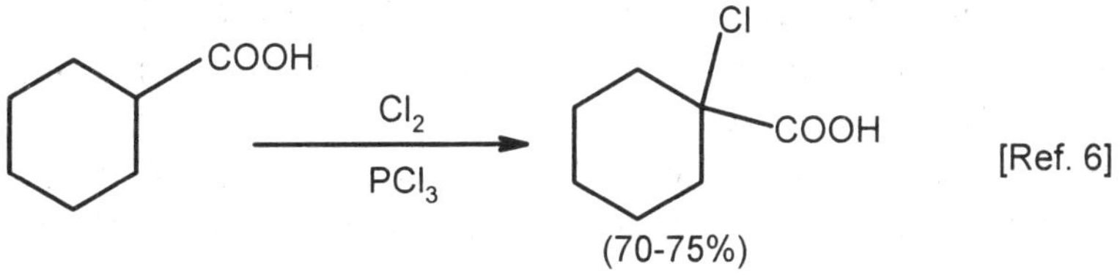

cyclohexyl-COOH $\xrightarrow[PCl_3]{Cl_2}$ 1-chloro-cyclohexane-1-COOH (70-75%)   [Ref. 6]

[Ref. 7]

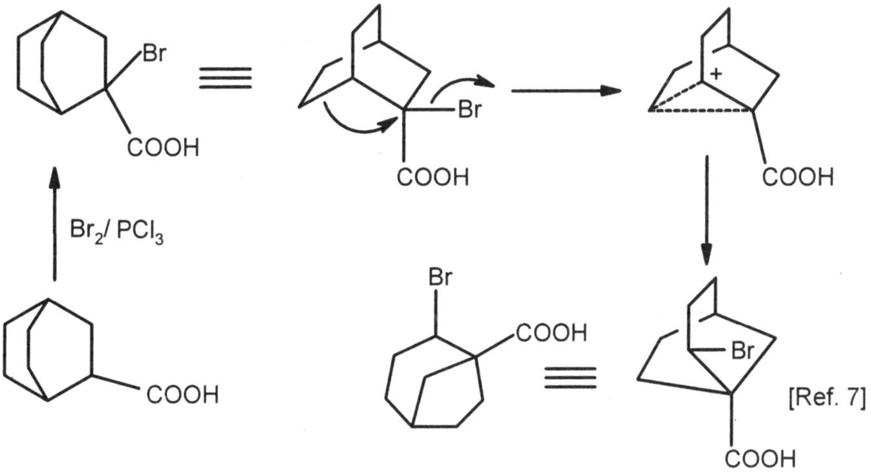

References

1. Hell, C. (1881), *Ber.*, **14**, 891; Volhard, J. (1887), *Ann.*, **242**, 141; Zelinsky, N. (1887), *Ber.*, **20**, 2026.
2. Ward, C. F. (1922), *J. Chem. Soc.*, 1164.
3. Kwart, H. and Scalzi, F. V. (1964), *J. Am. Chem. Soc.*, **86**, 5496.
4. Zhang, L. H., Duan, J., Xu, Y. and Dolbier Jr., W. R. (1998), *Tetrahedron Lett.*, **39**, 9621.
5. Ogata, Y. and Watanabe, S. (1979), *J. Org. Chem.*, **44**, 2768; (1980), *ibid*, **45**, 2831.
6. Little, J. C., Sexton, A. R., Tong, Y. –L. C. and Zurawic, T. E. (1969), *J. Am Chem Soc.*, **91**, 7098.
7. Chow, A. W., Jakas, D. R. and Hoover, J. R. E. (1966), *Tetrahedron Lett.*, 5427.

- Harwood, H. J. (1962), *Chem. Rev.*, **99**, see pp. 102-03.
- Lange, G. L. and Otulakowski, J. A. (1982), *J. Org. Chem.,* **47**(26), 5093.
- Ogata, Y. and Sugimoto, T. (1978), *J. Org. Chem.,* **43**(19), 3684.
- Chatterjee, N. R. (1978), *Indian J. Chem.*, **16B**, 730.
- Ogata, Y. and Tomizawa, K. (1979), *J. Org. Chem.,* **44**, 2768.
- Kolasa, T. and Miller, M. J. (1990), *J. Org. Chem.*, **55**, 4246.
- Krasnov, V. P., Bukrina, I. M., Zhdanova, E. A., Kodess, M. I. and Korolyova, M. A. (1994), *Synthesis*, 961.
- Sharma, A. and Chattopadhyay, S. (1999), *J. Org. Chem.*, **64**, 8059.
- Stack, D. E., Hill, A. L., Differdaffer, C. B. and Burns, N. M. (2002), *Org. Lett.*, **4**, 4487.

31

Hofmann Rearrangement

Introduction

On treatment of halogen (chlorine or bromine) in presence of a strong base, an unsubstituted amide undergoes an interesting reaction that leads to a primary amine with one carbon less than the starting amide; hypochlorites or bromites may also be used as the reagent. This reaction was discovered by the German chemist August W. Hofmann over 120 years ago and is called as *Hofmann rearrangement*[1]. The latter facet of the reaction (loss of one carbon atom in the product), often called as *Hofmann degradation*, makes this rearrangement a distinct reaction in synthesis.

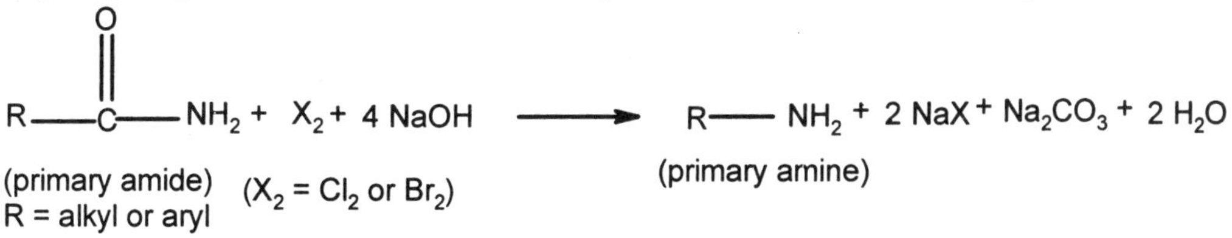

$$R-\underset{\underset{O}{\|}}{C}-NH_2 + X_2 + 4\,NaOH \longrightarrow R-NH_2 + 2\,NaX + Na_2CO_3 + 2\,H_2O$$

(primary amide) $(X_2 = Cl_2$ or $Br_2)$ (primary amine)
R = alkyl or aryl

The *Hofmann rearrangement* can produce primary amines with 1^0, 2^0, or 3^0 alkyl groups or aryl amines.

$$n\text{-}C_4H_9CH_2CONH_2 \xrightarrow[H_2O]{Cl_2,\ NaOH} n\text{-}C_4H_9CH_2NH_2$$

(n-hexamide)   (1-pentanamine, 90%)

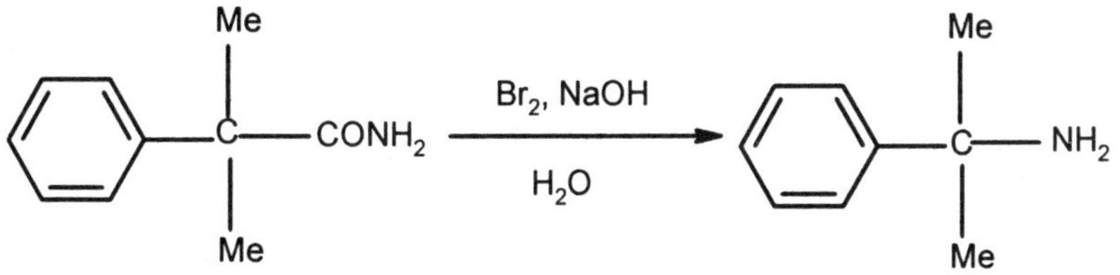

$$Ph\underset{\underset{Me}{|}}{\overset{\overset{Me}{|}}{C}}-CONH_2 \xrightarrow[H_2O]{Br_2,\ NaOH} Ph\underset{\underset{Me}{|}}{\overset{\overset{Me}{|}}{C}}-NH_2$$

(2-methyl-2-phenyl propanamide)   (2-phenyl-2-propanamine)

Mechanism

The *Hofmann rearrangement* takes place in several steps:

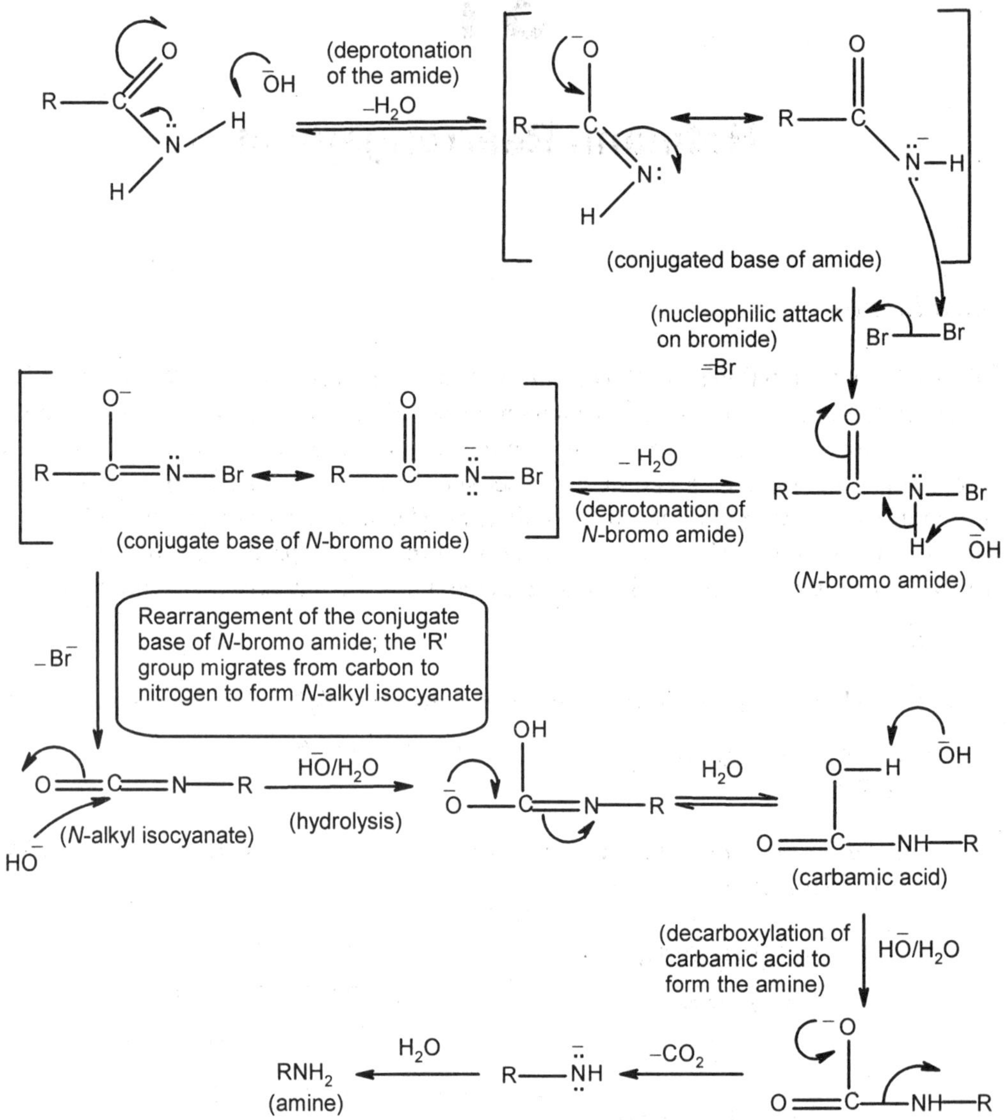

Formation of isocyanate from the conjugate base of N-bromo amide may be supposed to involve two steps — loss of bromide to form nitrene (RCON:), followed by the actual migration. This possibility has been excluded considering the fact that no hydroxamic acid (RCONHOH) is isolated in the *Hofmann rearrangement*, since nitrenes are known to react with water to give hydroxylamine. The concerted mechanism has been supported by several workers[2]. Moreover, Wright and Fry noticed an

appreciable kinetic isotope effect on this rearrangement using phenyl-1-C^{14}-labelled
N-bromobenzamide that supports the concerted mechanism[3].

Critical Views

(a) The *Curtius rearrangement*[4] (Theodor Curtius, 1890) and the *Lossen rearrangement*[5] are of similar to that of the *Hofmann rearrangement*.

$$R-\underset{\underset{O}{\|}}{C}-N_3 \xrightarrow[\text{(pyrolysis)}]{\text{heat}} R-N=C=O + N_2 \quad (\textit{Curtius rearrangement})$$
$$(\textit{N-alkyl isocyanate})$$

$$\begin{array}{c} R-\underset{\underset{O}{\|}}{C} \\ \diagdown \\ NH \\ \diagup \\ R'-\underset{\underset{O}{\|}}{C}-O \end{array} \xrightarrow{\text{base}} R-N=C=O + R'CO\bar{O} \quad (\textit{Lossen rearrangement})$$
$$(\textit{N-alkyl isocyanate})$$

(b) In case of an alkyl group bearing more than six or seven carbons, low yields are obtained unless Br_2 and NaOH are substituted by Br_2 and sodium methoxide[6] or NBS/sodium methoxide[7]. Under these conditions, the product of addition to the isocyanate is the carbamate RNHCOOMe, which is easily isolated or can be hydrolyzed to the amine.

(c) It is interesting to note that the *Hofmann degradation* of optically active amides with a chiral carbon bound to the carbonyl groups yields products with *retention of configuration*[8]. Thus —

$$\underset{\substack{(S)\text{-}(+)\text{-}2\text{-Methyl-3-} \\ \text{phenyl propanamide}}}{\overset{H}{\underset{Me}{\overset{C_6H_5CH_2}{\diagdown}}}\overset{*}{C}-CONH_2} \xrightarrow[\substack{H_2O \\ (\textit{retention of} \\ \textit{configuration})}]{Br_2,\ NaOH} \underset{(S)\text{-}(+)\text{-}1\text{-phenyl-2-propanamine}}{\overset{H}{\underset{Me}{\overset{C_6H_5CH_2}{\diagdown}}}\overset{*}{C}-NH_2}$$

(d) That the rearrangement step of the reaction is an *intramolecular process* in which the departure of the halide ion and the shift of the migrating group are *synchronous* is best evidenced from the following examples:

(i) [β-camphoramic acid structure with CONH₂ and COOH] →[Hofmann rearrangement] [1-aminodihydro-α-campholytic acid with NH₂ and COOH] →[heat] [lactam structure with NH-CO]

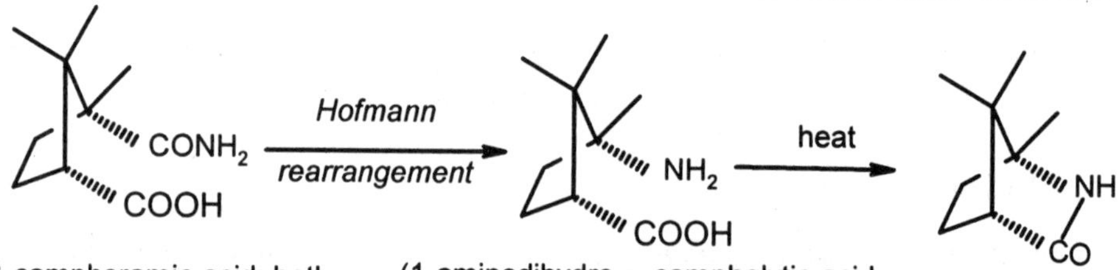

(β-camphoramic acid; both the –CONH₂ and –COOH groups lie *cis* to each other)

(1-aminodihydro-α-campholytic acid; both the –NH₂ and –COOH groups lie *cis* to each other; hence gives lactam on heating – otherwise cannot)

(lactam)

(ii) [Structure A: 2,4-dinitrophenyl-naphthyl with CONH₂] →[Hofmann rearrangement] [Structure B: 2,4-dinitrophenyl-naphthyl with NH₂]

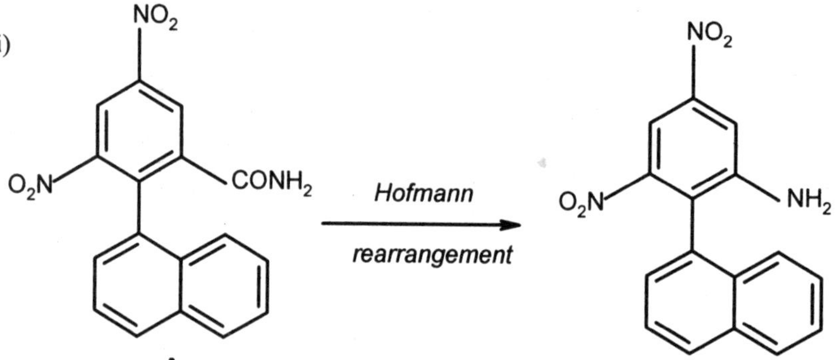

A (optically active)

B (optically active)

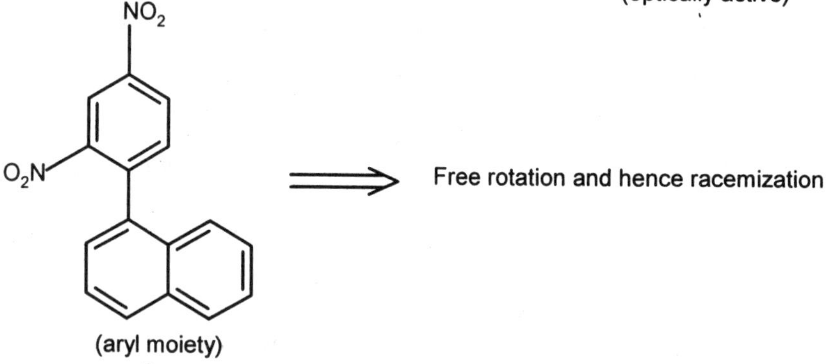

[Structure C: 2,4-dinitrophenyl-naphthyl aryl moiety] ⟹ Free rotation and hence racemization

(aryl moiety)
C

The amide (**A**) shows optical activity due to restricted rotation (presence of two bulky –NO₂ and –CONH₂ substituents) around single bond joining the phenyl and the naphthyl groups. If the migratory aryl moiety (**C**) becomes free during the rearrangement, the rotation would be no more restricted and would have resulted in racemization. The product, amine (**B**) has, on the other hand, been found to be optically pure, and hence the rearrangement proceeds through a transtition state in which the migratory group is partially bound to both the migration origin and the terminus (*i.e. synchronous process*)

Applications

The *Hofmann rearrangement* finds immense applications in organic syntheses. Few instances are cited here:

i) Formation of amines–

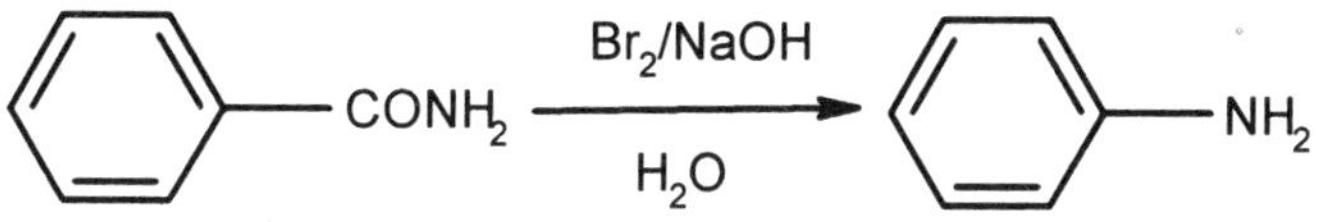

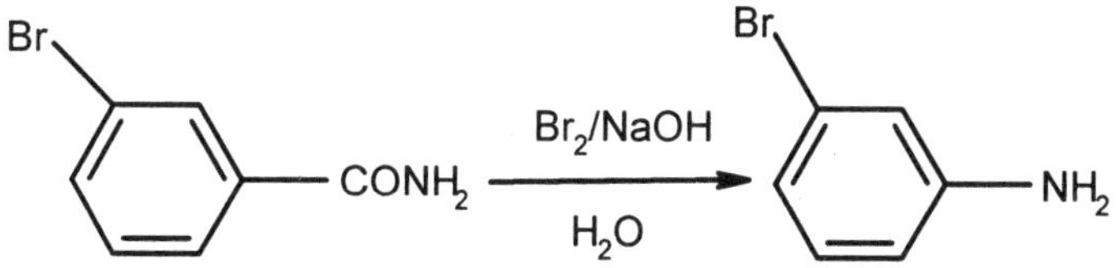

Ph–CONH$_2$ $\xrightarrow{\text{Br}_2/\text{NaOH}, \text{H}_2\text{O}}$ Ph–NH$_2$

(*m*-bromobenzamide) $\xrightarrow{\text{Br}_2/\text{NaOH}, \text{H}_2\text{O}}$ (*m*-bromoaniline)

benzocyclobutane-CONH$_2$ $\xrightarrow[\text{MeOH}, -40°\text{C, then }-15°\text{C}]{\text{Br}_2/\text{NaOMe}}$ benzocyclobutane-NHCOOMe $\xrightarrow{\text{hydrolysis}}$ benzocyclobutane-NH$_2$ [Ref. 6]

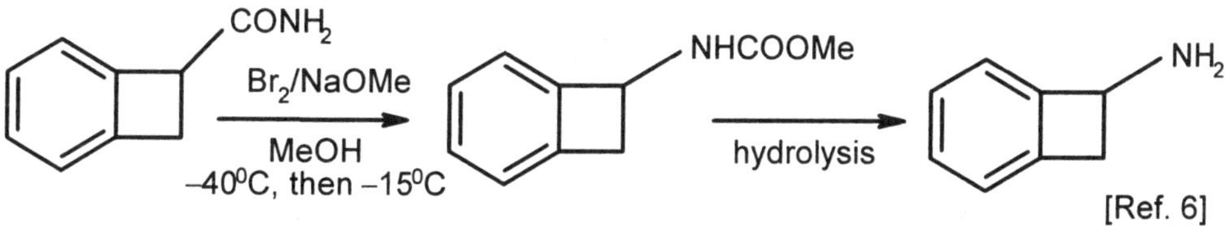

1-methylcyclohexane-CONH$_2$ $\xrightarrow{\text{Br}_2/\text{NaOH}, \text{H}_2\text{O}}$ 1-methylcyclohexane-NH$_2$ [Ref. 9]

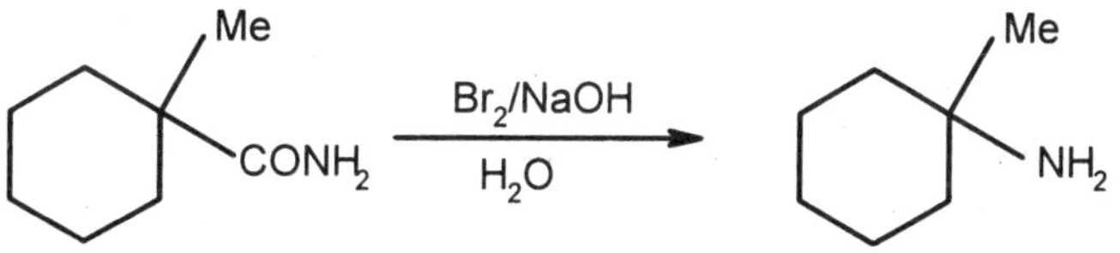

ii) Hydrazine and anthranilic acid, the two important chemicals, are prepared by using this reaction:

H$_2$NCONH$_2$ $\xrightarrow{\text{NaOBr}}$ H$_2$NNH$_2$
(urea) → (hydrazine)

phthalimide $\xrightarrow{\text{Br}_2/\text{NaOH}, \text{H}_2\text{O}}$ anthranilic acid (o-NH$_2$-C$_6$H$_4$-COOH)

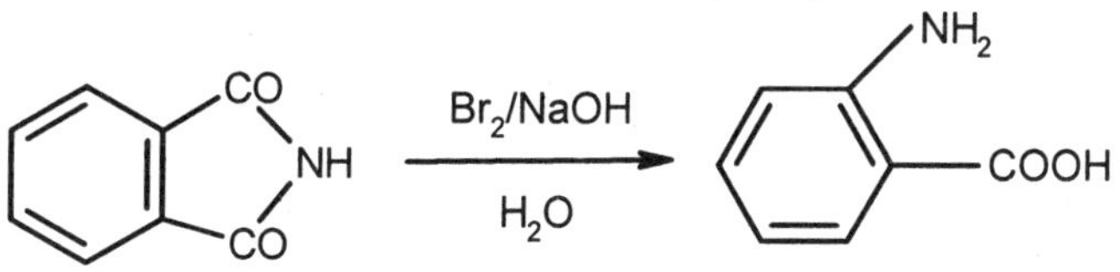

iii) Preparation of β-amino pyridine:

$$\underset{\text{(fluoronicotinamide)}}{\text{3-CONH}_2\text{-2-F-pyridine}} \xrightarrow[\text{H}_2\text{O}]{\text{Br}_2/\text{KOH}} \underset{\text{(β-amino pyridine)}}{\text{3-NH}_2\text{-2-F-pyridine}} \quad \text{[Ref. 10]}$$

iv) A compound may be descended to its lower homologue by this reation:

$$RCH_2CH_2OH \xrightarrow{[O]} RCH_2COOH \xrightarrow[\text{2. NH}_3/\Delta]{\text{1. SOCl}_2} RCH_2CONH_2 \xrightarrow[\text{H}_2\text{O}]{\text{Br}_2/\text{NaOH}} RCH_2NH_2 \xrightarrow{HNO_2} RCH_2OH \xrightarrow{[O]} RCOOH$$

References

1. Hofmann, A. W. (1881), *Ber. Dtsch. Chem. Ges.*, **14**, 2725; (1882), *Ber. Dtsch. Chem. Ges.*, **15**, 762.
2. Imamoto, T., Tsuno, Y. and Yukawa, Y. (1971), *Bull. Chem. Soc. Jpn.*, **44**, 1632, 1639, 1644; Imamoto, T., Kim, S., Tsuno, Y. and Yukawa, Y. (1971), *Bull. Chem. Soc. Jpn.*, **44**, 1776.
3. Wright, J. C. and Fry, A. (1968), *Chem. Engg. News.*, No. 1, 28.
4. Abramovitch, R. A. and Davis, B. A. (1964), *Chem. Rev.*, **64**, 149 [Review]; Banthorpe, D. V. in Patai, *The Chjemistry of the Azido Group*, Wiley, NY, 1971, p. 397.
5. Lossen, W. (1872), *Ann.*, 161, 347; Bauer, L. and Exner, O. (1974), *Angew. Chem. Int. Ed. Engl.*, **13**, 376.
6. Radlick, P. and Brown, L. R. (1974), *Synthesis*, 290.
7. Huang, X. and Keillor, J. W. (1997), *Tetrahedron Lett.*, **38**, 313.
8. Wallis, E. S. and Moyer, W. W. (1933), *J. Am. Chem. Soc.*, **55**, 2598.
9. Sy, A. O. and Raksis, J. W. (1980), *Tetrahedron Lett.*, 2223.
10. Finger, G. C., Starr, L. D., Roe, A. and Link, W. J. (1962), *J. Org. Chem.*, **27**, 3965.

- Maryanoff, B. E. and Reitz, A. B. (1989), *Chem. Rev.*, **89**, 863. [Review]
- Ando, K. (1999), *J. Org. Chem.*, **64**, 6815.
- Reiser, U. and Jauch, J. (2001), *Synlett*, 90.
- Comins, D. L. and Ollinger, C. G. (2001), *Tetrahedron Lett.*, **42**, 4115.
- Harusawa, S., Koyabu, S., Inoue, Y., Sakamopto, Y., Araki, L. and Kurihara, T. (2002), *Synthesis*, 1072.
- Lattanzi, A., Orelli, L. R., Barone, P., Massa, A., Iannece, P. and Scettri, A. (2003), *Tetrahedron Lett.*, **44**, 1333.

32

Houben – Hoesch Reaction

Introduction

Friedel-Crafts type acylation using nitriles and HCl in the presence of a Lewis acid (commonly zinc chloride or aluminium chloride) is called the *Hoesch* or the *Houben-Hoesch reaction*[1]; the reaction is usually applicable to phenols, phenolic ethers and some reactive heterocyclic compounds (*e.g.* pyrrole). The reaction, however, is not successful toward monohydric phenols due to the formation of imino-ether hydrochloride [Ar–O–C(R)=NH$_2^+$Cl$^-$] as almost the exclusive product resulting from the electrophilic attack on the hydroxyl oxygen[2].

The reaction is carried out by passing dry HCl gas through an equimolecular mixture of nitrile and the substrate in dry ether containing ZnCl$_2$. The resulting ketimine (as hydrochloride) on hydrolysis yields the aromatic hydroxyl ketone. Thus, resorcinol gives 2,4-dihydroxyacetophenone:

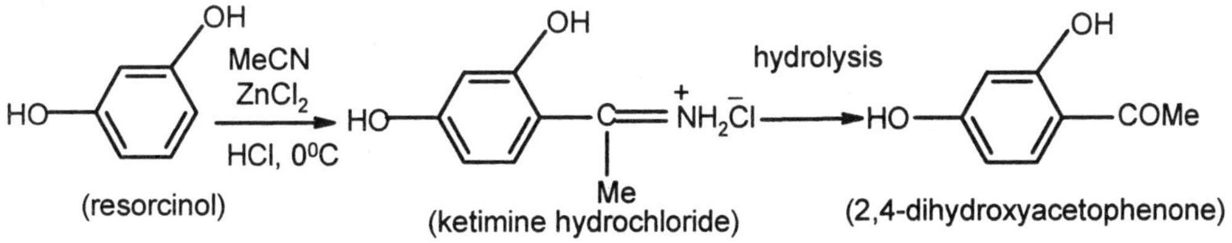

The reaction is very successful with polyhydroxy phenols specially, the *m*-polyhydroxy phenols. A variety of aliphatic nitriles (*e.g.* acetoniriles, mono- and trichloroacetonitriles) are in use; even aryl nitriles give good yields in many cases[3].

Mechanism

The reaction mechanism is complex and still not completely settled[4]. The generally represented form is mentioned here taking resorcinol as the substrate:

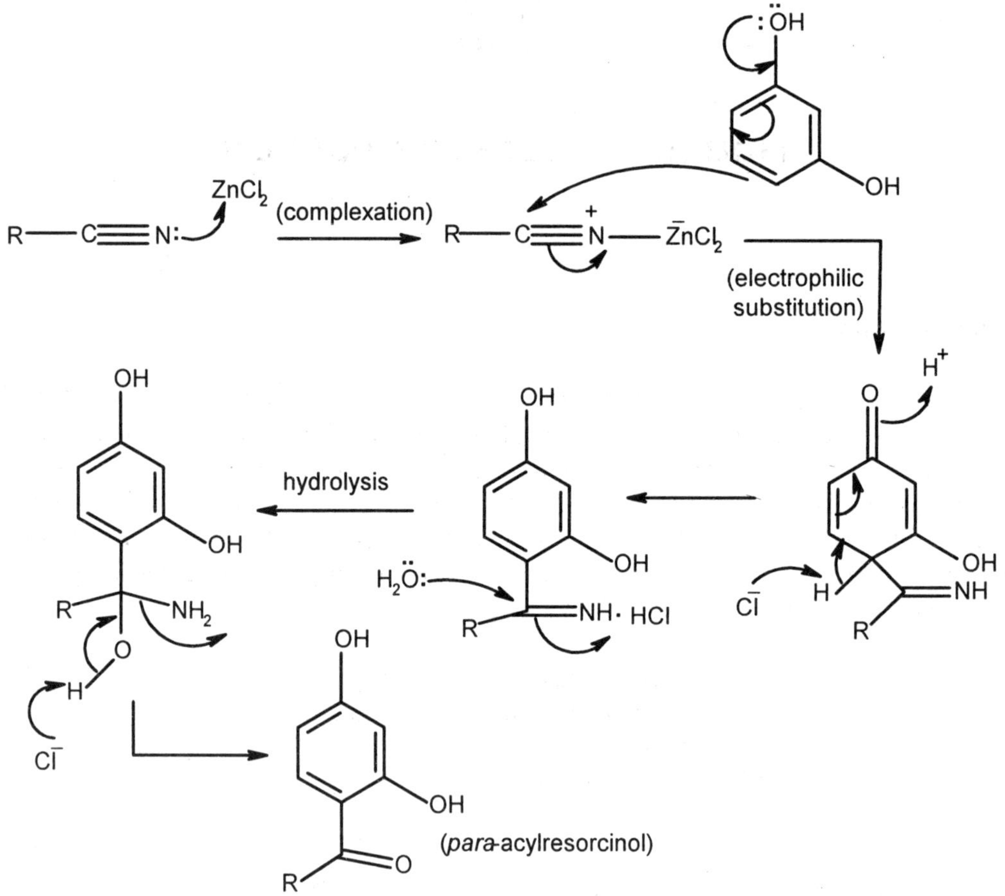

Critical Views

(a) The reaction can be extended to aromatic amines by the use of boron trichloride[5], and the acylation is regioselectively *ortho*.
(b) Ketones can also be obtained by treating the substarte with a nitrile in the presence of F_3CSO_2OH, but the mechanism in this case is different[6].
(c) When hydrogen cyanide is employed, aromatic aldehyde may be obtained; thus the *Gattermann reaction* is a special case of the *Hoesch reaction*.

Applications

The reaction provides a very useful method for the synthesis of polyhydroxy acetophenones and polyhydroxy benzophenones. A few examples are cited:

i) 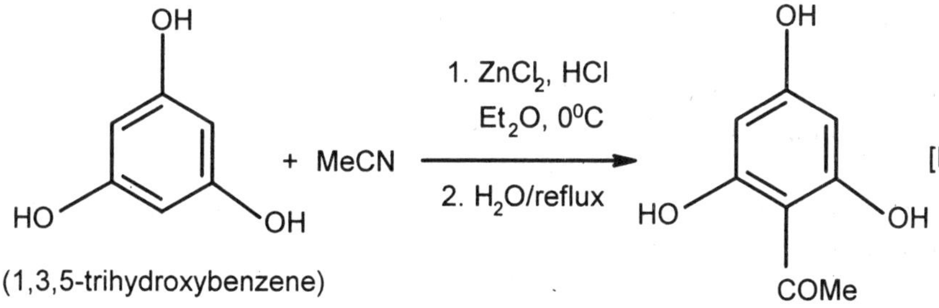 1,3,5-trihydroxybenzene + MeCN $\xrightarrow[\text{2. H}_2\text{O/reflux}]{\text{1. ZnCl}_2, \text{HCl, Et}_2\text{O, 0°C}}$ 2,4,6-trihydroxyacetophenone [Ref. 7]

ii) 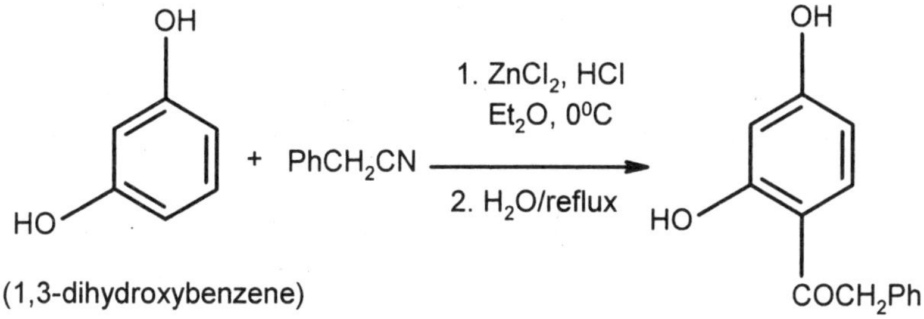 1,3-dihydroxybenzene + PhCH$_2$CN $\xrightarrow[\text{2. H}_2\text{O/reflux}]{\text{1. ZnCl}_2, \text{HCl, Et}_2\text{O, 0°C}}$ 2,4-dihydroxy-ω-phenylacetophenone

iii) 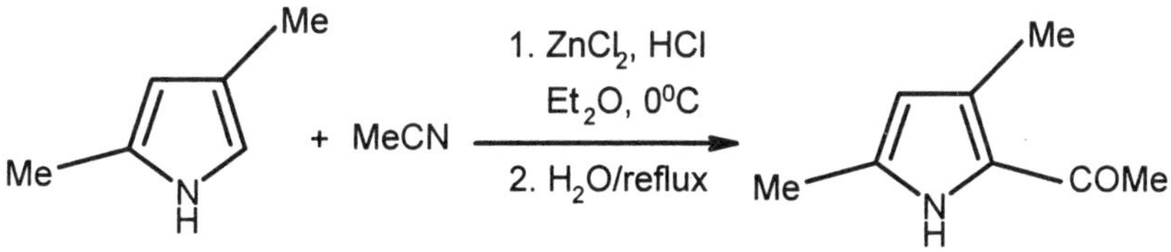 2,4-dimethylpyrrole + MeCN $\xrightarrow[\text{2. H}_2\text{O/reflux}]{\text{1. ZnCl}_2, \text{HCl, Et}_2\text{O, 0°C}}$ 2,4-dimethyl-5-acetyl pyrrole

iv) Important organic molecules, coumarin derivatives, can be prepared by using this reaction:

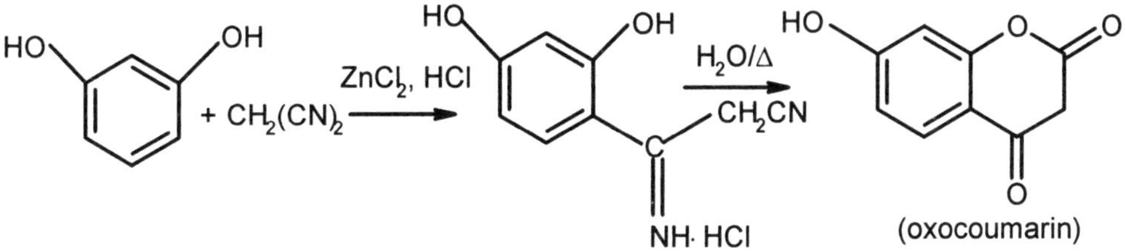

resorcinol + CH$_2$(CN)$_2$ $\xrightarrow{\text{ZnCl}_2, \text{HCl}}$ [imine intermediate with NH·HCl and CH$_2$CN] $\xrightarrow[\text{-CH}_2\text{CN}]{\text{H}_2\text{O}/\Delta}$ oxocoumarin

References

1. Hoesch, K., *Ber. Dtsch. Chem. Ges.*, **48**, 1122; Ruske, W. in Olah, *Friedel-Crafts and Related Reactions*, vol. 3, Wiley, NY, 1964, p. 383. [Review]

2   For an exception, see Toyoda, T., Sasakura, K. and Sugasawa, T. (1981), *J. Org. Chem.*, **46**, 189.
3   Zil'berman, E. N. and Rybakova, N. A. (1960), *J. Gen. Chem. USSR*, **30**, 1972.
4   Jeffery, E. A. and Satchell, D. P. N. (1966), *J. Chem. Soc.* **B**, 579.
5   Sugasawa, T., Toyoda, T., Adachi, M. and Sasakura, K. (1978), *J. Am. Chem. Soc.*, **100**, 4842; Sugasawa, T., Adachi, M., Sasakura, K. and Kitagawa, A. (1979), *J. Org. Chem.*, **44**, 578.
6   Amer, M. I., Booth, B. L., Noori, G. F. M. and Proenca, M. F. J. R. P. (1983), *J. Chem. Soc., Perkin Trans. 1*, 1075.
7   Gulati, K. C., Seth, S. R. and Venkataraman, K. (1943), *Org. Syn.*, **Coll. Vol. 2**, 522.
- Yato, M., Ohwada, T. and Shudo, K. (1991), *J. Am. Chem. Soc.*, **113**, 691.
- Sato, Y., Yato, M., Ohwada, T., Saito, S. and Shudo, K. (1995), *J. Am. Chem. Soc.*, **117**, 3037.
- Kawecki, R., Mazurek, A. P., Kozerski, L. and Maurin (1999), *Synthesis*, 751.
- Udwary, D. W., Casillas, L. K. and Townsend, C. A. (2002), *J. Am. Chem. Soc.*, **124**, 5294.

33

Hunsdiecker Reaction

Introduction

The conversion of silver salts of carboxylic acids into organic halides by treating the silver carboxylates with halogens in a refluxing inert solvent (such as carbon tetrachloride, chloroform or ether) under anhydrous conditions is known as *Hunsdiecker reaction*[1]. The reaction offers a way of decreasing the length of carbon chain by one unit[2]. Although bromine is the most often used halogen, chlorine and iodine have also been used.

$$RCOOAg + Br_2 \xrightarrow[reflux]{CCl_4} R\text{—}Br + CO_2 + AgBr$$

The reaction may be looked upon as decarboxylative bromination. The 'R' group may be both alkyl and aryl. Good yields are obtained for *n*-alkyl (bearing 2 to 18 carbons) and also for branched alkyl groups, producing primary, secondary and tertiary bromides. The reaction, however, does not go well if 'R' contains unsaturation.

Mechanism

The reaction is not catalyzed by acids but is catalyzed by irradiation and it shows induction period. Hence the mechanism of the reaction is believed to proceed through free-radical pathway[3]. The widely accepted mechanistic scheme of *Hunsdiecker reaction* is depicted below:

Step-I: In the first step the halogen and the silver carboxylate react to produce an acyl hypohalite as an intermediate.

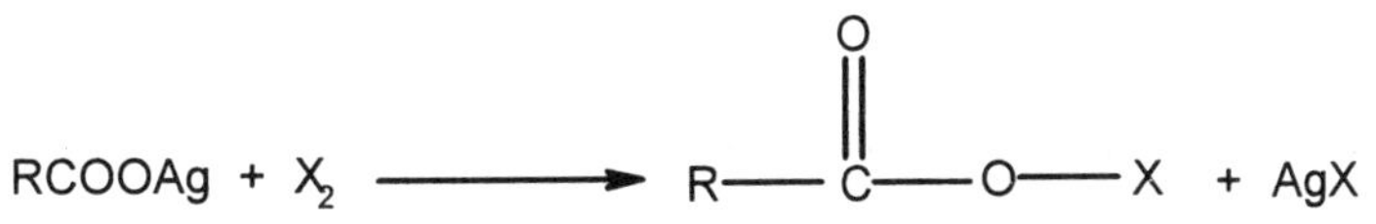

$$RCOOAg + X_2 \longrightarrow R\text{—}\underset{\underset{\text{(acyl hypohalite)}}{}}{\overset{\overset{O}{\|}}{C}}\text{—}O\text{—}X + AgX$$

Step-II: This is the initiation step; the acyl hypohalite undergoes thermal decoposition into acyl and halogen free radicals by means of homolytic fission.

$$R-\underset{\underset{O}{\|}}{C}-O-X \xrightarrow{\text{homolytic cleavage}} R-\underset{\underset{O}{\|}}{C}-\ddot{O}\cdot + X\cdot$$

Step-III: Decarboxylation of the acyl radical to form alkyl or aryl free radical.

$$R-\underset{\underset{O}{\|}}{C}-\ddot{O}: \longrightarrow R\cdot + CO_2$$

Step-IV: Formation of alkyl or aryl halide –

$$R-\underset{\underset{O}{\|}}{C}-O-X + \cdot R \longrightarrow R-X + R-\underset{\underset{O}{\|}}{C}-\ddot{O}:$$

The acyl radical propagates radical chain reaction.

Critical Views

(a) The free radical pathway for *Hunsdiecker reaction* is evidenced from the facts –
  (i) isolation of the side product, R-R, which is formed presumably by radical recombination, is consistent with a free-radiocal mechanism.
  (ii) except in a few cases, an optically active carboxylic acid leads to a racemic halide[2,4]. Besides, if 'R' is neopentyl, there is no rearrnagemnt, which would certainly happen with a carbocation. These stereochemical results support the radical nature of the reaction.

(b) *Cristol-Firth modification:* Cristol and Firth[5] suggested an inproved procedure for the reaction that consists in refluxing a solution of the carboxylic acid with mercuric oxide in carbon tetrachloride, followed by treatment with bromine. The mechanism in this case also involves a mercury salt of the acid[6].

(c) A 1:1 ratio of the carboxylate to iodine (when is the reagent), the product becomes the usual halide —— but a 2:1 ratio forms the ester, RCOOR. This ester-forming reaction is termed as *Simonini reaction*[7]; the mechanism of this reaction is similar to that of the *Hunsdiecker reaction*.

$$2\ RCOOAg + I_2 \longrightarrow RCOOR + 2\ AgI + CO_2$$

Applications

The reaction is of wide scope, producing alkyl (primary, secondary, and tertiary) and aryl halides. A few examplers are cited below:

(i) $CH_3O_2C(CH_2)_4COOAg \xrightarrow[\substack{CCl_4,\ \text{reflux} \\ -AgBr}]{Br_2} CH_3O_2C(CH_2)_3CH_2Br$ [Ref. 8]
                                                       (65-68%)

(ii) $n\text{-}C_{17}H_{35}COOH \xrightarrow[CCl_4,\ reflux]{HgO,\ Br_2} n\text{-}C_{17}H_{35}Br$ (93%)   [Ref. 9]

(iii) The reaction is applicable for preparing bridgehead halides –

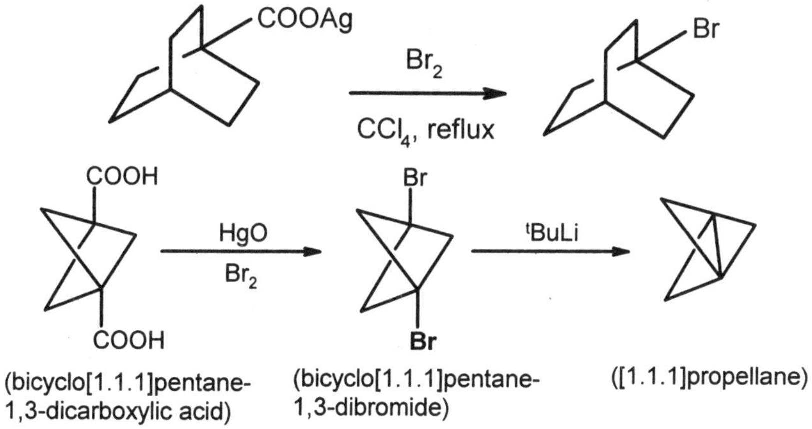

bicyclo[2.2.2]octane-COOAg $\xrightarrow[CCl_4,\ reflux]{Br_2}$ bicyclo[2.2.2]octane-Br

bicyclo[1.1.1]pentane-1,3-dicarboxylic acid (COOH, COOH) $\xrightarrow{HgO,\ Br_2}$ bicyclo[1.1.1]pentane-1,3-dibromide (Br, Br) $\xrightarrow{^tBuLi}$ [1.1.1]propellane

(iv) $\triangleright\!-\!COOH \xrightarrow{HgO,\ Br_2} \triangleright\!-\!Br$ (41–46%)   [Ref. 10]

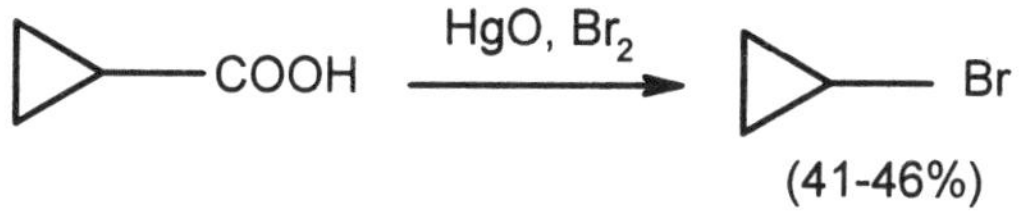

(v) 4-Cl-C$_6$H$_4$-COOH $\xrightarrow[2.\ Br_2]{1.\ HgO/light}$ 4-Cl-C$_6$H$_4$-Br   (80%)   [Ref. 11]

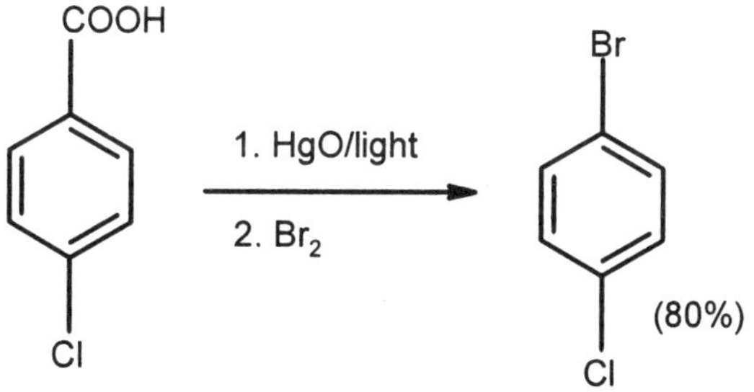

(vi) $2\ CH_3(CH_2)_nCOOH \xrightarrow{Tl_2CO_3} 2\ CH_3(CH_2)_nCOOTl \xrightarrow{Br_2} 2\ CH_3(CH_2)_nBr$
    (thalium modification)[12]   (80–875%) [Ref. 13]

References

1   Borodin, B. (1861), *Justus Liebigs Ann. Chem.*, **119**, 121; Hunsdiecker, H. and Hunsdiecker, C. (1942), *Ber. Dtsch. Chem. Ges.*, **75**, 291.

2   For reviews: Johnson, R. G. and Ingham, R. K. (1956), *Chem. Rev.*, **56**, 219; Naskar, D., Chowdhury, S. and Roy, S. (1998), *Tetrahedron Lett.*, **39**, 699.
3   Pundit, U. K. and drik, I. P. (1963), *Tetrahedron Lett.*, 891; Chelmers, D. J. and Thomson, R. H. (1968), *J. Chem. Soc., (C)*, 848.
4   Arnold, R. T. and Morgan, P. (1948), *J. Am. Chem. Soc,* **70**, 4248.
5   Cristol, S. J. and Firth, W.C. (1961), *J. Org. Chem.*, **26**, 280.
6   Cason, J. and Walba, D. M. (1972), *J. Org. Chem.*, **37**, 669.
7   Simonini, A. (1893), *Monatsh.*, **14**, 81; Bachman, G. B., Kite, G. F., Tuccarbasu, S. and Tullman, G. M. (1970), *J. Org. Chem.*, **35**, 3167.
8   Wilson, C. V. (1957), *Org. React.*, **9**, 332.
9   Gunstone, F. D. (1960), *Adv. Org. Chem.*, **1**, 117.
10 Meek, J. and Osuga, D. (1973), *Org. Synth.*, **V**, 126.
11 Meyers, A. and Fleming, M. (1979), *J. Org. Chem.*, **44**, 3402.
12 McKillop, A., Bromley, D. and Taylor, E. C. (1969), *J. Org. Chem.*, **34**, 1172.
13 Cambie, R., Hayward, R., Jurlina, J., Rutledge, P. and Woodgate, P. (1981), *J. Chem. Soc., Perkin Trans. 1*, 2608.

- Sheldon, R. A. and Kochi, J. K. (1972), *Org. React.*, **19**, 326. [Review]
- Barton, D. H. R., Crich, D. and Motherwell, W. B. (1983), *Tetrahedron Lett.*, **24**, 4979.
- Wilberg, K. B. (1984), *Acc. Chem. Res.*, **17**, 379.
- Crich, D. in *Comprehensive Organic synthesis* (Eds. Trost, B. M. and Steven, V. L.), Pergamon, 1991, **vol. 7**, pp.723-34.
- Camps, P., Lukach, A. E., Pujol, X. and Vazquez, S. (2000), *Tetrahedron*, **56**, 2703.
- De Luca, L., Giacomelli, G., Porcu, G. and Taddei, M. (2001), *Org. Lett.*, **3**, 855.
- Das, J. P. and Roy, S. (2002), *J. Org. Chem.*, **67**, 7861.

34

Knoevenagel Reaction

Introduction

The base-catalyzed aldol-type condensation of aldehydes or ketones, usually without any α-hydrogen, with compounds having active methylene group is known as *Knoevenagel reaction*[1]; in most cases the intermediate aldol is dehydrated to the α,β–unsaturated product. A general scheme for the reaction may be represented as:

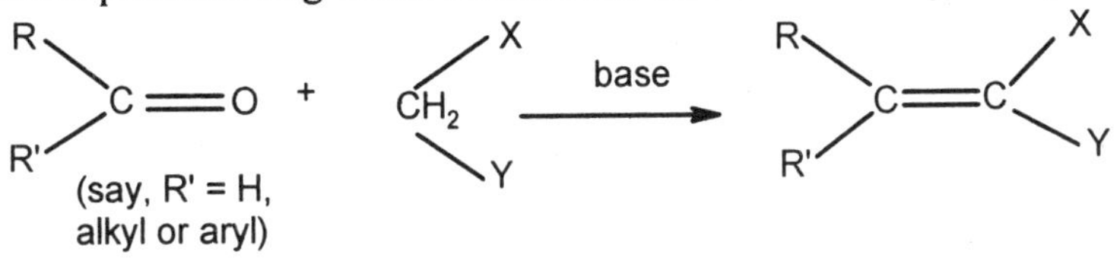

$$\underset{\substack{\text{(say, R' = H,}\\ \text{alkyl or aryl)}}}{\overset{R}{\underset{R'}{>}}C=O} \;+\; \overset{X}{\underset{Y}{>}}CH_2 \;\xrightarrow{\text{base}}\; \overset{R}{\underset{R'}{>}}C=C\overset{X}{\underset{Y}{<}}$$

Generally, organic bases are often used as catalysts — e.g. mono-, di- or tri-alkylamines, aniline, pyridine, piperidine, etc. Alkoxides are also common catalysts. In X-CH$_2$-Y, both X and Y may be strong electron-withdrawing groups like — CHO, COR, COOH, COOR, CN, NO$_2$, SOR, SO$_2$R, SO$_2$OR, or similar groups. Actually, any compound that contains a C-H bond the hydrogen of which can be removed by a base undergoes the condensation with carbonyl compounds; thus, chloroform, 2-methylpyridine, terminal acetylenes, cyclopentadienes, and other such type of molecules can take part in the condensation. On the other hand, if sufficiently strong base is employed, the reaction may also proceed with the either compound having X or Y. It is important to note that ketones are less reactive than aldehydes due to streic and electronic reasons. Few reaction schemes are shown below:

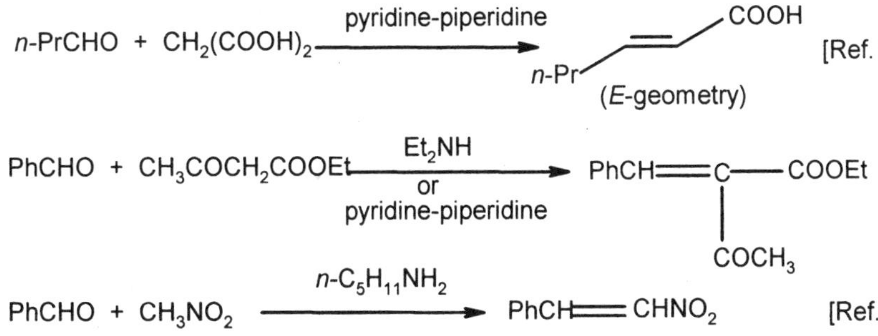

$$n\text{-PrCHO} + CH_2(COOH)_2 \xrightarrow{\text{pyridine-piperidine}} \underset{(E\text{-geometry})}{\overset{n\text{-Pr}}{>}C=C\overset{}{<}COOH} \quad [\text{Ref. 2}]$$

$$PhCHO + CH_3COCH_2COOEt \xrightarrow[\text{or pyridine-piperidine}]{Et_2NH} PhCH=\underset{\underset{COCH_3}{|}}{C}-COOEt$$

$$PhCHO + CH_3NO_2 \xrightarrow{n\text{-}C_5H_{11}NH_2} PhCH=CHNO_2 \quad [\text{Ref. 3}]$$

Mechanism

The reaction proceeds through the following steps:

Step-I: Reversible abstraction of the methylenic proton by base forming a resonance-stabilized carbanion.

$$\underset{Y}{\overset{X}{>}}CH_2 + :B \rightleftharpoons \underset{Y}{\overset{X}{>}}\overset{-}{C}H + \overset{+}{B}H$$

Step-II: Nucleophilic attack of the carbanion at the electron-deficient carbonyl carbon.

$$\underset{R'}{\overset{R}{>}}C=O + \overset{-}{C}H\underset{Y}{\overset{X}{<}} \rightleftharpoons \underset{R'}{\overset{R}{>}}\overset{O^-}{\underset{}{C}}-CH\underset{Y}{\overset{X}{<}} \quad (I)$$

Step-III: Protonation of (I) by BH^+ to form the aldol (II).

$$\underset{R'}{\overset{R}{>}}\overset{O^-}{\underset{}{C}}-CH\underset{Y}{\overset{X}{<}} + \overset{+}{B}H \rightleftharpoons \underset{R'}{\overset{R}{>}}\overset{OH}{\underset{}{C}}-CH\underset{Y}{\overset{X}{<}}$$
(I)   (II)

Step-IV: Further abstraction of the acidic proton of –CHXY moiety to form a resonance-stabilized carbanion (III).

$$\underset{R'}{\overset{R}{>}}\overset{OH}{\underset{}{C}}-CH\underset{Y}{\overset{X}{<}} \underset{-BH^+}{\overset{:B}{\rightleftharpoons}} \underset{R'}{\overset{R}{>}}\overset{OH}{\underset{}{C}}-\overset{-}{C}\underset{Y}{\overset{X}{<}}$$
(II)   (III)

Step-V: Elimination of the –OH group as hydroxide by $E1_{cb}$ pathway to form α,β-unsaturated compound.

$$\underset{R'}{\overset{R}{>}}\overset{OH}{\underset{}{C}}-\overset{-}{C}\underset{Y}{\overset{X}{<}} \xrightarrow{-OH^-} \underset{R'}{\overset{R}{>}}C=C\underset{Y}{\overset{X}{<}}$$
(III)   (IV)

Critical Views

(a) Cope modification: The *Knoevenagel reaction* is reversible favouring the equilibrium to the left, and hence yield of the product is not satisfactory; to improve the yield of the reaction Cope introduced a modified procedure by performing the reaction in benzene and removing the water produced, as an azeotropic mixture in Dean and Stark apparatus, thus shifting the equilibrium completely to the right. That's why the modified technique is very often called as *Cope-Knoevenagel reaction*.

(b) Doebner modification: When the catalyst is pyridine (to which piperidine may or may not be added), the reaction is known as *Doebner modification* of the *Knoevenagel reaction*. Condensation with malonic acid or its monoester under this condition leads to decarboxylation[4] in the reaction mixture yielding α,β-unsaturated carboxylic acids or its derivatives — the carboxylic or the ester functionality takes up *anti*-orientation to the larger β-substituent.

$$RCHO + CH_2(COOH)(COOR') \xrightarrow[\Delta, -CO_2]{\text{pyridine-piperidine}} \underset{\text{trans-oriented product}}{\underset{(R' = H \text{ or alkyl})}{R(H)C=C(H)(COOR')}}$$

(R = alkyl or aryl)   (R' = H or alkyl)

(c) In some cases (where 2 equivalents of active methylene compounds are used) the *Knoevenagel product* underogoes *Michael addition* with the second molecule of active methylene component; the addition product bears high synthetic value.

$$CH_3CHO + CH_2(COOEt)_2 \underset{}{\overset{R_2NH}{\rightleftharpoons}} \underset{\text{Knoevenagel product}}{CH_3CH=C(COOEt)_2}$$

$$CH_3CH=C(COOEt)_2 \underset{CH_2(COOEt)_2}{\overset{R_2NH}{\rightleftharpoons}} CH_3-CH(COOEt)-CH(COOEt)_2$$
Michael addition

Applications

The *Knoevenaggel reaction* is a potential tool for the synthesis of a wide variety of condensation products because the carbonyl component as well as the active-methylene component can be varied. A few reaction schemes are cited:

i) 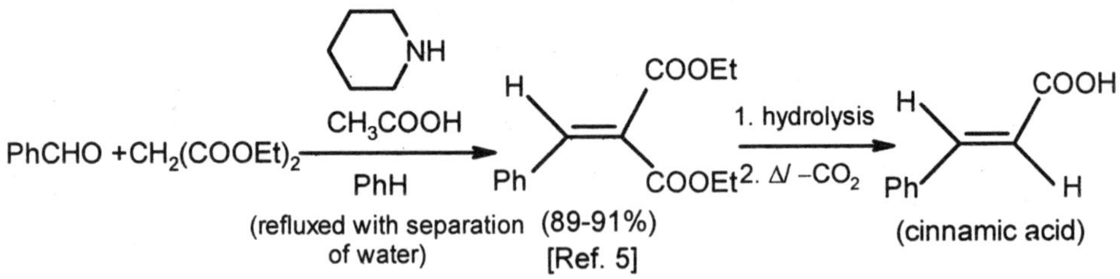 PhCHO + CH$_2$(COOEt)$_2$ →[piperidine NH / CH$_3$COOH / PhH (refluxed with separation of water)] Ph−CH=C(COOEt)$_2$ (89–91%) [Ref. 5] →[1. hydrolysis 2. Δ / −CO$_2$] Ph−CH=CH−COOH (cinnamic acid)

ii) 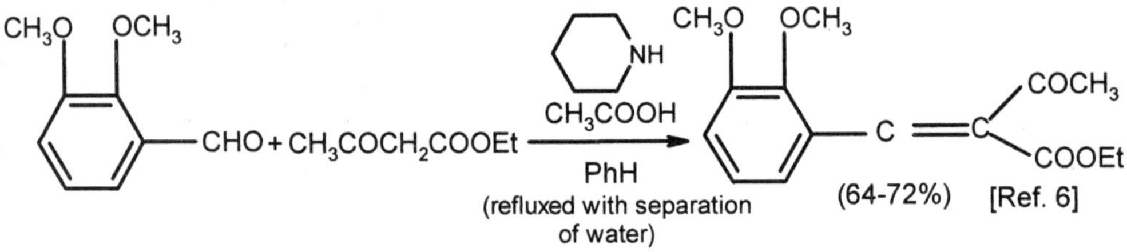 (2,3-dimethoxyphenyl)−CHO + CH$_3$COCH$_2$COOEt →[piperidine NH / CH$_3$COOH / PhH (refluxed with separation of water)] (2,3-dimethoxyphenyl)−C=C(COCH$_3$)(COOEt) (64–72%) [Ref. 6]

iii) CH$_3$CH=CH−CHO + CH$_2$(COOH)$_2$ →[1. pyridine/100°C 2. work-up] CH$_3$CH=CH−CH=CH−COOH (28–32%) [Ref. 7]

iv) 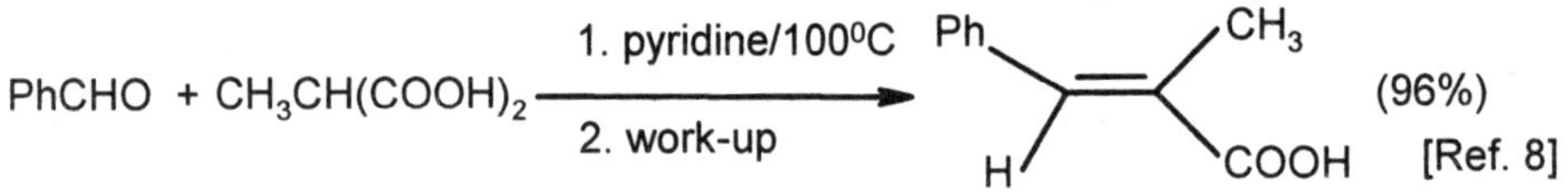 PhCHO + CH$_3$CH(COOH)$_2$ →[1. pyridine/100°C 2. work-up] Ph(H)C=C(CH$_3$)(COOH) (96%) [Ref. 8]

v) 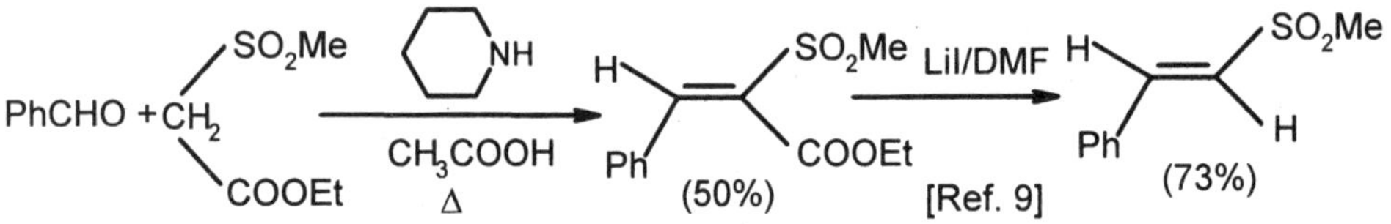 PhCHO + CH$_2$(SO$_2$Me)(COOEt) →[piperidine NH / CH$_3$COOH / Δ] Ph(H)C=C(SO$_2$Me)(COOEt) (50%) →[LiI/DMF] Ph(H)C=C(SO$_2$Me)(H) (73%) [Ref. 9]

vi) 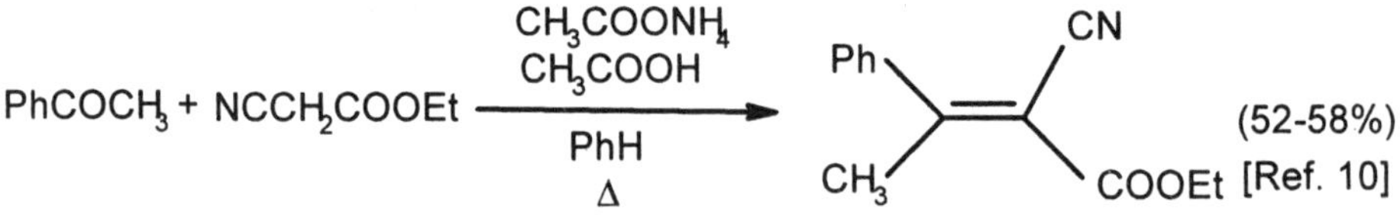 PhCOCH$_3$ + NCCH$_2$COOEt →[CH$_3$COONH$_4$ / CH$_3$COOH / PhH / Δ] Ph(CH$_3$)C=C(CN)(COOEt) (52–58%) [Ref. 10]

vii) 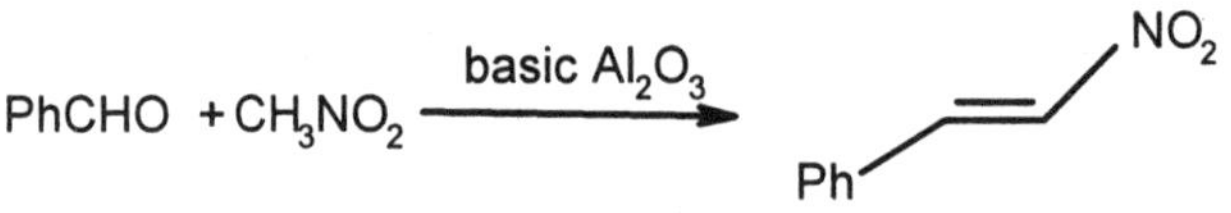 PhCHO + CH$_3$NO$_2$ →[basic Al$_2$O$_3$] Ph−CH=CH−NO$_2$

viii) R−CHO (1.3 eq.) + NC−CH$_2$−Y →[0.2 eq. PPh$_3$ / 75–80°C, 2.5–5.5 h or MW, 450W, 2–5 min.] R−CH=C(Y)(CN) [Ref. 12]

Y = −CN, −COOEt

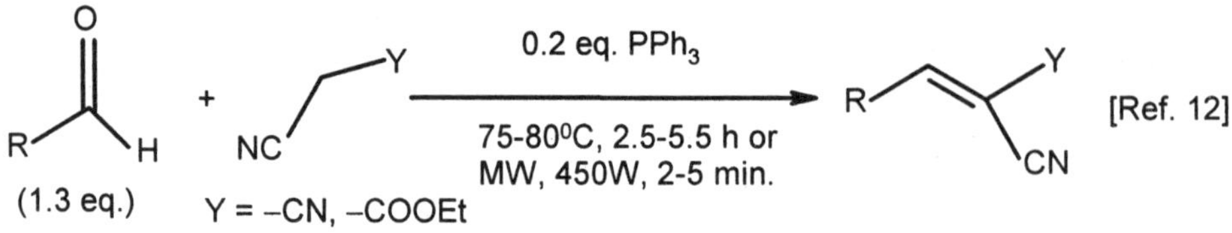

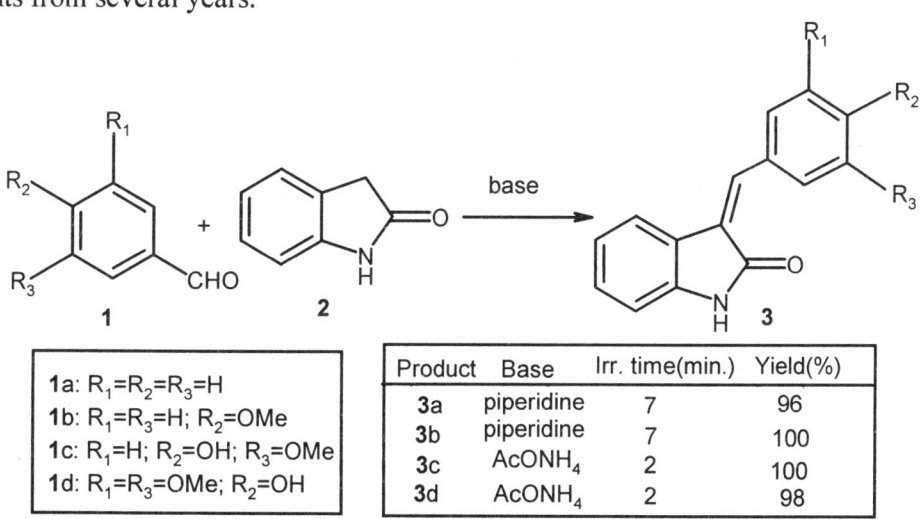

x) Seijas et al.[13] carried out microwave enhanced *Knoevenagel reaction* to prepare 3-benzylidene-1,3-dihydroindol-2-ones, which have been known as pharmacological agents from several years.

References

1. Knoevenagel, E. (1898), *Ber. Dtsch. Chem. Ges.*, **31**, 2596; Jones, G. (1967), *Org. React.*, **15**, 204 [Review]; Wilk, B. K. (1997), *Tetrahedron*, **53**, 7097.
2. Tanaka, M., Oota, O., Hiramatsu, H. and Fujiwara, K. (1988), *Bull. Chem. Soc. Jpn.*, **61**, 2473.
3. Matsumoto, K. (1984), *Angew. Chem. Int. Ed. Engl.*, **23**, 617; Barrett, A. G. M. and Graboski, G. G. (1986), *Chem. Rev.*, **86**, 751.

4   Corey, E. J. (1953), *J. Am. Chem. Soc.*, **75**, 1163; Corey, E. J. and Fraenkel, G. (1953), *J. Am. Chem. Soc.*, **75**, 1168; Klein, J. and Meyer, A. Y. (1964), *J. Org. Chem.*, **29**, 1038.
5   Allen, C. F. H. and Spangler, F. W. (1955), *Org. Syn.*, **Coll. Vol. 3**, 377.
6   Horning, E. C., Koo, J., Fish, M. S. and Walker, G. N. (1963), *Org. Syn.*, **Coll. Vol. 4**, 408.
7   Allen, C. F. H. and VanAllen, J. (1955), *Org. Syn.*, **Coll. Vol. 3**, 783.
8   Gensier, W. J. and Berman, E. (1958), *J. Am. Chem. Soc.*, **80**, 4949.
9   Happer, D. A. R. and Steenson, B. E. (1980), *Synthesis*, 806.
10 McElvain, S. M. and Clemens, D. H. (1963), *Org. Syn.*, **Coll. Vol. 4**, 463; Sakurai, A. and Midorikawa, H. (1969), *J. Org. Chem.*, **34**, 3612.
11 Cope, A. C., D'Addieco, A., Whyte, D. E. and Glickman, S. A. (1963), *Org. Syn.*, **Coll. Vol. 4**, 234.
12 Yadav, J. S., Reddy, B. S. S., Basak, A. K., Visali, B., Narsaiah, A. V., Nagaiah, K. (2004), *Eur. J. Org. Chem.,* 546.
13 Seijas, J. A., Vázquez Tato, M. P., Fernández, M. C. and Nazaret, A. (1999), *Proceedings of the Third International Electronic Conference on Synthetic Organic Chemistry (ECSOC-3),* http://www.mdpi.org/ecsoc-3.htm, September 1-30.

- Angeletti, E., Canepa, C., Martinetti, G. and Venturello, P. (1989), *J. Chem. Soc., Perkin Trans. 1*, 105.
- Pyne, S. G. and Boche, G. (1989), *J. Org. Chem.*, **54**, 2663.
- Niwa, S. and Soai, K. (1990), *J. Chem. Soc., Perkin Trans. 1*, 937.
- Chandrasekar, S., Yu, J., Falck, J. R. and Mioskowski, C. (1994), *Tetrahedron Lett.*, **35**, 5441.
- Prajapati, D., Lekhok, K. C., Sandhu, J. S. and Ghosh, A. C. (1996), *J. Chem. Soc., Perkin Trans. 1*, 959.
- Reddy, T. I. and Varma, R. S. (1997), *Tetrahedron Lett.*, **38**, 1721.
- McNulty, J., Steeve, J. A. and Wolf, S. (1998), *Tetrahedron Lett.*, **39**, 8013.
- Villemin, D. and Martin, B. (1998), *Synth. Commun.*, **28**, 3201.
- Bogdal, D. (1998), *J. Chem. Research (S)*, 468.
- Paquette, L. A., Kern, B. E. and Mendez-Andino, J. (1999), *Tetrahedron Lett.*, 40, 4129.
- Balalaie, S. and Nemati, N. (2000), *Synth. Commun.*, **30**, 869.
- Pearson, A. J. and Mesaros, E. F. (2002), *Org. Lett.*, **4**, 2001.
- Kourouli, T., Kefalas, P., Ragoussis, N. and Ragoussis, V. (2002), *J. Org. Chem.*, **67**, 4615.
- Wada, S. and Suzuki, H. (2003), *Tetrahedron Lett.*, **44**, 399.

35

Mannich Reaction

Introduction

The three-component condensation reaction between an active-methylene compound, formaldehyde and an amine to form a β-aminocarbonyl compound (called as *Mannich base*) is known as *Mannich reaction*[1].

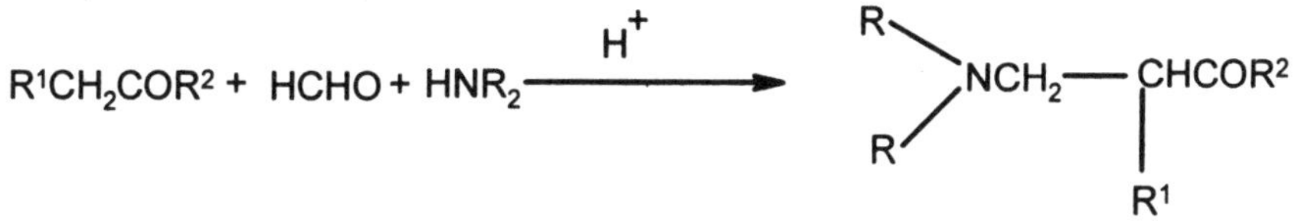

$$R^1CH_2COR^2 + HCHO + HNR_2 \xrightarrow{H^+} R_2NCH_2-CHCOR^2$$
$$\text{(with } R^1 \text{ substituent)}$$
(Mannich base)

This is one of the most widely used reactions for the formation of carbon-carbon bonds. The reaction is usually carried out in water, methanol, ethanol, or acetic acid. The amine is normally employed as its hydrochloride, and several drops of hydrochloric acid are frequently added into the reaction mixture. Usually secondary amines (*e.g.,* dimethylamine, diethylamine, piperidine, morpholine, pyrrolidine) are selected to avoide side reactions. Typical examples of the *Mannich reaction* are:

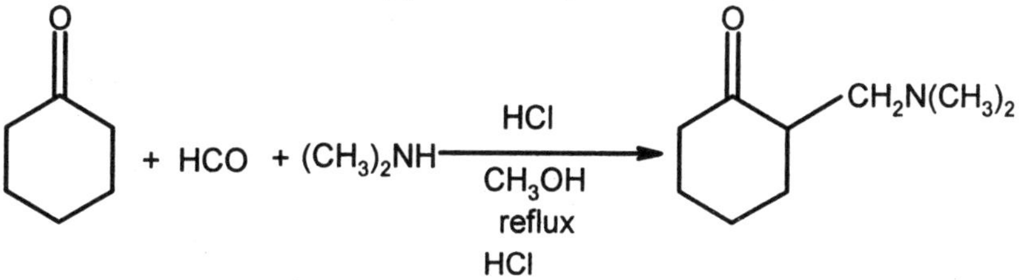

cyclohexanone + HCO + $(CH_3)_2NH$ $\xrightarrow[\text{reflux}]{\text{HCl}, CH_3OH}$ 2-((dimethylamino)methyl)cyclohexanone

$PhCOCH_3 + HCHO + (CH_3)_2NH \xrightarrow[\text{reflux}]{\text{HCl}, CH_3OH} PhCOCH_2N(CH_3)_2$ [Ref. 2]

The scope of the reaction has already been extended widely.

Mechanism

Under the usual acidic conditions, the mechanism of the *Mannich reaction* is believed[3] to proceed through electrophilc attack of an iminium salt (**1**) on the enolic tautomer of the acive-methylene compound. The intermediacy of the iminium salt has been supported from kinetic evidence[4].

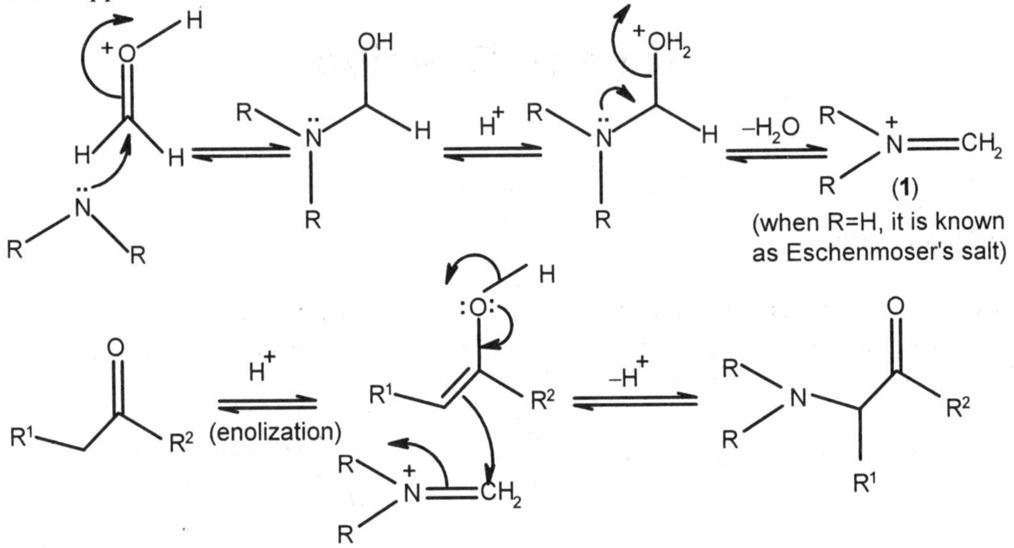

(when R=H, it is known as Eschenmoser's salt)

Critical Views

(a) The *Mannich reaction* can also be catalyzed by base; the reaction goes through enolate of the avtive-methylene compound:

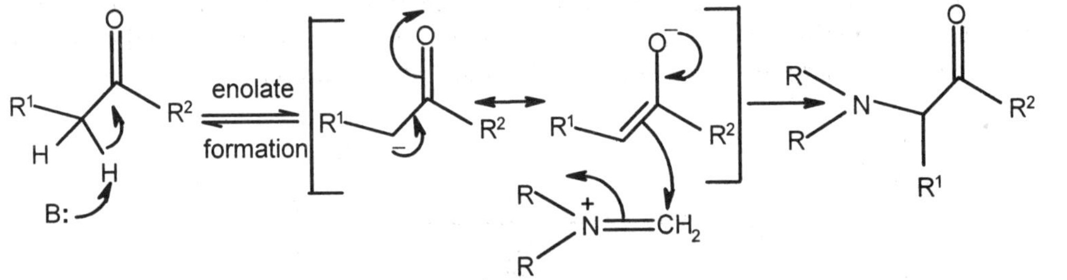

(b) Since the ease of enol formation is increased by the presence of alpha-alkyl substituents (corresponds to greater enol-stability), it is very much logical to understand that unsymmetrical ketones will react predominantly at the more highly substituted alpha-position, as observed in the following cases:

$$CH_3COCH(CH_3)_2 + HCHO + (CH_3)_2NH \xrightarrow[\substack{CH_3OH \\ reflux}]{HCl} CH_3COC(CH_3)_2CH_2N(CH_3)_2 \; [76\%]$$
$$+ (CH_3)_2NCH_2CH_2CO(CH_3)_2 \; [22\%]$$

[Ref. 5]

[Reaction scheme: 2-methylcyclohexanone + HCHO + (CH₃)₂NH, HCl, H₂O, reflux → 2-methyl-2-(dimethylaminomethyl)cyclohexanone [30%] + 2-methyl-6-(dimethylaminomethyl)cyclohexanone [70%] [Ref. 6]]

However, presence of steric interaction into the reacting molecule(s) directs the product to orient in the reverse:

$$(CH_3)_2CHCH_2COCH_3 + HCHO + (C_2H_5)_2NH \xrightarrow[CH_3OH/reflux]{HCl, H_2O} (CH_3)_2CHCH_2COCH_2CH_2N(C_2H_5)_2 \quad (63\%) \quad [Ref. 7]$$

Hence, regioselectivity can be attained by treating appropriate preformed iminium salts with the active-hydrogen component — for instance, the use of methylene dimethylammonium trifluoroacetate [$(CH_3)_2N^+=CH_2CF_3COO^-$] in CF_3COOH gives substitution at the more substituted position, while addition at the less substituted position takes place with methylene di-isopropylammonium perchlorate [$(i\text{-propyl})_2N^+=CH_2ClO_4^-$][8].

(c) Aryl amines do not normally undergo the reaction. Besides the enolizable carbonyl compounds, the Mannich reaction also applicable to other compounds having active-hydrogens like esters, nitriles, nitroalkanes, terminal alkynes, etc. — apart from those some aromatic and heterocyclic compounds (phenols, pyrroles, furans, indoles, etc.) also participate in the reaction.

Applications

The synthetic importance of the *Mannich reaction* lies in the facile conversion of *Mannich bases* into products of interests (particularly, a number of natural products having significant pharmacological applications); actually, they are used primarily as synthetic intermediates. Few examples are cited:

i) $CH_3COCH_2CH_2N(CH_3)_2$ (Mannich base)
- $\xrightarrow{Ni/H_2}$ $CH_3COCH_2CH_3$ (saturated ketone)
- $\xrightarrow{heating}$ $CH_3COCH_2=CH_2$ (α,β-unsaturated ketone)
- $\xrightarrow{CH_3I}$ $CH_3COCH_2CH_2N(CH_3)_3I$ (quaternary salt) $\xrightarrow[heating]{base}$ $CH_3COCH_2=CH_2$ (α,β-unsaturated ketone)

ii) 2-methylfuran + HCHO + (CH₃)₂NH →(H₂O, AcOH, 100°C)→ 5-methyl-2-(dimethylaminomethyl)furan (69-76%) [Ref. 9]

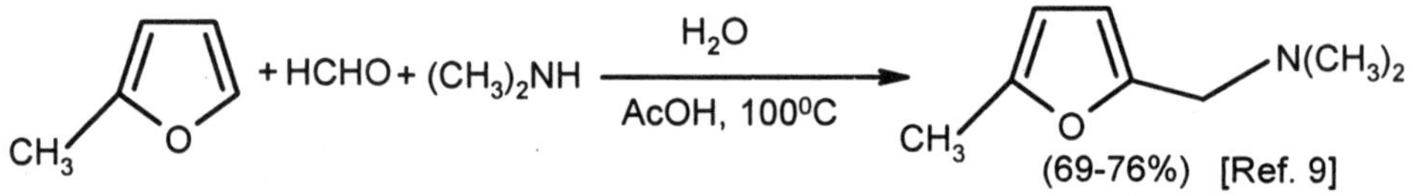

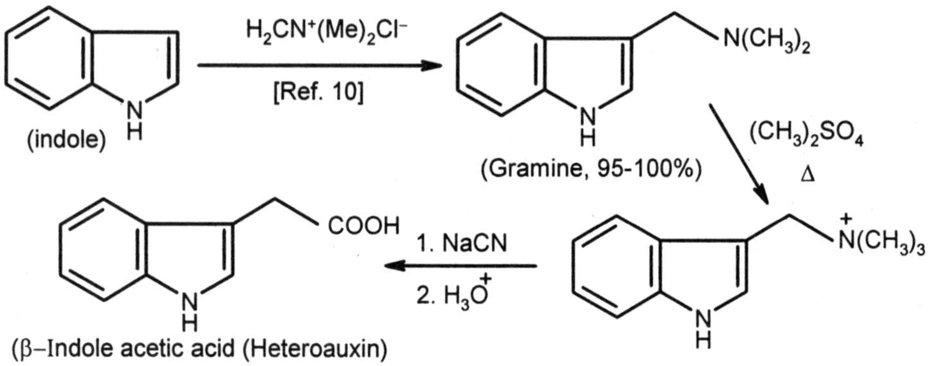

iii) (indole) →(H₂CN⁺(Me)₂Cl⁻, [Ref. 10])→ (Gramine, 95-100%) →((CH₃)₂SO₄, Δ)→ 3-(trimethylammoniomethyl)indole →(1. NaCN, 2. H₃O⁺)→ (β–Indole acetic acid (Heteroauxin))

iv) 4-methoxyphenol →(H₂CN⁺(Me)₂Cl⁻)→ (Mannich base) →(CH₃I)→ trimethylammonium iodide →(NaCN/DMF)→ (2-hydroxy, 5-dimethoxy benzylcyanide) →(CH₃I, K₂CO₃)→ (2,5-dimethoxy benzylcyanide) →(NiCl₂, NaBH₄)→ (2,5-dimethoxyphenyl-ethylamine)

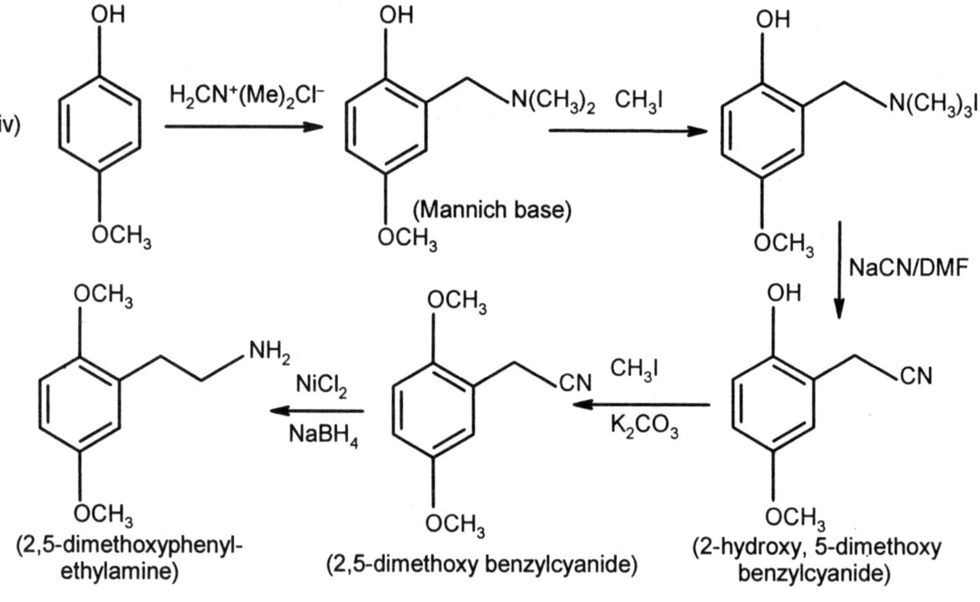

v) OHC-CH₂-CH₂-CHO + H₂NCH₃ + HOOC-CH₂-CO-CH₂-COOH → tropinone dicarboxylic acid → tropinone → (atropine)

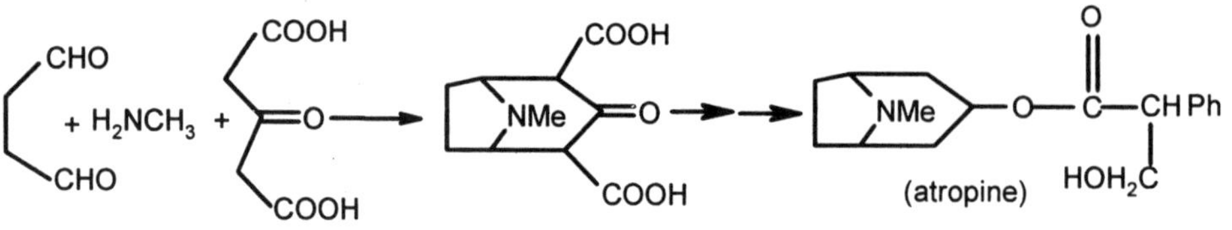

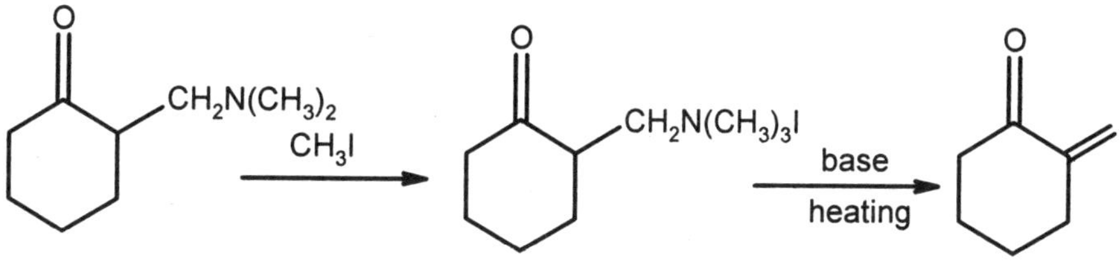

ix) Benjamin List[12] has reported a direct catalytic asymmetric three-component *Mannich reaction*:

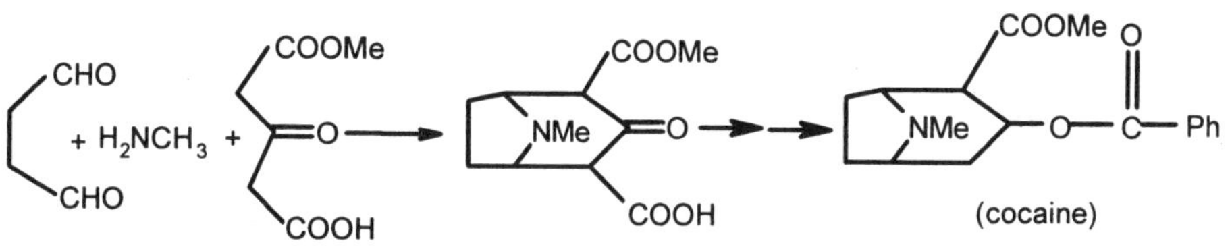

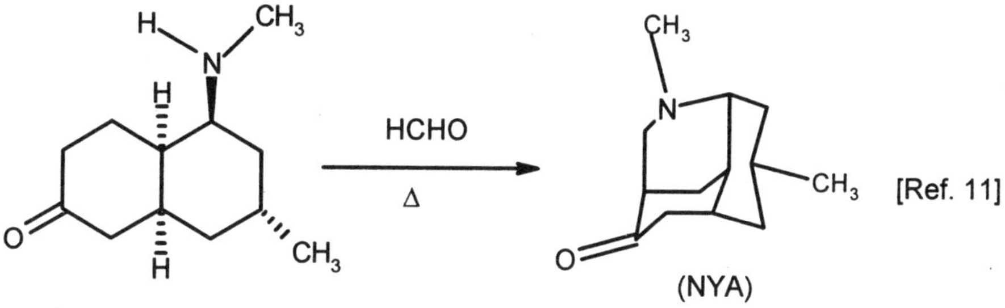

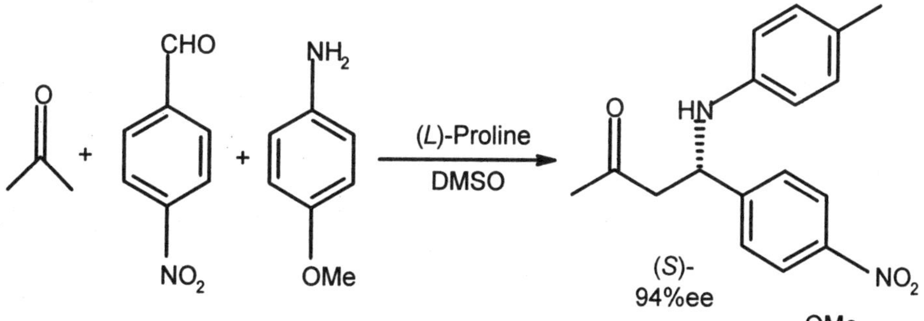

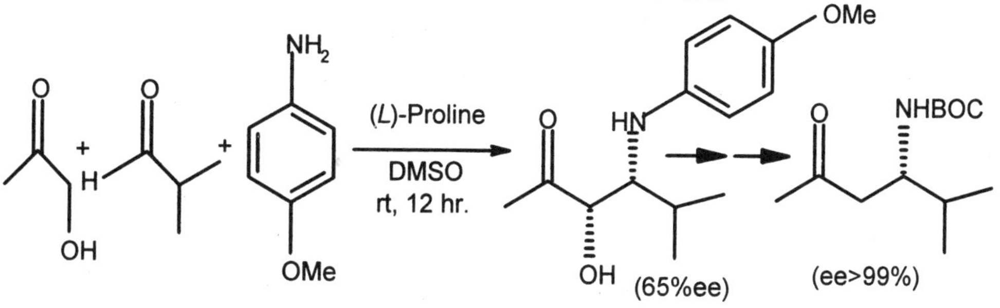

x) Nimavat *et al.*[13] prepared the following Mannich bases bearing antimicrobial properties:

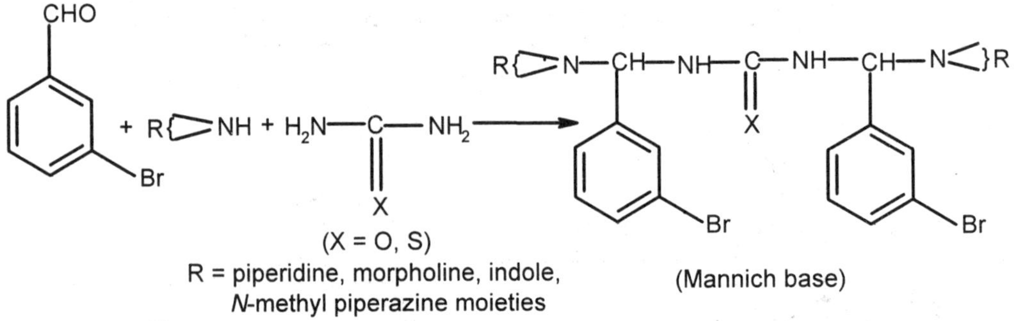

CHO
+ R{ NH + H_2N—C—NH_2
X
Br
R{ N—CH—NH—C—NH—CH—N }R
X
Br
Br
(X = O, S)
R = piperidine, morpholine, indole,
N-methyl piperazine moieties
(Mannich base)

xi) Robinson[14] synthesized tropinone, a precursor of atropine and related compounds, by a *Mannich reaction* of succindialdehyde and methylamine with acetone; better yields were obtained at the use of calcium salt of acetonedicarboxylic acid instead of simple acetone.

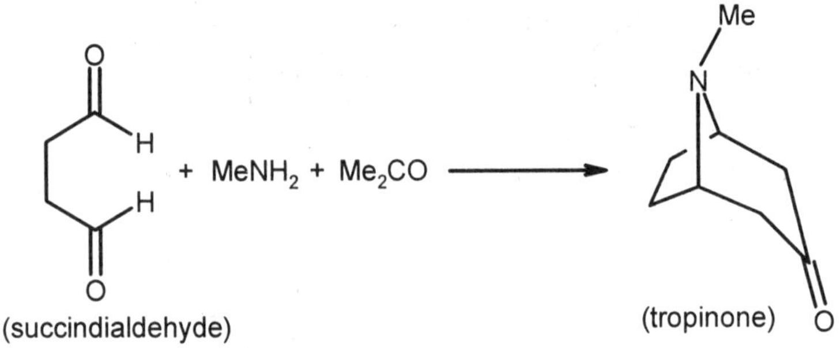

O
H
H
O
+ $MeNH_2$ + Me_2CO
Me
N
O
(succindialdehyde)
(tropinone)

References

1 Mannich, C. and Krosche, W. (1912), *Arch. Pharm.*, **250**, 647; Blicke, F. F. (1942), *Org. React.*, **1**, 3030; Reichert, B., *Die Mannich Reaktion*, Springer Verlag, Berlin, 1959; Gevorgyan, G. A., Agababyan, A. G. and Mndzhoyan, O. L. (1985), *Russ. Chem. Rev.*, **54**, 495 [Review]; Tramontini, M. and Angiolini, L. (1990), *Tetrahedron*, **46**, 1791 [Review].
2 Maxwell, C. E. (1955), *Org. Syn.*, **Coll. Vol. 3**, 305.
3 Hellmann, H. and Opitz, G. (1956), *Angew. Chem.*, **68**, 265; Cummings, T. F. and Shelton, J. R. (1960), *J. Org. Chem.*, **25**, 419; Smissman, E. E., Sorenson, J. R. J., Albrecht, W. A. and Creese, M. W. (1970), *J. Org. Chem.*, **35**, 1357.
4 Benkovic, S. J., Benkovic, P. A. and Comfort, D. R. (1969), *J. Am. Chem. Soc.*, **91**, 1860.

5 Brown, M. and Johnson, W. S. (1962), *J. Org. Chem.*, **27**, 4706; Buchanan, G. L., Curran, A. C. W. and Wall, R. T. (1969), *Tetrahedron*, **25**, 5503.
6 House, H. O. and Trost, B. M. (1964), *J. Org. Chem.*, **29**, 1339.
7 Spencer, T. A., Watt, D. S. and Friary, R. J. (1967), *J. Org. Chem.*, **32**, 1234.
8 Jasor, Y., Luche, M., Gaudry, M. and Marquet, A. (1974), *J. Chem. Soc., Chem. Commun.*, 253; Gaudry, M., Jasor, Y. and Khac, T. B., *Org. Synth.*, **VI**, 474.
9 Eliel, E. L. and Fisk, M. T. (1963), *Org. Syn.*, **Coll. Vol. 4**, 816.
10 Kozikowski, A. and Ishida, H. (1980), *Heterocycles*, **4**, 55.
11 Scot, W. L. and Evans, D. A. (1972), *J. Am. Chem. Soc.*, **94**, 4779.
12 List, B. (2000), *J. Am. Chem. Soc.*, **122**, 9336.
13 Nimavat, K. S., Popat, K. H., Vasoya, S. L. and Joshi, H. S. (2003), *J. Indian Chem. Soc.*, **80**(7), 711.
14 Robinson, R. (1917), *J. Chem. Soc.*, **111**, 762.

- Wender, P. A. and lechleeiter, J. C. (1980), *J. Am. Chem. Soc.*, **102**, 6340.
- Katritzky, A. R. and Harris, P. A. (1990), *Tetrahedron*, **46**, 987.
- Overman, L. (1992), *Acc. Chem. Res.*, **25**, 352.
- Texier-Boulett, F., Latouche, R. and Hamelin, J. (1993), *Tetrahedron Lett.*, **34**, 2123.
- Kobayashi, S. and Ishitani, H. (1995), *J. Chem. Soc., Chem. Commun.*, 1379.
- Loh, T. –P. and Wei, L. L. (1998), *Tetrahedron Lett.*, **39**, 323.
- Arned, M., Westermann, B. and Risch, N. (1998), *Angew. Chem. Int. Ed., Engl.*, **37**, 1045.
- Pawda, A. and Waterson, A. G. (2000), *J. Org. Chem.*, **65**, 235.
- Atlan, V., Bienayme, H., El Kaim, L. and Majee, A. (2000), *Chem. Commun.*, 1585.
- Schlienger, N, Bryce, M. R. and Hansen, T. K. (2000), *Tetrahedron*, **56**, 10023.
- Bur, S. K., Martin, S. F. (2001), *Tetrahedron*, **57**, 3221. [Review]
- Vicario, J. L., Badia, D. and Carrillo, L. (2001), *Org. Lett.*, **3**, 773.
- Ranu, B. C., Samanta, S. and Guchhait, S. K. (2002), *Tetrahedron*, **58**, 983.
- Martin, S. F. (2002), *Acc. Chem. Res.*, **35**, 985. [Review]
- Pawda, A., Bur, S. K., Danca, D. M., Ginn, J. D. and Lynch, S. M. (2002), *Synlett* 851. [Review]
- Yang, X. –F., Wang, M. and Varma, R. S., Li, C. –J. (2003), *Org. Lett.*, **5**, 657.

36

Meerwein – Ponndorf – Verley Reduction

Introduction

The conversion of aldehydes and ketones into the corresponding alcohols by the treatment of aluminium isopropoxide (catalyst) in excess of isopropyl alcohol is called the *Meerwein-Ponndorf-Verley reduction*[1].

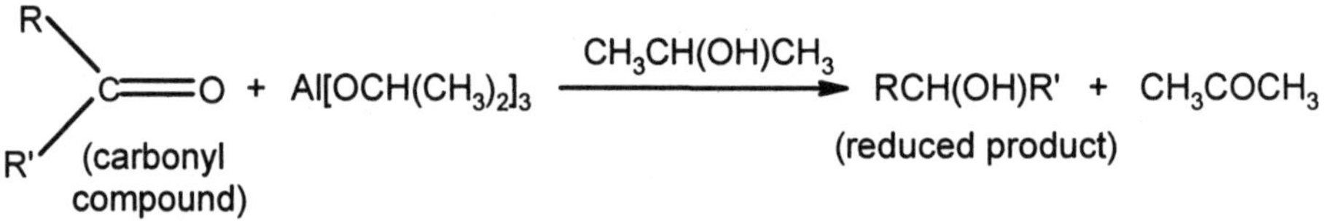

$$\underset{\text{(carbonyl compound)}}{\overset{R}{\underset{R'}{>}}C=O} + Al[OCH(CH_3)_2]_3 \xrightarrow{CH_3CH(OH)CH_3} \underset{\text{(reduced product)}}{RCH(OH)R'} + CH_3COCH_3$$

The reaction is reversible — the reverse reaction (*i.e.* oxidation of alcohol to carbonyl compound) is called the *Oppenauer oxidation*[2]. The reaction is controllable with an excess of isopropanol, and is shifted in the forward direction by distilling the acetone formed continuously out of the system. Thus, by keeping the reaction mixture at a temperature just above the boiling point of acetone, the reaction can then be driven to completion.

Mechanism

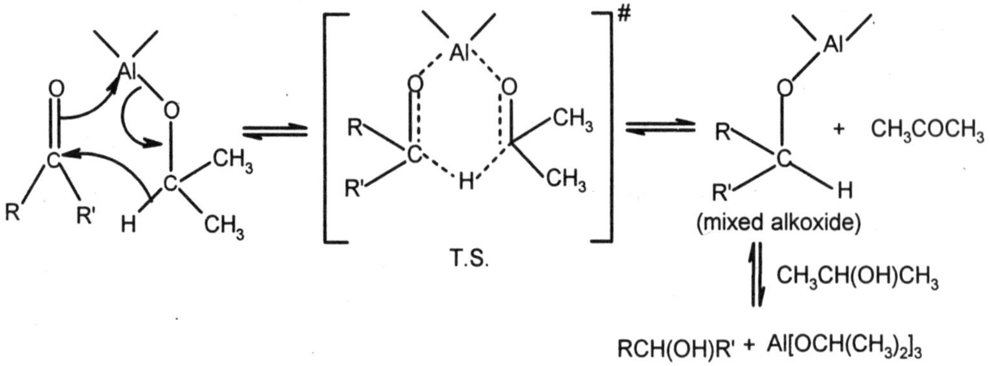

The reaction involves a six-membered cyclic transition state[3]. The aluminium atom within the Lewis-acidic reagent coordinates to the carbonyl oxygen, thereby enhancing the electron-deficiency at the carbonyl carbon so that hydride transfer from the isopropyl moiety is facilitated. The desired alcohol [RCH(OH)R'] is liberated on exchanging of isopropanol with the mixed alkoxide formed. Thus among the two hydrogens in RCH(OH)R', one is transferred from the aluminium reagent and the other comes from the solvent.

Critical Views

(a) Free-radical mechanism has been found to be involved in certain substrates[4]. It has also been established that the presence of bulky groups at the α-position of cyclic ketones retards the rate of reduction[5], and the formation of a less hindered alcohol to a greater extent is observed.

(b) Other Lewis-acidic alkoxides might also be employed; however use of aluminium isopropoxide is more advantageous. It is sufficiently soluble in organic solvents, and aluminium–oxygen bond being nearly covalent it undergoes very little dissociation to form alkoxide ions which usually cause some polymerization of the carbonyl compounds, particularly the sensitive aldehydes. Thus, the side reactions are negligible. Moreover, its relatively higher boiling point (140-50°C) enables to distill off the acetone (as formed during the reaction) out of the system so as to shift the reaction towards the forward direction. A useful substitute, lanthum isopropoxide was reported to be employed successfully[6].

(c) *Meerwein-ponndorf-Verley reduction* is specific for carbonyl group; other reducible functionalities such as C–C double or triple bonds, nitro, acetal, nitrile, *etc.* can be present safely in the substratre molecule without themselves being reduced. β-Keto esters, β-diketones, and other ketones and aldehydes with relatively high enol content do not respond the reaction.

Applications

The rection finds immense application in organic chemistry. A few examples are cited:

(i) $CH_2=CH-CHO$ (acraldehyde) $\xrightarrow[CH_3CH(OH)CH_3]{Al[OCH(CH_3)_2]_3}$ $CH_2=CH-CH_2OH$ (allyl alcohol)

(ii) cyclohexenyl methyl ketone $\xrightarrow[CH_3CH(OH)CH_3]{Al[OCH(CH_3)_2]_3}$ 1-(cyclohex-1-enyl)ethanol

(iii) Cl-C6H4-C(=O)-CH3 →[Al[OCH(CH3)2]3 / CH3CH(OH)CH3] Cl-C6H4-CH(OH)-CH3 (81%) [Ref. 7]

(iv) o-(NO2)C6H4-CHO →[Al[OCH(CH3)2]3 / CH3CH(OH)CH3] o-(NO2)C6H4-CH2OH

References

1. Meerwein, H. and Schmidt, R. (1925), *Justus Liebigs Ann. Chem.*, **444**, 221; Ponndorf, W. (1926), *Angew. Chem.*, **39**, 138; Verley, A. (1925), *Bull. Soc. Chim.*, **37**, 537; Wilds, A. L. (1944), *Org. React.*, **2**, 178. [Review].
2. Oppenauer, R. V. (1937), *Recl. Trav. Chim. Pays-Bas*, **56**, 137.
3. Warnhoff, E. W., Reynolds-Warnhoff, P. and Wong, M. Y. H. (1980), *J. Am. Chem. Soc.*, **102**, 5956; Shiner Jr., V. J. and Whittaker, D. (1963), *J. Am. Chem. Soc.*, **85**, 2337.
4. Yamataka, H. and Hanafusa, T. (1987), *Chem. Lett.*, 643; Nasipuri, D., Gupta, M. D. and Banerjee, S. (1984), *Tetrahedron Lett.*, **25**, 5551; Screttas, C. G. and Cazianis, C. T. (1978), *Tetrahedron*, **34**, 933.
5. Hach, V. (1973), *J. Org. Chem.*, **38**, 293.
6. Okano, T., Matsuoka, M., Konishi, H. and Kiji, J. (1987), *Chem. Lett.*, 181.
7. Marvel, C. S. and Schertz, G. L. (1943), *J. Am. Chem. Soc.*, **65**, 2055.

- Hutton, J. (1979), *Synth. Commun.*, **9**, 483.
- Namy, J. L., Souppe, J., Collin, J. and Kagan, H. B. (1984), *J. Org. Chem.*, **49**, 2045.
- Ashby, E. C. and Argyropoulos, J. N. (1986), *Tetrahedron Lett.*, **27**, 465; Ashby, E. C. and Argyropoulos, J. N. (1986), *J. Org. Chem.*, **51**, 3593.
- Ashby, E. C. (1988), *Acc. Chem. Res.*, **21**, 414. [Review]
- Aremo, N. and Hase, T. (2001), *Org. React.*, **42**, 3637. [Review]
- Campbell, E. J., Zhou, H. and Nguyen, S. T. (2002), *Angew. Chem. Int. Ed. Engl.*, **41**, 1020.
- Nishide, K. and Node, M. (2002), *Chirality*, **14**, 759.
- Jerome, J. E. and Sergent, R. H. (2003), *Chem Ind.*, **89**, 97.

37

Michael Reaction

Introduction

The base-catalyzed addition of a 'donor' compound possessing at least one active α-hydrogen atom to an 'acceptor' compound containing an activated double bond is classified as the *Michael reaction*[1]. The carbanion formed from the donor in alkaline solution attacks the more positive end of the polarized system of the acceptor yielding an anion (enolate), which on hydrolytic work-up affords the ultimate product.

$$\text{—CH}_2\text{—EWG} + \underset{\text{(acceptor)}}{\overset{\text{EWG'}}{\underset{}{>}}\text{C}=\text{C}\overset{}{<}} \xrightarrow[\text{ii. H}_2\text{O}]{\text{i. base}} \underset{\text{—CH—EWG}}{\overset{\text{EWG'}}{>}\text{C—C}\overset{}{<}\text{H}}$$

(donor)   (acceptor)

(EWG & EWG' both are electron-withdrawing groups)

Less specifically, it is the base-catalyzed addition of a *pseudo*-acidic carbonyl compound, ester, nitrile, nitro-compound or sulphone to the α,β-double bond of a conjugated unsaturated ketone, ester or nitrile. The anionic part of the *pseudo*-acidic addendum attaches itself to the β-end of the α,β-double bond.

$$\text{RCH}=\text{CHCOOR}^1 + \text{R}^2\text{CH(COOR}^3\text{)}_2 \xrightarrow[\text{ii. H}_2\text{O}]{\text{i. base}} \underset{\text{R}^2\text{C(COOR}^3\text{)}_2}{\text{RCH—CH}_2\text{COOR}^1}$$

R = R^2 = R^3 = H, alkyl, aryl, aralkyl, and the groups alike; R^1 = Me, CH$_2$Ph, OEt; CO$_2$R^1 may also be substituted by CN, NO$_2$, sulphones, *etc.*

$$\text{PhCH}=\text{CHCOOEt} + \text{CH}_2\text{(COOEt)}_2 \xrightarrow[\text{ii. H}_2\text{O}]{\text{i. } \overline{\text{O}}\text{Et}} \underset{\text{CH(COOEt)}_2}{\text{PhCH—CH}_2\text{COOEt}}$$

(ethyl cinnamate) (malonic ester)

The classical *Michael reaction* is carried out in a protic solvent (*e.g.* alcohol) by the use of a base, usually an alkali metal alkoxide such as sodium or potassium ethoxides, potassium *t*-butoxide, potassium isopropoxide, *etc.* Besides, mild bases like 2^0 and 3^0 amines, piperidine, pyridine have been reported to in use with success in some cases.

Mechanism

A resonance-stabilized carbanion is formed firstly as a result of acidic-proton abstraction from the 'donor' molecule by the base; in the second step the intermediate attacks the β-carbon of the α,β-unsaturated 'acceptor' molecule to produce an enolate, which ultimately yields the more stable ketonic product upon hydrolytic work-up.

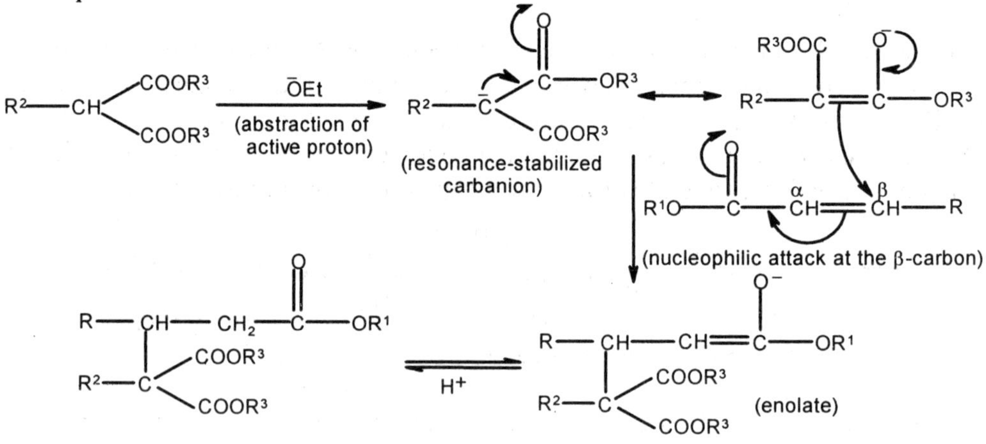

Critical Views

(a) If the donor molecule bears more than one active hydrogen atoms, then repetition of *Michael reaction* occurs[2].

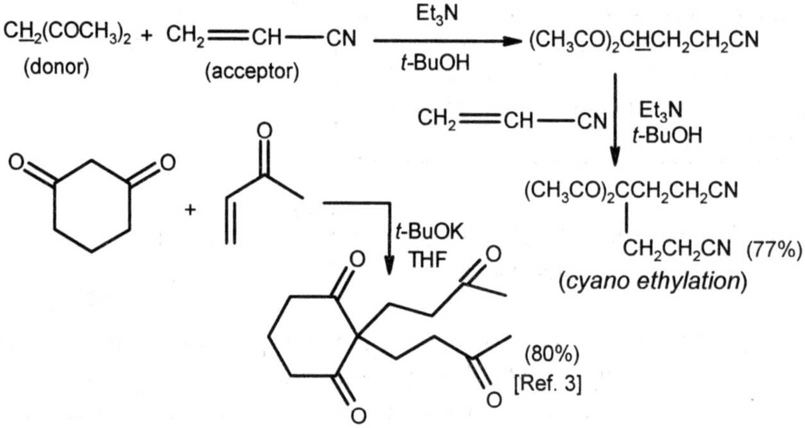

$$CH_2(COCH_3)_2 + CH_2=CH-CN \xrightarrow[t\text{-BuOH}]{Et_3N} (CH_3CO)_2CHCH_2CH_2CN$$

$$\xrightarrow[t\text{-BuOH}]{CH_2=CH-CN \quad Et_3N} (CH_3CO)_2\underset{\underset{\displaystyle CH_2CH_2CN}{|}}{C}CH_2CH_2CN \quad (77\%)$$

(cyano ethylation)

(80%)
[Ref. 3]

(b) *Intramolecular Michael condensations*[4] have been effected and cyclic products are obtained.

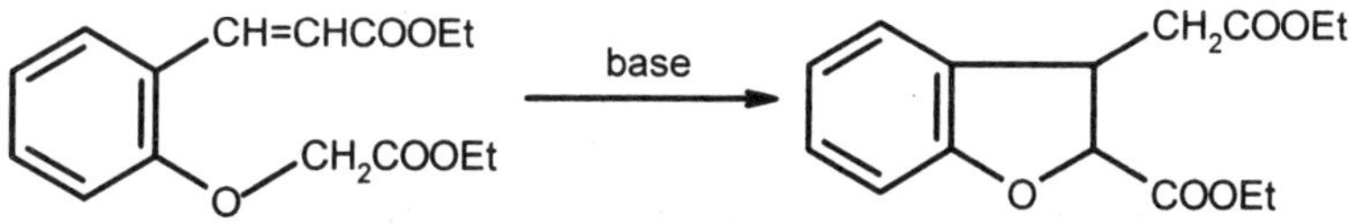

[benzofuran formation: o-substituted benzene with CH=CHCOOEt and OCH₂COOEt groups → base → bicyclic product with CH₂COOEt and COOEt substituents on the dihydrofuran ring]

(c) Amount of catalyst (base) used is very much important in carrying out the normal *Michael reaction*; otherwise '*abnormal Michael*' reaction may take place to give rearranged products.

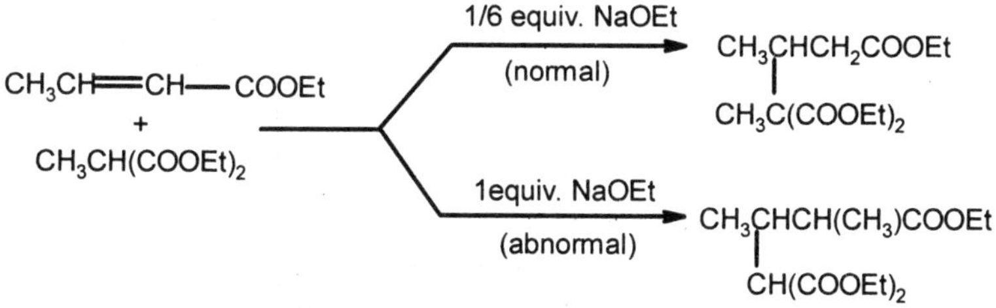

$CH_3CH=CH-COOEt$ + $CH_3CH(COOEt)_2$

→ 1/6 equiv. NaOEt (normal) → CH_3CHCH_2COOEt | $CH_3C(COOEt)_2$

→ 1 equiv. NaOEt (abnormal) → $CH_3CHCH(CH_3)COOEt$ | $CH(COOEt)_2$

(d) In certain cases, it has been observed that *Michael reaction* depends upon the extent of pressure applied during the reaction, as exemplified in the following scheme. There occurs no reaction under atmopheric pressure, but can be effected at 15 Kbar in 77% yield.

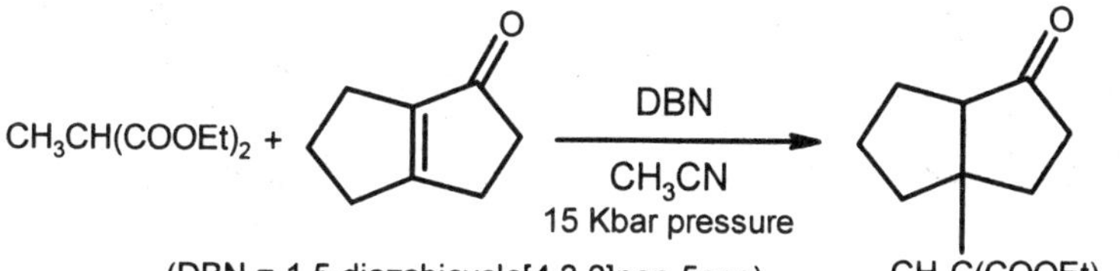

$CH_3CH(COOEt)_2$ + [bicyclic enone] → DBN, CH_3CN, 15 Kbar pressure → [bicyclic ketone with $CH_3C(COOEt)_2$ substituent]

(DBN = 1,5-diazabicyclo[4.3.0]non-5ene)

Applications

The *Michael reaction* is of great importance in organic synthesis. A very few examples are cited:

(i) 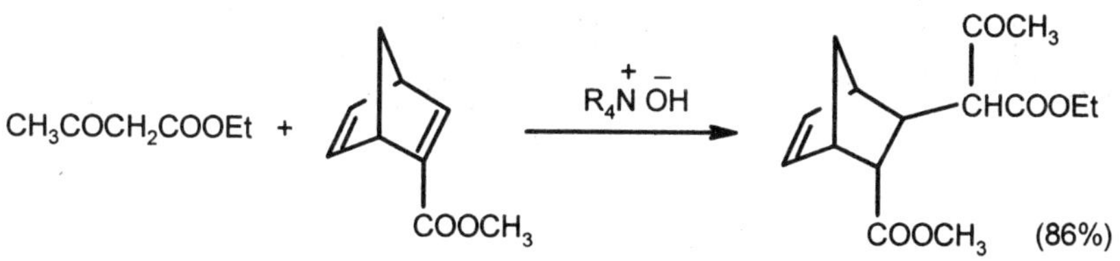 CH_3COCH_2COOEt + [norbornene with COOCH₃] → $R_4N^+ \bar{O}H$ → [norbornane product with COOCH₃ and CHCOOEt(COCH₃) substituents] (86%)

(ii) 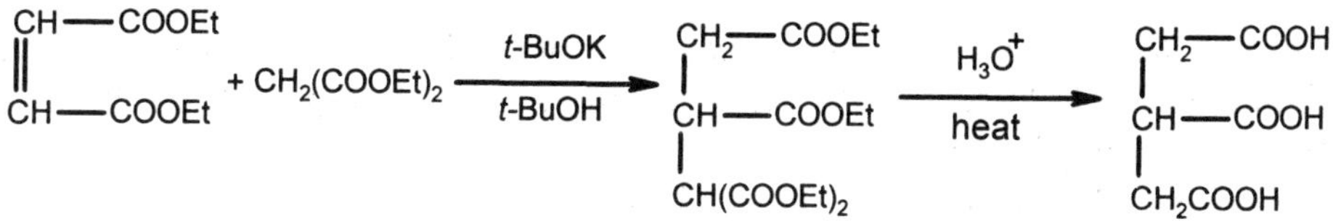 $\begin{array}{c}CH-COOEt\\ \| \\ CH-COOEt\end{array}$ + $CH_2(COOEt)_2$ → t-BuOK / t-BuOH → $\begin{array}{c}CH_2-COOEt\\ | \\ CH-COOEt \\ | \\ CH(COOEt)_2\end{array}$ → H_3O^+ / heat → $\begin{array}{c}CH_2-COOH\\ | \\ CH-COOH \\ | \\ CH_2COOH\end{array}$

(iii) $CH_3CH_2NO_2 + CH_2=CH-COCH_3 \xrightarrow{\overline{O}Et} CH_3CH(NO_2)CH_2CH_2COCH_3$

$\downarrow TiCl_3 | H_2O$

$CH_3COCH_2CH_2COCH_3$ (1,4-diketone)

(iv) One of the most important synthetic applications of *Michael addition* is in the *Robinson annulation*[6] that consists of two consecutive reactions — a *Michael addition* followed by an *aldol condensation* — producing a new six-membered ring.

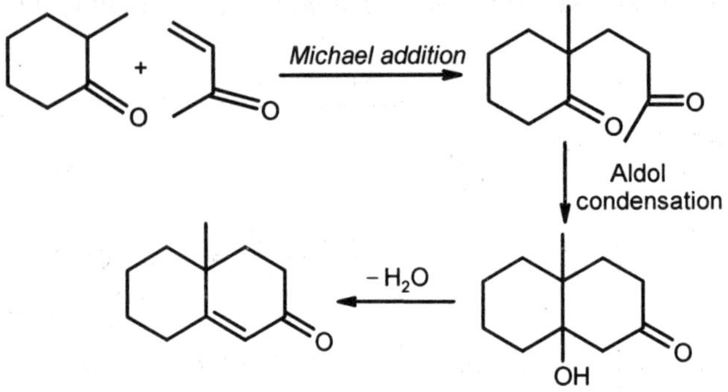

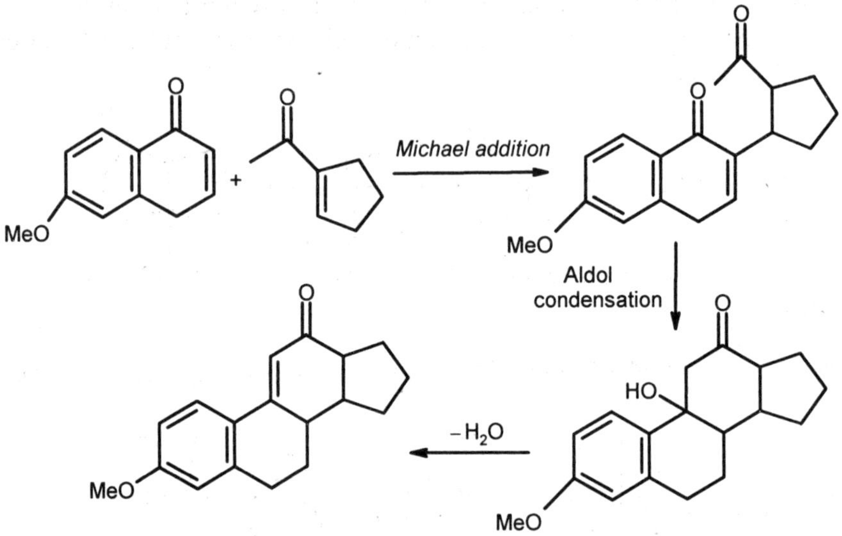

References

1. Michael, A. (1887), *J. Prakt. Chem.*, **35**, 349; Michael, A. (1887), *J. Prakt. Chem.*, **36**, 113; Bergman, E. D., Gunsburg, D. and Rappo, R. (1959), *Org. React.*, **10**, 179. [Review]
2. Van Allan, J. A. and Reynolds, G. A. (1968), *J. Org. Chem.*, **33**, 1102.

3 Houbrechb, Y., Laszlo, P. and Pennetreau, P. (1986), *Tetrahedron Lett.*, 705.
4 Ihara, M. and Fukumoto, K. (1993), *Angew. Chem.*, **105**, 1059; Little, R. D., Masjedizadeh, M. R., Wallquist, O. and McLoughliu, J. I. (1995), *Org. React.*, **47**, 315. [Review]
5 Dauben, W. G., Gerdes, J. M. and Look, G. C. (1986), *Synthesis*, 532.
6 Rapson, W. S. and Robinson, R. (1935), *J. Chem. Soc.*, 1285; Conforth, J. W. and Robinson, R. (1949), *J. Chem. Soc.*, 1855; Marshall, J. A. and Warren Jr., T. M. (1971), *J. Org. Chem.*, **36**, 178; Gawley, R. E. (1976), *Synthesis*, 777 [Review]; Jung, M. E. (1976), *Tetrahedron*, **32**, 1 [Review].

- Blarev, S. J., Schweizer, W. B. and Seebach, D. (1982), *Helv. Chimica Acta*, **65**, 1637.
- Oppolzer, W., Dudfield, P., Sevengon, T. and Godel, T. (1985), *Helv. Chimica Acta*, **68**, 212.
- Enders, D., Papadopanlos, K. and Rendenbach, E. M. (1986), *Tetrahedron Lett.*, 3491.
- Aoki, S., Sasaki, S. and Koga, K. (1989), *Tetrahedron Lett.*, 7229.
- Hoz, S. (1993), *Acc. Chem. Res.*, **26**, 69. [Review]
- Itoh, T. and Shirakami, S. (2001), *Heterocycles*, **55**, 37.
- Sundararajan, G. and Prabagaran, N. (2001), *Org. Lett.*, **3**, 389.
- Bolm, C., Kasyan, A., Heider, P., Saladin, S., Drauz, K., Gunther, K. and Wagner, C. (2002), *Org. Lett.*, **4**, 2265.
- Eilitz, U., Lebmann, F., Seidelmann and O., Wendisch, V. (2003), *Tetrahedron: Asymmetry*, **14**, 189.

38

Oppenauer Oxidation

Introduction

The reversal of the *Meerwein-Ponndorf-Verley reduction* involving particularly the oxidation of a secondary alcohol to the corresponding carbonyl compound on treatment with a base, usually metal alkoxides and an excess of a ketone is called the *Oppenauer oxidation*[1]. The ketones most commonly used are acetone, butanone, and cyclohexanone. Commonly used bases are aluminium *t*-butoxide, aluminium isopropoxide, potassium *t*-butoxide, *etc*.

$$R_2CHOH + CH_3COCH_3 \underset{}{\overset{Al(OPr-i)_3}{\rightleftharpoons}} R_2CO + (CH_3)_2CHOH$$
(2°-alcohol) (acetone) (carbonyl compound)

The oxidation takes place under mild conditions, and is highly selective for the oxidation of alcoholic hydroxyls. Although the method is usually employed for the preparation of ketones, it has also been used for aldehydes. By use of excess ketone (*e.g.* acetone) the equlibrium is forced to the right.

Mechanism

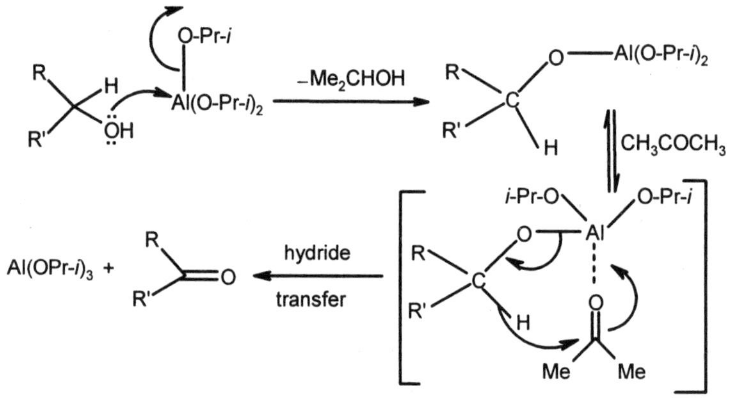

The aluminium alkoxide serves to form the alkoxide of the alcohol to be oxidized; the newly formed alkoxide then reacts with the ketone (acetone) to form a six-membered cyclic transition state that undergoes internal hydride ion transfer resulting in the oxidation of the alcohol into the corresponding carbonyl compound.

Critical Views

(a) The carbonyl compound in the reaction mixture acts as a carbonyl hydrogen acceptor; during oxidation of primary alcohol it is needed to use the carbonyl compound possessing better hydrogen acceptability. For this purpose benzoquinone, anisaldehyde, and cinnamaldehyde are employed.

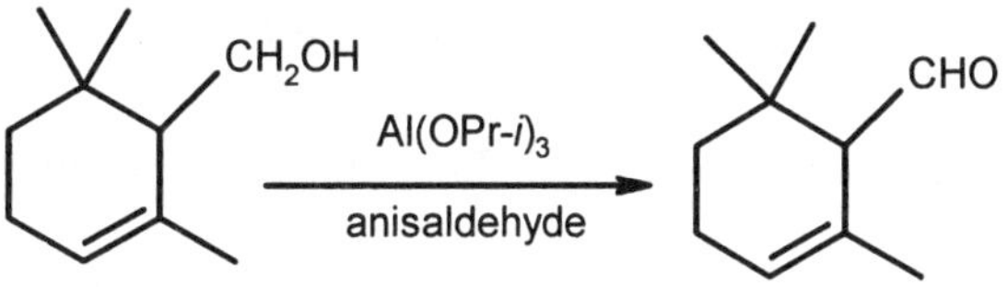

(b) Interesetingly, it is noticed that β,γ-double bonds generally migrate into conjugation with the carbonyl group under the conditions of the reaction.

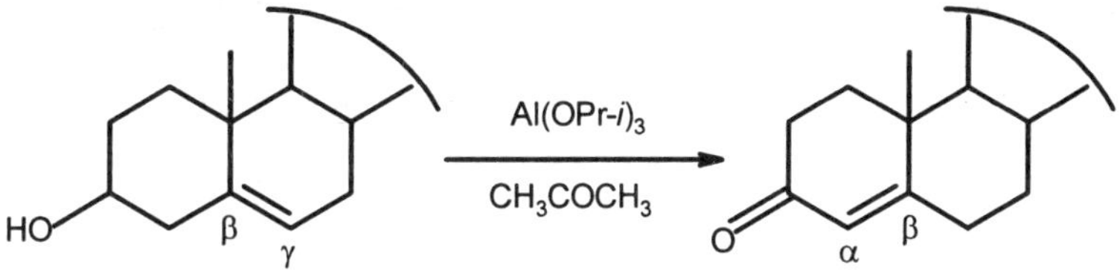

Applications

Since the oxidation is highly selective for the oxidation of alcoholic hydroxyls and takes place in mild conditions, it finds immense application in synthetic organic chemistry, particularly in field of terpenoid and steroid chemistry[2].

(i) $CH_3CH(OH)CH=CHCH=C(CH_3)CH=CH_2$ $\xrightarrow[\text{acetone/benzene}]{Al(O-Bu^t)_3}$ $CH_3COCH=CHCH=C(CH_3)CH=CH_2$

(ii)

(iii) Geraniol —[Al(O-Bu^t)$_3$, Quinone]→ Citral

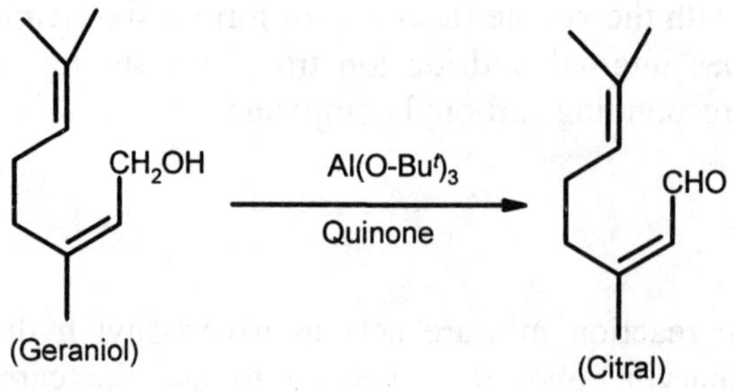

References

1. Oppenauer, R. V. (1937), *Rec. Trav. Chim.*, **56**, 137.
2. Djerassi, C. (1951), *Org. React.*, **6**, 207. [Review]

- Almedia, M. L. S., Kocovsky, P. and Backvall, J. –E. (1996), *J. Org. Chem.*, **61**, 6587.
- Akamanchi, K. G. and Chaudhari, B. A. (1997), *Tetrahedron Lett.*, **38**, 3285.
- Raja, T., Jyothi, T. M., Sreekumar, K., Talawar, M. B., Santhanalakshmi, J. and Rao, B. S. (1999), *Bull. Chem. Soc. Jpn.*, **72**, 2117.
- Nait Ajjou, A. (2001), *Tetrahedron Lett.*, **42**, 13.
- Ooi, T., Otsuka, H., Miura, T., Ichikawa, H. and Maruoka, K. (2002), *Org. Lett.*, **4**, 2669.
- Suzuki, T., Morita, K., Tsuchida, M. and Hiroi, K. (2003), *J. Org. Chem.*, **68**, 1601.

39

Perkin Reaction

Introduction

The aldol-type base-catalyzed condensation of an aromatic aldehyde with a carboxylic acid anhydride is referred to as the *Perkin reaction*[1]. When the anhydride contains two α-hydrogen atoms, the product is the α,β-unsaturated carboxylic acid. However, in some cases β-hydroxy carboxylic acid is isolated when the anhydride bears only one α-hydrogen, because dehydration cannot take place here.

$$Ar\text{-}CHO + (RCH_2CO)_2O \xrightarrow[\Delta]{RCH_2COONa} Ar\text{-}CH=C(R)COOH$$

Cinnamic acid is thus obtained by the condensation of benzaldehyde and acetic anihydride with sodium acetate:

$$PhCHO + (CH_3CO)_2O \xrightarrow[170\text{-}200°C]{CH_3COONa} Ph\text{-}CH=CHCOOH$$
(benzaldehyde)   (cinnamic acid)

The base used is a weak one, and usually an alkali salt of the carboxylic acid corresponding to the acid anhydride; other bases such as sodium carbonate, quinoline, pyridine and triethylamine may also be employed. Since a weak base is to react with a weak acid (anhydride), the general procedure is to heat the mixture of aldehyde and anhydride along with the base at 170-200°C for several hours (~ 5-6 hrs.).

$$Ph\text{-}CH=CHCHO + (CH_3CO)_2O \xrightarrow[\Delta]{CH_3COONa} Ph\text{-}CH=CH\text{-}CH=CHCOOH$$
(cinnamaldehyde)   (5-phenyl-2,4-dienoic acid)

Mechanism

The base abstracts an α-hydrogen atom of the anhydride to generate a carbanion, which in turn attacks the carbonyl carbon of the aldehyde followed by protonation of the anion formed. One molecule of water then eliminated in the presence of the anhydride under the reaction temperature; the final product (α,β-unsaturated carboxylic acid) is obtained on hydrolysis being effected during the work-up.

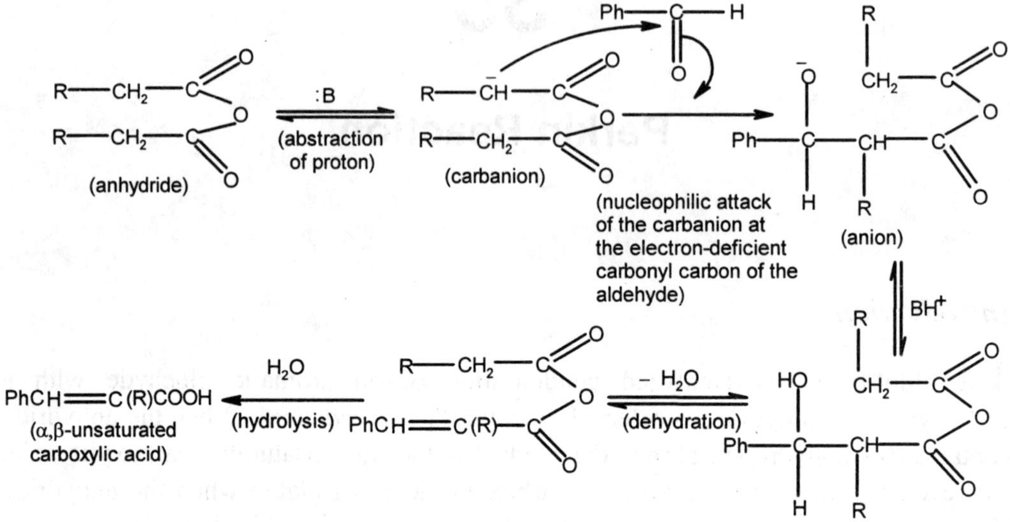

Critical Views

(a) A mixture of *E*- and *Z*-isomers of the α,β-unsaturated carboxylic acid is anticipated; however the formation of the *E*-isomer is preferred.
(b) It has been reported that use of cesium (Cs) salt of the carboxylic acid corresponding to the anhydride affords higher yields and shorter reaction time[2].
(c) In regard to substituted aromatic aldehydes, Crawford and Little[3] showed that the reactivity of the aromatic aldehydes containing *para*-substituents decreases in the order: $NO_2 > Cl > H > OCH_3 >$ alkyl.
(d) A variant of the *Perkin reaction* is the *Erlenmeyer-Plochl azalactone synthesis*[4] that involves the condensation of an aromatic aldehyde with an *N*-acyl glycine (**I**) in the presence of sodium acetate and acetic anhydride yielding the azalactone (**II**).

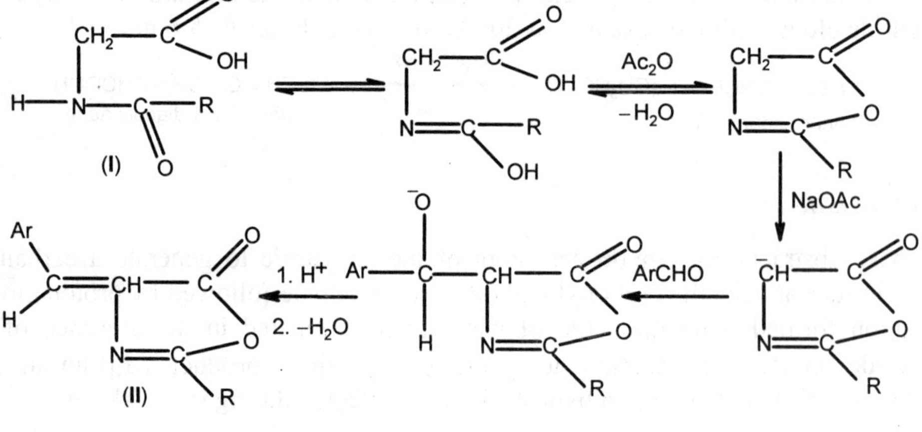

α-Amino acids and α-keto acids are prepared from azalactone derivatives.

Applications

(i) 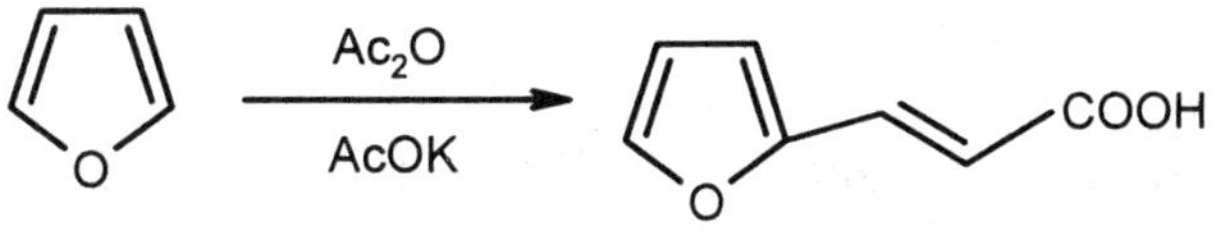 furan + Ac$_2$O / AcOK → 2-furyl-CH=CH-COOH [Ref. 5]

(ii) 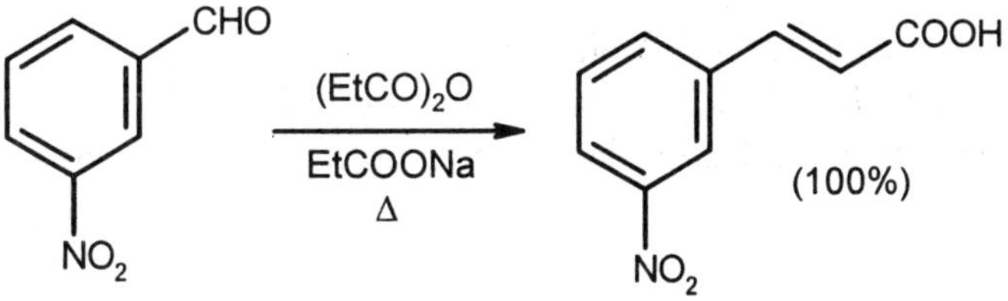 3-nitrobenzaldehyde + (EtCO)$_2$O / EtCOONa, Δ → 3-nitrocinnamic acid (100%) [Ref. 6]

(iii) 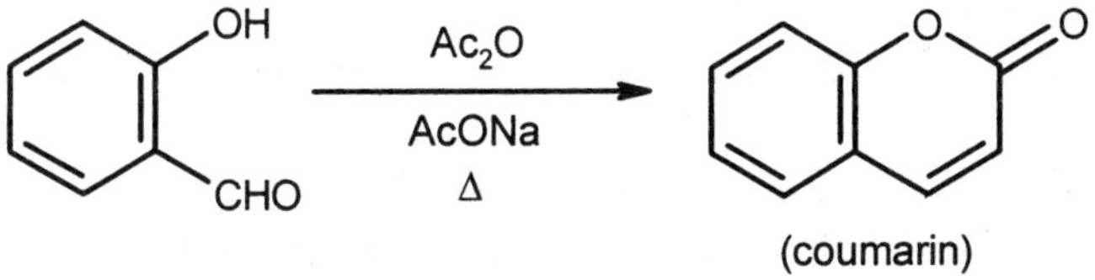 salicylaldehyde + Ac$_2$O / AcONa, Δ → coumarin

(iv) 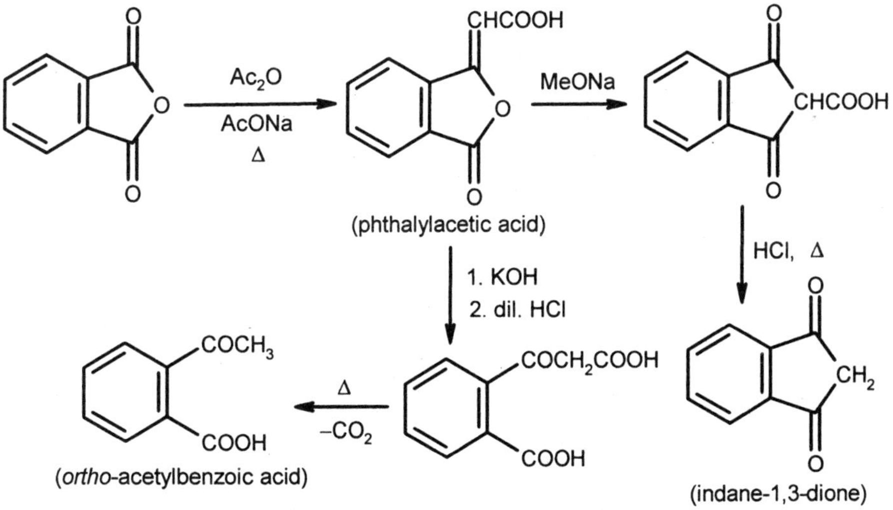 phthalic anhydride → (Ac$_2$O / AcONa, Δ) → phthalylacetic acid → (MeONa) → CHCOOH derivative → (HCl, Δ) → indane-1,3-dione

phthalylacetic acid → (1. KOH, 2. dil. HCl) → o-COCH$_2$COOH-benzoic acid → (Δ, −CO$_2$) → ortho-acetylbenzoic acid

(v) 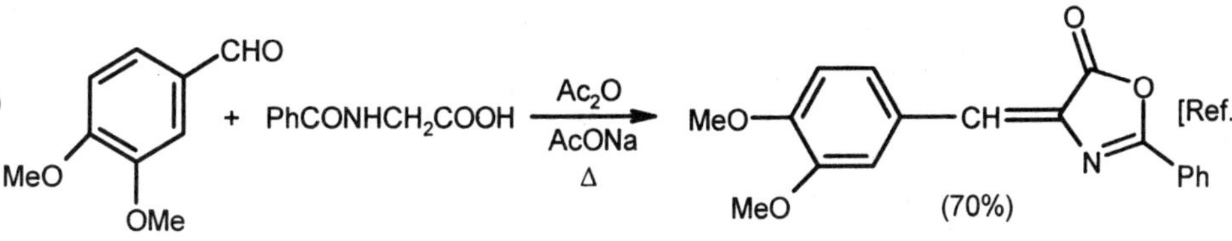 3,4-dimethoxybenzaldehyde + PhCONHCH$_2$COOH → (Ac$_2$O / AcONa, Δ) → azlactone (70%) [Ref. 7]

References

1 Perkin, W. H. (1868), *J. Chem. Soc.*, **21**, 53, 181; Perkin, W. H. (1877), *J. Chem. Soc.*, **31**, 388; Johnson, J. R. (1942), *Org. React.*, **1**, 210 (Review).
2 Koepp, E. and Vogtle, F. (1987), *Synthesis*, 177.
3 Crawford, M. and Little, W. T. (1959), *J. Chem. Soc.*, 722.
4 Erlenmeyer, E. (1893), *Justus Liebigs Ann. Chem.*, **275**, 1; Plochl, J. (1883), *Ber. Dtsch. Chem. Ges.*, **16**, 2815.
5 Johnson, J. R. (1955), *Org. Synth.*, **Coll. Vol. 3**, 426.
6 Maxwell, R. W. and Adams, R. (1930), *J. Am. Chem. Soc.*, **52**, 2907.
7 Vanderberg, G. E., Harrison, J. B., Carter, H. E. and Magerlein, B. J. (1967), *Org. Synth.*, **47**, 101.

- Pohjala, E. (1975), *Heterocycles*, **3**, 615.
- Poonia, N. S., Sen, S., Porwal, P. K. and Jayakumar, A. (1980), *Bull. Chem. Soc. Jpn.*, **53**, 3338.
- Gaset, A. and Gorrichon, J. P. (1982), *Synth. Commun.*, **12**, 71.
- Brady, W. T. and Gu, Y. –Q. (1988), *J. Heterocyclic Chem.*, **25**, 969.

40

Pinacol Rearrangement

Introduction

The acid-catalyzed rearrangement of *vic*-diols (1,2-glycols) to aldehydes or ketones is called the *pinacol rearrangement*[1].

$$R^1-\underset{\underset{OH}{|}}{\overset{\overset{R^2}{|}}{C}}-\underset{\underset{OH}{|}}{\overset{\overset{R^3}{|}}{C}}-R^4 \xrightarrow{H^+} R^1-\underset{\underset{O}{\|}}{C}-\underset{\underset{R^4}{|}}{\overset{\overset{R^2}{|}}{C}}-R^3$$

(R= alkyl, aryl or hydrogen)

The name is originated from the classical example of conversion of pinacol to pinacolone[2]:

$$Me-\underset{\underset{OH}{|}}{\overset{\overset{Me}{|}}{C}}-\underset{\underset{OH}{|}}{\overset{\overset{Me}{|}}{C}}-Me \xrightarrow{H^+} Me-\underset{\underset{O}{\|}}{C}-\underset{\underset{Me}{|}}{\overset{\overset{Me}{|}}{C}}-Me$$

(pinacol)   (migration of methyl group)   (pinacolone)

The migrating group may be alkyl, aryl, hydrogen, and even ethoxycarbonyl (COOEt)[3]. However, elimination of water to yield alkene — the normal reaction of alcohols— may be observed as a side-reaction.

Mechanism

In the initial step, one of the hydroxyl groups is protonated and thus, is converted into a good leaving group (water)[4]; protonation occurs to that hydroxyl so as to give more stable carbenium intermediate (**I**) formed on loss of water molecule from the protonated substrate. The next step is the 1,2-shift of a group/atom ('R' from the carbon adjacent to the +vely charged carbon atom) to give a hydroxycarbenium ion; the driving force of the migration is associated with the resonance stabilization of the rearranged carbenium ion (**II**) by the attached oxygen atom, and the species can immediately stabilze itself by losing a proton, thus yielding the desired carbonyl product.

[Reaction scheme: 1,2-diol → (protonation, H⁺) → protonated diol → (elimination of water, –H₂O) → carbenium intermediate (**I**) → (1,2-shift of group) → hydroxycarbenium ion (**II**) → (–H⁺) → carbonyl compound]

The intermediacy of carbenium ion (**I**) is evidenced from the fact that when pinacol [$Me_2C(OH)C(OH)Me_2$] is rearrnged in the presence of H_2O^{18} and the reaction is stopped before completion, the recovered pinacol is found to bear O^{18} in its molecule, thereby, indicating reversible formation of a carbenium ion (*i.e.* all the R groups in (**I**) are methyls). Besides, the intermediate arising out of pinacol has been trapped by the addition of tetrahydrothiophene[5]. The rearrangement is strickly *intramolecular* — the migrating group begins to attach itself with the carbocationic carbon before separating completely from its original position. The intramolecularity of the rearrangement is supported by the facts that (i) the migrating group, if chiral, retains its configuration, and (ii) formation of no cross-over products, but only of two rearranged products occurs when the reaction is carried out with a mixture of two different pinacols.

Critical Views

(a) *Migratory aptitude*: It is necessary to get an idea about the migratory aptitude of groups in the cases where glycols contain four different groups. Various experimental observations suggest the *realtive migratory aptitude* of groups in pinacol-type rearrangements as in the order of aryl > 3^0-alkyl > 2^0-alkyl > 1^0-alkyl.

[Reaction scheme: Me₂C(OH)–C(OH)(Ph)₂ → (H⁺) → carbocation intermediate → (1. migration of phenyl group, 2. –H⁺) → Ph–C(Me)₂–C(O)–Ph product (MeC(Ph)₂–C(=O)–Me)]

However, there is no clear-cut generalization in regard to relative migratory aptitude; the phenomenon largely depends upon the structure of the substrates as well as the experimental conditions[6]. Interestingly, the substrate (**1**) on treatment with cold and concentrated sulphuric acid yields mainly the ketone (**2**) [methyl migration], while the ketone (**2a**) [phenyl migration] is obtained predominanly when treated with acetic acid containing a trace of sulphuric acid[7].

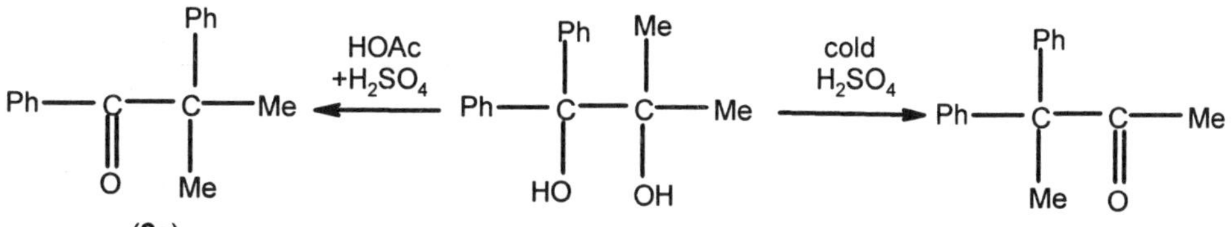

In some cases, migration of hydrogen is found to be more preferred to aryl migration[8].

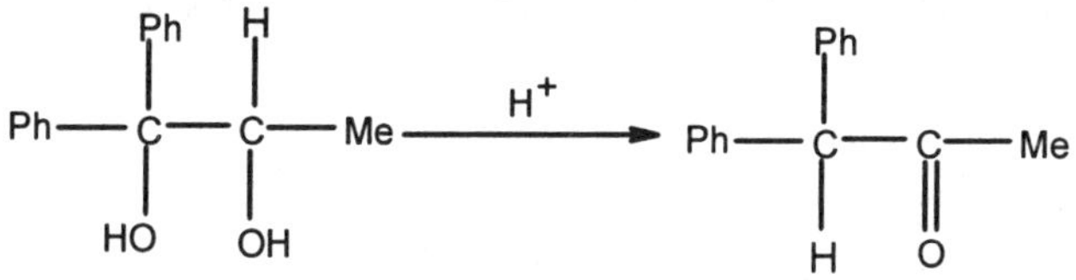

Again, between methyl and ethyl groups one is more preferred to the other in some cases, and in some other cases the reverse is true[9]. However, it may be generalized that among aryl migrating groups, electron-donating substituents in the *para* and *meta*- positions enhances the migratory aptitudes, while these substituents in the *ortho*-positions decrease the aptitude. Electron-withdrawing substituents in all the positions reduces the ability of migration. Bachmann and Ferguson[10] studied the relative migratory aptitudes of some aryl substituents as given below:

p-MeOC$_6$H$_4$ > p-MeC$_6$H$_4$ > m-MeC$_6$H$_4$ > C$_6$H$_5$ > p-ClC$_6$H$_4$ > o-MeOC$_6$H$_4$
  (500)   (15.7)   (1.95)   (1.0)   (0.7)   (0.3)

The poor migrating aptitude of o-MeOC$_6$H$_4$ group, despite of being electron-donating, is supposed to be associted with steric cause.

(b) In the case of unsymmetrically substituted glycols, the –OH that becomes protonated and ultimately is eliminated is the one whose loss gives rise to the more stable carbenium ion. Thus, 1,1-diphenyleth-1,2-diol (**3**) yields the pinacolone (**4**) involving hydrogen migration —— not (**5**), although it would involve the migration of phenyl group having greater migratory aptitude; formation of the more stable carbenium ion intermediate is a matter of prime consideration.

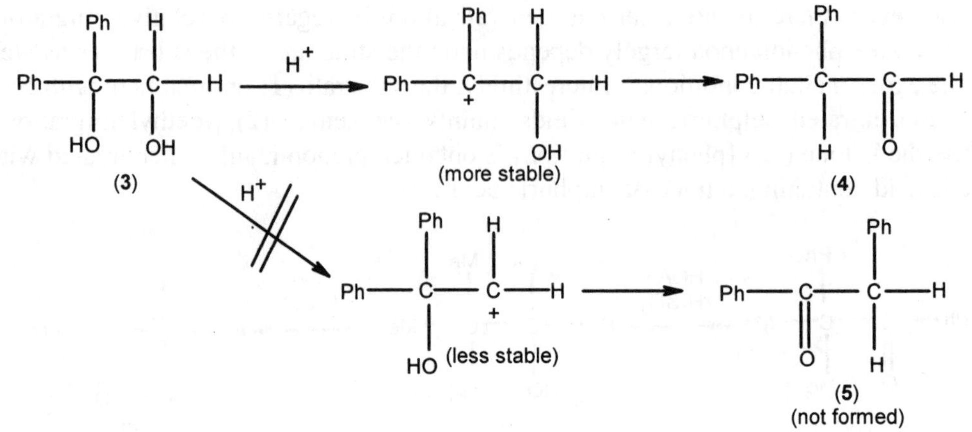

[Scheme showing rearrangement of compound (3) via H⁺ to more stable carbocation leading to (4), and less stable carbocation pathway leading to (5) (not formed).]

(c) *Piancol rearrangement* may also be catalyzed photochemically. Phtochemical *pinacol rearrangement* of 9,9'-bifluorene-9,9'-diol (6) into *spiro*[9H-fluorene-9,9'(10-H)-phenanthrene]-10'-one (7), for example, was reported by Hoang and his coworkers[11].

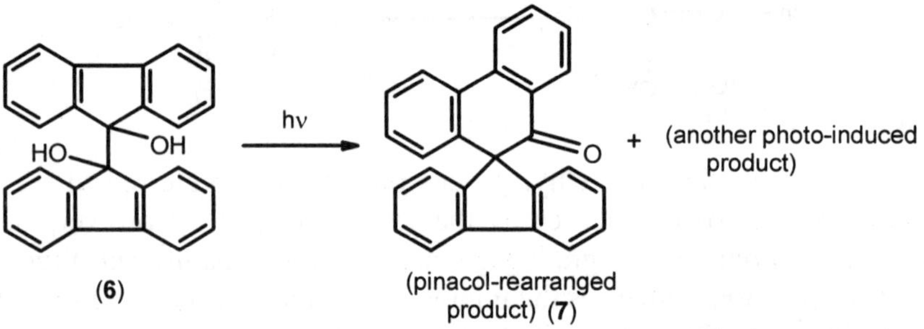

[Scheme: (6) ──hν──> (7) (pinacol-rearranged product) + (another photo-induced product)]

(d) *Pinacol rearrangement* can also be accomplished in the solid state — the method that usually involves mild conditions as well as improved yields[12]. Recently, Rashidi-Ranjbar and Kianmehr[13] developed a facile and efficient procedure for effecting *pinacol rearrangement* catalyzed by AlCl₃ in the absence of solvent. They studied the reaction with a number of substrates and found that benzylic pinacols rearrange under this condition at room temperature in a few minutes and in all most quantitative yield, while aliphatic pinacols do not react. A few are cited:

$$PhCH(OH)CH(OH)Ph \xrightarrow[\text{room temp., 2 min.}]{AlCl_3} PhCOCH_2Ph \quad (87\%)$$

$$(p\text{-}MeOC_6H_4)CH(OH)C(OH)Ph_2 \xrightarrow[\text{room temp., 2 min.}]{AlCl_3} (p\text{-}MeOC_6H_4)CHPhCOPh \quad (94\%)$$

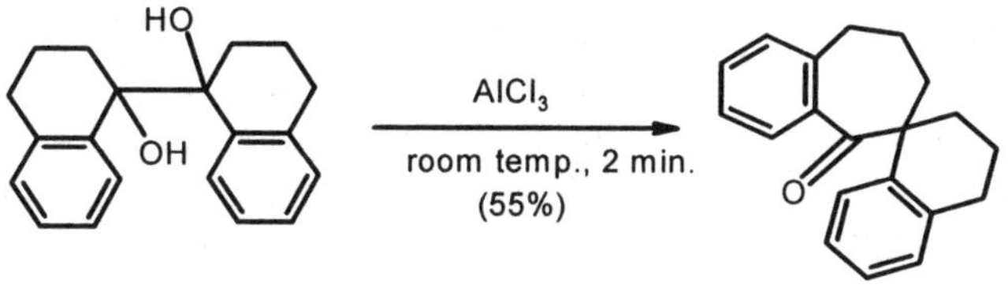

Applications

The *pinacol rearrangement reaction* is a useful alternative tool to the standard methods for synthesis of aldehydes and ketones.

$$Me_2C(OH)CH_2OH \xrightarrow{H^+} Me_2CHCHO$$

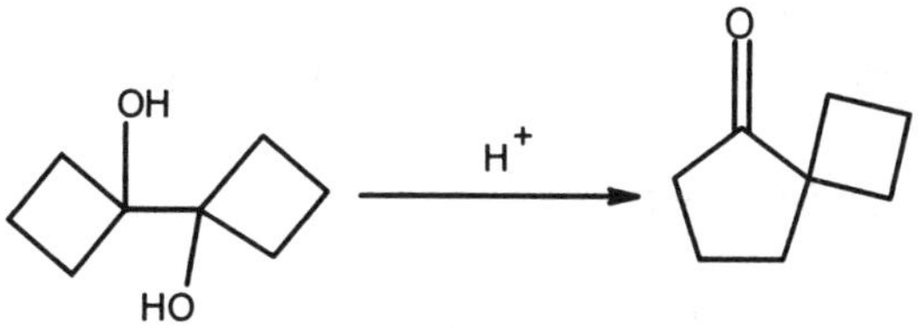

[Ref. 14]

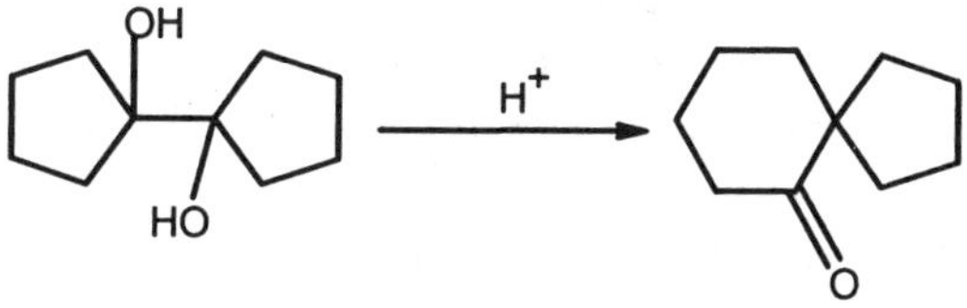

[Ref. 15]

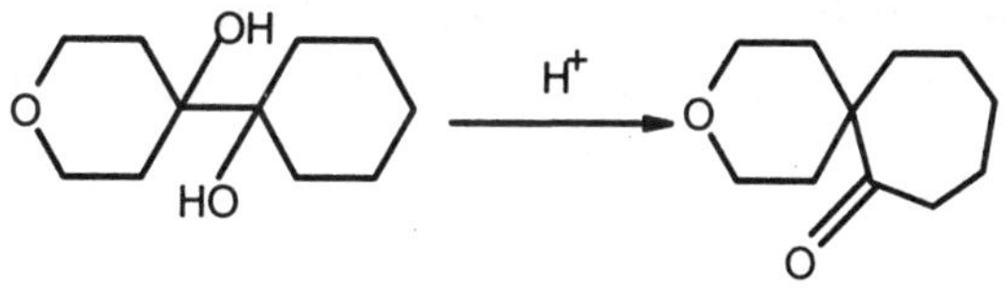

[Ref. 16]

References

1. Fittig, R. (1859), *Justus Liebigs Ann. Chem.*, **110**, 17; Fittig, R. (1860), *Justus Liebigs Ann. Chem.*, **114**, 54; Collins, C. J. (1960), *Quat. Rev.*, **14**, 357.
2. Bartok, M. and Molnar, A. in *The chemistry of Functional Groups, Suppliment E* (Ed. S. Patai), Wiley: NY, 1980, p. 722.
3. Kagan, J., Agdeppa Jr., D. A., Mayers, D. A., Singh, S. P., Walters, M. J. and Wintermute, R. D. (1976), *J. Org. Chem.*, **41**, 2355.
4. Collins, C. J. (1955), *J. Am. Chem. Soc.*, **77**, 5517.
5. Bosshard, H., Baumann, M. E. and Schetty, G. (1970), *Helv. Chim. Acta.*, **53**, 1271.
6. Botteron, D. G. and Wood, G. (1965), *J. Org. Chem.*, **30**, 3871.

7   Ramart-Lucas, P. and Salmon-Legagneur, F. C. R. (1928), *Acad. Sci.*, **188**, 1301.
8   Nakamura, K. and Osamura, Y. (1990), *Tetrahedron Lett.*, 251.
9   Wistuba, E. and Ruchardt, C. (1981), *Tetrahedron Lett.*, **22**, 4069; Pilkington, J. W. and Waring, A. J. (1976), *J. Chem. Soc., Perkin Trans 2*, 1349; Dubois, J. E. and Bauer, P. (1968), *J. Am. Chem. Soc.*, **90**, 4510; Heidke, R. L. and Saunders Jr., W. H. (1966), *J. Am. Chem. Soc.*, **88**, 5816.
10 Bachmann, W. E. and Ferguson, J. W. (1934), *J. Am. Chem. Soc.*, **56**, 2081.
11 Hoang, M., Gadosy, T., Ghazi, H., Hou, D.-F.,, Hopkinson, A. C., Johnston, L. and Lee-Ruff, E. (1998), *J. Org. Chem.*, **63**, 7168.
12 Toda, F. and Shigemasa, T. (1989), *J. Chem. Soc., Perkin Trans. 1*, 209.
13 Rashidi-Ranjbar, P. and Kianmehr, E. (2001), *Molecules*, **6**, 442.
14 Vogel, E. (1952), *Chem. Ber.*, **85**, 25.
15 Mundy, B. P., Srinivasa, R., Otzenberger, R. D. and DeBernardis, A. R. (1979), *Tetrahedron Lett.*, 2673.
16 Mundy, B. P., Kim, Y. and Warnet, R. J. (1983), *Heterocycles*, **20**, 1727.

- Nakamura, K. and Osamura, Y. (1993), *J. Am. Chem. Soc.*, **115**, 9112.
- Paquette, L. A., Lord, M. D. and Negri, J. T. (1993), *Tetrahedron Lett.*, **34**, 5693.
- Patra, D. and Ghosh, S. (1995), *J. Org. Chem.*, **60**, 2526.
- Magnus, P., Diorazio, L., Donohoe, T. J., Gils, M., Pye, P., Tarrant, J. and Thom, S. (1996), *Tetrahedron*, **52**, 14147.
- Bach, T. and Eilers, F. (1999), *J. Org. Chem.*, **64**, 8041.
- Razavi, H. and Polt, R. (2000), *J. Org. Chem.*, **65**, 5693.
- Marson, C. M., Oare, C. A., McGregor, J., Walsgrove, T., Grinter, T. J. and Adams, H. (2003), *Tetrahedron Lett.*, **44**, 141.

41

Reformatsky Reaction

Introduction

The reaction of α-halo esters with aldehydes or ketones in the presence of metallic zinc to form β-hydroxy esters is called the *Reformatsky reaction*[1].

$$\underset{R}{\overset{R'}{>}}C=O + BrCH_2COOEt \xrightarrow[\text{ether}]{Zn} \underset{R}{\overset{R'}{>}}C\underset{OZnBr}{\overset{CH_2COOEt}{<}} \xrightarrow{\text{hydrolysis}} \underset{R}{\overset{R'}{>}}C\underset{OH}{\overset{CH_2COOEt}{<}}$$

The reaction is carried out in an inert solvent such as diethyl ether, tetrahydrofuran or dioxane. Usually α-bromo esters are employed; vinylog of an α-halo ester (RCHBrCH=CHCOOEt) may also be used. The carbonyl compound may be aliphatic, aromatic, or heterocyclic, and may contain various functional groups. Use of activated zinc enhances both the reactivity and yields of the reaction[2]; activation of the metal involves removal of the oxide layer, and preparation of finely devided metal by means of treating with iodine or dibromomethane, or washing with dilute hydrochloric acid prior to use.

Mechanism

$$BrCH_2COOEt \xrightarrow[\text{(electron transfer from the metal)}]{Zn} \overset{+}{Br}Zn \left[\bar{C}H_2-\underset{\text{}}{\overset{O}{C}}-OEt \leftrightarrow CH_2=\underset{\text{}}{\overset{O^-}{C}}-OEt \right]$$

(nucleophilic addition of the enolate anion to the carbonyl group)

$$\underset{R}{\overset{R'}{>}}C\underset{OH}{\overset{CH_2COOEt}{<}} \xleftarrow[\text{hydrolysis}]{H_2O} \underset{R}{\overset{R'}{>}}C\underset{\bar{O}\overset{+}{Z}nBr}{\overset{CH_2COOEt}{<}}$$

(β-hydroxy ester)

Critical Views

(a) In certain cases (*e.g.* especially with aroamtic aldehydes) the hydrolysis product, β-hydroxy ester undergoes elimination reaction by removing one molecule of water from its molecule to afford an alkene. If tributylphosphine (an Lewis acid) is used along with zinc during carrying out the reaction, the predominant product obtained is the alkene[3].

(b) The advantage of using zinc instead of magnesium is that an organozinc comound is less reactive than organomagnesium one; thus the former (organozinc species) attacks to an ester carbonyl at much slower rate than that it adds to an aldehyde or ketone — although such addition of the organozinc derivative to the carbonyl group of unreacted α-halo ester is the most frequently observed side-reaction:

$$BrZnCH_2COOEt \xrightarrow{BrCH_2COOEt} \underset{BrZnO}{\overset{EtO}{>}}C\underset{CH_2Br}{\overset{CH_2COOEt}{<}} \xrightarrow{-EtOZnBr} O=C\underset{CH_2Br}{\overset{CH_2COOEt}{<}}$$
(β-keto ester)

(c) A modification of the *Reformatsky reaction* is the *Blaise reaction*[4], which recommends the use of nitriles to get β-keto esters.

$$R'-CN + BrZnCR_2COOEt \longrightarrow \underset{R'-C=NZnBr}{\overset{CR_2COOEt}{|}} \xrightarrow{H_2O \text{ hydrolysis}} \underset{R'-C=O}{\overset{CR_2COOEt}{|}}$$

(d) High reactivity of the *Reformatsky reaction* is achieved on application of ultrasound[5]. Recently, it has been reported that high-energy ultrasound (HIU)-promoted *Reformatsky reactions* afford several advantages over conventional thermal methods[6]. The work of Ross *et al.*[7] is cited here as for example:

(1) $\xrightarrow[I_2, \text{dioxane}]{BrCX_2COOEt, Zn, HIU}$ (2)

2a: X = H, 75%
2b: X = F, 89%

Applications

The reaction finds useful applications in organic synthesis:

(i) thiophene + BrCH$_2$COOEt $\xrightarrow[\text{ii. } H_3O^+]{\text{i. Zn/ether}}$ thiophene-CH=CH-COOEt

(ii) Reaction scheme:

CH₂(COOEt)—C(=O)—COOEt (oxaloacetate diester) + BrCH₂COOEt →(i. Zn/ether, I₂; ii. H₃O⁺)→ HO—C(CH₂COOH)(COOH)—CH₂COOH (citric acid)

(iii) Reaction scheme:

4-methylpentan-2-one + BrCH₂COOEt →(i. Zn/ether, I₂; ii. H₃O⁺)→ β-hydroxy ester (61%) [Ref. 8]

References

1. Reformatsky, S. (1887), *Ber. Dtsch. Chem. Ges.*, **20**, 1210; Rathke, M. W. (1975), *Org. React.*, **22**, 423 (review); Furstner, A. (1989), *Synthesis*, 571 (review); Furstner, A. in Organozinc Reagents (Eds. P. Knochel and P. Jones), Oxford University Press: New York, 1999, p. 287 (Review).
2. Rieke, R. D. and Uhm, S. J. (1975), *Synthesis*, 452; Furstner, A. (1989), *Synthesis*, 571; Bouhlel, E. and Rathke, M. W. (1991), *Synth. Comm.*, **21**, 133; Furstner, A. (1993), *Angew. Chem. Int. Ed. Engl.*, **32**, 164.
3. Shen, Y., Xin, Y. and Zhao, J. (1988), *Tetrahedron, Lett.*, **29**, 6119.
4. Hannick, S. M. and Kishi, Y. (1983), *J. Org. Chem.*, **48**, 3833; Cason, J., Rinehart Jr., K. L. and Thornton, S. D. (1953), *J. Org. Chem.*, **18**, 1594.
5. Han, B. N. and Boudjouk, P. (1982), *J. Org. Chem.*, **47**, 5030.
6. Ross, N. A. and Bartsch, R. A. (2001), *J. Heterocyclic Chem.*, **38**, 1255; Ross, N. A. and Bartsch, R. A. (2003), *J. Org. Chem.*, **68**, 360.
7. Ross, N. A., Bartsch, R. A. and Marchand, A. P. (2003), *ARKIVOC*, **xii**, 27.
8. Mattes, H. and Benezra, C. (1985), *Tetrahedron Lett.*, 5697.

- Hirashita, T., Kinoshita, K., Yamamura, H., kawai, M. and Araki, S. (2000), *J. Chem. Soc., Perkin Trans. 1*, 825.
- Ocampo, R., Dolbier, W. R., Abboud, K. A. and Zuluga, F. (2002), *J. Org. Chem.*, **67**, 72.
- Obringer, M., Colobert, F., Neugnot, B. and Solladie, G. (2003), *Org. Lett.*, **5**, 629.

42

Reimer – Tiemann Reaction

Introduction

The formylation of an activated aromatic ring compound with chloroform in alkaline solution is known as the *Reimer-Tiemann reaction*[1]. The method is useful only for phenols and certain heterocyclic compounds such as pyrroles and indoles. It leads preferentially to the formation of an *ortho*-formylated product; when both the *ortho*-positions are blocked, the incoming group occupies the *para*-position.

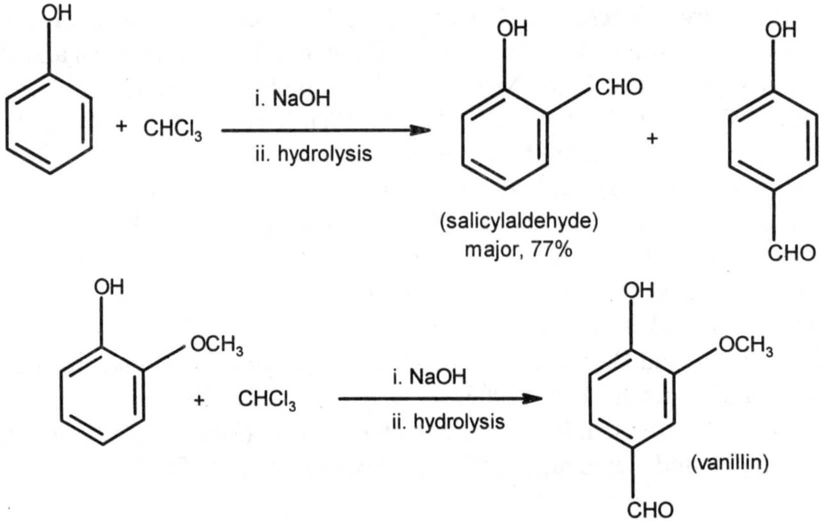

The *Reimer-Tiemann reaction* is mainly used for the synthesis of *ortho*-hydroxy aromatic aldehydes; yields are generally low (not more than 50%). However, it has been reported that application of ultrasound leads to shorter reaction times and improved yields[2].

If CCl_4 is used instead of $CHCl_3$, carboxylation occurs under the reaction conditions.

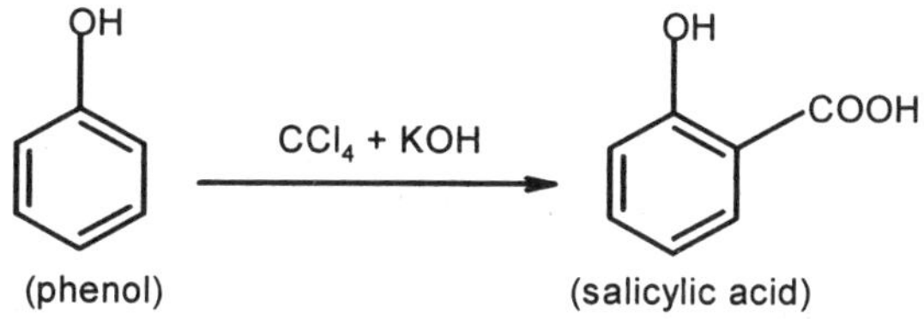

Mechanism

The reaction is believed to proceed in the following manner[3]; firstly the dichlorocarbene ($:CCl_2$), an electron-deficient reactive species, is formed by the reaction of chloroform with strong alkali. In the second step, it attacks the electron-rich *ortho*-position (if this position remains free) of the aromatic ring to form *ortho*-dichloromethylphenolate, which on hydrolysis yields the final product, *ortho*-hydroxyarylaldehyde.

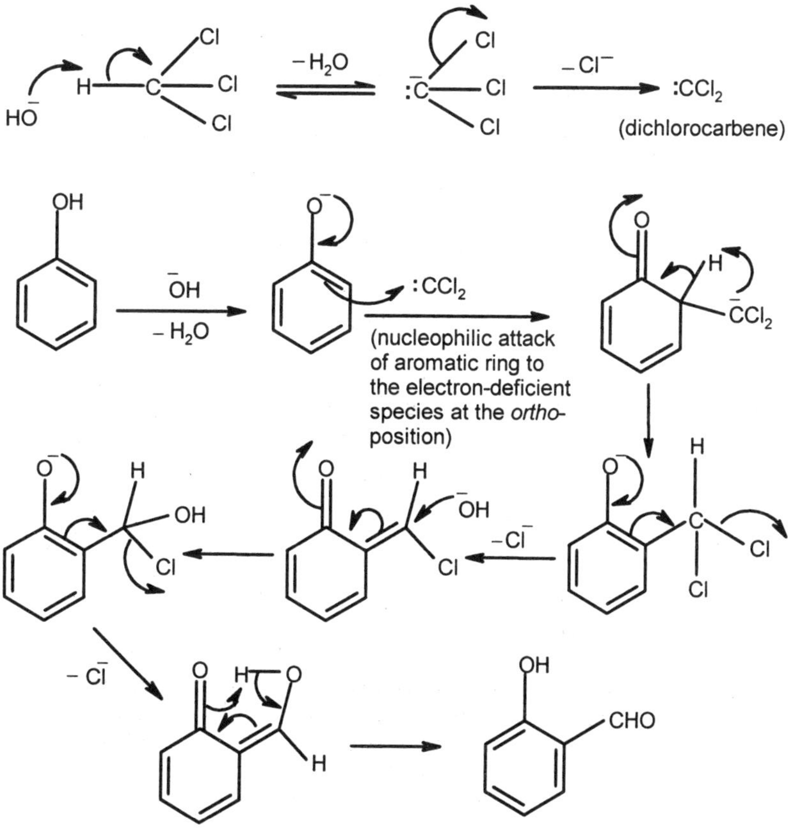

Critical Views

(a) *Ortho*-formylated product usually predominates in the *Reimer-Tiemann reaction*, however it has been reported that enhanced *para*-selectivity can be achieved by the use of polyethylene glycol[4].

(b) The involvement of dichlorocarbene as the reaction intermediate receives support from the fact that certain substrates, *e.g.* pyrrole, *p*-cresol, etc., under the reaction condition give side-products (called as *abnormal products*) along with the normal products or instead of those.

For example, pyrrole on *Reimer-Tiemann reaction* affords an 'abnormal product' (3-chloropyridine) developed due to ring-expansion along with the normal product, 2-formylpyrrole.

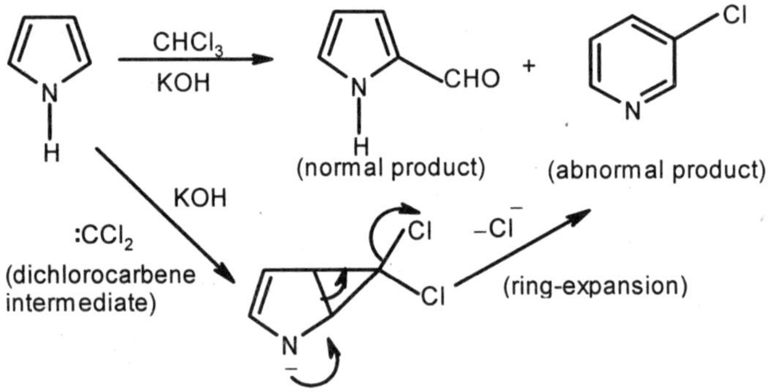

p-Cresol also forms an abnormal product, the dienone as the side-product.

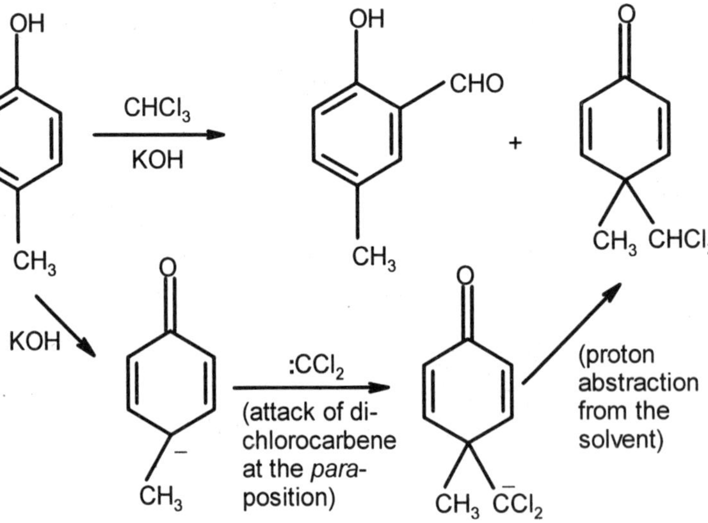

Applications

The reaction offers a useful method for introducing formyl (and also carboxyl function) group in phenolic compounds as well as certain heterocycles. Few examples are cited:

i) 2-naphthol → 1-formyl-2-naphthol (1. CHCl$_3$/KOH, 2. hydrolysis)

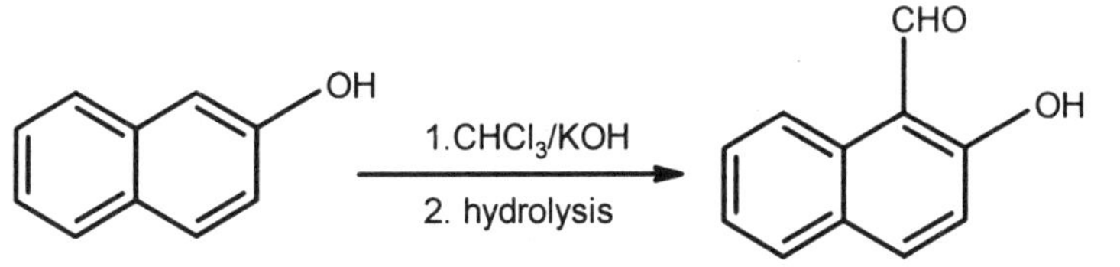

ii) indole → 3-formylindole (1. CHCl$_3$/KOH, 2. hydrolysis)

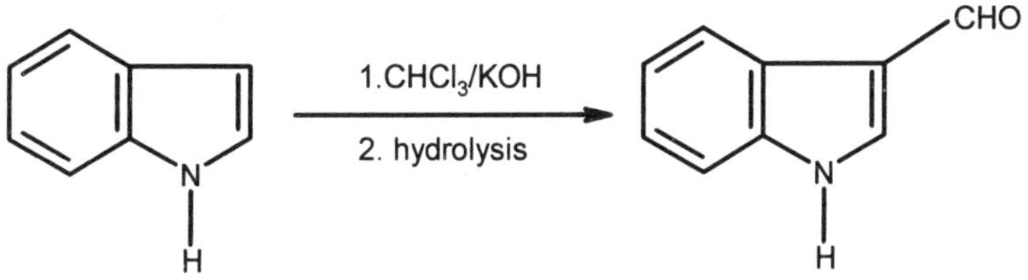

References

1. Reimer, F. and Tiemann, F. (1876), *Ber. Dtsch. Chem. Ges.*, **9**, 423, 824, 1268, 1285; Gilman, H., Arntzen, C. E. and Webb, F. J. (1945), *J. Org. Chem.*, **10**, 374; Ferguson, L. N. (1946), *Chem. Rev.*, **38**, 229; Wynberg, H. (1960), *Chem. Rev.*, **60**, 169.
2. Cochran, J. C. and Melville, M. G. (1990), *Synth. Commun.*, **20**, 609.
3. Robinson, E. A. (1961), *J. Chem. Soc.*, 1663; Hine, J. and van der Veen, J. M. (1959), *J. Am. Chem. Soc.*, **81**, 6446.
4. Neumann, R. and Sasson, Y. (1986), *Synthesis*, 569.
- Kobayashis, S., Tagawas, S. and Nakajima, S. (1963), *Chem. Pharm. Bull.*, **11**, 123.
- Wynberg, H. and Meijer, E. W. (1982), *Org. React.*, **28**, 1. [Review]
- Thoer, A., Denis, G., Delmas, M. and Gaset, A. (1988), *Synth. Commun.*, **18**, 2095.
- Langlois, B. R. (1991), *Tetrahedron Lett.*, **32**, 3691.
- Jimenez, M., Miranda, M. A. and Tormos, R (1995), *Tetrahedron*, **51**, 5825.
- Jung, M. E. and Lazarova, T. I. (1995), *J. Org. Chem.*, **62**, 1553.

43

Sandmeyer Reaction

Introduction

The reaction leading to the formation of aromatic halogeno or cyano compounds from the corresponding diazonium salts under the catalytic action of cuprous halide or cyanide is called as the *Sandmeyer reaction*[1].

$$Ar\overset{+}{N_2}\overset{-}{X} + CuX \longrightarrow Ar-X$$

The reaction is not useful for the direct preparation of fluorides or iodides, but for bromides and chlorides it is of wide scope. The yields become usually high. During preparation of benzonitrile derivatives (ArCN), the reaction is generally conducted in neutral solution to avoid evolution of HCN.

$$Ar\overset{+}{N_2}\overset{-}{X} \xrightarrow[50^\circ C]{CuCN + KCN} Ar-CN$$

The *Sandmeyer reaction* can thus generally be represented as:

$$Ar\overset{+}{N_2}\overset{-}{X} + \overset{+1}{CuNu} \longrightarrow Ar-Nu + N_2 + \overset{+1}{CuX}$$

$$(Nu = Cl, Br, CN)$$

The Cu(I)-species is regenerated and is thus a true catalyst.

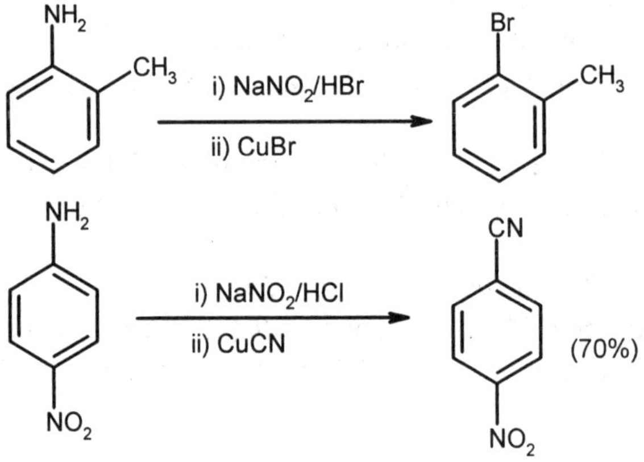

o-toluidine →(i) NaNO$_2$/HBr; ii) CuBr)→ 2-bromotoluene

p-nitroaniline →(i) NaNO$_2$/HCl; ii) CuCN)→ 4-nitrobenzonitrile (70%)

Mechanism

The mechanism of the *Sandmeyer reaction* is not rigorously known, however is believed to proceed in the following manner[2]. In the first two steps, the arenediazonium ion species is reduced by Cu(II) salt to give an aryl radical species. But two alternative mechanisms are possible for the third step — either the resulting Cu(II) salt binds to the aryl radical forming the intermediate Ar–Cu(III)NuX followed by its dissociation into Cu(I)X and the substitution product Ar–Nu, or the aryl radical reacts with the Cu(II) salt giving the substitution product Ar–Nu through ligand transfer and Cu(I) through concomitant reduction.

$$Ar-\overset{+}{N}\equiv N\ \ \overset{-}{X} + \overset{+1}{Cu}Nu \longrightarrow Ar-N\equiv N\cdot + \overset{+2}{Cu}NuX$$

$$Ar-N\equiv N\cdot \longrightarrow Ar\cdot + N_2$$

$$Ar\cdot + \overset{+2}{Cu}NuX \longrightarrow Ar-\overset{+3}{Cu}NuX \text{ (intermediate)} \longrightarrow Ar-Nu + \overset{+1}{Cu}X$$

(alternative path) one step
$$\downarrow$$
$$Ar-Nu + \overset{+1}{Cu}X$$

Critical Views

(a) In case of *in situ* preparation of required diazonium salt from an aryl amine by means of *diazotization reation*, an acid HX is used, that corresponds the halo substituent (X) to be introduced onto the aromatic ring; otherwise — *e.g.* when using HCl/CuBr — a mixture of aryl chloride and bromide will be obtained.

(b) Aryl iodides and fluorides can also be prepared from the corresponding diazonium salts. The cuprous catalyst is not necessary for the preparation of aryl iodide because iodide ion is sufficient enough to decompose the diazonium salt.

$$Ar-\overset{+}{N}\equiv N\ \ \overset{-}{X} \xrightarrow{KI} Ar\text{-}I + N_2 + KX$$

$$\downarrow I^-$$

$$Ar-N\equiv N\cdot + I\cdot \xrightarrow{I^-} Ar-N\equiv N\cdot + I_2^{\cdot-}$$

$$\downarrow Ar-\overset{+}{N}\equiv N$$

$$Ar\cdot + N_2 \longleftarrow Ar-N\equiv N\cdot$$

$$Ar\text{-}I + I_2^{\cdot-} \longleftarrow \qquad\qquad I_3^- \xleftarrow{I^-} I_2$$

Aryl fluoride can be prepared by careful heating of diazonium fluoroborate (Schiemann salt); careful handling is required (using proper equipment) because the Schiemann diazonium salt is potentially explosive.

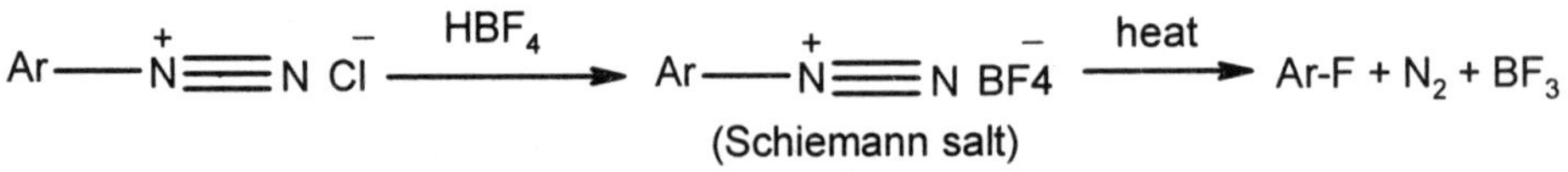

$$Ar-\overset{+}{N}\equiv N\ \overset{-}{Cl} \xrightarrow{HBF_4} \underset{\text{(Schiemann salt)}}{Ar-\overset{+}{N}\equiv N\ \overset{-}{BF_4}} \xrightarrow{heat} Ar\text{-}F + N_2 + BF_3$$

However, the advantageous method of preparing aryl fluoride that does not require forming and isolating potentially explosive Schiemann diazonium fluoroborate is as follows:

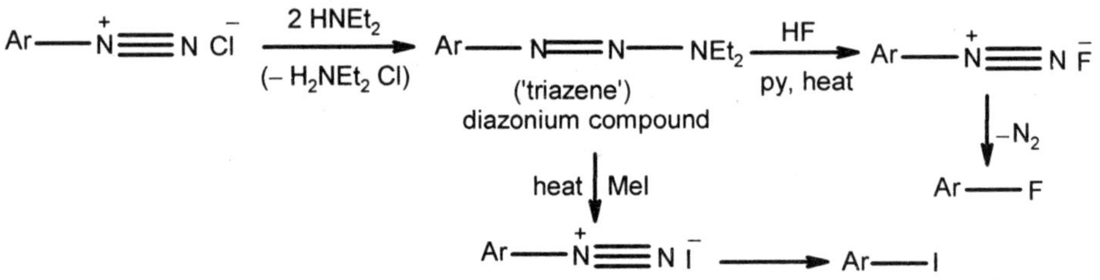

$$Ar-\overset{+}{N}\equiv N\ \overset{-}{Cl} \xrightarrow[(-H_2NEt_2\ Cl)]{2\ HNEt_2} \underset{\substack{\text{('triazene')}\\ \text{diazonium compound}}}{Ar-N=N-NEt_2} \xrightarrow[py,\ heat]{HF} Ar-\overset{+}{N}\equiv N\ \overset{-}{F}$$

heat ↓ MeI , ↓ $-N_2$

$$Ar-\overset{+}{N}\equiv N\ \overset{-}{I} \longrightarrow Ar-I \qquad Ar-F$$

Applications

The *Sandmeyer reaction* is a very useful synthetic tool in organic synthesis. Few examples are:

i) 2-aminonaphthalene $\xrightarrow[\text{ii) CuCl}]{\text{i) NaNO}_2\text{/HCl}}$ 2-chloronaphthalene

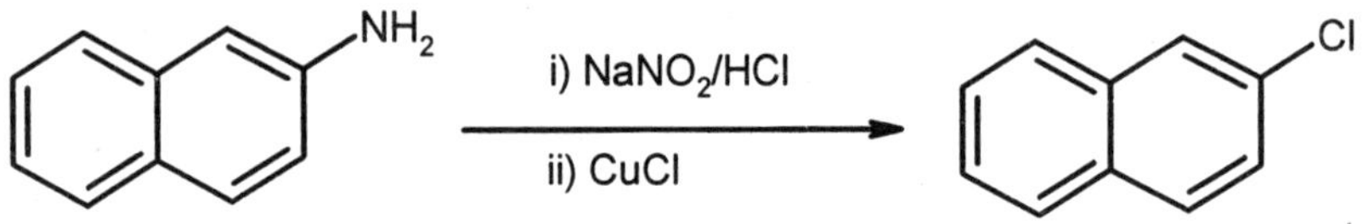

ii) 3-aminophenyl propyl ketone $\xrightarrow[\text{ii) KI}]{\text{i) NaNO}_2\text{/HCl}}$ 3-iodophenyl propyl ketone

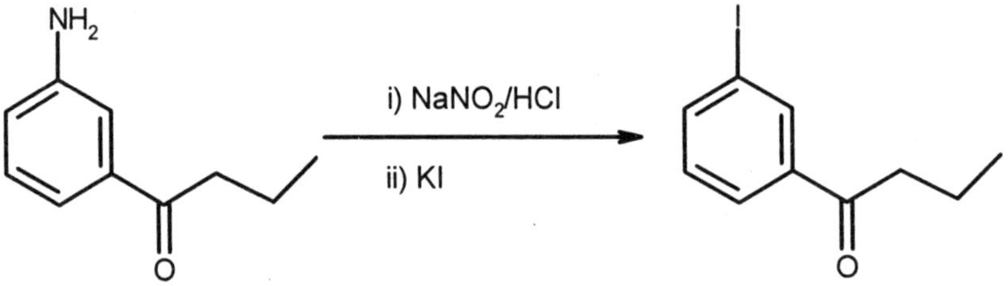

iii) 2,6-dibromo-4-nitrobenzenediazonium hydrogensulfate $\xrightarrow[-N_2]{KI}$ 1,3-dibromo-2-iodo-5-nitrobenzene

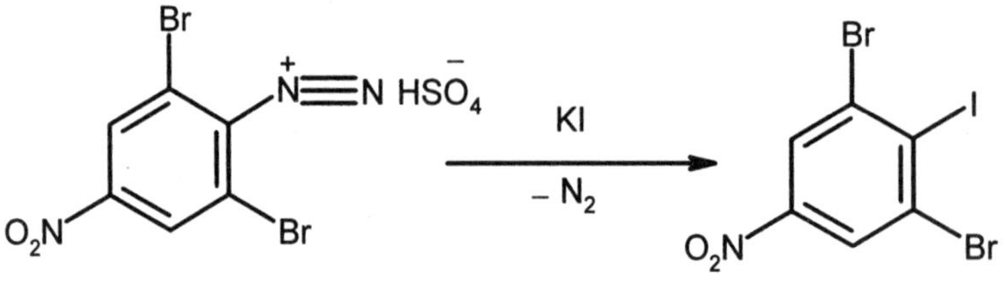

iv) 2-nitroaniline →(i) NaNO₂/HCl; ii) CuCN)→ 2-nitrobenzonitrile (87%)

v) 2-methylaniline →(i) NaNO₂/HCl; ii) CuCN)→ 2-methylbenzonitrile [Ref. 3]

vi) 2-bromoaniline →(i) NaNO₂/HCl; ii) KI)→ 1-iodo-2-bromobenzene [Ref. 4]

References

1. Sandmeyer, T. (1884), *Ber. Dtsch. Chem. Ges.*, **17**, 1633.
2. Dickermann, S. C., Weiss, K. and Ingbermann, A. K. (1958), *J. Am. Chem. Soc.*, **80**, 1904; Dickermann, S. C., DeSouza, D. J. and Jacobson, N. (1969), *J. Org. Chem.*, **34**, 710; Galli, C. (1981), *J. Chem. Soc., Perkin Trans. 2*, 1459; Galli, C. (1982), *J. Chem. Soc., Perkin Trans. 2*, 1139; Galli, C. (1984), *J. Chem. Soc., Perkin Trans. 2*, 897; Hanson, P., Jones, J. R., Gilbert, Timms, A. W. (1991), *J. Chem. Soc., Perkin Trans. 2*, 1009.
3. Clarke, H. T. and Read, R. R. (1941), *Org. Synth.*, **1**, 514.
4. Heaney, H. and Millar, I. T. (1960), *Org. Synth.*, **40**, 105.

- Pfeil, E. (1953), *Angew. Chem.*, **65**, 155.
- Merkushev, E. B. (1988), *Synthesis*, 923.
- Obushak, M. D., Lyakhovych, M. B. and Ganushchak, M. I. (1998), *Tetrahedron Lett.*, **39**, 9567.
- Chandler, St. A., Hanson, P., Taylor, A. B., Walton, P. H. and Timms, A. W. (2001), *J. Chem. Soc., Perkin Trans. 2*, 214.
- Hanson, P., Rowell, S. C., Taylor, A. B., Walton, P. H. and Timms, A. W. (2002), *J. Chem. Soc., Perkin Trans. 2*, 1126.
- Hanson, P., Jones, J. R., Taylor, A. B., Walton, P. H. and Timms, A. W. (2002), *J. Chem. Soc., Perkin Trans. 2*, 1135.

44

Sharpless Asymmetric Epoxidation

Introduction

Sharpless asymmetric epoxidation[1] is a beautiful example of asymmetric synthesis. In this reaction, achiral allylic alcohols are converted to chiral 2,3-epoxy alcohols with high enantiomeric excess (usually more than 90% ee). This enantioselective epoxidation is effected by *tert*-butylhydroperoxide (oxidizing agent) in presence of titanium tetraisopropoxide and the chiral additive, an enantiomerically pure (+)- or (−)-dialkyltartrate (usually diethyltartrate, DET)[2]. The titanium tetraisopropoxide and DET can be used in catalytic amounts (10-15 mol%) if molecular sieves are present[3]. Since both (+)- and (−)-DET are readily available, and the reaction is practically stereospecific, either enantiomer of the product can be synthesized[4].

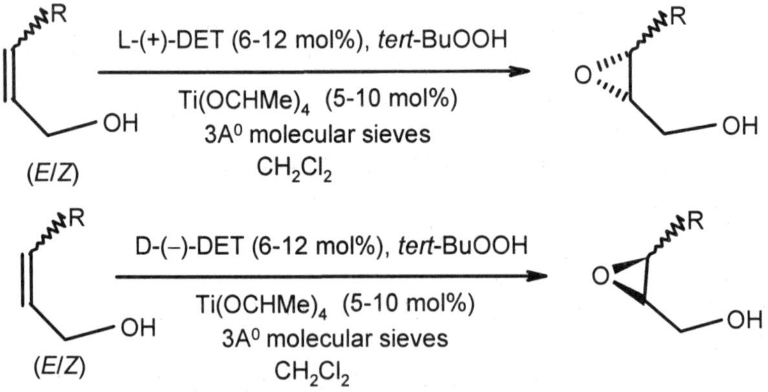

(*note:* to remember the stereochemistry of the product the following mnemonic may be applied: **L**, from lower face and **D**, doesn't attack from downface)

Titanium tetraisopropoxide is added so that the oxidizing agent, the chiral ligand, and the substrate can assemble to form an enantiomerically pure chiral complex; no epoxidation occurs at all in the absence of the metal isopropoxide.

Sharpless Asymmetric Epoxidation

Mechanism

The mechanism of this reaction[5] is a complex one; however a lucid representation may be sketched as follows. Firstly, a dimeric complex (**I**) may be formulated developed due to the reaction of titanium tetraisopropoxide with the chrial additive; at each titanium centre two isopropoxide groups of the original tetraisopropoxytitanium have been replaced by the chiral tartrate ligand.

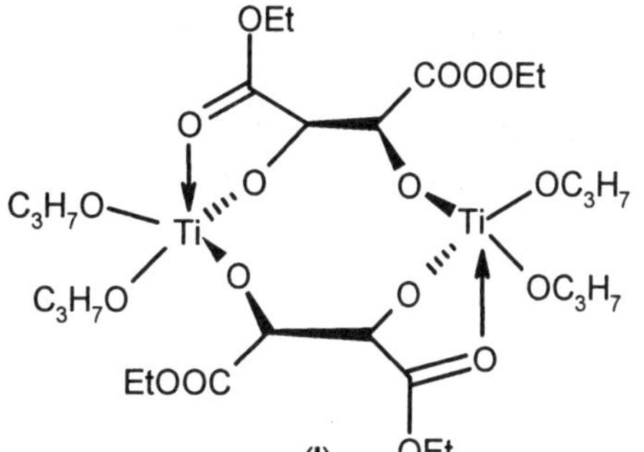

The dimer (**I**), in turn, is attacked by the substrate molecule (allylic alcohol) and the oxidizing agent (*tert*-butylhydroperoxide) to give the complex (**II**). The titanium atom acts as a template for the reactants — they are geometrically oriented in such a manner to allow a facial selection, resulting in an enantioselective epoxidation step. The metal atom behaves also as a Lewis acid to facilitate the course of the reaction as shown below:

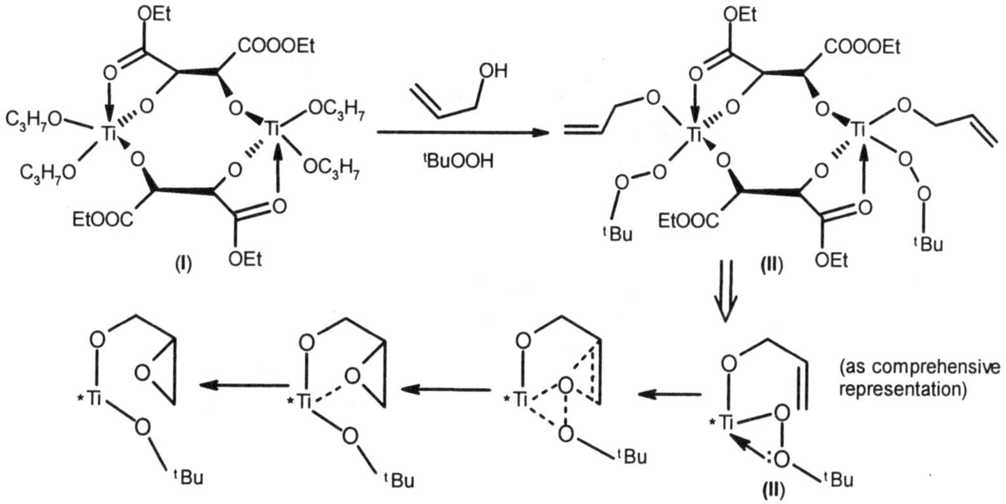

Critical Views

The reaction should be carried out under water-free condition; water has adverse effect on the enantioselectivity of the reaction. The work of Sharpless et al.[6], for example, may be mentioned here — the investigators showed that in case of the epoxidation of (E)-α-phenylcinnamyl alcohol, the addition of one equivalent of water led to a decrease in enantioselectivity from 99% ee to 48% ee. Thus, the reaction mixture must be completely devoid of water.

This procedure has established itself as one of the most important methods of asymmetric synthesis, and has been used to prepare a large number of optically active natural products and other compounds. The method has been found successful for a wide range of primary allylic alcohols, where the double bond is mono-, di-, tri- or tetrasubstituted[7].

Applications

The *Sharpless asymmetric epoxidation* is one of the most important among the newer reactions, and finds wide application in organic chemistry. For this novel work, K. B. Sharpless was awarded the Nobel Prize in Chemistry in 2001.

i) [geranyl-type allylic alcohol] $\xrightarrow{\text{L-(+)-DET (6-12 mol\%), tert-BuOOH}}_{\text{Ti(OCHMe)}_4 \text{ (5-10 mol\%)}, \text{3A}^0 \text{ molecular sieves}, \text{CH}_2\text{Cl}_2}$ [2,3-epoxy product] (77%)

ii) $C_{10}H_{19}$-CH=CH-CH$_2$OH $\xrightarrow{\text{L-(+)-DET, tert-BuOOH}}_{\text{Ti(OCHMe)}_4, \text{3A}^0 \text{ molecular sieves}, \text{CH}_2\text{Cl}_2}$ [epoxy alcohol with $C_{10}H_{19}$]

iii) [acetonide allylic alcohol] $\xrightarrow{\text{D-(−)-DET, tert-BuOOH}}_{\text{Ti(OCHMe)}_4}$ [epoxy product] (90%)

References

1. Katsuki, T. and Sharpless, K. B. (1980), *J. Am. Chem. Soc.*, **102**, 5974.
2. Sharpless, K. B., Woodard, S. S. and Finn, M. G. (1983), *Pure Appl. Chem.*, **55**, 1823.
3. Gao, Y., Hanson, R. M., Klunder, J. M., Ko, S. Y., Masamune, H. and Sharpless, K. B. (1987), *J. Am. Chem. Soc.*, **109**, 5765; Wang, Z. and Zhou, W. (1987), *Tetrahedron*, **43**, 2935.
4. Pfenninger, A. (1986), *Synthesis*, 89 (review); Sharpless, K. B. (1985), *Chemtech*, **15**, 692; Schinzer, D. in *Organic Synthesis Hightlights II* (Ed. H. Waldmann), VCH, Weinheim, Germany, 1995, p.3-9.
5. Williams, I. D., Pederson, S. F., Sharpless, K. B. and Lippard, S. (1984), *J. Am. Chem. Soc.*, **106**, 6430; Jorgensen, K. A., Wheeler, R. A. and Hoffmann, R. (1987), *J. Am. Chem. Soc.*, **109**, 3240; Carlier, P. R. and Sharpless, K. B. (1989), *J. Org. Chem.*, **54**, 4016; Corey, E. J. (1990), *J. Org. Chem.*, **55**, 1693; Finn, M. G. and Sharpless, K. B. (1991), *J. Am. Chem. Soc.*, **113**, 113.
6. Hill, J. G., Rossiter, B. E. and Sharpless, K. B. (1983), *J. Org. Chem.*, **48**, 3607.
7. Schweiter, M. J. and Sharpless, K. B. (1985), *Tetrahedron Lett.*, **26**, 2543.

- Yamamoto, K., Kawanami, Y. and Miyazawa, M. (1993), *J. Chem. Soc., Chem. Commun.*, 436.
- Katsuki, T. and Martin, V. S. (1996), *Org. React.*, **48**, 1. [Review]
- Honda, T., Ohta, M. and Mizutani, H. (1999), *J. Chem. Soc., Perkin Trans.* **1**, 23.
- Black, P. J., Jenkins, K. and Williams, J. M. (2002), *Tetrahedron: Asymmetry*, **13**, 317.
- Ghosh, A. K. and Lei, H. (2003), *Tetrahedron: Asymmetry*, **14**, 629.

45

Stevens Rearrangement

Introduction

When a quarternary ammonium species bearing an electron-withdrawing group on one of the carbons attached to the nitrogen is heated with a strong base (such as NaOR, NaNH$_2$), a rearranged tertiary amine is obtained in a reaction known as the *Stevens rearrangement*[1].

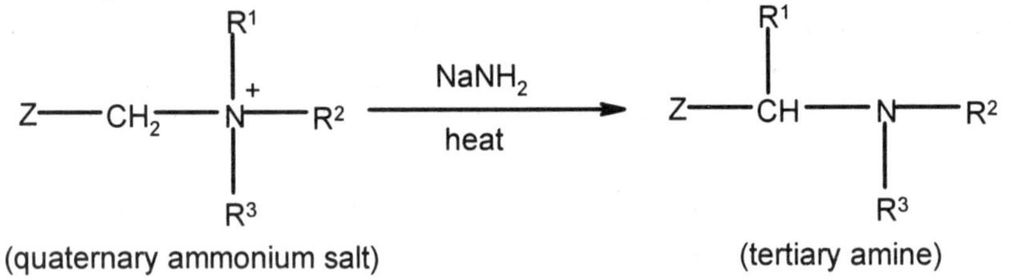

$$Z-CH_2-\overset{R^1}{\underset{R^3}{\overset{+}{N}}}-R^2 \xrightarrow[\text{heat}]{NaNH_2} Z-CH-\overset{R^1}{\underset{R^3}{N}}-R^2$$

(quaternary ammonium salt)   (tertiary amine)

The electron-withdrawing group (Z) may be RCO, ROOC, phenyl, and so on[2]. The most common migrating groups are allylic, benzylic, benhydryl, phenacyl, *etc.*[3] An illustrative example[4] is cited below:

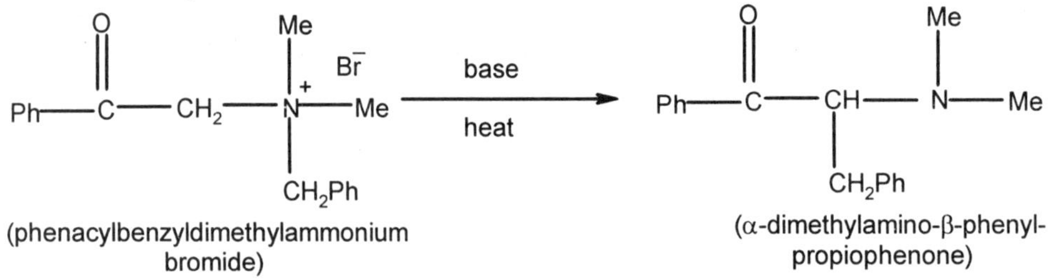

$$Ph-\overset{O}{\overset{\|}{C}}-CH_2-\overset{Me}{\underset{CH_2Ph}{\overset{+}{N}}}-Me \;\; Br^- \xrightarrow[\text{heat}]{\text{base}} Ph-\overset{O}{\overset{\|}{C}}-\overset{}{\underset{CH_2Ph}{CH}}-\overset{Me}{\underset{}{N}}-Me$$

(phenacylbenzyldimethylammonium bromide)   (α-dimethylamino-β-phenyl-propiophenone)

The order of migration is observed as phenacyl > propargyl > allylic > benzyl > alkyl[5].

Mechanism

The mechanism of the *Stevens rearrangement* is a matter of much discussion[6]. However it was established that the rearrangement is intramolecular in nature (by crossover experiment[4,7]), and the migrating group retains its configuration[8]. In the initial step, a strong base abstracts a proton from the α-carbon of quaternary ammonium species resulting a nitrogen ylide (that has been isolated[9]).

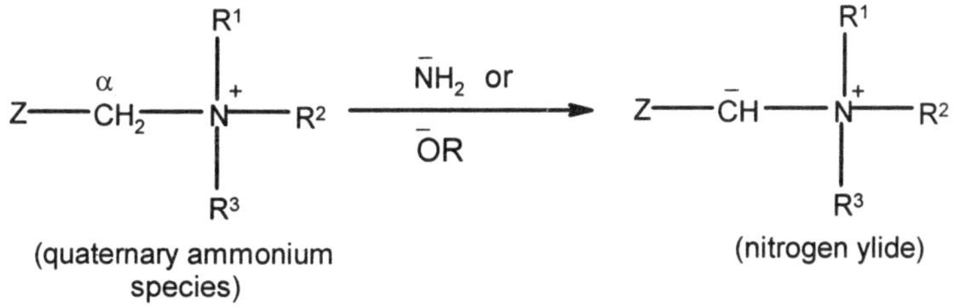

$$Z-CH_2-\overset{R^1}{\underset{R^3}{\overset{+}{N}}}-R^2 \quad \xrightarrow{\overset{-}{N}H_2 \text{ or } \overset{-}{O}R} \quad Z-\overset{-}{C}H-\overset{R^1}{\underset{R^3}{\overset{+}{N}}}-R^2$$

(quaternary ammonium species)   (nitrogen ylide)

The rearrangement step may proceed *via* either through ion-pair or radical-pair intermediate. Stevens and Johnstone[7] suggested that the initially formed nitrogen ylide is cleaved to an ion-pair residing in solvent cage, followed by their subsequent recombination to give the tertiary amine.

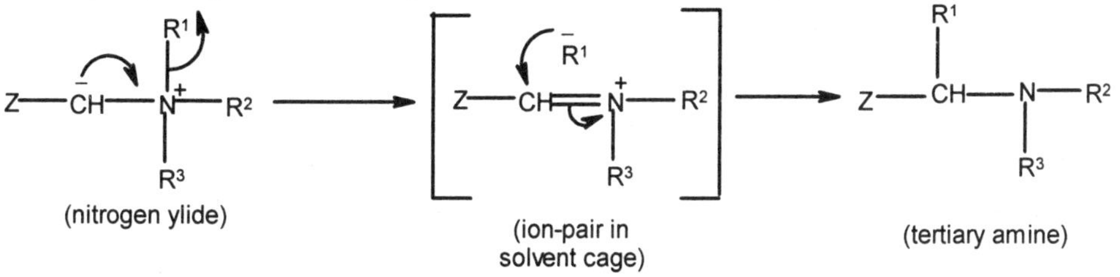

(nitrogen ylide)   (ion-pair in solvent cage)   (tertiary amine)

However, in many cases involvement of radical-pair intermediate has been evidenced[10]; the proposed scheme[11] is given as:

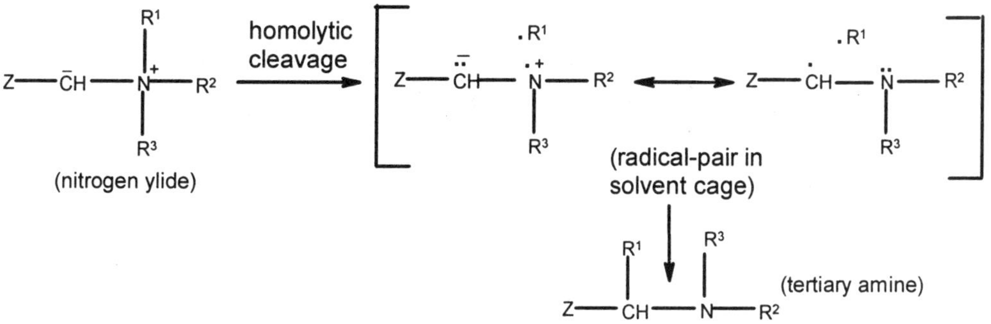

(nitrogen ylide)   (radical-pair in solvent cage)   (tertiary amine)

The formation of radical-pair intermediate was evidenced by CIDNP spectra, and also from the fact that in some cases small amounts of coupling products (R^1–R^1) have been isolated[12], formation of which is rationalized from the leakage of some ·R^1 (radical) out of the solvent cage.

Critical Views

Sometimes 1,4-shift is also observed along with the normal 1,2-shift in the *Stevens rearrangement*. Hill and Chan[13] noticed that allylbenzylmethylphenylammonium iodide under Stevens conditions yields a 1,2-rearranged product (15%) along with a 1,4-rearranged one.

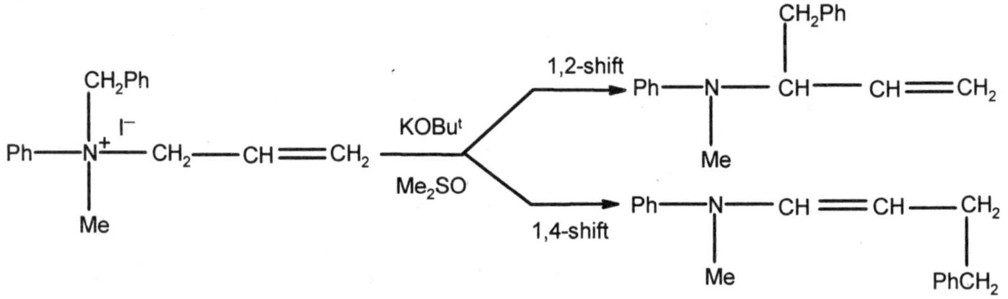

Sulphur ylides[14] bearing electron-withdrawing groups at the α-carbons are also reported to undergo *Stevens rearrangement*.

Applications

(i) [Ref. 15]

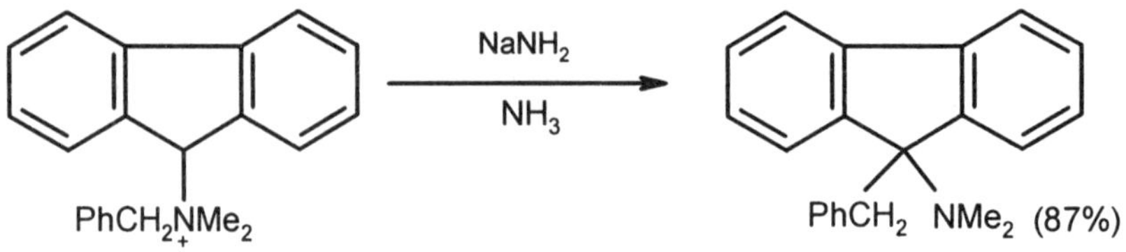

(ii) [Ref. 16]

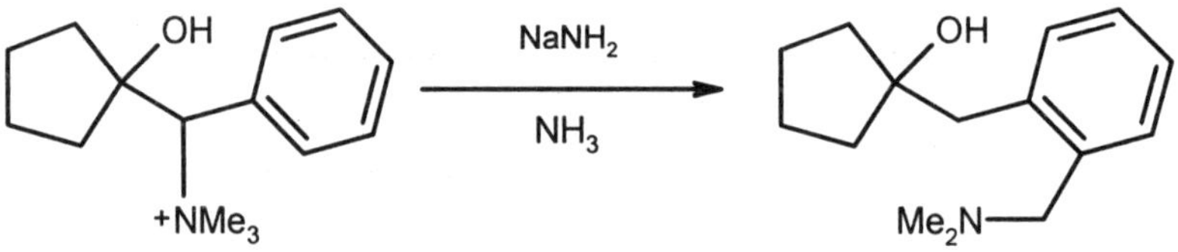

(iii) (ring expansion) (90%) [Ref. 17]

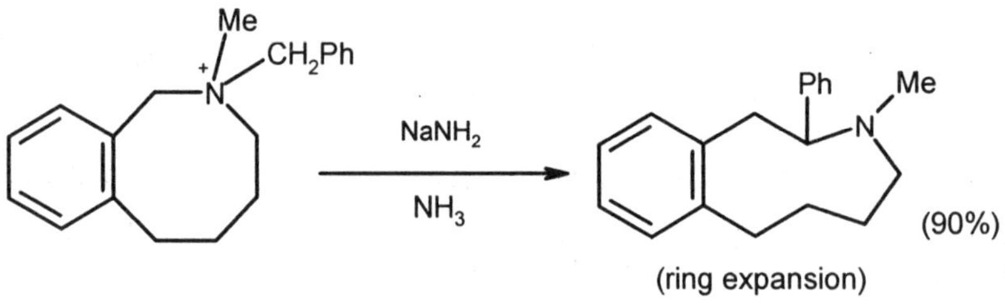

iv) [Diethyl-diallyl-ammonium bromide] →(KOBuᵗ, MeCN)→ Et₂N-CH(CH=CH₂)-CH₂-CH=CH₂ (74%) [Ref. 18]

References

1. Stevens, T. S., Creighton, E. M., Gordon, A. B. and MacNicol, M. (1928), *J. Chem. Soc.*, 3193; Thomson, T. and Stevens, T. S. (1932), *J. Chem. Soc.*, 55.
2. Schuster, H. F. and Coppola, G. M., *Allens in Organic Synthesis*, Wiley: NY, 1984, p57.
3. Brown, J. M. (1987), *Angew. Chem. Int. Engl.*, 26, 190.
4. Stevens, T. S. (1930), *J. Chem. Soc.*, 2107.
5. Pine, S. H. (1970), *Org. React.*, 18, 403 (review).
6. Pine, S. H. (1971), *J. Chem. Educ.*, 48, 99; Yates, G. L. and Yates, B. F. (1994), *Aust. J. Chem.*, 47, 1685.
7. Johnstone, R. A. W. and Stevens, T. S. (1955), *J. Chem. Soc.*, 4487.
8. Schollkopf, U., Ludwig, U., Ostermann, G. and Patsch, M. (1969), *Tetrahedron Lett.*, 3415; Brewster, J. H. and Kline, M. W. (1952), *J. Am. Chem. Soc.*, 74, 5179.
9. Jemison, R. W., Mageswaran, S., Ollis, W. D., Potter, S. E., Pretty, A. J. Sutherland, I. O. and Thebtaranonth, Y. (1970), *Chem. Commun.*, 1201.
10. Lepley, A. R., Becker, R. H. and Giumanini, A. G. (1971), *J. Org. Chem.*, 36, 1222; Baldwin, J. E. and Brown J. E. (1969), *J. Am. Chem. Soc.*, 91, 3646.
11. Schollkopf, U. and Ludwig, U. (1968), Chem. Ber., 101, 2224; Schollkopf, U., Ludwig, U., Ostermann, G. and Patsch, M. (1969), *Tetrahedron Lett.*, 3415.
12. Schollkopf, U., Ludwig, U., Ostermann, G. and Patsch, M. (1969), *Tetrahedron Lett.*, 3415; Hennion, G. F. and Shoemaker, M. F. (1970), *J. Am. Chem. Soc.*, 92, 1769.
13. Hill, R. K. and Chan, T. (1966), *J. Am. Chem. Soc.*, 88, 866.
14. Olsen, R. K. and Currie Jr., J. O. in Patai, *The Chemistry of the Thiol Group*, **pt. 2**, Wiley: NY, 1974, p. 561.
15. Hauser, C. R., Manyik, R. M., Brasen, W. R. and Bayless, P. L. (1955), *J. Org. Chem.*, 20, 1119.
16. Puterbaugh, W. H. and Hauser, C. R. (1964), *J. Am. Chem. Soc.*, 86, 1394.

17 Elmasmodi, A., Cotelle, P., Barbry, D., Hasiak, B. and Couturier, D. (1989), *Synthesis*, 327.
18 Allin, S. M., Button, M. A. C. and Shuttleworth, S. J. (1997), *Synlett,* 725.

- Makita, K., Koketsu, J., Ando, F., Ninomiya, Y. and Koga, N. (1998), *J. Am. Chem. Soc.*, **120**, 5764.
- Feldman, K. S. and Wrobleski, M. L. (2000), *J. Org. Chem.*, **65**, 8659.
- Kitagaki, S., Yanamoto, Y., Tsutsui, H., Anada, M., Nakajima, M. and Hashimoto, S. (2001), *Tetrahedron Lett.*, **42**, 6361.
- Knapp, S., Morriello, G. J. and Doss, G. A. (2002), *Tetrahedron Lett.*, **43**, 5797.
- Hanessian, S., Parthasarathy, S., Mauduit, M. and Payza, K. (2003), *J. Med. Chem.*, **46**, 34.

46

Stobbe Condensation

Introduction

The base catalyzed condensation of aldehydes and ketones with diethyl succinate and its derivatives to form monoesters of an α-alkylidene- (or arylidene) succinic acid is called the *Stobbe condensation*[1]. The bases generally used are NaOEt, NaH, KOBu^t, *etc*. One of the ester groups becomes hydrolyzed in the course of the reaction.

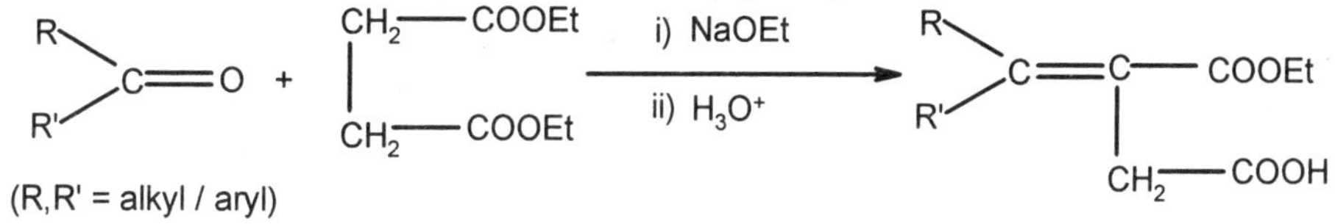

$$\underset{(R,R' = alkyl / aryl)}{\overset{R}{\underset{R'}{>}}C=O} + \begin{matrix} CH_2-COOEt \\ | \\ CH_2-COOEt \end{matrix} \xrightarrow[\text{ii) }H_3O^+]{\text{i) NaOEt}} \underset{R'}{\overset{R}{>}}C=C\underset{CH_2-COOH}{\overset{COOEt}{<}}$$

Mechanism

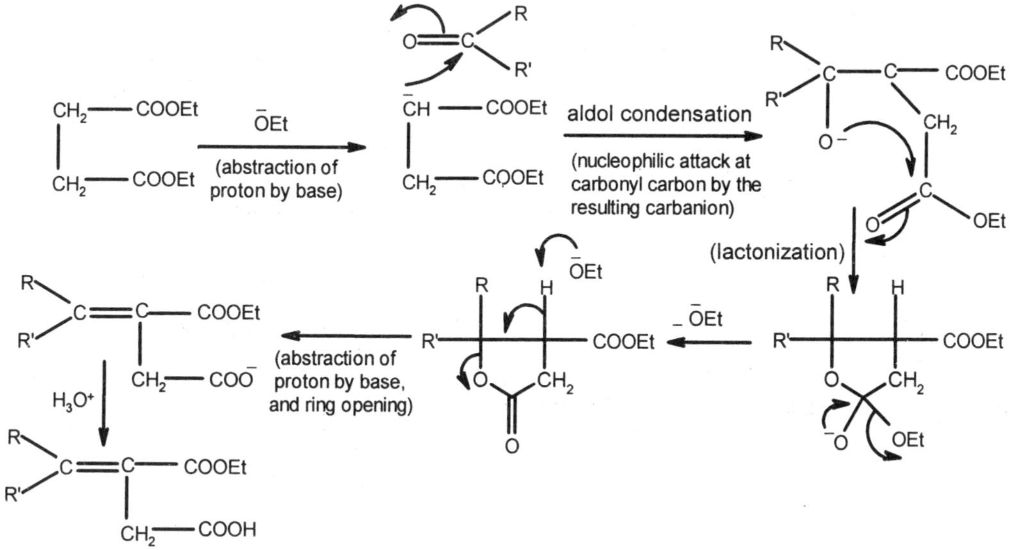

The reaction is believed to proceed by initial aldol condensation and subsequent intramolecular formation of a lactone intermediate[2], which then undergoes base-catalyzed elimination (E1 or E2) to yield the product.

Critical Views

(a) The *Stobbe condensation* has been extended to di-*tert*-butyl esters of glutaric acid[3].

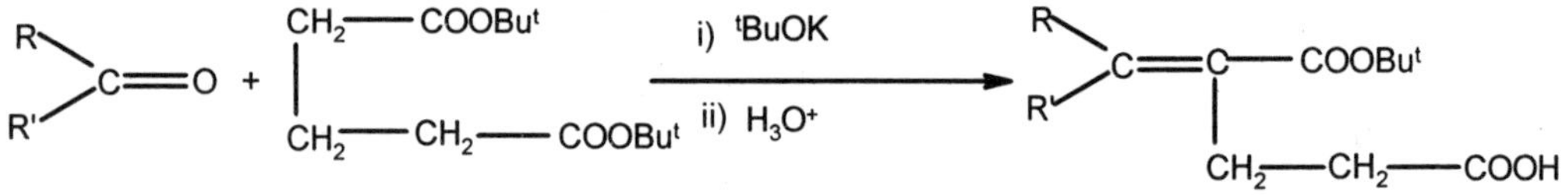

(b) It is observed that the unsaturated *Stobbe product* may consist of mixtures — both of stereoisomers and of structural isomers in which the double bond can take up several possible orientations[4].

(mixture of products)

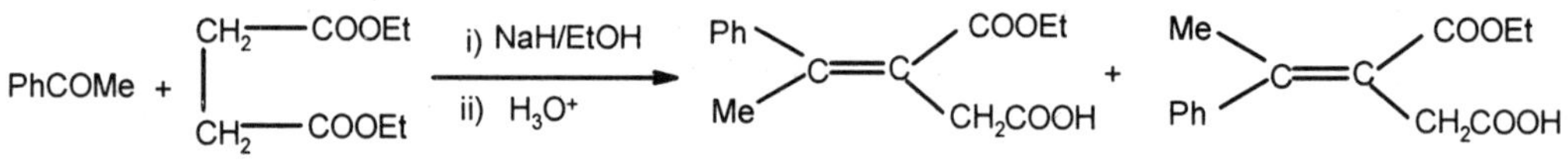

Applications

(i) Synthetically important cyclic ketones can be prepared starting from the *Stobbe condensation product*.

[Ref. 5]

ii)

[Ref. 6]

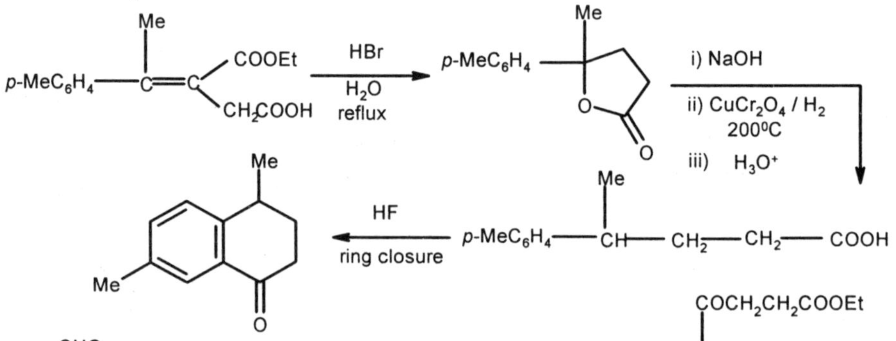

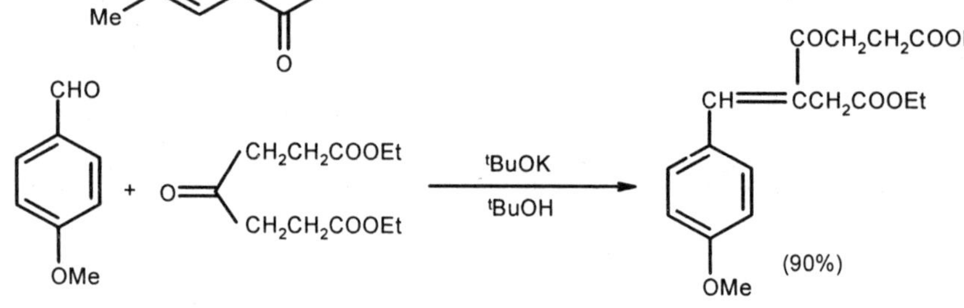

iii) Dimethyl succinate →
i) PhCHO, LiOMe, MeOH
ii) CH$_2$N$_2$-ether
→ Ph-CH=C(COOMe)-CH$_2$-COOMe [Ref. 7]

References

1. Stobbe, H. (1893), *Ber. Dtsch. Chem. Ges.*, **26**, 2312; Stobbe, H. (1894), *Ann.* **282**, 280; Billet (1949), *Bull. Soc. Chim. France*, 297.
2. Robinson, R. and Seijo, E. (1941), *J. Chem. Soc.*, 582; Johnson, W. S. and Schneider, W. P. (1963), *Org. Syn.*, **Coll. Vol. 4**, 132.
3. Puterbaugh, W. H. (1962), *J. Org. Chem.*, 27, 4010; El-Newaihy, M. F., Salem, M. R., Enayat, E. I. And El-Bassiouny, F. A. (1982), *J. Prakt. Chem.*, **324**, 379.
4. Johnson, W. S. and Daub, G. H. (1951), *Org. React.*, **6**, 1 (review); Overberger, C. G. and Roberts, C. W. (1949), *J. Am. Chem. Soc.*, **71**, 3618; Heller, H. G. and Swinney, B. (1967), *J. Chem. Soc., C*, 2452; House, H. O. and Larson, J. K. (1968), *J. Org. Chem.*, **33**, 448.
5. Johnson, W. S., Peterson, J. W. and Schneider, W. P. (1947), *J. Am. Chem. Soc.*, **69**, 74.
6. Bagavant, G. and Swaminathan, K. B. V. (1975), *Current Science*, **44**, 661.
7. Gordaliza, M., del Corral, J. M. M., Castro, M. A., Salinero, M. A., San Feliciano, A., Dorado, J. M. and Valle, F. (1996), *Synlett,* 1201.

- Baghos, V. B., Nasr, F. H. and Gindy, M. (1979), *Helv. Chim. Acta*, **62**, 90.
- Moldvai, I., Temesvari-Major, E., Balazs, M., Gacs-Baitz, E., Egyed, O. and Szantay, C. (1999), *J. Chem. Res. (S)*, 3018.
- Liu, J. and Brooks, N. R. (2002), *Org. Lett.*, **4**, 3521.
- Moldvai, I., Temesvari-Major, E., Incze, M., Platthy, T., Gacs-Baitz, E. and Szantay, C. (2003), *Heterocycles*, **60**, 309.

47

Williamson Ether Synthesis

Introduction

The *Williamson ether synthesis*[1] is the most reliable and versatile method for the preparation of unsymmetrical as well as symmetrical ethers. The normal method involves the treatment of an alkoxide or aroxide ion prepared from the corresponding alcohol or phenol (aromatic alcohol) with an unhindered alkyl halide.

$$RO^- Na^+ + R'-X \longrightarrow R-O-R' + NaX$$

(R = alkyl/ aryl)   (R' = alkyl, allyl, or benzyl)   (ether)

The reaction is very useful when primary alkyl halides are used; iodides are the most reactive than the corresponding bromides or chlorides. Alkyl tosylates are also in use. The alkoxide or phenoxide is commonly made by adding Na, K, or NaH to the substrate. An inert solvent such as benzene, toluene, or xylene, or an excess of the alcohol corresponding to the alkoxide is often used as solvent.

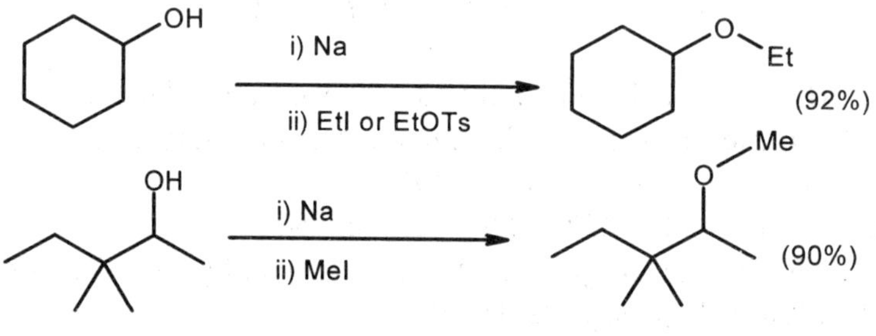

Cyclohexanol —i) Na; ii) EtI or EtOTs→ Cyclohexyl ethyl ether (92%)

3,3-dimethyl-2-butanol —i) Na; ii) MeI→ 2-methoxy-3,3-dimethylbutane (90%)

Mechanism

In most of the cases the alkoxide or phenoxide ion reacts with the alkyl halide following a bimolecular nucleophilic substitution (S_N2) path.

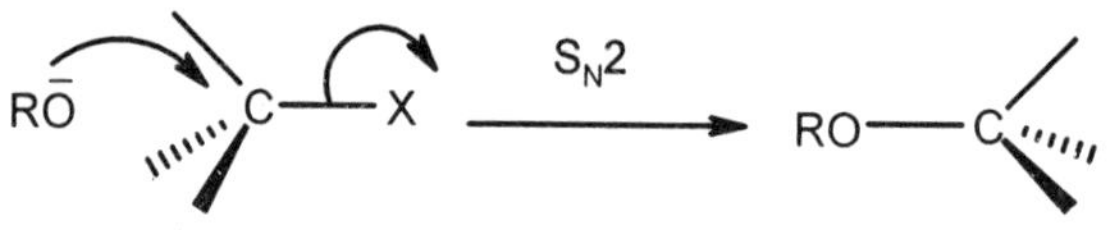

$$RO^- + C-X \xrightarrow{S_N2} RO-C + X^-$$

However, there is evidence that in some cases, particularly with alkyl iodides, the SET (single electron transfer) mechanism can take place[2].

Critical Views

(a) The reaction can also be carried out by mixing the halide and alcohol or phenol directly with Cs_2CO_3 in acetonitrile[3] or with solid KOH in dimethylsulphoxide[4].

(b) Secondary and tertiary alkyl halides are not suitable substrates, because they tend to undergo elimination rather than substitution under the reaction conditions — the counter-reactant alkoxide behaves as a strong base. The reaction is not successful with tertiary alkyl halide, while only low yields are obtained with secondary ones. An ether having one tertiary moiety may successfully be synthesized by the use of a less hindered alkyl group (primary alkyl halide) as the S_N2 substrate and the alkoxide of the more hindered alkyl group (tertiary alkoxide).

$$Me_3C\text{---}\bar{O}\ \overset{+}{Na} + EtCH_2\text{---}I \longrightarrow Me_3C\text{---}O\text{---}CH_2Et$$

(sodium *tert*-butoxide) (*n*-propyliodide)   (*tert*-butyl propyl ether)

(c) *C*-Alkylation is a major concern in cases where phenoxides are used as the nucleophiles. The extent of undesired *C*-alkylation and desired *O*-alkylation depends upon the nature of solvent used. For instance, β-naphthoxide reacts with benzylbromide in DMF (*N,N*-dimethylformamide) to yield β-naphthyl benzylether almost exclusively; but the *C*-alkylated product, 1-benzyl-2-naphthol, is obtained as the major one when the reaction is carried out in water[5].

(β-napthoxide) + (benzyl bromide) →

- DMF → (β-naphthyl benzylether)
- H_2O → (1-benzyl-2-naphthol)

Applications

i) [cyclopentyl]–ONa + EtBr ⟶ [cyclopentyl]–OEt

ii) PhOH —NaOH, MeOSO₂OMe→ PhOMe [Ref. 6]

iii) diol —TsCl, Py, heat→ mono-tosylate —−TsOH→ cyclic ether [Ref. 7]

References

1. Williamson, W. (1851), *Justus Liebigs Ann. Chem.*, **77**, 37; Williamson, W. (1852), *J. Chem. Soc.*, **4**, 229; Dermer, O. C. (1934), *Chem. Rev.*, **14**, 409 (review); Ullmann, F. and Sponagel, P., *Ber.* **38**, 2211 (1905); Allen, C.F.H. and Gates, J.W. (1945), *Organic Syntheses*, **25**, 9; Baker, R. H. and Martin, W. B. (1960), *J. Org. Chem.*, **25**, 1496; Feuer, H. in *The Chemistry of the Ether Linkage* (Ed., S. Patai), Wiley: NY, 1967, p. 445.
2. Ashby, E. C., Bae, D., Park, W., Depriest, R. N. and Su, W. (1984), *Tetrahedron Lett.*, **25**, 5107.
3. Lee, J. C., Yuk, J. Y. and Cho, S. H. (1995), *Synth. Commun.*, **25**, 1367.
4. Benedict, D. A., Bianchi, T. A. and Cate, L. A. (1979), *Synthesis*, 428; Johnstone, R. A. W. and Rose, M. E. (1979), *Tetrahedron*, **35**, 2169.
5. Kornblum, N., Seltzer, R. and Haberfield, P. (1963), *J. Am. Chem. Soc.*, **85**, 1148.
6. Hiers, G. S. and Hager, F. D. (1941), *Org. Synth.*, **Coll. Vol. 1**, 58.
7. Mundy, B. P. and Wilkening, D. (1984), *J. Org. Chem.*, **49**, 3379.

- Jursic, B. (1988), *Tetrahedron*, **44**, 6677.
- Silva, A. L., Quiroz, B. and Maldonado, L. A. (1998), *Tetrahedron Lett.*, **39**, 2055.
- Peng, Y. and Song, G. (2002), *Green Chem.*, **4**, 349.
- Stabile, R. G. and Dicks, A. P. (2003), *J. Chem. Educ.*, **80**, 313.

48

Wittig Reaction

Introduction

The *Wittig reaction*[1] finds immense application in the synthetic organic chemistry, and because of its wide usefulness the Inventor received the Nobel Prize in chemistry in 1979. In this method, an aldehyde or ketone reacts with a phosphorus ylide (*i.e.* alkylidene phosphorane) to give an alkene; thus this reaction is sometimes called as the *Wittig olefination reaction*. The beauty of the reaction lies in its regiospecificity — the double bond is formed between the carbonyl carbon of the aldehyde or ketone and the negatively charged carbon of the ylide.

$$\begin{matrix} \diagdown \\ \diagup \end{matrix} C=O + \left[Ph_3\overset{+}{P}-\overset{-}{C}\diagup^{R}_{R'} \longleftrightarrow Ph_3P=C\diagup^{R}_{R'} \right] \longrightarrow \begin{matrix} \diagdown \\ \diagup \end{matrix} C=C\diagup^{R}_{R'} + Ph_3P=O$$

(aldehyde or ketone)   (alkylidene phosphorane) phosphorus ylide   (alkene) + $Ph_3P=O$ (phosphine oxide)

Phosphorus ylides are usually prepared by the reaction of triphenylphosphine with varying alkyl halides to give triphenylphosphonium salts, which on treatment with a base form the corresponding ylides. However, when phosphonium fluorides are used, no base is required since these react directly with the substrate to give the alkene[2]. *Wittig reactions* may be carried out in a number of different solvents; tetrahydrofuran (THF) and dimethylsulphoxide (DMSO) are most oftenly used.

$$Ph_3P + X-CHR(R') \longrightarrow \underset{\text{(phosphonium salt)}}{Ph_3\overset{+}{P}\diagdown^{H \text{ acidic}}_{\underset{R'}{\diagdown}R} X^-} \xrightarrow[\text{of proton}]{BuLi \atop \text{(abstraction}} \left[Ph_3\overset{+}{P}-\overset{-}{C}\diagup^{R}_{R'} \longleftrightarrow Ph_3P=C\diagup^{R}_{R'} \right]$$

(alkylidene phosphorane) phosphorus ylide

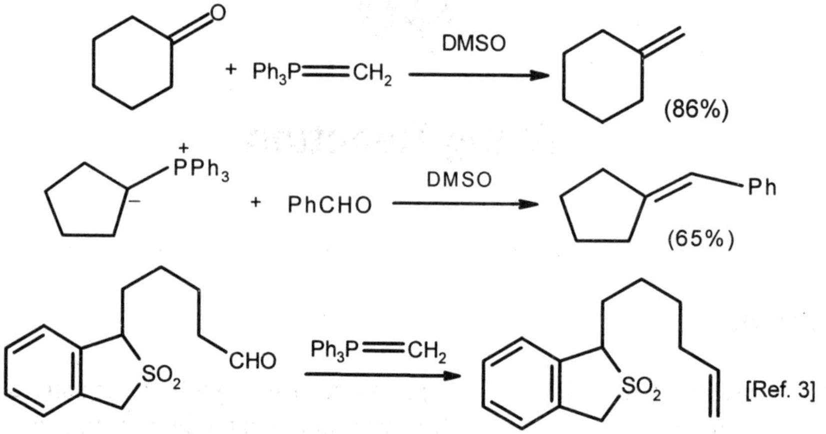

Mechanism

The ylide as a nucleophile attacks the carbonyl group of an aldehyde or ketone generating a four-membered ring called *oxaphosphetane intermediate*. The *oxaphosphetane intermediate* is unstable, and quickly undergoes dissociation to give the alkene and phosphine oxide. Triphenylphosphine oxide is extremely stable; the phosphorus – oxygen double bond in the oxide has been estimated to be greater than 540 kJ/mole. Formation of this double bond between phosphorus and oxygen is, thus, the driving force for the *Wittig reaction*.

Mechanism of the *Wittig reaction* is of much discussion[4]; the reaction is still undergoing mechanistic investigation. Another possibility, which suggests that the *oxaphosphetane intermediate* is formed by a two-step process involving 'betaine' instead of being formed directly, cannot be completely discarded[5]. Here, the following mechanistic scheme for the *Wittig reaction* is depicted:

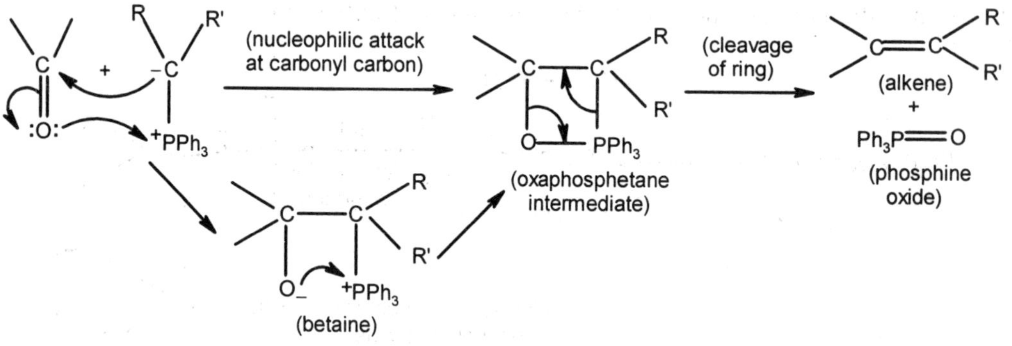

There is little evidence in favour of 'betaine' formation[5,6]; on the contrary, *oxaphosphetane intermediate* receives strong evidences[7] including its isolation in certain cases[8] and also ^{31}P-NMR spectral behaviour[9].

Critical Views

(a) The reactivity of the phosphorus ylide largely depends on the substituents (R and R') attached with the carbonyl carbon; when these groups are electron-withdrawing, the negative charge can be delocalized over several centres reducing the reactivity of the ylide carbon. The reactivity of the carbonyl compound towards addition with the ylide increases with the increase in electrophilic character of the carbonyl group of the substrate.

(b) It has been observed that some *Wittig reactions* give (Z)-selective alkene; some the (E), and others give the diastereomeric mixtures of alkenes. The course of stereoselectivity has been studied much[10], and seems to depend upon many factors. A change in solvent or addition of salts can alter the (E/Z) ratio of the product[11].

However, it is generally observed that with *stabilized* ylides (*i.e.* ylides whose anion is stabilized by further conjugation) *Wittig reaction* is *E*-selective[12], while it is *Z*-selective with *unstabilized* ylides[13].

Geometry of the alkene formed is determined by the stereoselectivity of the oxaphosphetane-forming step; in case of unstabilized ylide *syn*-diastereoisomer of the oxaphosphetane is formed preferentially as the kinetically controlled product, stereospecific dissociation of which leads to the formation of *Z*-selective alkene **Scheme-1**). On the contrary, when the cycloaddition reaction takes place between a stabilized ylide and the carbonyl compound, the stereoselectivity of oxaphosphetane-forming step is no longer kinetically controlled but is reversible and thermodynamically controlled. Less stable *syn*-diastereoisomer of the oxaphosphetane intermediate is converted to the more stable *anti*-form that leads to the formation of *E*-alkene predominantly (**Scheme-2**).

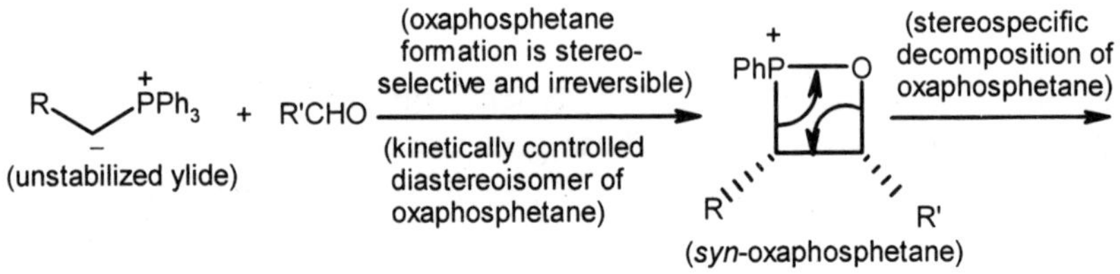

(**Scheme-1**)

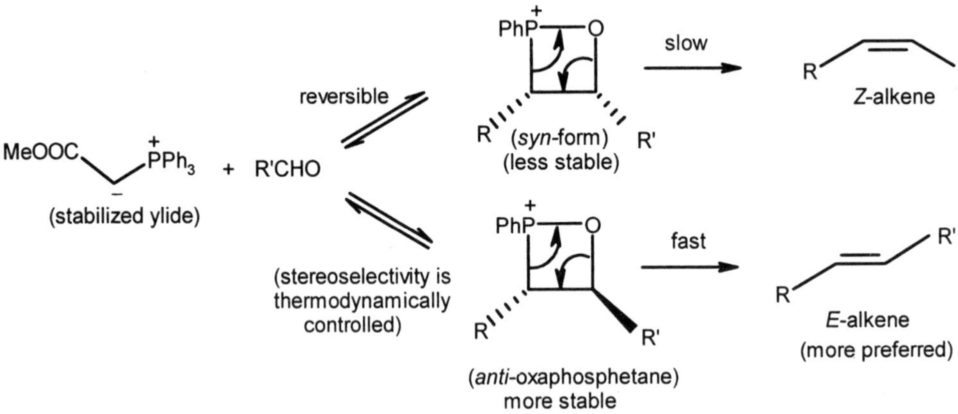

(**Scheme-2**)

(c) *Horner-Wadsworth-Emmons modification*[14], which involves the use of phosphonate ester ylides, has found application in many cases for the synthesis of α,β-unsaturated esters, α,β-unsaturated ketones and related conjugated systems.

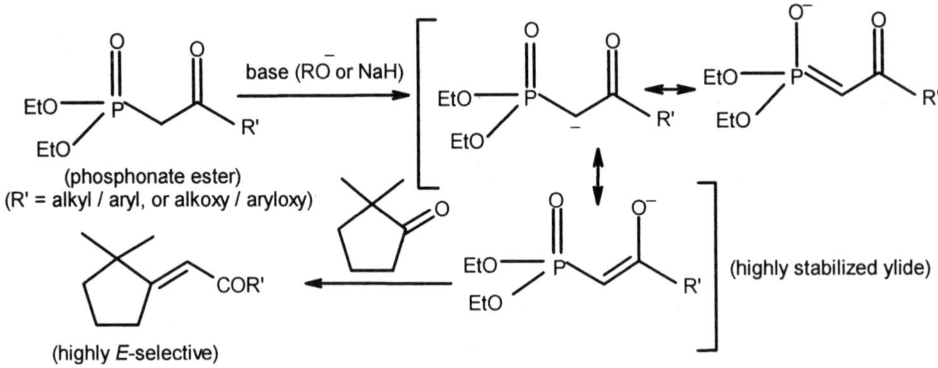

Applications

The *Wittig reaction* bears significant impact in organic synthesis, particularly in the synthesis of many target molecules, for example in natural product synthesis.

i) [Reaction scheme showing 2 equivalents of a β-ionylidene ylide + dialdehyde → β-carotene] [Ref. 15]

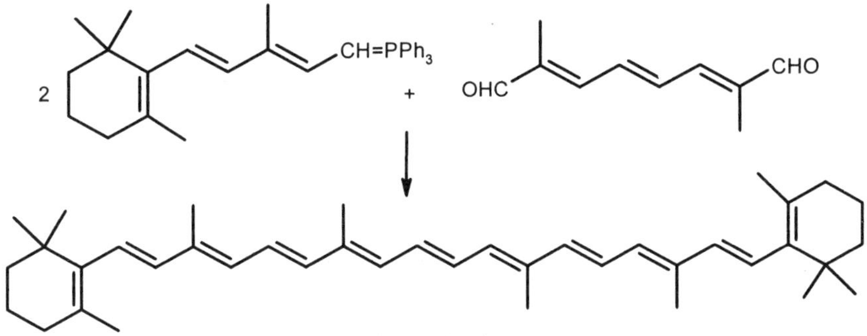

ii) [Reaction scheme: OHC-(CH₂)₈-COOMe + OHC-CH=PPh₃ ylide → aldehyde-ester (96% E), yield 52%; then + butyl-PPh₃ ylide → Z,E-diene ester, 79% yield; then LAH → bombykol, 92% yield (an E,Z-diene, a pheromone produced by female silkworm moth)]

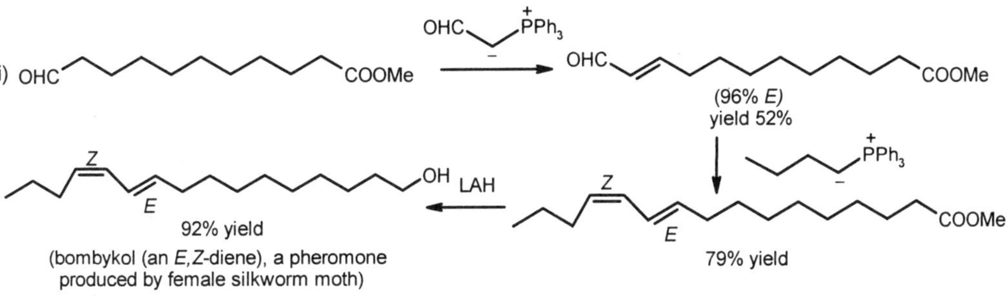

iii) [Reaction scheme: acetonide-protected glyceraldehyde + Ph₃P=CH-COOMe → α,β-unsaturated ester]

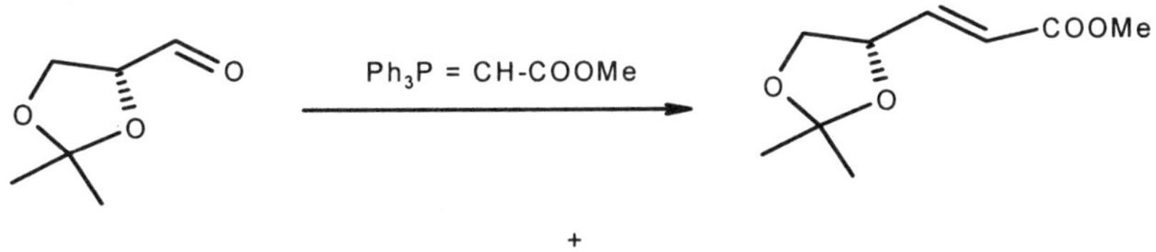

iv) [Reaction scheme: aldehyde-lactone + R-CH=PPh₃ → Z-alkene lactone (Z:E = 90:10)]

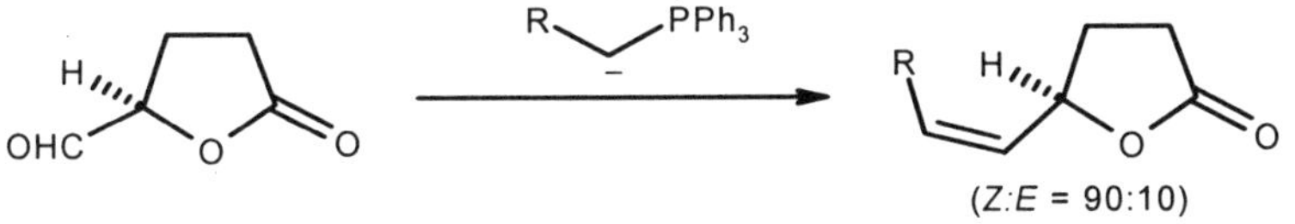

References

1   Wittig, G. and Geissler, G. (1953), *Justus Liebigs Ann. Chem.*, **580**, 44; Wittig, G. and Schollkopf, U. (1954), *Ber. Dtsch. Chem. Ges.*, **87**, 1318; Maercker, A. (1965), *Org. React.*, **14**, 270 (review); Pommer, H. (1977), *Angew. Chem. Int. Ed. Engl.*, **16**, 423; Murphy, P. J. and Brennan (1988), *J. Chem. Soc. Rev.*, **17**, 1 (review).

2   Schiemenz, G. P., Becker, J. and Stockigt, J. (1970), *Ber. Dtsch. Chem. Ges.*, **103**, 2077.
3   Nicolaou, K. C., Barnette, W. E. and Ma, P. (1980), *J. Org. Chem.*, **45**, 1463.
4   Cockerill, A. F. and Harrison, R. G. in *The Chemistry of Functional Groups: Suppliments A, pt. 1* (Ed. S. Patai), Wiley: NY, 1977, p.288; Vedejs, E. and Marth, C. F. (1988), *J. Am. Chem. Soc.*, **110**, 3948.
5   Vedejs, E. and Marth, C. F. (1990), *J. Am. Chem. Soc.*, **112**, 3905.
6   Wittig, G., Weigmann, H. and Schlosser, M. (1961), *Ber. Dtsch. Chem. Ges.*, **94**, 676; Schlosser, M. and Christmann, K. F. (1967), *Liebigs Ann. Chem.*, **708**, 1; Neumann, R. A. and Berger, S. (1998), *Eur. J. Org. Chem.*, 1085.
7   McEwen, W. E., Kumli, K. F., Blade-Font, A., Zanger, M. and VaderWerf, C. A. (1964), *J. Am. Chem. Soc.*, **86**, 2378; Maryanoff, B. E., Reitz, A. B., Mutter, M. S., Inners, R. R., Almond Jr., H. R., Whittle, R. R. and Olofson, R. A. (1986), *J. Am. Chem. Soc.*, **108**, 7664.
8   Mazhar-Ul-Haque, Caughlan, C. N., Ramirez, F., Pilot, J. F. and Smith, C. P. (1971), *J. Am. Chem. Soc.*, **93**, 5229; Birum, G. H. and Matthews, C. N. (1967), *Chem. Commun.*, 137.
9   Vedejs, E., Meier, G. P. and Snoble, K. A. J. (1981), *J. Am. Chem. Soc.*, **103**, 2823.
10 Maryanoff, B. E. and Reitz, A. B. (1989), *Chem. Rev.*, **89**, 863 (review); Schlosser, M. (1970), *Top. Stereochem.*, **5**, 1; Takeuchi, K., Paschal, J. W. and Loncharich, R. J. (1995), *J. Org. Chem.*, **60**, 156.
11 Reitz, A. B., Nortey, S. O., Jordan Jr., A. D., Mutter, M. S. and Maryanoff, B. E. (1986), *J. Org. Chem.*, **51**, 3302.
12 Le Bigot, Y., El Gharbi, R., Delmas, M. and Gaset, A. (1986), *Tetrahedron*, 42, 3813; Maryanoff, B. E., Reitz, A. B. and Duhl-Emswiler, B. A. (1985), *J. Am. Chem. Soc.*, **107**, 217.
13 Schlosser, M., Schaub, B., de Oliveira-Neto, J. and Jeganathan, S. (1986), *Chimica*, **40**, 244.
14 Horner, L., Hoffmann, H., Wippel, H. G. and Klahre, G. (1959), *Chem. Ber.*, **92**, 2499; Wadsworth Jr., W. S. and Emmons, W. D. (1961), *J. Am. Chem. Soc.*, **83**, 1733; Wadsworth Jr., W. S. (1977), *Org. React.*, **25**, 73.
15 Wittig, G. and Pommer, H. (1959), *Chem Abstr.*, **53**, 2279.

- Vedejs, E. and Peterson, M. (1994), *Top. Stereochem.*, **21**, 1. [Review]
- Heron, B. M. (1995), *Heterocycles*, **41**, 2357.
- De Luca, L., Giacomelli, G. and Porcheddu, A. (2002), *Org. Lett.*, **4**, 533.
- Blackburn, L., Kanno, H. and Taylor, R. J. K. (2003), *Tetrahedron Lett.*, **44**, 115.

49

Wolff – Kishner Reduction

Introduction

The *Wolff-Kishner reduction*[1] is an efficient method of conversion of a carbonyl group of an aldehyde or ketone to methylene group by heating the corresponding carbonyl compound with hydrazine hydrate and a strong base (usually NaOH or KOH) at high temperature (~200°C).

$$\underset{\text{(aldehyde or ketone)}}{\diagdown\!\!\!\diagup\!\!\text{C}=\text{O}} \xrightarrow{\text{NH}_2\text{NH}_2} \underset{\text{(hydrazone derivative)}}{\diagdown\!\!\!\diagup\!\!\text{C}=\text{NNH}_2} \xrightarrow[\Delta]{\text{KOH}} \underset{\text{(reduced product)}}{\diagdown\!\!\!\diagup\!\!\text{CH}_2}$$

The original procedure of the reduction involved the heating of the preformed hydrazone with potassium hydroxide in the absence of solvent or heating a solution of either the preformed hydrazone or the carbonyl compound and hydrazine in ethanolic sodium ethoxide to 160 - 200°C in a sealed tube or an autoclave. Latter on, Huang-Minlon[2] modified the reaction procedure in which a mixture of the carbonyl compound, excess hydrazine hydrate, sodium hydroxide or potassium hydroxide, and diethylene glycol is refluxed for several hours; then the mixture is heated to 190 - 200°C so that the water and excess hydrazine are distilled out from the reaction flask. The reaction mixture is then heated to 190 - 200°C until the evolution of nitrogen is complete. The method is known as *Huang-Minlon modification*.

PhCOEt $\xrightarrow[\Delta]{\text{NH}_2\text{NH}_2,\ \text{KOH},\ \text{diethylene glycol}}$ PhCH$_2$Et

(camphore) $\xrightarrow[\Delta]{\text{NH}_2\text{NH}_2,\ \text{KOH},\ \text{diethylene glycol}}$ (camphane)

Mechanism

The initial step is the hydrazone formation. The actual reduction step involves two tautomeric proton transfers from nitrogen to carbon. In strogly basic medium, it is expected that a proton transfer from nitrogen to carbon to occur by loss of a proton from the nitrogen, followed by reprotonation on carbon. A second deprotonation gives an azo anionic species, from which loss of nitrogen gas (N_2) leads to a carbanion intermediate that is then protonated quickly by the solvent to yield the final product.

Critical Views

The *Wolff-Kishner reaction* can also be accomplished with the semicarbazones of carbonyl compounds. A further improvement of the reaction conditions was introduced by Cram et al.[3]; the reaction can be made to proceed at room temperature by the use of potassium *tert*-butoxide as the base and dimethylsulphoxide (DMSO) as solvent.

Particularly for sterically hindered ketones, an alternate procedure[4] is followed. The method consists of forming the hydrazone under acid-catalyzed conditions using the mixture of ketone, hydrazine and hydrazine dihydrochloride in tritheylene glycol at 130 °C; the mixture is then made basic with potassium hydroxide and heated to 210-220°C.

The reaction is useful for the reduction of carbonyl compounds containing acid-sensitive functionalities, and also for the reduction of high-molecular substrates. The reduction is specific for the carbonyl functionality, and thus other reducible groups (such as C – C double or triple bond, carboxylic, ester, nitro, cyano, *etc.*) present in the carbonyl substrates remain intact.

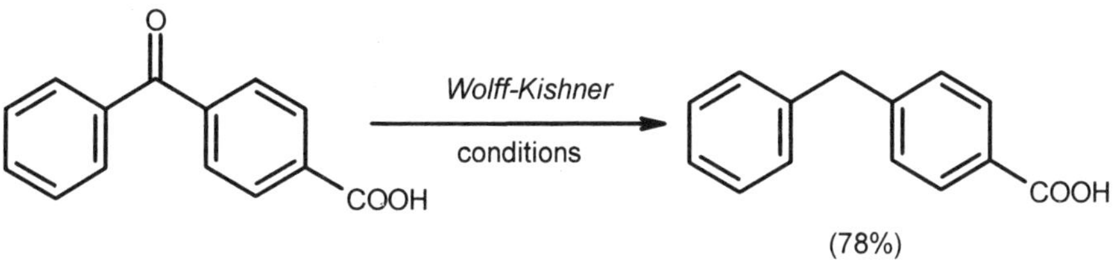

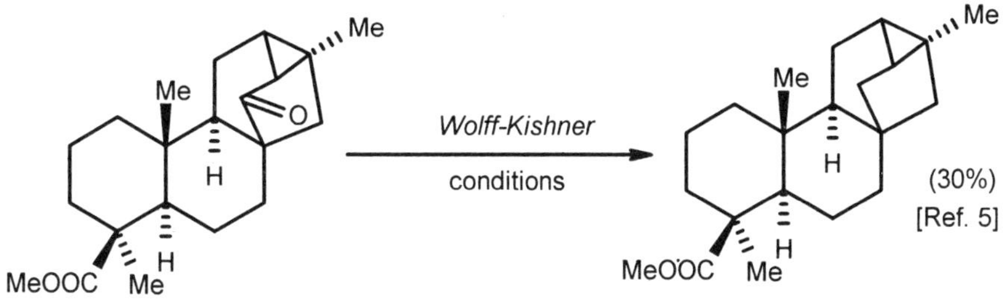

The reduction method is not suitable for α,β-unsaturated carbonyl compounds; it may involve a shift in the position of the double bond, or in some other cases pyrazoline derivatives may be formed that decompose giving cyclopropanes along with the expected alkene.

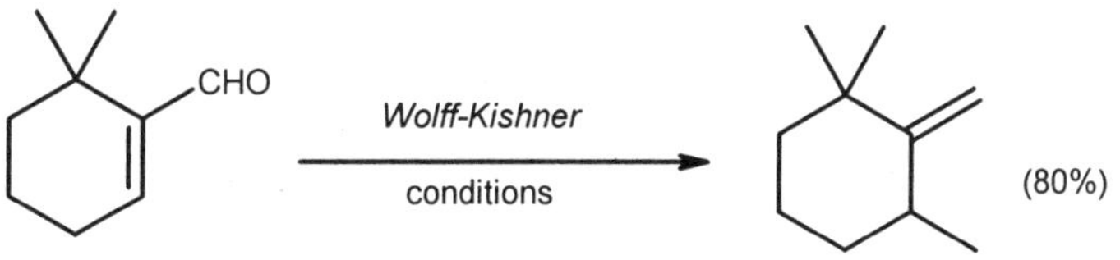

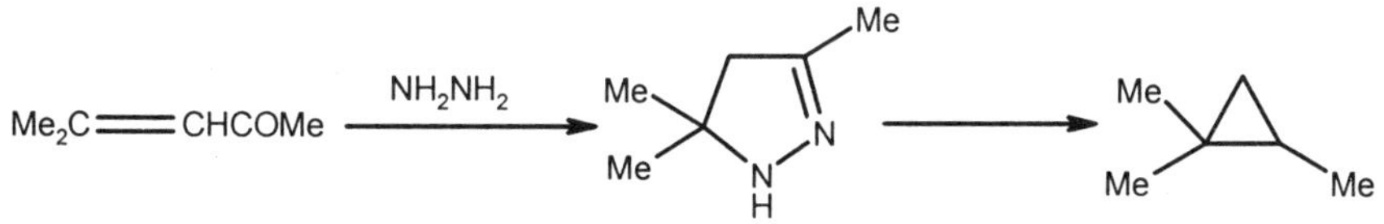

Applications

The *Wolff-Kishner reaction* is a useful method for the specific reduction of carbonyl group into methylene moiety. Few examples are cited:

i)

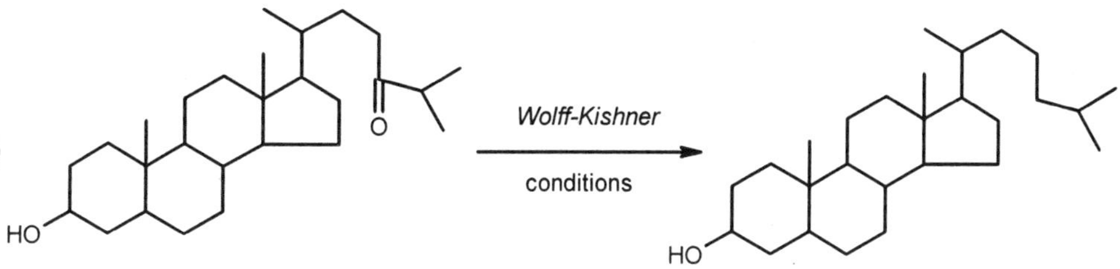

ii) [Structure: cyclohexanone ethylene ketal with -CH₂-C(=O)-CH₃ substituent] →(Wolff-Kishner conditions)→ [Structure: cyclohexanone ethylene ketal with -CH₂CH₂CH₃ (propyl) substituent]

iii) [Steroid-like structure with Me, Me, ketal, C=O, Me, H, Me, Me groups] →(NH₂NH₂, KOH, diethylene glycol, 120-220°C)→ [Reduced steroid-like structure] (92%) [Ref. 6]

References

1. Kishner, N. (1911), *J. Russ. Phys. Chem. Soc.*, **43**, 582; Wolff, L. (1912), *Justus Liebigs Ann. Chem.*, **394**, 85; Todd, D. (1948), *Org. React.*, **4**, 378 (review).
2. Huang-Minlon (1946), *J. Am. Chem. Soc.*, **68**, 2487; *ibid*, 1949, **71**, 3301.
3. Cram, D. J., Sahyun, M. R. V. and Knox, G. R. (1962), *J. Am. Chem. Soc.*, **84**, 1734.
4. Audrieth, L. F. and Ogg, B. A., *The Chemistry of Hydrazine*, Wiley: NY, 1951; Nagata, W. and Itazaki, H. (1964), *Chem. Ind. (London),* 1194.
5. Cory, R., Chan, D., Naquib, Y., Rastall, M. and Renneboog, R. (1980), *J. Org. Chem.*, **45**, 1852.
6. Abad, A., Agulló, C., Arnó, M., Marín, M. L. and Zarazozá, R. J. (1997), *Synlett,* 574.

- Szmant, H. H. (1968), *Angew. Chem. Int. Ed. Engl.*, **7**, 120.
- Akhila, A. and Banthorpe, D. V. (1980), *Indian J. Chem.*, **19B**, 998.
- Taber, D. F. and Stachel, S., J. (1992), *Tetrahedron Lett.*, **33**, 903.
- Gadhwal, S., Baruah, M. and Sandhu, J. S. (1999), *Synlett*, 1573.
- Chattopadhyay, S., Banerjee, S. K. and Mitra, A. K. (2002), *J. Indian Chem. Soc.*, **79**, 906.
- Bashore, C. G., Samardjiev, I. J., Bordner, J. and Coe, J. W. (2003), *J. Am. Chem. Soc.*, **125**, 3268.

50

Yamaguchi Esterification

Introduction

Yamaguchi esterification[1] offers a method for useful esterification of carboxylic acids using 2,4,6-trichlorobenzoyl chloride (often called as the *Yamaguchi reagent*) in the presence of triethylamine (base) and 4-N,N-dimethylaminopyridine (DMAP).

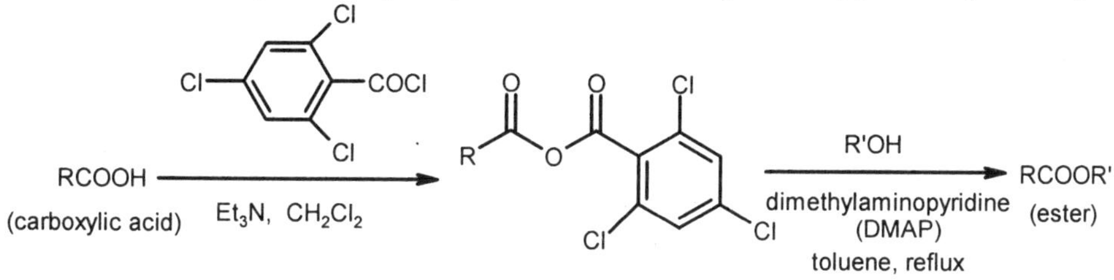

Mechanism

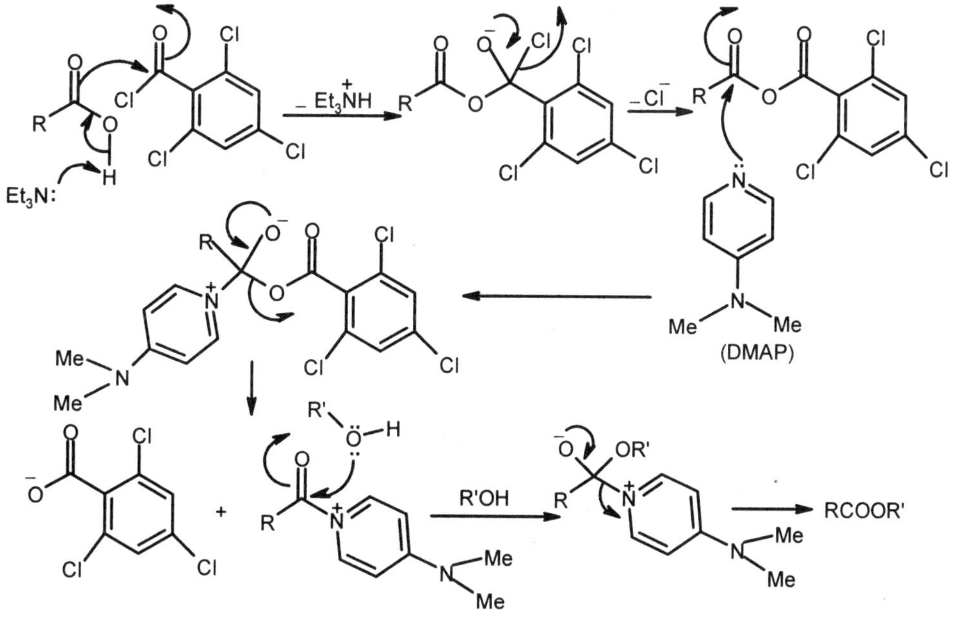

The esterification occurs in the steps as depicted in the afore-mentioned mechanistic scheme.

Critical Views

It is interesting to note that DMAP cann't attack the benzoyl carbonyl because of the steric hindrance developed due to the presence of two chloro substituents *ortho* to the benzoyl carbobyl. Thus only the desired attack by DMAP at the carboxylic carbon takes place exclusively.

However, in some cases modified *Yamaguchi reaction* is applied where 2,6-dichlorobenzoyl chloride (DCBC) is used instead of 2,4,6-trichlorobenzoyl chloride. For instance, the most general procedure for the synthesis of cyclocholates (macrocyclic polyesters with two to six steroid units formed by head-to-tail cyclization of bile acids) from monomeric or linear dimeric derivatives of bile acids by modified *Yamaguchi esterification (macrolactonization)* using DCBC and DMAP as coupling reagents[2].

(R¹=R²=H, lithocholic acid) *Yamaguchi macrolactonization procedure for cyclocholates*

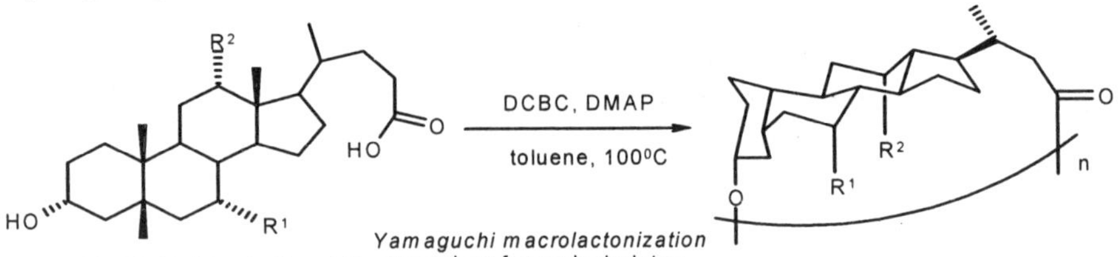

Applications

The *Yamaguchi esterification* is an extremely useful reaction in organic syntheses, particularly in the macrocyclic lactonization, which is the key step for the total synthesis of a number of natural and unnatural macrocylic compounds, *e.g.* technique used in the stereoselective synthesis of macrolide and polyether antibiotics[3], total synthesis of (–) baconipyrone-C (a marine polypropionate)[4],total synthesis of filipin III (a methylpentaene macrolide antibiotic)[5], *etc.*

(Filipin III)

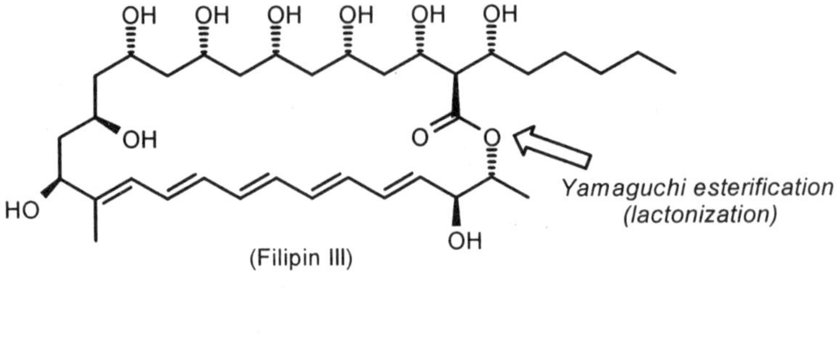

One more example[6] of using *Yamaguchi macrolactonization* to synthesize epothilone analogues is cited below:

(Troc = 2,2,2-trichloroethyloxylcarbonyl)

(epothilone analogue)

References

1. Inanaga, J., Hirata, K., Saeki, H., Katsuki, T. and Yamaguchi, M. (1979), *Bull. Chem. Soc. Jpn.*, **52**, 1989. Kawanami, Y., Dainobu, Y., Inanaga, J., Katsuki, T. and Yamaguchi, M. (1981), *Bull. Chem. Soc. Jpn.*, **54**, 943.
2. Lappalainen, K., Kolehmainen, E. and Saman, D. (1995), *Spectrochem. Acta, Part A*, **51**, 1543; Lappalainen, K., Kolehmainen, E. and Kotoneva, J. (1996), *Magn. Res. Chem.*, **34**, 316; Lappalainen, K., Kolehmainen, E. (1997), *Liebigs Annalen-Recueil*, 1965; Li, Y. and Dias, J. R. (1997), *Synthesis*, 425; Li, Y. and Dias, J. R. (1997), *Synth. Commun.*, 27, 757; Li, Y. and Dias, J. R. (1998), *New J. Chem.*, 579; Li, Y. and Dias, J. R. (1998), *Eur. J. Org. Chem.*, 719; Tamminen, J. and Kolehmainen, E. (201), *Molecules*, **6**, 21 (review).
3. Yonemitsu, O. (1990), *Yakugaku Zasshi*, **110**(8), 523. [Review]
4. Paterson, I., Chen, D. Y-K., Acena, J. L. and Franklin, A. S. (2000), *Org. Lett.*, **2**(11), 1513.
5. Richardson, T. I. and Rychnousky, S. (1999), *Tetrahedron*, **55**(29), 8977.
6. Lapeva, T. (2002), *Ph.D. Thesis: A New Synthetic Pathway to Epithilone Analogues and an Efficient Approach to Cyclohexenylamines as Precursors for the Total Synthesis of Epibatidine*, Eberhard-Karls-Universitat Tubingen, Germany, p.80.

- Bartra, M. and Vilarrasa, J. (1991), *J. Org. Chem.*, **56**, 5132.
- Berger, M. and Mulzer, J. (1999), *J. Am. Chem. Soc.*, **121**, 8393.
- Hamelin, O., Wang, Y., Depres, J. –P. and Greene, A. E. (2000), *Angew. Chem. Int. Ed. Engl.*, **39**, 4314.
- Tian, Z., Hong, C. and Wang, Y. (2002), *Synth. Commun.*, **32**, 3821.
- Muñoz, D. M., Passey, S. C., Simpson, T. J., Willis, C. L., Campbell, J. B. and Richard Rosser, R. (2004), *Australian J. Chem.*, **57**(7), 645.

Allan – Robinson Reaction

Allan–Robinson reaction is one of the most useful synthetic methods for preparing flavones and their derivatives. Flavones are obtained by this method in one step from the condensation of an *o*-hydroxyacetophenone with the anhydride of an aromatic acid in the presence of the salt of the same acid, or in the presence of triethylamine or pyridine as catalyst at oil bath temperature. To get 3-hydroxyflavones or 3-methoxyflavones, it is required to use ω-benzyloxyacetophenone or ω-methoxyacetophenone as the starting material.

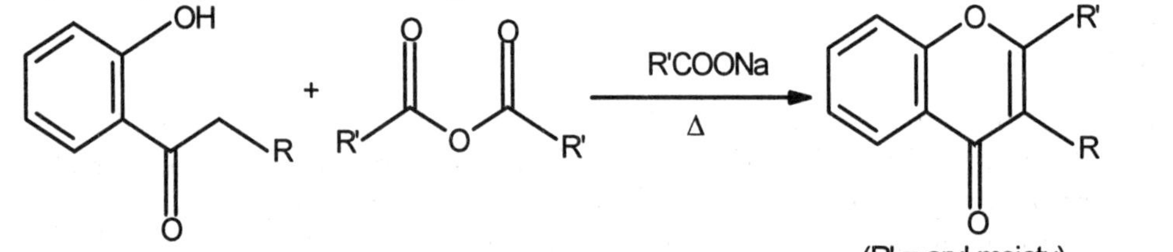

(R = H, –OCOPh or –OMe)   (R' = aryl moiety)   (R' = aryl moiety) (flavone derivative)

Mechanism

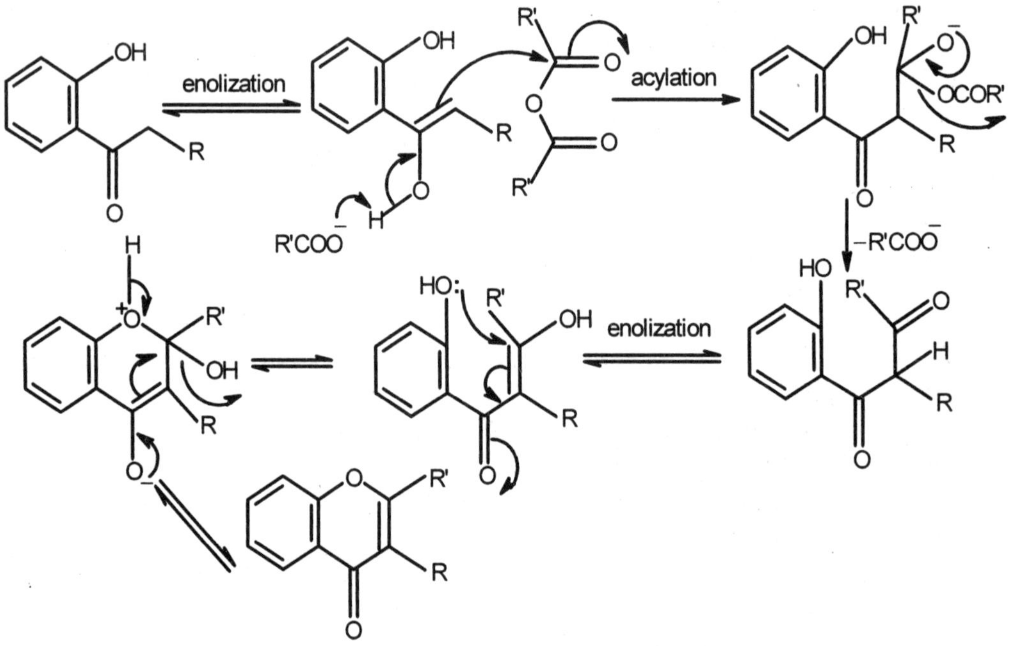

Few examples involving the reaction are cited below:

1) [reaction: 2'-hydroxyacetophenone + (C₆H₅CO)₂O, C₆H₅COONa, 180° → flavone]

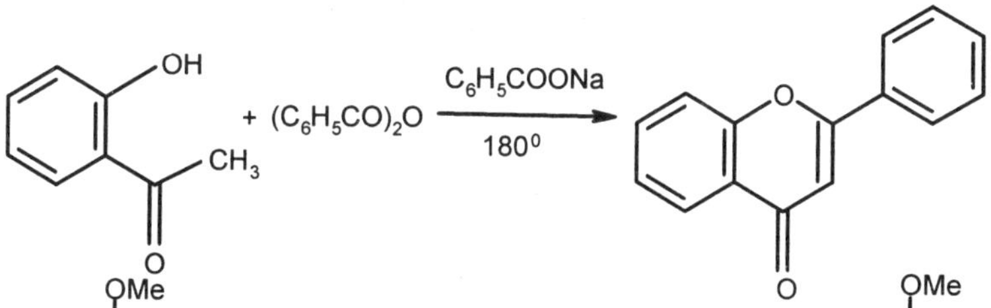

2) [reaction: polymethoxy/hydroxy acetophenone + (R-C₆H₄-CO)₂O, i) Et₃N/Δ, ii) alc. KOH → product]

R = H; araneol
R = OMe; araneosol

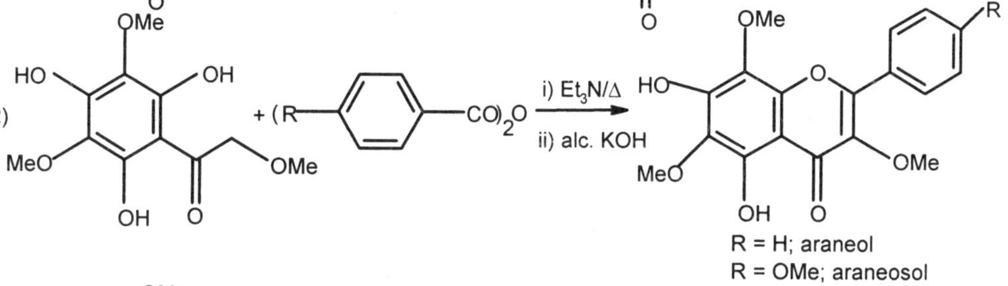

3) [reaction: methoxy/hydroxy acetophenone-OBz + (C₆H₅CO)₂O, PhCOONa/Δ → Prudomestin]

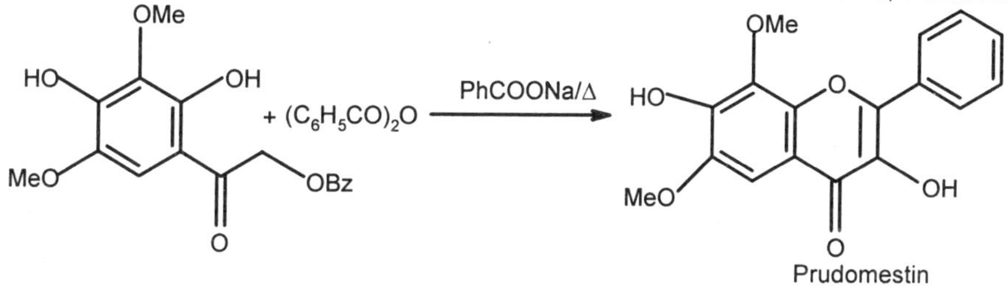

References

- Allan, J. and Robinson, R. (1924), *J. Chem. Soc.*, **125**, 2192.
- Krishnamurti, M., Seshadri, T. R. and Shankaran, P. R. (1966), *Tetrahedron*, **22**, 941.
- Krishnamurti, M., Seshadri, T. R. and Shankaran, P. R. (1967), *Indian J. Chem., Sec. B,* **5B**, 137.
- Wagner, H., Maurer, I., Farkas, L. and Strelisky, J. (1977), *Tetrahedron*, **33**, 1405.
- Dutta, P. K., Bagchi, D. and Pakrashi, S. C. (1982), *Indian J. Chem., Sec. B,* **21B**, 1037.
- Patwardhan, S. A. and Gupta, A. S. (1984), *J. Chem. Res.*, (S), 395.
- Horie, T., Kawamura, Y., Tsukayama, M. and Yoshizaki, S. (1989), *Chem. Pharm. Bull.*, **37**, 1216.

Amadori Rearrangement

Amadori rearrangement involves the conversion of *N*-glycosides of aldoses (or say aldosylamines) to *N*-glycosides of the corresponding ketoses (ketosylamines) in the presence of an acid or a base.

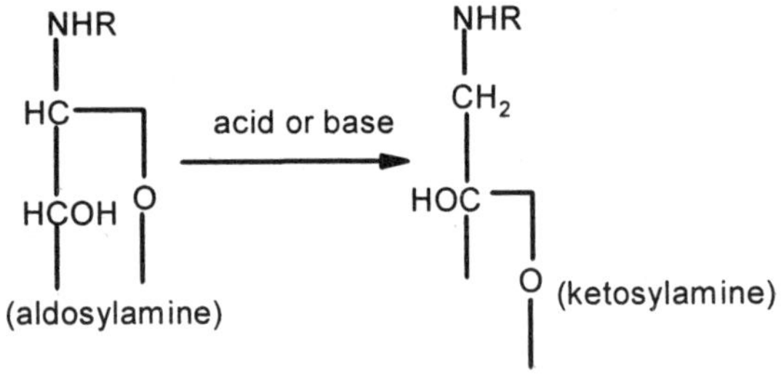

Mechanism

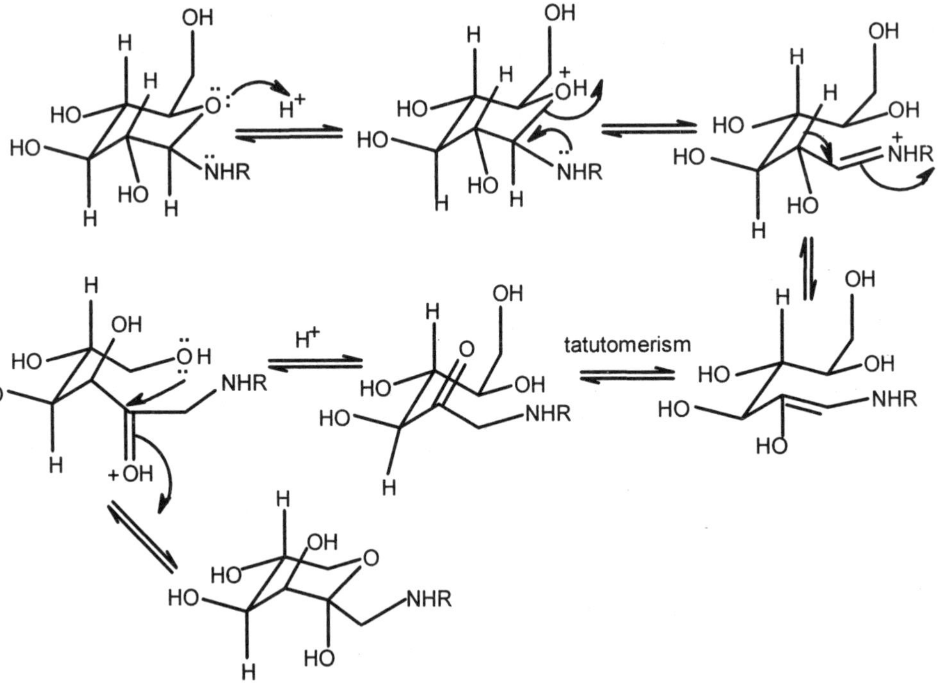

Amadori rearrangement finds immense applications in carbobohydrate chemistry. Recently, Kadokawa *et al.* (1998) have reported a new method of synthesis of aminopolysaccharide that involves this rearrangement reaction as the key step.

(6-amino-6-deoxy-D-glucose hydrochloride)

(aminopolysaccharide having aminoketose structure)

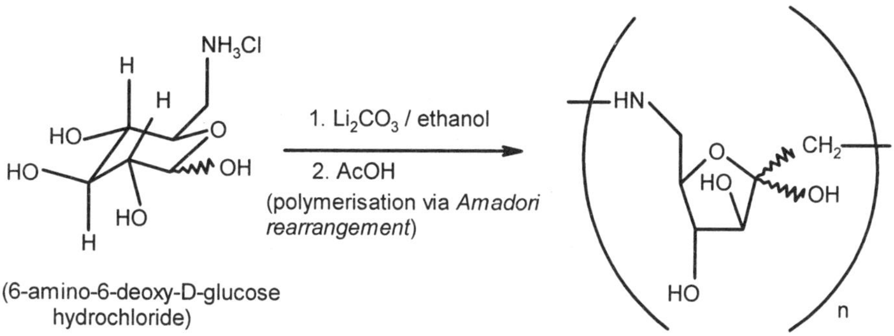

References

- Amadori, M. (1925) *Atti Accad. Nazl. Lincei*, **2**(6), 337, *Chemical Abstr.*, **20**, 902 (1926); *ibid.* **9**(6), 68, 226 (1929), *Chemical Abstr.*, **23**, 3211, 3443 (1929).
- Isbell and Frusch (1958), *J. Org. Chem.*, **23**, 1309.
- Hodges, J. E. (1955), *Adv. Carbohydrate Chem.*, **10**, 169. [Review]
- Lemieux R. U. in *Molecular Rearrangements* Part 2, de Mayo, P. (Ed.), Wiley-Interscience, New York, 1964, p 753.
- Wrodnigg, T. M., Stutz, A. E. and Withers, S. G. (1997), *Tetrahedron Lett.*, **38**, 5463.
- Kadokawa, J.–I., Hino, D., Karasu, M., Tagaya, H. and Chiba, L. M. (1998), *Chem. Lett.*, 383.
- Turner, J. J., Wilschut, N., Overkleeft, H. S., Klaffke, W., van der Marel, G. A. and van Boom, J. H. (1999), *Tetrahedron Lett.*, **40**, 7039.
- Liu, Z. and Sayre, L. M. (2003), *Chem. Res. Tox.*, **16**, 232.

Angeli–Rimini Hydroxamic Acid Synthesis

Hydroxamic acids can effectively be synthesized by means of the reaction of aldehydes with *N*-hydroxybenzenesulphonamide (called *Piloty's acid*) under basic conditions. This reaction constitutes the basis for a well-known spot test used in the qualitative identification of aldehydes. This test, known as the *Angeli–Rimini test*, involves the formation of a hydroxamic acid which forms characteristically coloured complex with ferric ions.

$$RCHO + C_6H_5SO_2NHOH \xrightarrow{\text{NaOH or NaOMe}} RCONHOH$$

(R = alkyl or aryl)

Mechanism

Mechanistic path of *Angeli–Rimini hydroxamic acid* formation may be accomplished from route-a; but Hassner *et al.* suggested the possibility of route-b and at the same time they rejected the postulate of formation of nitrenes as intermediate since they observed that *N*-hydroxybenzenesulphonamide does not undergo reaction with olefins under the basic conditions.

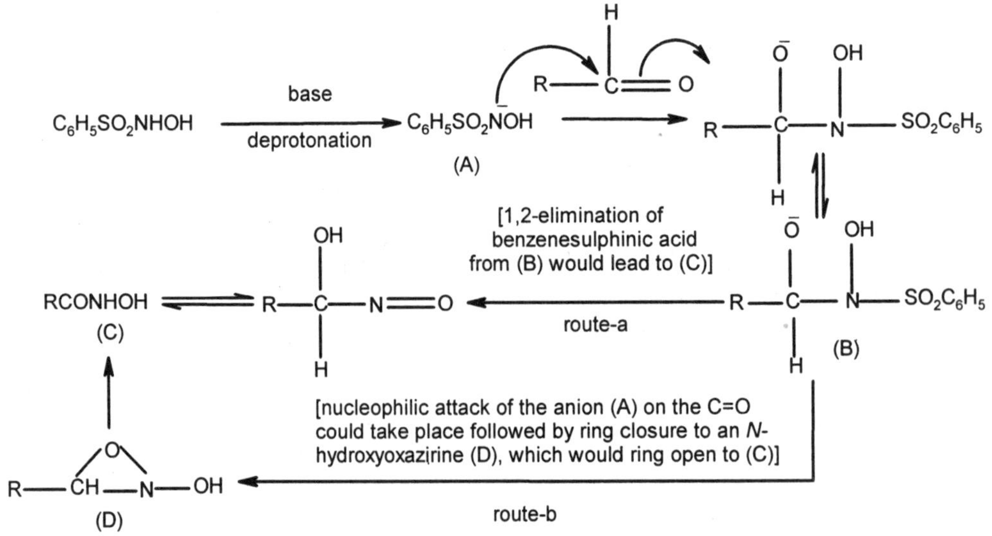

References

- Angeli, A. (1896), *Gazz. Chim. Ital.*, **26**(II), 17; Rimini, E. (1901), *Gazz. Chim. Ital.*, **31**(II), 84.
- Yale H. L. (1943), *Chem. Rev.*, **33**, 209. [Review]
- Hassner, A., Wiederkehn, R. and Kaschores, A. J. (1970), *J. Org. Chem.*, **38**(6), 1962.
- Feigl, F., *Spot Tests In Organic Chemistry*, 2nd ed., Elsevier Publishing Co., New York, 1966, p.196.
- Zhou, S., Xie, F., Xu, Z. and Ni, S. (2001), *Huaxue Shiji*, **23**, 154.

Baker-Venkataraman Rearrangement

Isomerization of *o*-benzoyloxyacetophenones into *o*-hydroxy-β-diketones by the action of a base is known as *Baker–Venkataraman rearrangement*. In this method *o*-hydroxyacetophenones are acylated at oil-bath temperature with aromatic acid chlorides in acetone-potassium carbonate or pyridine, and the resulting esters are converted into β-diketones with potassium hydroxide in pyridine or with NaH.

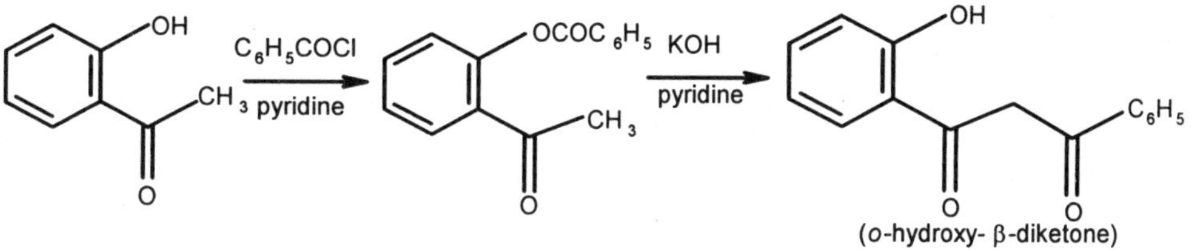

Mechanism

The reaction involves an intramolecular acyl transfer.

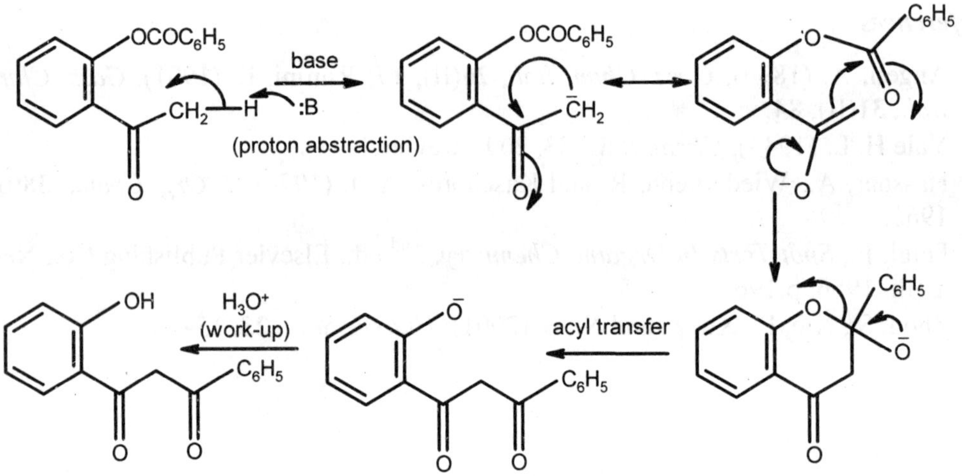

The reaction is used as a key step for synthesizing an important natural pigment, flavones. *o*-Hydroxy-β-diketone on ring closure yields flavone nucleus. A huge number of flavones were reported to be synthesized by using *Baker-Venkataraman method*.

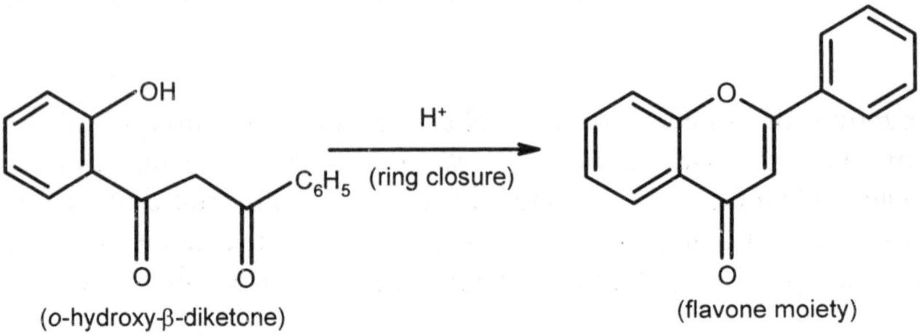

References

- Baker, W. (1933), *J. Chem. Soc.*, 1381.
- Farkas, L., Nogradi, M., Sudarsanam, V. and Herz, W. (1966), *J. Org. Chem.*, **31**, 3228.
- Farkas, L., Vermes, B. and Nogradi, M. (1967), *Tetrahedron*, **23**, 741.
- Govindachari, T. R., Parthasarathy, P. C., Pai, B. R. and Subranium, P. S. (1968), *Tetrahedron*, **24**, 7027.
- Finnegan, R. A., Bachmann, P. L. and Knutson, D. (1973), *Lloydia*, **35**, 457.
- Bowden, K. and Chehel-Amiran, A. (1986), *J. Chem. Soc., Perkin Trans.2*, 2039.
- Makrandi, J. K. and Kumari, V. (1989), *Synth. Commun.*, **19**, 1919.
- Reddy, B. P. and Krupadanam, G. L. D. (1996), *J. Heterocycl. Chem.*, **33**, 1561.

- Kalinin, A. V., Da Silva, A. J. M., Lopes, C. C., Lopes, R. S. C. and Snieckus, V. (1998), *Tetrahedron Lett.*, **39**, 4995.
- Kalanin, A. V. and Snieckus, V. (1998), *Tetrahedron Lett.*, **39**, 4999.
- Pinto, D. C. G. A., Silva, A. M. S. and Cavaleiro, J. A. S. (2000), *New J. Chem.*, **24**, 85.
- Thasana, N. and Ruchirawat, S. (2002), *Tetrahedron Lett.*, **43**, 4515.

Bamberger Rearrangement

B*amberger rearrangement* involves acid-catalyzed rearrangement reaction of aryl hydroxylamines to *para*-aminophenols.

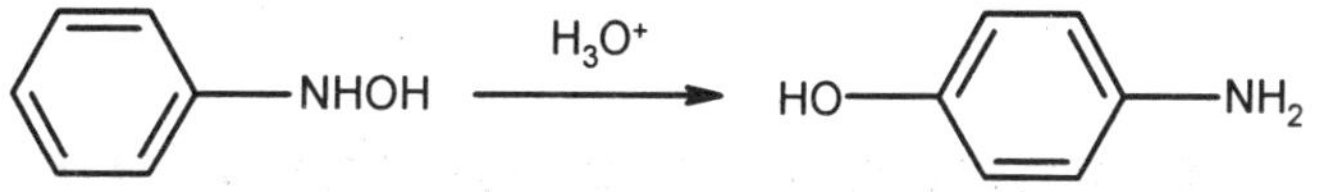

$$C_6H_5\text{-NHOH} \xrightarrow{H_3O^+} HO\text{-}C_6H_4\text{-}NH_2$$

(*N*-phenylhydroxylamine) → (*p*-aminophenol)

The rearrangement is intermolecular, and nucleophilic attack occurs at the ring; available evidences supports the following mechanism for the reaction.

Mechanism

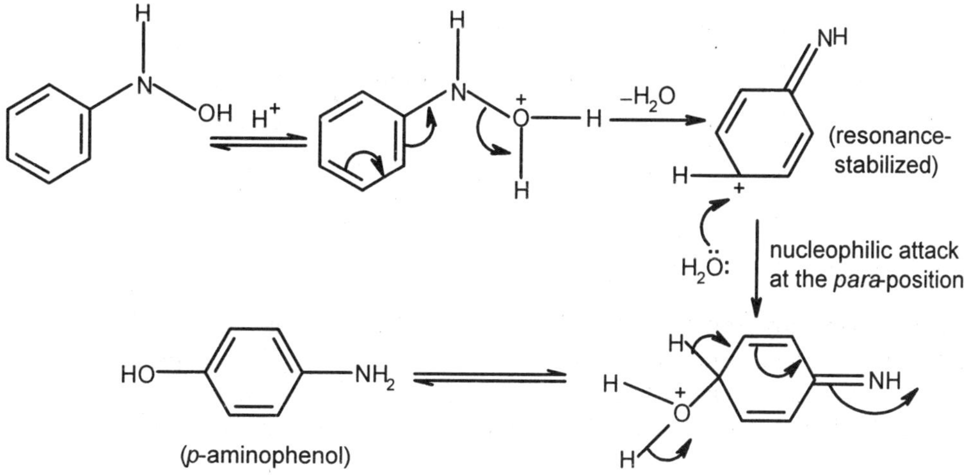

References

- Bamberger, E. (1894), *Ber. Dtsch. Chem. Ges.*, **27**, 1548.
- Shine, H. J. in *Aromatic Rearrangements*, Elsevier, NY, 1967, pp. 182-90.
- Sone, T., Tokuda, Y., Sakai, T., Shinkai, S. and Manabe, O. (1981), *J. Chem. Soc., Perkin 2*, 298.
- Kohnstam, G., Petch, W. A. and Williams D. L. H. (1984), *J. Chem. Soc., Perkin 2*, 423.
- Sternson, L. A. and Chandrasakar, R. (1984), *J. Org. Chem.*, **49**, 4295.
- Fishbein, J. C. and McClelland, R. A. (1987), *J. Am. Chem. Soc.*, **109**, 2824.
- Fishbein, J. C. and McClelland, R. A. (1996), *Can. J. Chem.*, **74**, 1321.
- Pirrung, M. C., Wedel, M. and Zhao, Y. (2002), *Synlett*, 143.

Bamford - Stevens Reaction

Treatment of tosylhydrazones of an aldehyde or a ketone with a strong base, such as Na/ethyleneglycol, NaH, NaOMe or $NaNH_2$, leads to the formation of an alkene; the reaction is known as *Bamford-Stevens reaction*.

$$\underset{H}{\overset{|}{\underset{|}{C}}}-\underset{N-NH-Ts}{\overset{|}{\underset{\|}{C}}}-\quad\xrightarrow{\text{base}}\quad \underset{}{>}C=C\underset{}{<}$$

(alkene)
(more substituted alkenes are generally predominated)

Mechanism

The reaction may involve both the possible mechanisms — a carbenoid and a carbocation mechanism. In protic solvent the reaction mainly proceeds through carbocation mechanism, while in aprotic solvent the carbenoid mechanism may operate. Both the routes involve the formation of a diazo-compound (**A**):

In protic solvent

In aprotic solvent

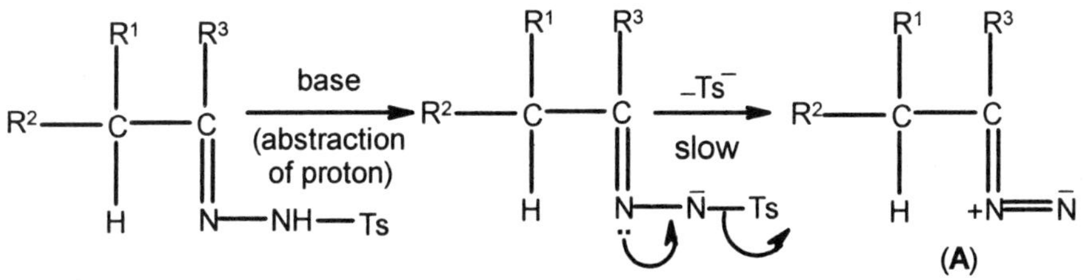

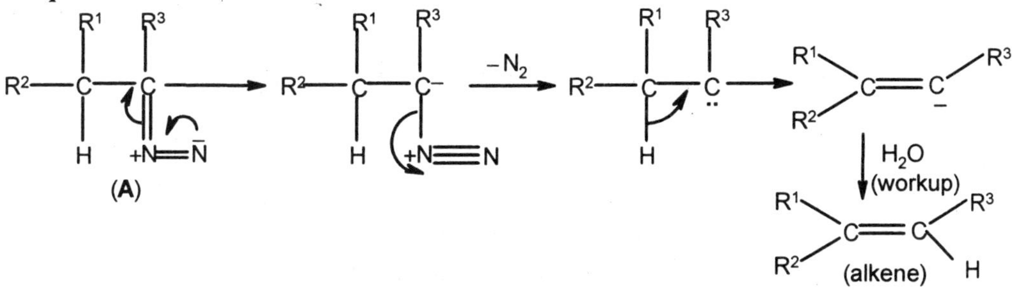

The reaction is called the *Shapiro reaction* when the bases used are alkyllithiums and Grignard reagents; the alkenes formed are generally the less-substituted kinetic products. Thus, major diference between the two reactions is the base employed.

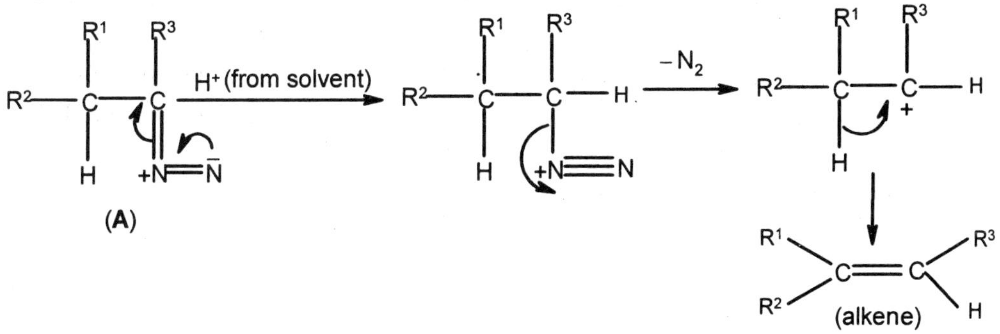

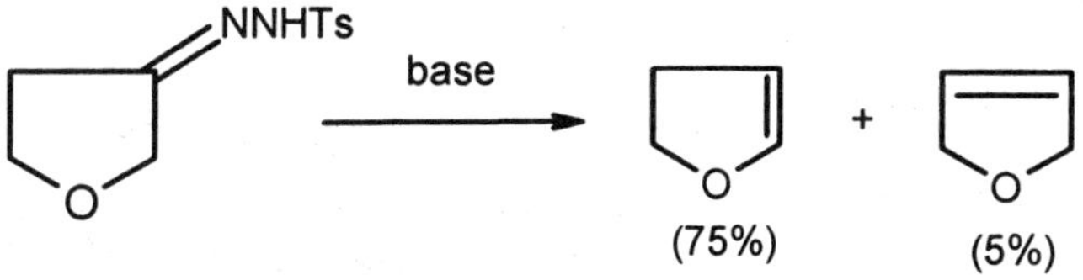

References

- Bamford, W. R. and Stevens, T. S. (1952), *J. Chem. Soc.*, 4735.
- Powell, J. W. and Whiting, M. C. (1959), *Tetrahedron Lett.*, **7**, 305; 1961, **12**, 168; DePuy, C. H. and Froemsdorf, D. H. (1960), *J. Am. Chem. Soc.*, **82**, 634;

Bayless, J. H., Friedman, L., Cook, F. B. and Shechter, H. (1968), *J. Am. Chem. Soc.*, **90**, 531; Nickon, A. and Werstiuk, N. H. (1972), *J. Am. Chem. Soc.*, **94**, 7081.
- Gianturco, M. A., Fridel, P. and Flanagan, V. (1965), *Tetrahedron Lett.*, 1847.
- Shapiro, R. H. (1976), *Org. React.*, **23**, 405. [Review]
- Grieco, P. A., Oguri, T., Wang, C.-L. J. and Williams, E. (1977), *J. Org. Chem.*, **42**, 4113
- Regitz, M. and Maas, G., Diazo Compounds, Academic Press, NY, 1986, p.257.
- Wulfman, D. S., Yousefian, S. and White, J. M. (1988), *Synth. Commun.*, **18**, 2349.
- Adlington, R. M. and Barrett, A. G. M. (1983), *Acc. Chem. Res.*, **16**, 55. [Review]
- Sarkar, T. K. and Ghorai, B. K. (1992), *J. Chem. Soc., Chem. Commun.*, **17**, 1184.
- Olmstead, K. K. and Nickon, A. (1999), *Tetrahedron*, **55**, 7389.
- May, J. A. and Stoltz, B. M. (2002), *J. Am. Chem. Soc.*, **124**, 12426.

Bardhan – Sengupta Synthesis

In 1932 Bardhan and Sengupta reported a novel method for synthesizing phenanthrene derivatives. The method involves the formation of octahydrophenanthrenes first by cyclodehydration of 2β-phenylethylcyclohexanols with the treatment of phosphorus pentoxide or anhydrous hydrogen fluoride. Octahydrophenanthrenes are, then, aromatized to phenanthrenes by heating with selenium. The reaction scheme is given as:

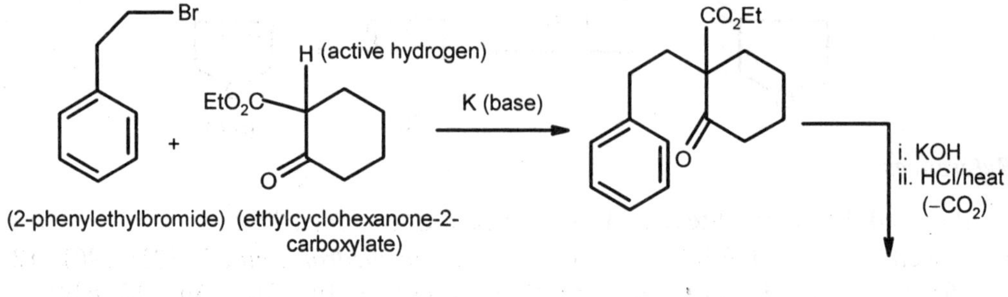

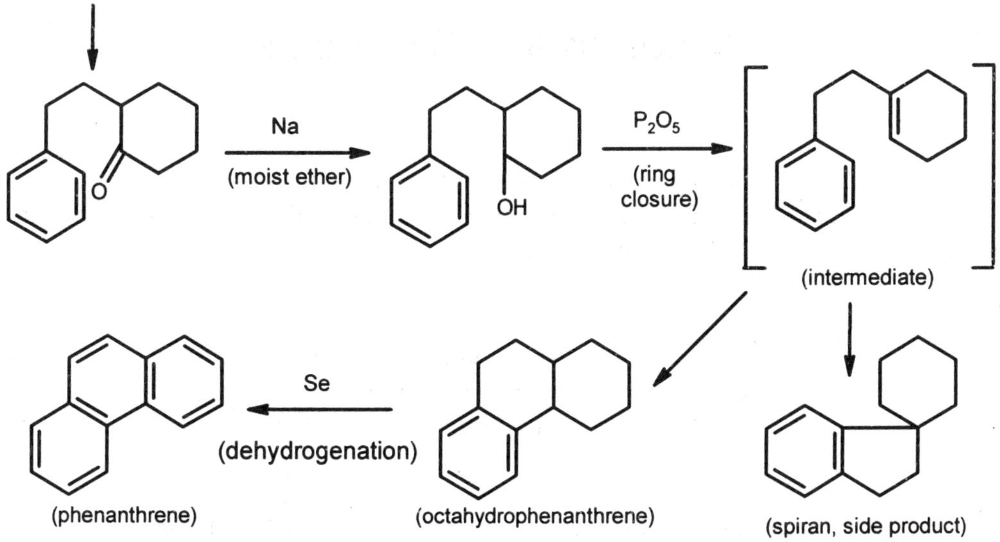

The method can be applied to 2β-phenylethylcyclopentanols also.

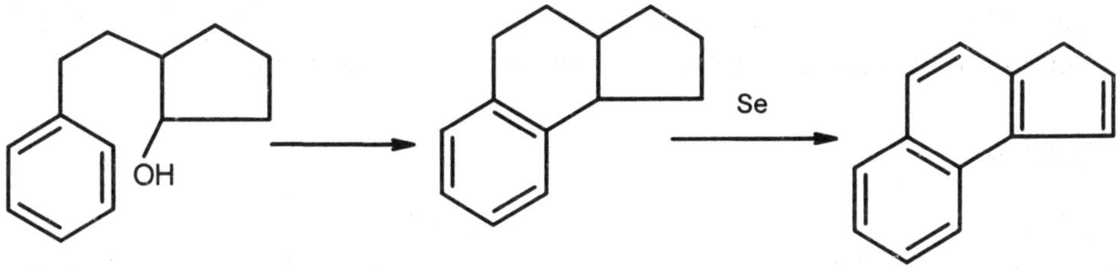

References

- Bardhan and Sengupta (1932), *J. Chem. Soc.*, 2520
- Kon (1933), *J. Chem. Soc.*, 1081.
- Renfrow, W. B., Renfrow, A., Shoun, E. and Sears, C. A. (1951), *J. Am. Chem. Soc.*, **73**, 317.

Bartoli Indole Synthesis

Bartoli synthesis offers an efficient and very simple method of synthesizing 7-substituted indoles on treatment of *ortho*-substituted nitrobenzenes with three equivalents of vinylmagnesium bromide.

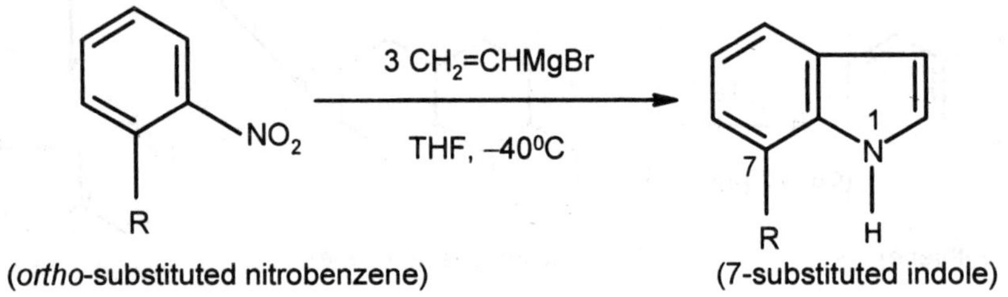

The process works best when the *ortho*- substitution in the nitrobenzene derivative is large. It is supposed that initial attack by the vinyl Grignard is at the nitro group oxygen with subsequent elimination of magnesium enolate yielding the nitroso equivalent of the starting compound; thus it seems that this step is encouraged by non-planarity of the nitro group facilitated by the bulky *ortho*-substituent.

Mechanism

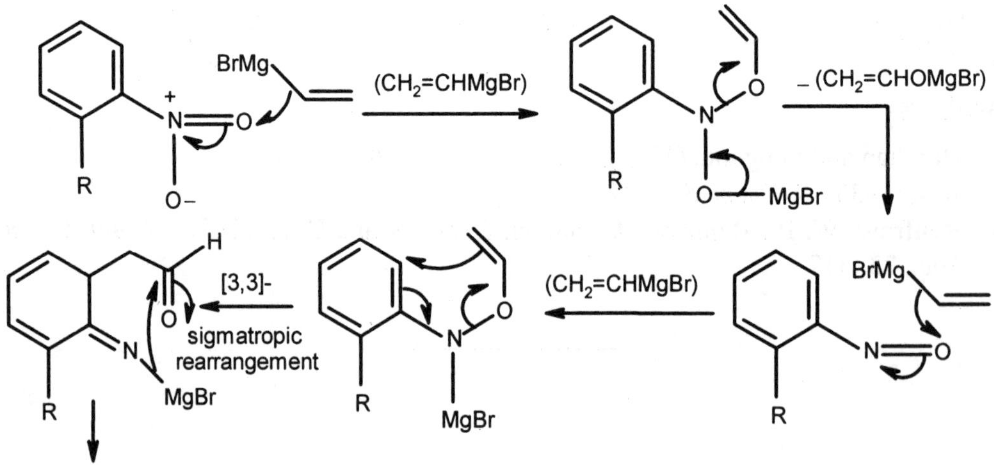

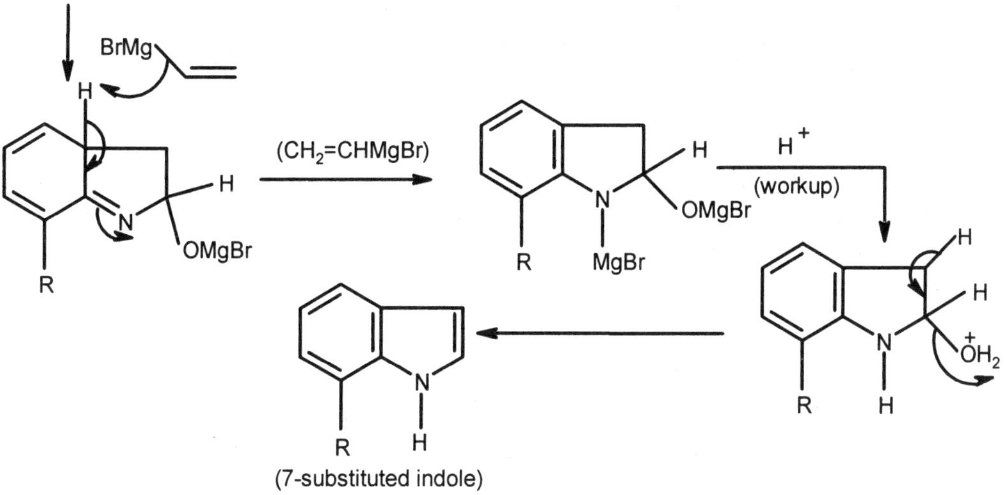

(7-substituted indole)

References

- Bartoli, G., Leardini, R., Medici, A. and Rosini, G. (1978), *J. Chem. Soc., Perkin Trans 1*, 892.
- Bartoli, G., Bosco, M., Dalpozzo, R. and Todesco, P. E. (1988), *J. Chem. Soc., Chem. Commun.*, 807.
- Bartoli, G., Palmieri, G., Bosco, M. and Dalpozzo, R. (1989), *Tetrahedron Lett.*, **30**, 2129.
- Bosco, M., Dalpozzo, R., Bartoli, G., Palmieri, G. and Petrini, M. (1991), *J. Chem. Soc., Perkin Trans 2*, 657.
- Dodson, D., Todd, A. and Gilmore, J. (1991), *Synth. Commun.*, **21**, 611.
- Dodson, D., Gilmore, J. and Long, D. A. (1992), *Synlett*, 79.
- Dobbs, A. P., Voyle, M. and Whitall, N. (1999), *Synlett*, 1594.
- Dobbs, A. P. (2001), *J. Org. Chem.*, **66**, 638.
- Pirrung, M. C., Wedel, M. and Zhao, Y. (2002), *Synlett*, 143.
- Garg, N. K., Sarporg, R. and Stoltz, B. M. (2002), *J. Am. Chem. Soc.*, **124**, 13179.

Baylis – Hillman Reaction

Baylis-Hillman reaction offers a method for carbon–carbon bond formation between α-carbon of a conjugated carbonyl system and an aldehydic carbon in presence of a suitable base (as catalyst) such as DABCO (1,4-diazabicyclo[2.2.2]octane) or trialkylphosphins. Here is a typical example:

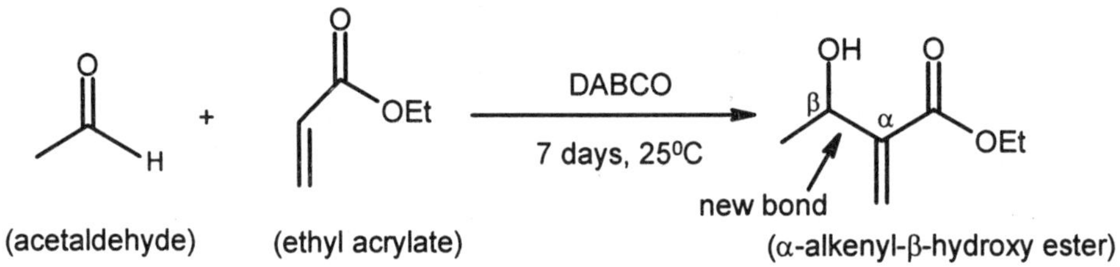

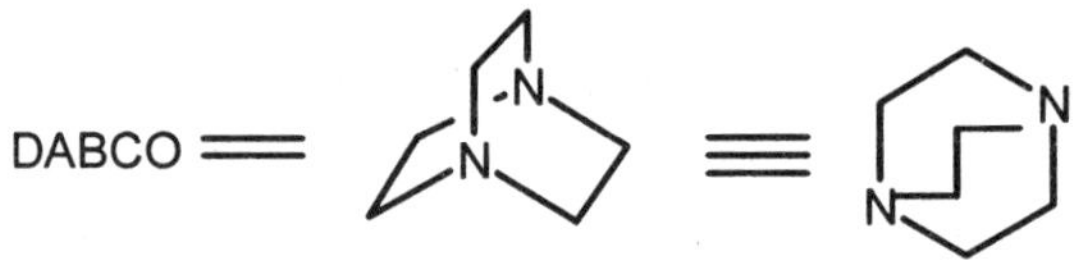

The reaction may be extended to a number of substrates, and a general scheme is shown:

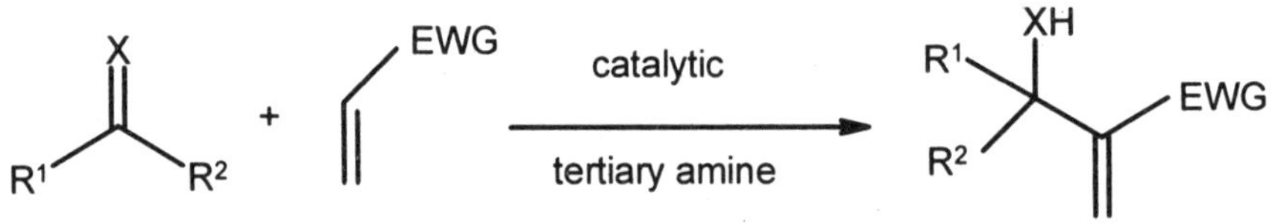

X=O, NR$_2$
EWG= COOR, COR, CHO, CN, SOOR, SO$_3$R, CONR$_2$, PO(OEt)$_2$

Mechanism

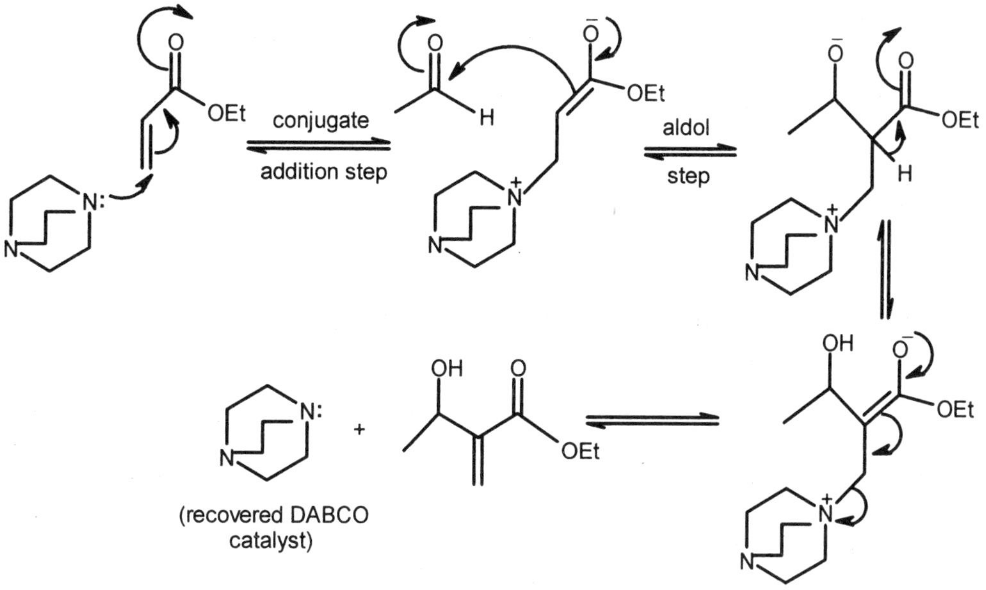

[Mechanism scheme showing DABCO-catalyzed conjugate addition to ethyl acrylate, aldol step with aldehyde, proton transfer, and elimination to regenerate DABCO catalyst and give the Baylis–Hillman product (ethyl 2-(1-hydroxyethyl)acrylate).]

(recovered DABCO catalyst)

A disadvantage of this reaction is that the rate is low — several days reaction time are required. Under certain conditions (*e.g.* under pressure, microwave irradiation), rate enhancements have been observed. As a catalyst, DABCO is the best because it is a good nucleophile as well as a good leaving group — DABCO's combination of nucleophilicity and leaving group ability is best suited here.

References

- Baylis, A. B. and Hillman, M. E. D. (1972), *Ger. Pat., 2*, **155**, 113; *Chem. Abstr.*, 1972, **77**, 34174q.
- Drewes, S. E. and Roos, G. H. P. (1988), *Tetrahedron*, **44**, 4653.
- Basavaiah, D., Rao, P. D. and Hyma, R. S. (1996), *Tetrahedron*, **52**, 8001. [Review]
- Ciganek, E. (1997), *Org. React.*, **51**, 201. [Review]
- Rafel, S. and Leahy, J. W. (1997), *J. Org. Chem.*, **62**, 1521.
- Brzezinski, L. S., Rafel, S. and Leahy, J. W. (1997), *J. Am. Chem. Soc.*, **119**, 4317.
- Auge, J., Lubin, N. and Lubineau, A. (1994), *Tetrahedron Lett.*, **35**, 7947.
- Kundu, M. K., Mukherjee, S. B., Balu, N., Padmakumar, R. and Bhat, S. V. (1994), *Synlett*, 444.

- Shi, M. and Feng, Y.–S.(2001), *J. Org. Chem.*, **66**, 406.
- Yu, C. and Hu, L. (2002), *J. Org. Chem.*, **67**, 219.
- Shi, M., Li, C.–Q. and Jiang, J.–K. (2003), *Tetrahedron*, **59**, 1181.

Blanc - Quelet Chloromethylation Reaction

Replacement of a hydrogen atom in an aromatic hydrocarbon by chloromethyl (–CH$_2$Cl) group on treatment with formaldehyde [generated from *trioxan (metaformaldehyde)*, (CH$_2$O)$_3$] and hydrogen chloride usually in the presence of a catalyst such as zinc chloride, aluminium chloride, stannic chloride or sulphuric acid, is known as *Blanc-Quelet chloromethylation reaction*.

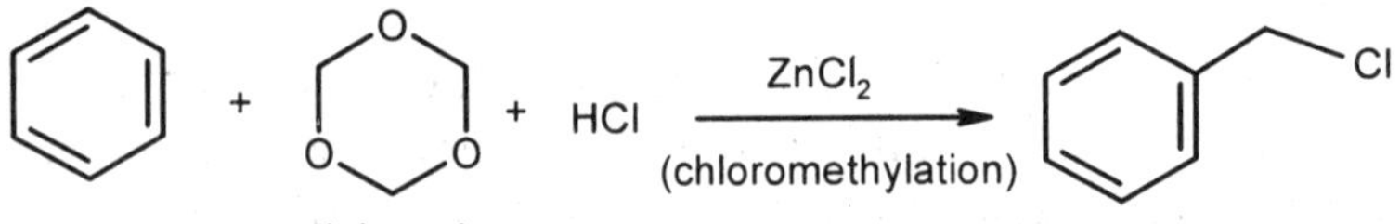

Mechanism

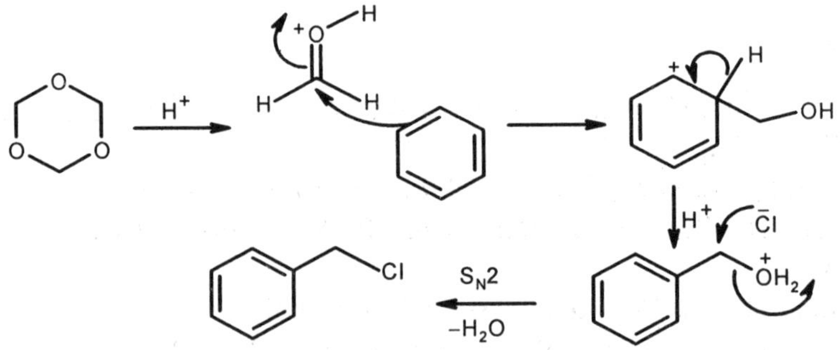

References

- Blanc, G. (1923), *Bull. Soc. Chim. Fr.*, **33**, 313.
- Fuson, R. C. and McKeever, C. H. (1942), *Org. React.*, **1**, 63. [Review]
- Sekine, Y. and Boekelheide, V. (1981), *J. Am. Chem. Soc.*, **103**, 1777.

- Mallory, F. B., Rodolph, M. J. and Oh, S. M.(1989), *J. Org. Chem.*, **54**, 4619.
- Tashiro, M., Tsuge, A., Sawada, T., Makishima, T., Horie, S., Arimura, T., Mataka, S. and Yamato, T. (1990), *J. Org. Chem.*, **55**, 2404.
- Ito, K., Ohba, Y., Shinagawa, E., Nakayama, S., Takahashi, S., Honda, K., Nagafuji, H., Suzuki, A. and Sone, T. J. (2000), *J. Hetercyclic Chem.*, **37**, 1479.
- Qiao, K. and Deng, Y.-Q. (2003), *Huaxue Xuebao*, **61**, 133.

Boord Reaction

B*oord reaction* offers a method for the synthesis of olefins from β-halo ethers upon simultaneous removal of the alkoxide group and halogen on treatment with zinc.

$$\underset{\underset{X\quad OR}{|\quad |}}{-\overset{|}{C}-\overset{|}{C}-} \xrightarrow{\text{Zn dust}} -\overset{|}{C}=\overset{|}{C}- \quad\text{(olefin)}$$

(X=Br, I)

Mechanism

$$\underset{R^1}{\overset{Br}{\diagup}}\underset{OEt}{\overset{}{\diagdown}}R^2 \xrightarrow[\text{(oxidative addition)}]{Zn} \underset{R^1}{\overset{ZnBr}{\diagup}}\underset{OEt}{\overset{}{\diagdown}}R^2 \xrightarrow[\text{elimination}]{\text{reductive}} R^1CH=CHR^2$$

Besides zinc, magnesium, sodium, or certain other reagents may also be used. The reaction may be extended to the compounds having general formula:

$$X-\overset{|}{\underset{|}{C}}-\overset{|}{\underset{|}{C}}-Z\text{, where } X = Br, I, \text{ and } Z = OCOR, OTs, NR_2, SR$$

β-Halo acetals readily yield vinylic ethers:

$$X-\overset{|}{\underset{|}{C}}-\overset{|}{\underset{|}{C}}(OR)_2 \xrightarrow{Zn} -\overset{|}{C}=\overset{|}{C}(OR)$$

References

- Swallen, L. C. and Boord, C. E. (1930), *J. Am. Chem. Soc.*, **52**, 651.
- Schmitt Claude G. and Boord, C. E. (1931), *J. Am. Chem. Soc.*, **53**, 2427.
- Amstutz, E. D. (1944), *J. Org. Chem.*, **9**, 310.
- House, H. O. and Ro, R. S. (1958), *J. Am. Chem. Soc.*, **80**, 332.
- Cristol, S. J. and Rademacher, L. E. (1959), *J. Am. Chem. Soc.*, **81**, 1600.
- Gurien, H. (1963), *J. Org. Chem.*, **28**, 878.
- Bruck, P. (1970), *J. Org. Chem.*, **35**, 2222.
- Reeve, W., Brown, R. and Steckel, T. F. (1971), *J. Am. Chem. Soc.*, **93**, 4607.
- Larock, R. C., *Comprehensive Organic Transformations*, VCH:NY, 1989, p.136.
- Abramovitch, R. A. and Bulman, A. (1992), *Synlett*, 795.
- Lin, G. and Zhang, A. (2000), *Tetrahedron*, **56**, 7163.
- Rebeiro, G. L. and Khadikar, B. M. (2001), *Synthesis*, 370.
- Bremner, J. B., Coates, J. A., Keller, P. A., Pyne, S. G. and Witchard, H. M. (2002), *Synlett*, 219.

Bouveault Aldehyde Synthesis

Alkyl or aryl halides can be formylated to homologous aldehydes by their transformation to the corresponding organometallic reagent, followed by the addition to dialkylformamide. The reaction is called *Bouveault aldehyde synthesis*.

$$R-X \xrightarrow[\text{iii. } H_3O^+]{\text{i. M; ii. DMF}} R-CHO$$

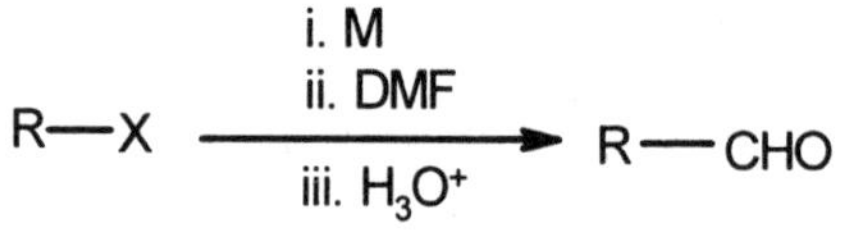

Mechanism

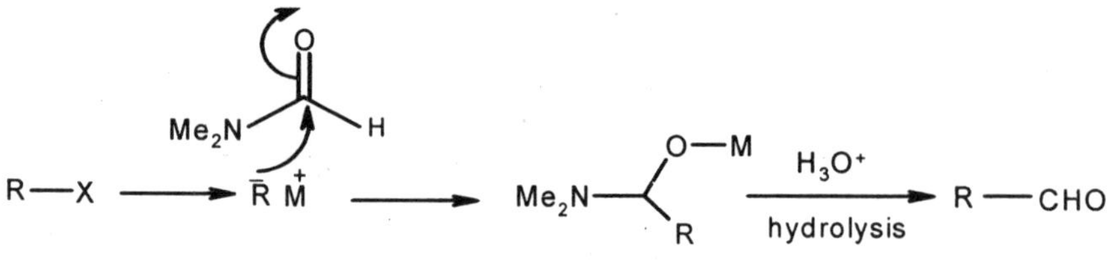

The reaction was modified by using Grignard reagents that react with the disubstituted formamides to yield aldehydes upon subsequent acidic hydrolysis.

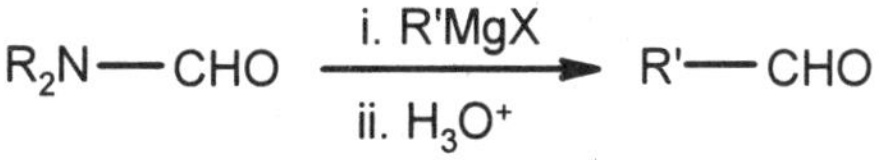

$$R_2N-CHO \xrightarrow[\text{ii. } H_3O^+]{\text{i. R'MgX}} R'-CHO$$

References

- Bouveault, L. (1904), *Bull. Soc. Chim. Fr.*, **31**, 1306.
- Smith, L. I. And Baylis, M. (1941), *J. Org. Chem.*, **6**, 437, 489.
- Spialter, L. and Pappalardo, J. A., *The Acyclic Aliphatic Tertiary Amines*, Macmillan : NY, 1965, p. 59.
- Petrier,, C., Gemal, A. L. and Luche, J. L. (1982), *Tertahedron Lett.*, **23**, 3361.
- Comins, D. L. and Brown, J. D. (1984), *J. Org. Chem.*, **49**, 1078.
- Einhorn, J. and Luche, J. L. (1986), *Tetrahedron Lett.*, **27**, 1791.
- Denton, S. M. and Wood, A. (1999), *Synlett*, 55.

Brook Rearrangement

When compounds having an OH and a silyl group on the same carbon atom are treated with a catalytic amount of base, migration of the silyl group from carbon to oxygen takes place; the reaction is called as the *Brook rearrangement*.

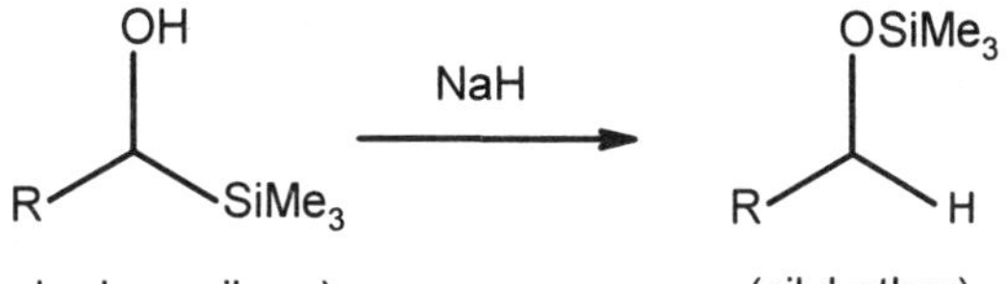

R-C(OH)(SiMe_3) →[NaH] R-CH(OSiMe_3)
(α-hydroxysilane) (silyl ether)

The reaction may be regarded as a nucleophilic substitution at silicon; the substitution goes through a three-membered cyclic intermediate bearing a pentacovalent silicon, where a linear arrangement of nucleophile and leaving group is not required.

Mechanism

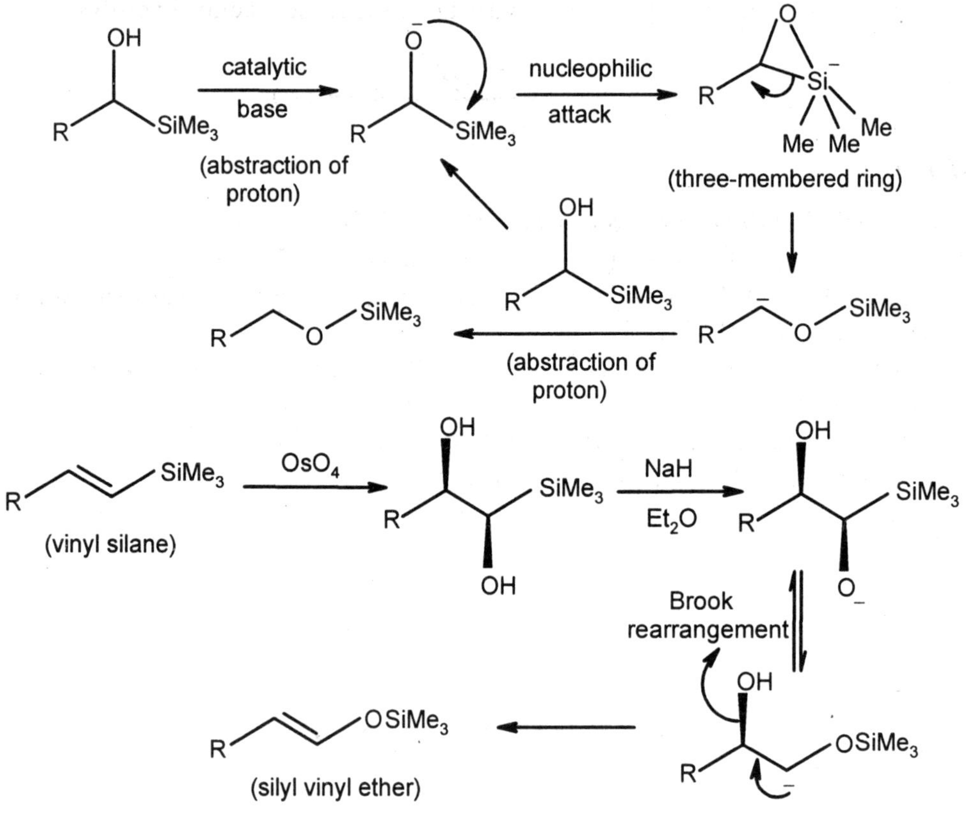

[Mechanism scheme showing the Brook rearrangement: an α-hydroxysilane is deprotonated by catalytic base, the alkoxide undergoes nucleophilic attack on silicon forming a three-membered ring intermediate, which opens to give a carbanion stabilized as a silyl ether; proton abstraction completes the rearrangement.]

[Second scheme: A vinyl silane R–CH=CH–SiMe₃ is dihydroxylated with OsO₄ to give the diol, which upon treatment with NaH/Et₂O undergoes Brook rearrangement to give the silyl vinyl ether R–CH=CH–OSiMe₃.]

References

- Brook, A. G. (1958), *J. Am. Chem. Soc.*, **80**, 1886.
- Brook, A. G. (1974), *Acc. Chem. Res.*, **7**, 77. [Review]
- Page, P. C. B., Klair, S. S. and Rosenthal, S. (1990), *Chem. Soc. Rev.*, **19**, 147. [Review]
- Fleming, I. And Ghosh, U. (1994), *J. Chem. Soc., Perkin Trans. 1*, 257.
- Takeda, K., Takeda, K. and Ohnishi, Y. (2000), *Tetrahedron Lett.*, **41**, 4169.
- Moser, W. H. (2001), *Tetrahedron*, **57**, 2065.
- Takeda, K., Sawada, Y. and Sumi, K. (2002), *Org. Lett.*, **4**, 1031.

Bucherer Reaction

B*ucherer reaction* offers a method for the conversion of naphthols into naphthylamines and *vice versa* using aqueous solution of a sulphite or bisulphite. Naphthols are converted into corresponding amines in presence of ammonia; primary amines can be used instead of ammonia, in which case *N*-substituted naphthylamines are formed.

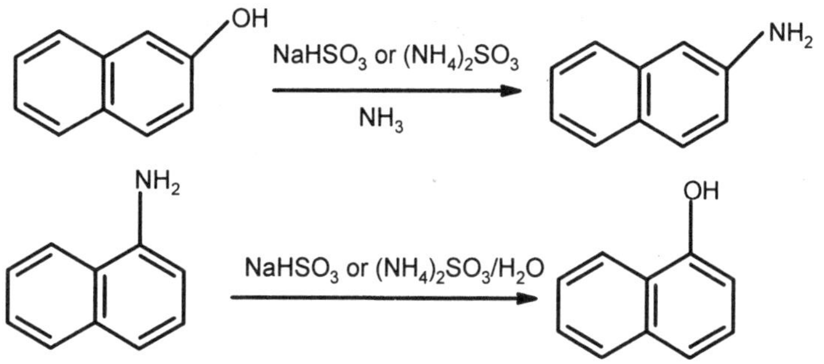

Mechanism

The meachanistic path of *Bucherer reaction* involves an overall addition-elimination:

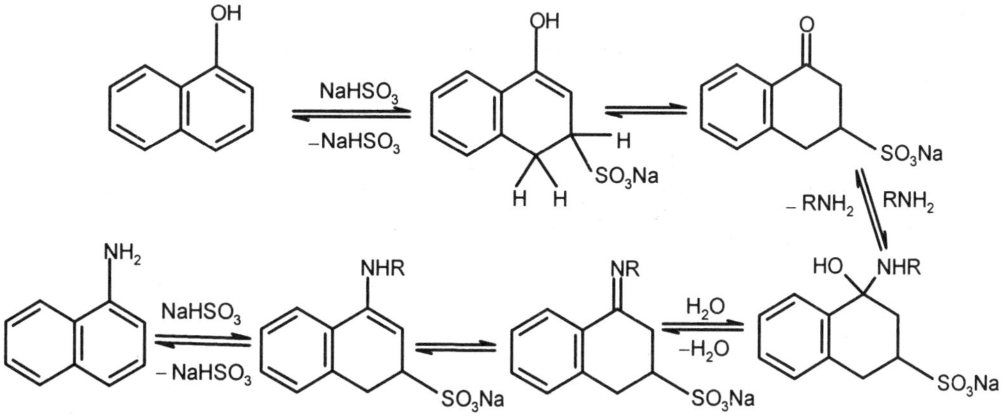

The reaction can also be employed in the quinoline and isoquinoline series.

References

- Bucherer, H. T. (1904), *J. Prakt. Chem.*, **69**, 49.
- Kozlov, V. V. and Veselovskaia, I. K. (1958), *J. Gen. Chem. USSR*, **28**, 3359.
- Rieche, A. and Seeboth, H. (1960), *Liebigs Ann. Chem.*, **638**, 66.
- Rieche, A. and Seeboth, H. (1960), *Liebigs Ann. Chem.*, **638**, 43, 57.
- Gilbert, E. E., *Sulfonation and Related Reactions,* Wiley : NY, 1965, p. 166.
- Seeboth, H. (1967), *Angew. Chem. Int. Ed. Engl.*, **6**, 307.
- Belica, P. S. and Manchand, P. S. (1990), *Synthesis*, 539.
- Singer, R. A. and Buchwald, S. L. (1999), *Tetrahedron Lett.*, **40**, 1095.
- Canete, A., Melendrez, M. X., Saitz, C. and Zanocco, A. L. (2001), *Synth. Commun.*, **31**, 2143.

Carroll Rearrangement

An allylic alcohol and a β-ketoester react under base-catalyzed condition to yield γ-keto-olefin; the reaction is said to be the *Carroll rearrangement*.

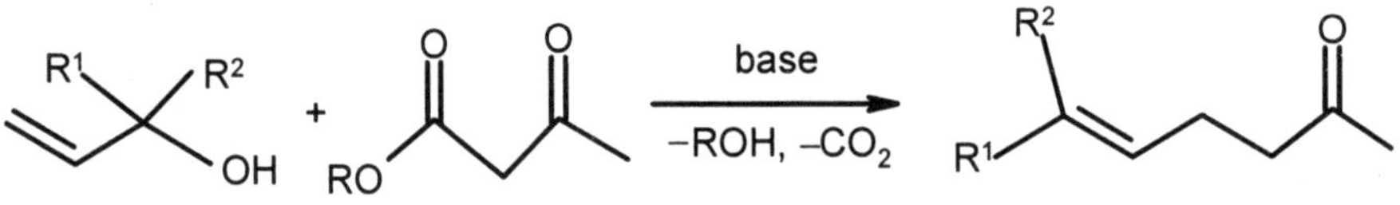

Mechanism

The reaction involves *anion-assisted Clasien reaction* as an intermediate step.

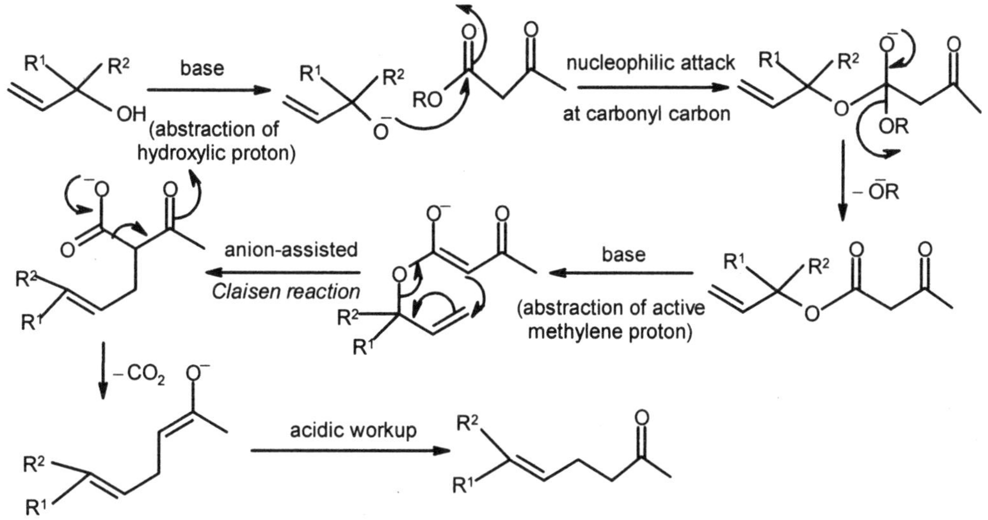

Pseduionone can also be obtained in a similar reaction:

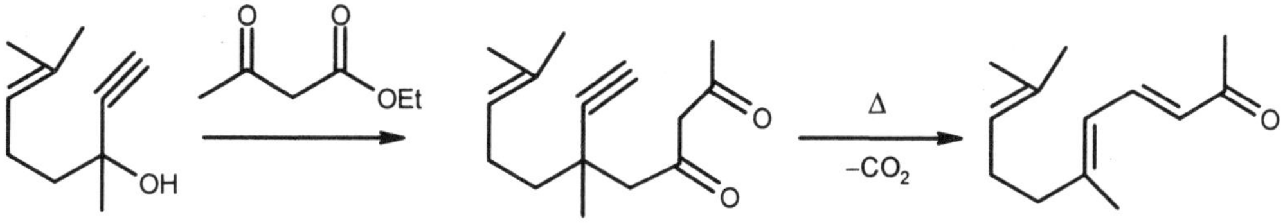

References

- Carroll, M. F. (1940), *J. Chem. Soc.*, 704.
- Carroll, M. F. (1941), *J. Chem. Soc.*, 507.
- Kimel, W. and Cope, A. C. (1943), *J. Am. Chem. Soc.* **65**, 1992.
- Kimel *et al.* (1957) *J. Org. Chem.*, **23**, 153.
- Hoffmann, W. (1973), *Chem. Ztg.*, **97**, 23.
- Gilbert, J. C. and Kelly, T. A. (1988), *Tetrahedron*, **44**, 7587.
- Teissere, P. J., *Chemistry of Fragrant Substances*, VCH Publishers Inc.: New York, 1994,.
- Enders, D., Knopp, M., Runsink, J. and Raabe, G. (1996), *Justus Liebigs Ann. Chem.*, 1095.
- Hatcher, M. A. and Posner, G. H. (2002), *Tetrahedron Lett.*, **43**, 5009.

Chapman Rearrangement

Thermal aryl migration of aryliminoethers leading to the formation of *N,N*-diarylamides is called the *Chapman rearrangement*.

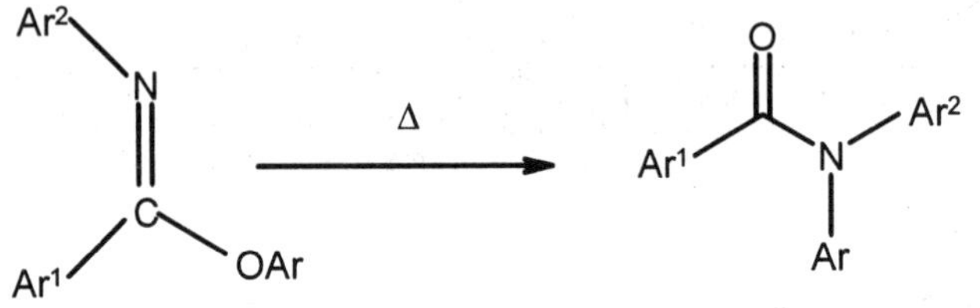

Diarylamines can easily be prepared by using this technique; *N,N*-diarylamides as formed produce the corresponding diarylamines on hydrolysis.

Though the reaction may occur without the presence of any solvent, it has been observed that best yields are obtained when the reaction is carried out in refluxing tetraethylene glycol dimethyl ether (tetraglyme). Aryl moieties may bear substituents such as alkyl, halogen, OR, CN, COOR, *etc*. Presence of electron-withdrawing substituents in the migrating aryl moiety enhances the reactivity, while presence of such groups in any of the other two or in both usually decreases the rate.

Mechanism

The mechanism of the reaction is supposed to be an intramolecular aromatic nucleophilic substitution involving a 1,3-shift of aryl group from oxygen to nitrogen *via* oxazete intermediate.

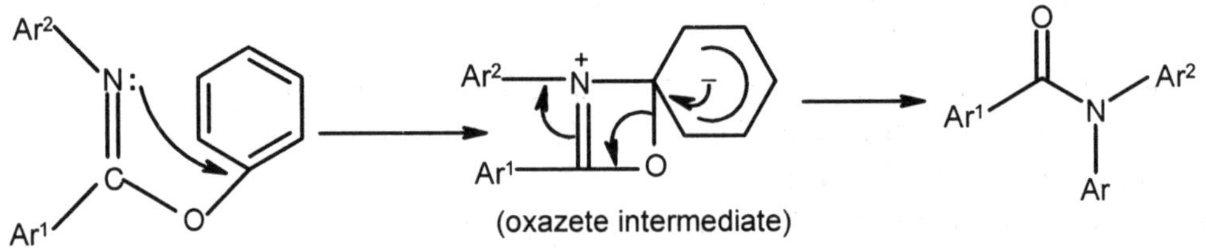

(oxazete intermediate)

References

- Chapman, A. W. (1925), *J. Chem. Soc.*, **127**, 1992.
- Wiberg, K. B. and Rowland, B. I. (1955), *J. Am. Chem. Soc.*, **77**, 2205.
- Schulenberg, J. W. and Archer, S. (1965), *Org. React.*, **14**, 1. [Review]

- Wheeler, O. H., Roman, F. and Rosado, O. (1969), *J. Org. Chem.*, **34**, 966.
- McCarty, C. G. and Garner, L. A. in Patai *The Chemsitry of Amidines and Imidates*, Wiley : NY, 1975, p. 189.
- Kimura, M. (1987), *J. Chem. Soc., Perkin Trans. 2*, 205.
- Dessolin, M., Eisenstein, O., Golfier, M., Prange, T. and Sautet, P. (1992), *J. Chem. Soc., Chem Commun.*, 132.

Chugaev Reaction

Thermal decomposition of xanthates prepared from alcohols involving stereospecific elimination (*cis*-elimination) to yield alkenes is called the *Chugaev reaction*. The reaction is advantageous because it requires relatively lower temperature and leads to the predominant formation of unrearranged terminal alkenes.

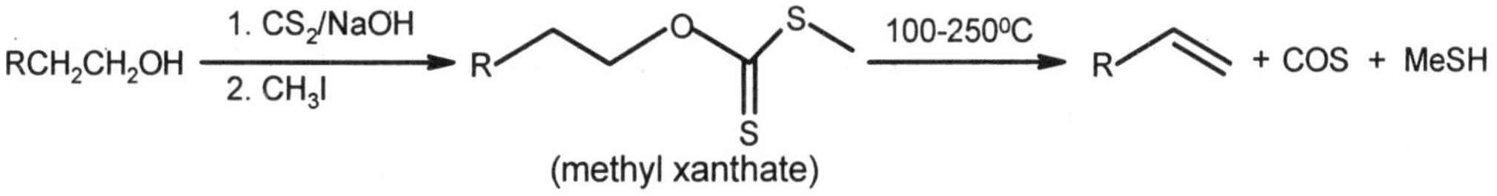

$$RCH_2CH_2OH \xrightarrow[\text{2. CH}_3\text{I}]{\text{1. CS}_2/\text{NaOH}} R\text{-CH}_2\text{CH}_2\text{-O-C(=S)-SMe (methyl xanthate)} \xrightarrow{100\text{-}250°C} R\text{-CH=CH}_2 + COS + MeSH$$

Mechanism

The reaction involves a six-membered cyclic transition state, and proceeds through E_i mechanistic pathway:

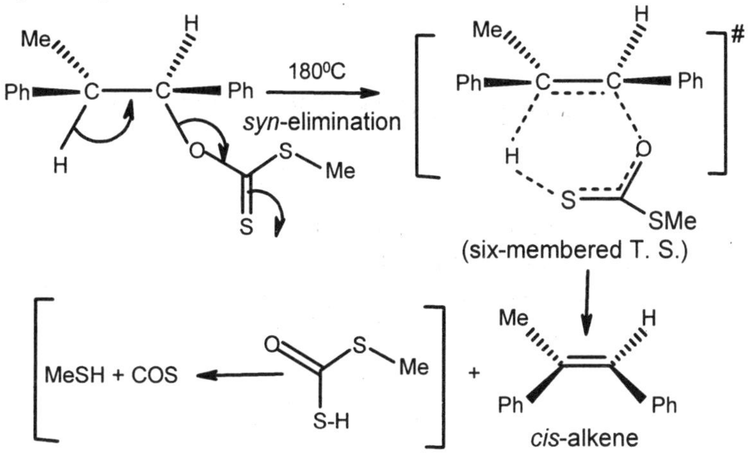

[Mechanism diagram showing *syn*-elimination at 180°C via six-membered T.S., producing *cis*-alkene + MeSH + COS]

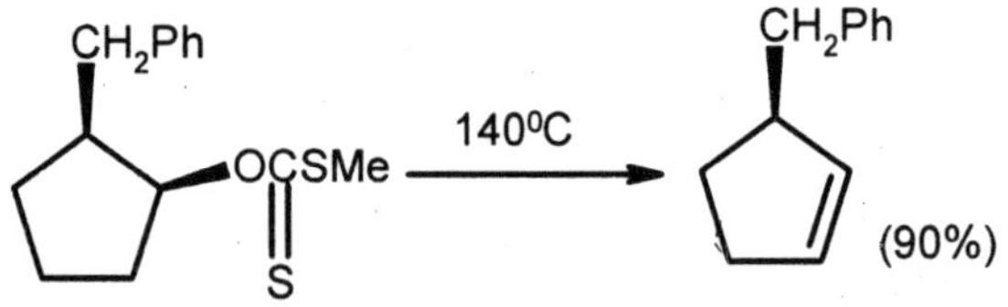

References

- Chugaev, L. (1899), *Ber. Dtsch. Chem. Ges.*, **32**, 3332.
- O'Connor, G. L. and Nace, H. R. (1953), *J. Am. Chem. Soc.*, **75**, 218.
- Bader, R. F. W. and Bourns, A. N. (1961), *Can. J. Chem.*, **39**, 348.
- Nace, H. R. (1962), *Org. React.*, **12**. 57. [Review]
- McNamara, L. S. and Price, C. C. (1962), *J. Org. Chem.*, **27**, 1230.
- Chande, M. S. and Pranjpe, S. D. (1973), *Indian J. Chem.*, **11**, 1206.
- Kim, M. and White, J. D. (1975), *J. Am. Chem. Soc.*, **97**, 451.
- Harano, K. and Taguchi, T. (1975), *Chem. Pharm. Bull.*, **23**, 467.
- Cernigliano, G. and Kocienski, P. (1977), *J. Org. Chem.*, **42**, 3622.
- Lee, A. W. M., Chan, W. H., Wong, H. C. and Wong, M. S. (1989), *Synth. Commun.*, **19**, 547.
- Meulemans, T. M., Stork, G. A., Macaev, F. Z., Jansen, B. J. M. and de Groot, A. (1999), *J. Org. Chem.*, **64**, 9178.
- Nakagawa, H., Sugahara, T. and Ogasawara, K. (2000), *Org. Lett.*, **2**, 3181.
- Nakagawa, H., Sugahara, T. and Ogasawara, K. (2001), *Tetrahedron Lett.*, **42**, 4523.

Corey – Kim Oxidation

Oxidation of alcohols to the corresponding aldehydes or ketones by means of *N*-chlorosuccinamide (NCS) and dimethylsulphide (DMS) followed by treatment with a base (usually NEt_3) is known as *Corey-Kim oxidation*.

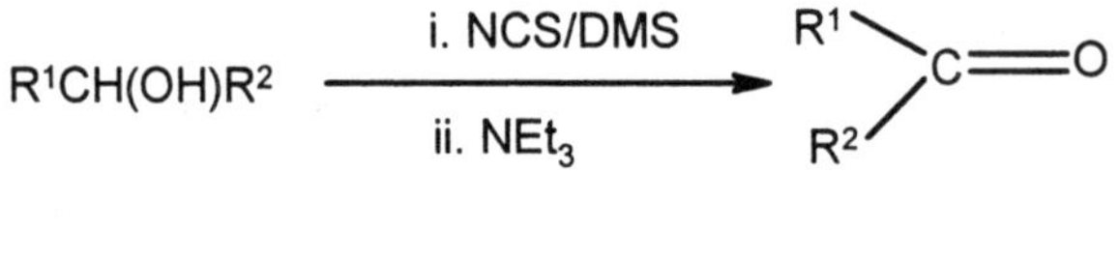

Mechanism

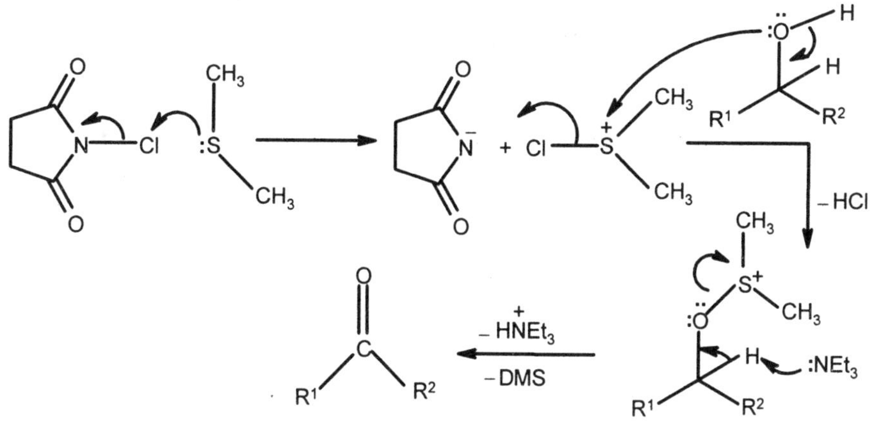

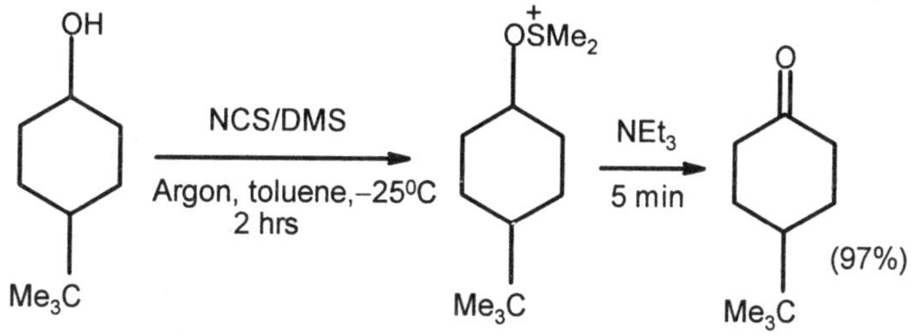

[Mechanism scheme showing NCS reacting with dimethyl sulfide to form a chlorosulfonium intermediate, which reacts with an alcohol R¹R²CH-OH to give an alkoxysulfonium ion (with loss of HCl), followed by deprotonation by NEt₃ to yield the ketone/aldehyde with loss of HNEt₃⁺ and DMS.]

[Example: 4-tert-butylcyclohexanol → alkoxysulfonium intermediate → 4-tert-butylcyclohexanone]

OH → (NCS/DMS, Argon, toluene, –25 °C, 2 hrs) → O$\overset{+}{S}$Me₂ → (NEt₃, 5 min) → ketone (97%)

If the base is not applied in the second step, the reaction provides a technique for converting an alcohol to an alkyl chloride.

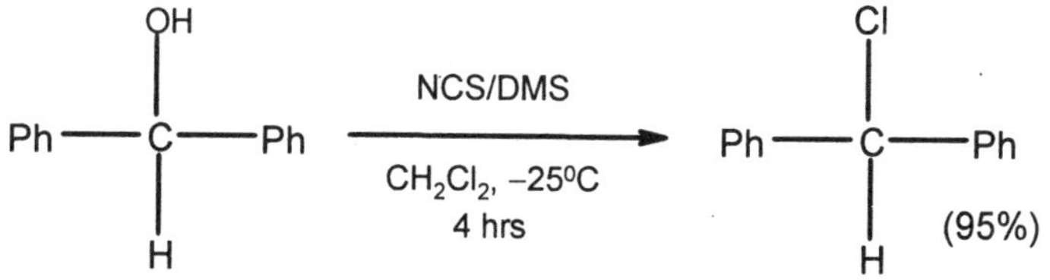

Ph–C(OH)(H)–Ph → (NCS/DMS, CH₂Cl₂, –25 °C, 4 hrs) → Ph–C(Cl)(H)–Ph (95%)

References

- Corey, E. J. and Kim, C. U. (1972), *J. Am. Chem. Soc.*, **94**, 7586.
- Corey, E. J., Kim, C. U. and Takeda, M. (1972), *Tetrahedron Lett.*, 4339.
- Tamura, Y., Chen, L. C., Fujita, M., Kiyokawa, H. and Kita, Y.(1979), *Chem. Ind.*, 668.
- Katayama, S., Fukada, K., Watanabe, T. and Yamauchi, M.(1988), *Synthesis*, 178.
- Shapiro, G. and Lavi, Y. (1990), *Heterocycles*, **31**, 2099.
- Pulkkinen, J. T. and Vepsalainen, J. J. (1996), *J. Org. Chem.*, **61**, 8604.
- Crich, D. and Neelamkavil, S. (2002), *Tetrahedron Lett.*, **58**, 3865.

Corey – Winter Reaction

Olefination reaction of 1,2-diols carried out by sequential treatment with 1,1'-thiocarbonyldiimidazole (TCDI) and trimetylphosphite is known as the *Corey-Winter reaction*. Cyclic thionocarbamate formed in the first step is cleaved to alkene.

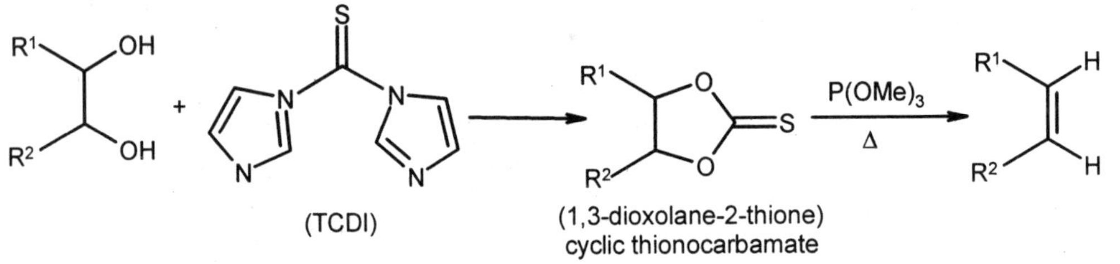

TCDI was originally suggested by Corey, but other reagents such as thiophosgene and 4-dimethylaminopyridine (DMAP) can also be used.

Mechanism

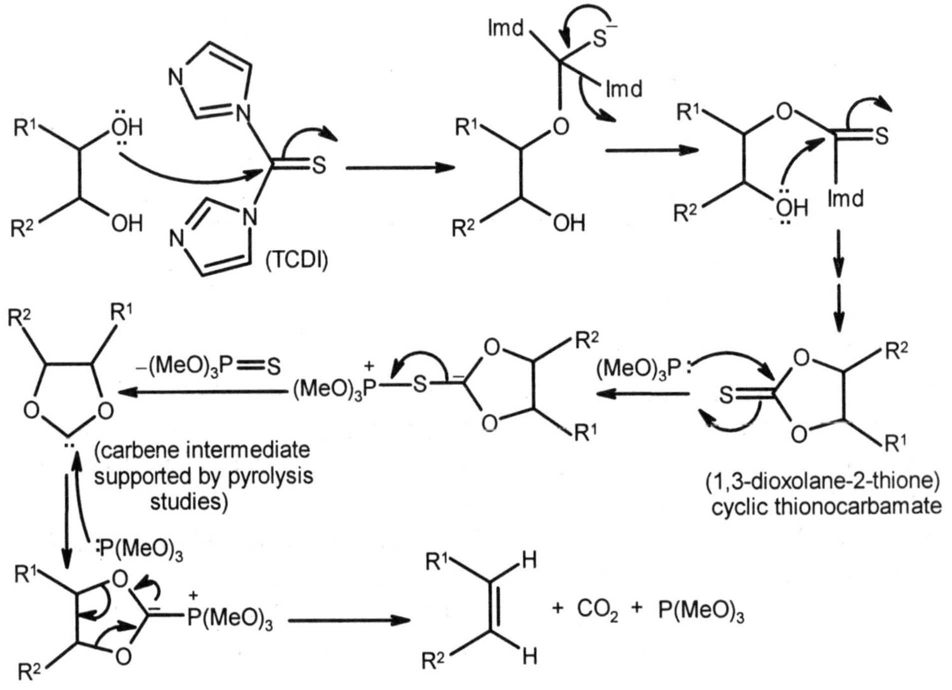

The elimination is *syn*, and the product becomes sterically controlled.

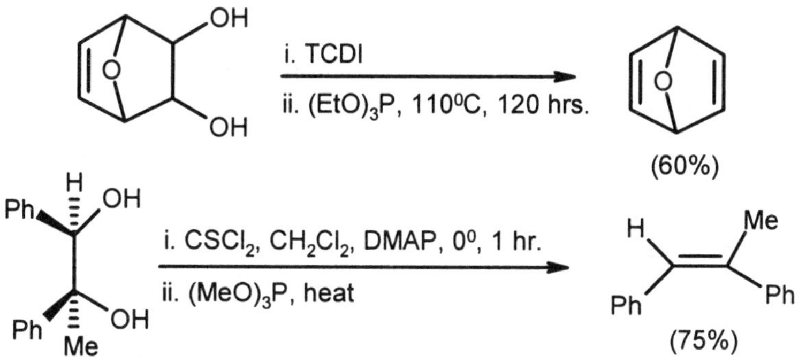

References

- Corey, E. J. and Winter, R. A. E. (1963), *J. Am. Chem. Soc.*, **85**, 2677.
- Corey, E. J., Carey, F. A. and Winter, R. A. E. (1965), *J. Am. Chem. Soc.*, **87**, 934.
- Corey, E. J. (1967), *Pure Appl. Chem.*, **14**, 19.
- Horton, D. and Tindall, C. G., Jr. (1970), *J. Org. Chem.*, **35**, 3558.
- Hartmann, W., Fischler, H. M. and Heine, H. G. (1972), *Tetrahedron Lett.*, 853.
- Davis, J., Trantz, V. and Erhardt, U. (1972), *Tetrahedron Lett.*, 4435.
- McGahren, W. J., Ellestad, G. A., Morton, G. O., Kunstmann, M. P. and Mullen, P. (1973), *J. Org. Chem.*, **38**, 3542.
- Prinzbach, H. and Babsh, H. (1975), *Angew Chem. Int. Ed. Engl.*, **14**, 753.
- Corey, E. J. and Hopkins, P. B. (1982), *Tetrahedron Lett.*, 1979.
- Sonnet, P. E. (1980), *Tetrahedron*, **36**, 557. [Review]
- Block, E. (1984), *Org. React.*, **30**, 457. [Review]
- Crich, D., Pavlovic, A. B. and Wink, D. J. (1999), *Synth. Commun.*, **29**, 359.
- Palomo, C., Oiarbide, M., Landa, A., Esnal, A. and Linden, A. (2001), *J. Org. Chem.*, **66**, 4180.

de Mayo Reaction

de Mayo reaction is a [2 + 2]-photocycloaddition between the enol form of 1,3-diketones and olefins followed by a cyclobutane-cleavage.

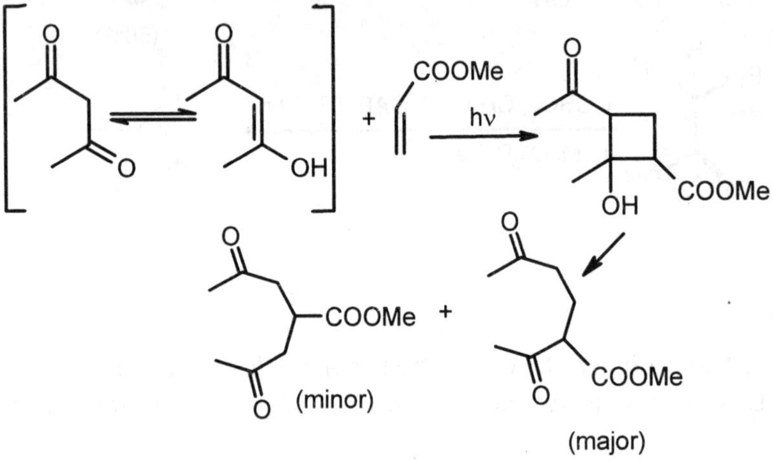

Mechanism

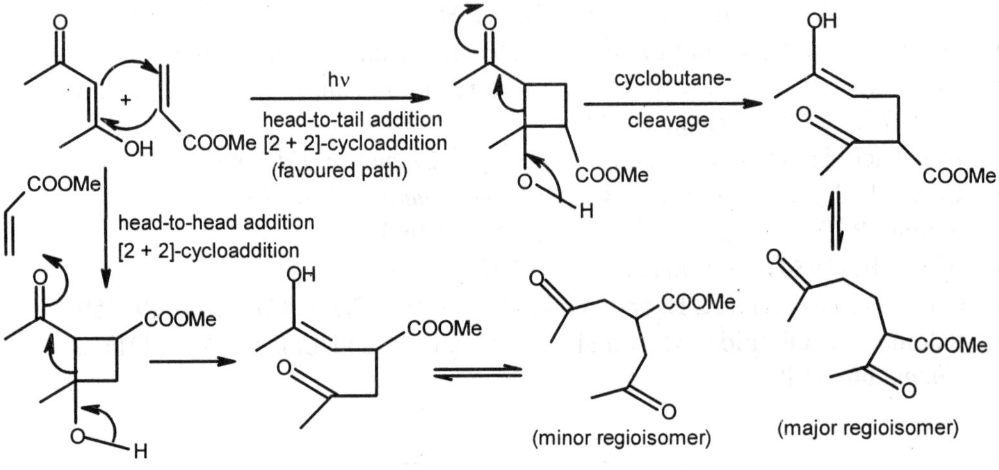

References

- de Mayo, P., Takeshita, H. and Sattar, A. B. M. A. (1962), *Proc. Chem. Soc., London*, 119.

- de Mayo, P. (1971), *Acc. Chem. Res.*, **4**, 49. [Review]
- Pearlman, B. A. (1979), *J. Am. Chem. Soc.*, **101**, 6398.
- Disanayaka, B. W. and Weedon, A. C. (1987), *J. Org. Chem.*, **52**, 2905.
- Sato, M., Sunami, S., Kogawa, T. and Kaneko, C. (1994), *Chem. Lett.*, 2191.
- Quevillon, T. M. and Weedon, A. C. (1996), *Tetrahedron Lett.*, **37**, 3939.

Demjanov Rearrangement

Rearrangement reaction of cabocations formed by diazotization of primary amines leading to the formation of rearranged alcohols is termed as *Demjanov rearrangement.* Thus, *n*-propylamine on treatment with nitrous acid yields minor amount of *n*-propyl alcohol together with a major proportion of isopropyl alcohol.

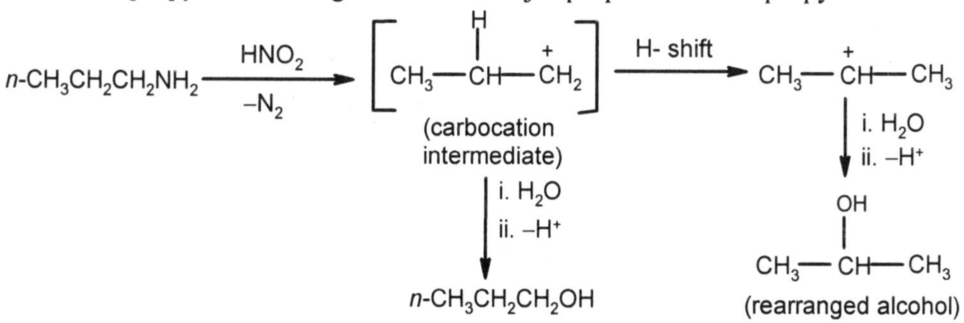

This reaction makes the basis of *Demjanov's method* of contracting and expanding alicyclic ring systems. Ring contraction takes place when a positive charge is formed on an alicyclic carbon, and ring expansion occurs when the positive charge is placed on a carbon α to an alicyclic ring.

(ring expansion and contraction products)

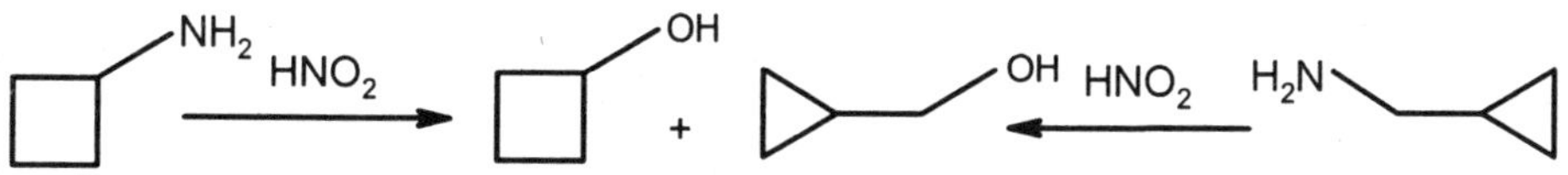

Mechanism

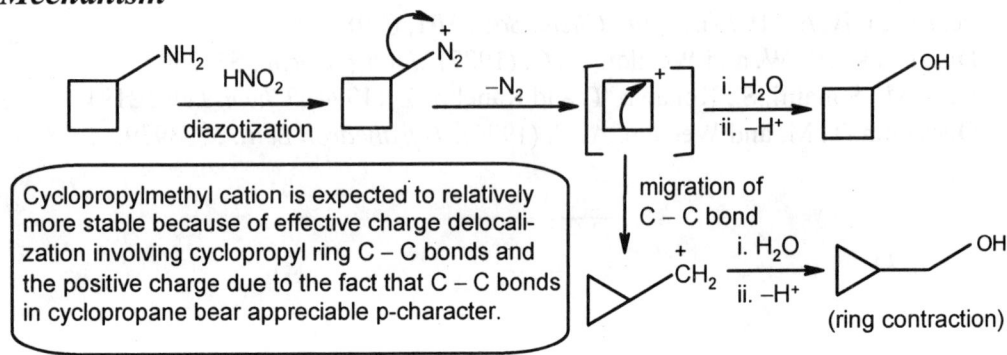

Cyclopropylmethyl cation is expected to relatively more stable because of effective charge delocalization involving cyclopropyl ring C – C bonds and the positive charge due to the fact that C – C bonds in cyclopropane bear appreciable p-character.

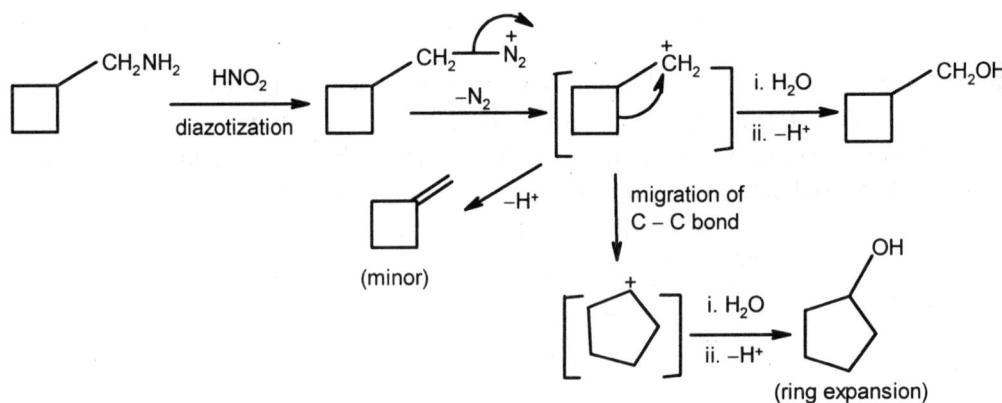

References

- Demjanov, N. J. and Lushnikov, M. (1903), *J. Russ. Phys. Chem. Soc.*, **35**, 26.
- Smith, P. A. S. and Baer, D. R.(1960), *Org. React.*, **11**, 157. [Review]
- Kotani, R. (1965), *J. Org. Chem.*, **30**, 350.
- Diamond, J., Bruce, W. F. and Tyson, F. T. (1965), *J. Org. Chem.*, **30**, 1840.
- Nakazaki, M., Naemura, K. and Hashimoto, M. (1983), *J. Org. Chem.*, **48**, 2289.
- Wong, H. N. C., Hon, M., Tse, C., Yip, Y., Tanko, J. and Hudlicky, T. (1989), *Chem. Rev.*, **89**, 165. [Review]
- Fattori, D., Henry, S. and Vogel, P. (1993), *Tetrahedron*, **49**, 1649.
- Wiberg, K. B., Shobe, D. and Nelson, G. C. (1993), *J. Am. Chem. Soc.*, **115**, 10645.
- Boeckman, R. K. (1999), *Org. Synth.*, **77**, 141.

Dienone - Phenol Rearrangement

***D**ienone-phenol rearrangement* involves 1,2-migration of an alkyl group of a 4,4-dialkyl cyclohexadienone under acid-catalyzed condition leading to the formation of 3,4-disubstituted phenol; it is quite a type of *retro* pinacol reaction.

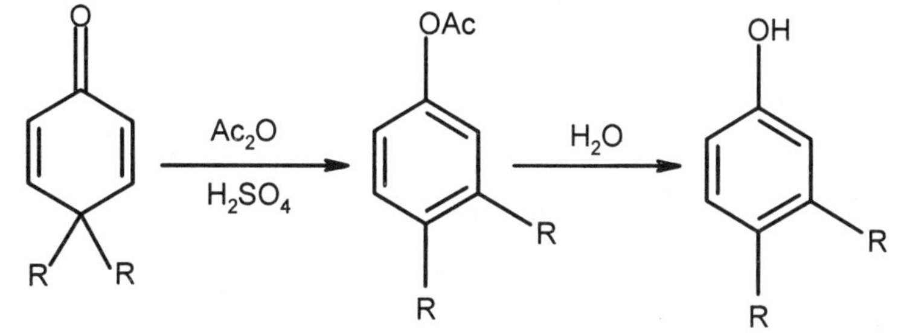

Mechanism

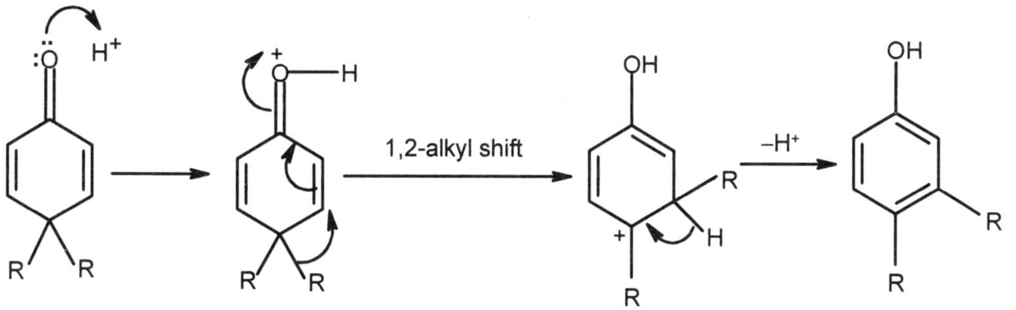

The driving force of the transformation is an outcome of the process of aromatization in the final product.

When one of the alkyl groups becomes the part of the cyclic system, either the alkyl or the ring methylene moiety may migrate — the actual course of migration depends upon the structural or the electronic as well as on the reaction condition.

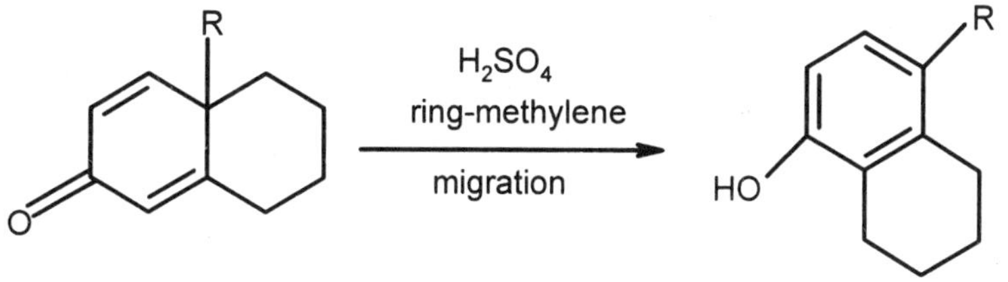

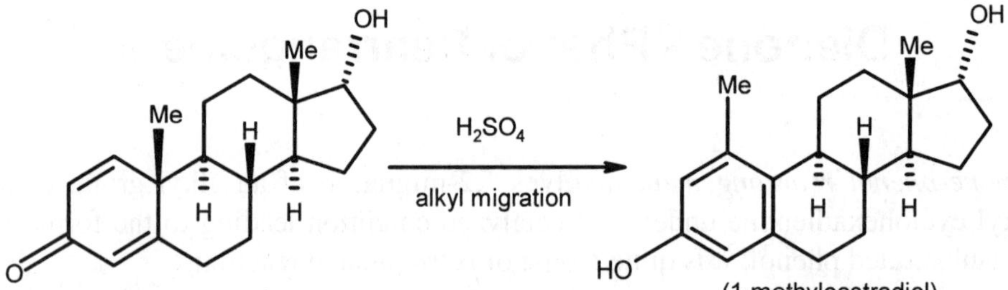

References

- Shine, H. J. in *Aromatic Rearrangements*, Elsevier: NY, 1967, pp. 55-68.
- Schultz, A. G. and Hardinger, S. A. (1991), *J. Org. Chem.*, **56**, 1105.
- Schultz, A. G. and Green, N. J. (1991), *J. Am. Chem. Soc.*, **114**, 1824.
- Frimer, A. A., Marks, V., Sprecher, M. and Gilinsky-Sharon, P. (1994), *J. Org. Chem.*, **59**, 1831.
- Banerjee, A. K., Castillo-Melendez, J. A., Vera, W., Azocar, J. A. and Laya, M. S. (2000), *J. Chem. Res. (S)*, 324.
- Zimmermann, H. E. and Cirkva, V. (2001), *J. Org. Chem.*, **66**, 1839.

Eglinton Reaction

Eglinton reaction offers a method for producing symmetrical diynes by oxidative coupling of terminal alkynes on heating with stoichiometric amount of cuprous acetate in pyridine or a similar base.

$$2\ R-C\equiv C-H \xrightarrow[\text{pyridine}]{Cu(OAc)_2} R-C\equiv C-C\equiv C-R$$

The oxidation is specific for triple-bond hydrogen; hence other functional groups, if present in the starting alkyne, remain unchanged.

Mechanism

The first step of the reaction involves the formation of acetylide ion, and the last step is probably the coupling of two radicals:

$$R-C\equiv C-H \xrightarrow[-C_5H_5NH^+]{\text{pyridine}} R-C\equiv C^- \xrightarrow[\text{(oxidation)}]{Cu(OAc)_2} R-C\equiv C\cdot \text{ (free radical)}$$

$$R-C\equiv C-C\equiv C-R \xleftarrow{\text{coupling between two radicals}} R-C\equiv C\cdot$$

References

- Eglinton, G. and Galbraith, A. R. (1956), *Chem. Ind.*, 737.
- Eglinton, G. and McRae, W. (1963), *Adv. Org. Chem.*, **4**, 225. [Review]
- Clifford, A. A. and Waters, W. A. (1963), *J. Chem. Soc.*, 3056.
- Fedenok, L. G., Berdnikov, V. M. and Shvartsberg, M. S. (1973), *J. Org. Chem. USSR*, **9**, 1806.
- Altmann, M., Friedrich, J., Beer, F., Reuter, R., Enkelmann, V. and Bunz, U. H. F. (1997), *J. Am. Chem. Soc.*, **119**, 1427.
- Srinivasan, R., Devan, B., Shanmugam, P. and Rajagopalan, K. (1997), *Indian J. Chem, Sec. B*, **36B**, 123.
- Nakanishi, H., Sumi, N., Aso, Y. and Otsubo, T. (1998), *J. Org. Chem.*, **63**, 8632.
- Siemsen, P., Livingston, R. C. and Diederich, F. (2000), *Angew. Chem. Int. Ed. Engl.*, **39**, 2632. [Review]
- Inouchi, K., Kabashi, S., Takimiya, K., Aso, Y. and Otsubo, T. (2002), *Org. Lett.*, **4**, 2533.

Elbs Persulphate Oxidation

Elbs persulphate oxidation is undergone by monohydric phenols leading to their oxidation to dihydric phenols by alkaline potassium persulphate. Hydroxylation usually takes palce in the *para*-position with respect to that already present; but if this position is blocked, the *ortho*-position is favoured.

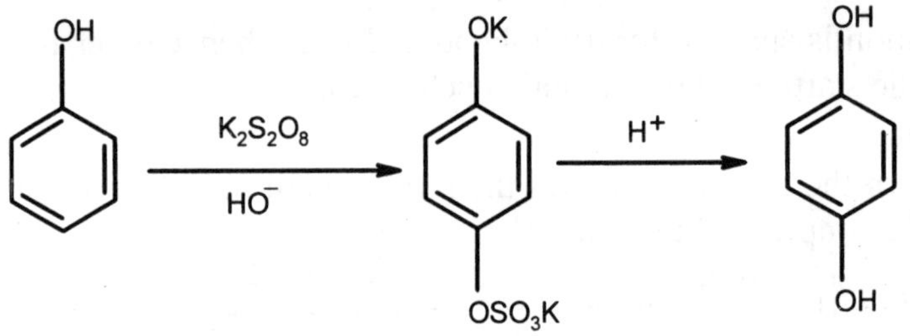

Mechanism

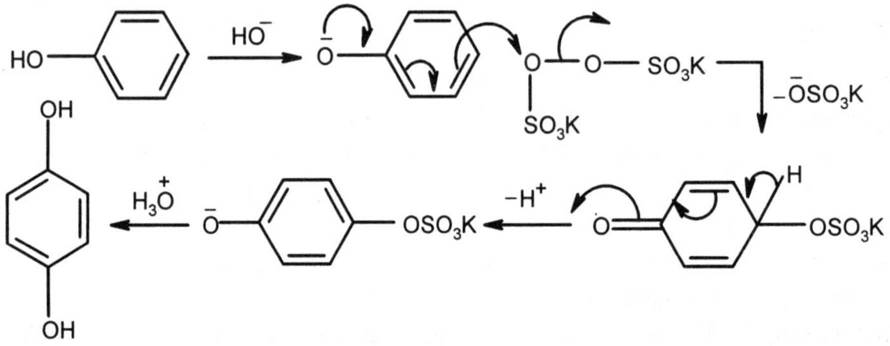

References

- Elbs, K. (1993), *J. Prakt. Chem.* **48**, 179.
- Sethna, S. M. (1951), *Chem. Rev.*, **49**, 91.

- Capdevielle, P. and Maumy, M. (1982), *Tetrahedron Lett.*, **23**, 1573, 1577.
- Behrman, E. J. (1988), *Org. React.*, **35**, 421. [Review]
- Watson, K. G. and Serban, A. (1995), *Aust. J. Chem.* **48**, 1503.

Fischer – Hepp Rearrangement

When *N*-nitroso derivatives of secondary aromatic amines are treated with acid (preferably by hydrochloric acid), migration of the nitroso group occurs usually at the *para*-position yielding the corresponding *para*-nitroso secondary aromatic amines. The reaction, known as the *Fischer-Hepp rearrangement*, is of importance because *para*-nitroso secondary aromatic amines cannot generally be prepared directly by *C*-nitrosation of the parent compounds.

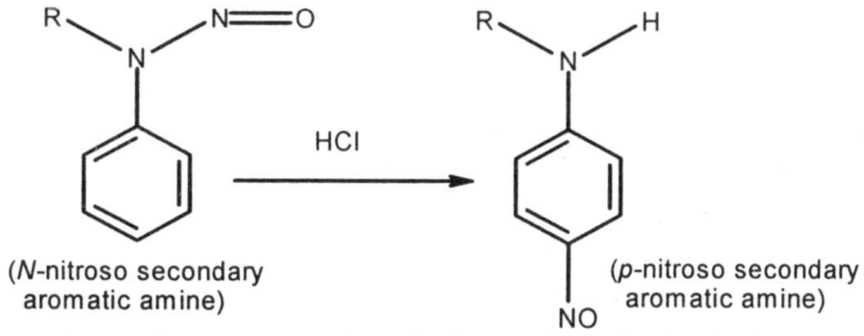

(*N*-nitroso secondary aromatic amine) → (*p*-nitroso secondary aromatic amine)

In benzene systems the *para*-product is formed exclusively; but in other aromatic systems where *para*-position is not free for attack, *ortho*-migration may also take place. For example:

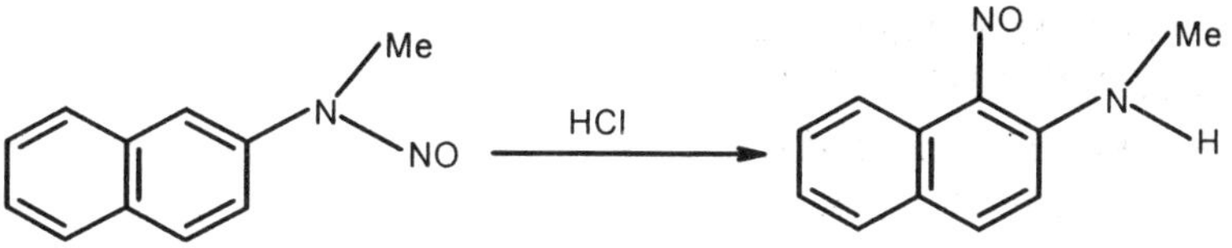

Mechanism

The mechanism of the reaction is not completely understood. However, it is intramolecular in nature as evidenced from the fact that a large excess of urea cannot freeze the reaction — since, if NO⁺, NOCl or some other similar species would become free during the reaction course, urea would capture them preventing the rearrangement.

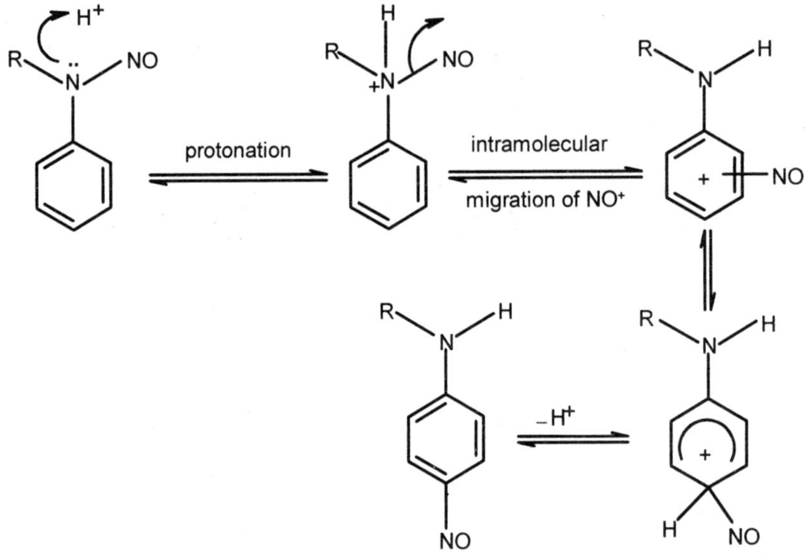

References

- Fischer, O. and Hepp, E. (1886), *Ber. Dtsch. Chem. Ges.*, **19**, 2991.
- Baliga, B. T. (1970), *J. Org. Chem.*, **35**, 2031.
- Belyaev, E. Yu, Kumarev, V. P. and Porai-Koshits, B. A. (1971), *Org. React. USSR*, **7**, 165.
- Morgan, T. D. B. and Williams, D. L. H. (1972), *J. Chem. Soc, Perkin Trans. 2*, 74.
- Williams, D. L. H. (1975), *Tetrahedron*, **31**, 1343.
- Williams, D. L. H. (1982), *J. Chem. Soc, Perkin Trans. 2*, 801.
- Williams, D. L. H. in Patai's *The Chemistry of Functional Groups, Supplement F, pt. 1*, Wiley: NY, 1982, p. 113.
- Lunn, G., Sansone, E. B. and Keefer, L. K. (1984), *J. Org. Chem.*, **49**, 3470.
- Titova, S. P., Arinich, A. K. and Gorelik, M. V. (1986), *J. Org. Chem. (USSR)*, **22**, 1407.
- Morris, P. I. (1999), *Chem. Ind.*, 968.

Friedlander Synthesis

Friedlander synthesis is a useful method for synthesizing quinoline and its derivatives. In this method quinolines can be prepared by condensing *ortho*-aminobenzaldehyde or *ortho*-acylarylamines with an aldehyde or ketone containing an active α-methylene group followed by ring closure in refluxing alcoholic sodium hydroxide solution. Acids are also effective catalyst for this reaction.

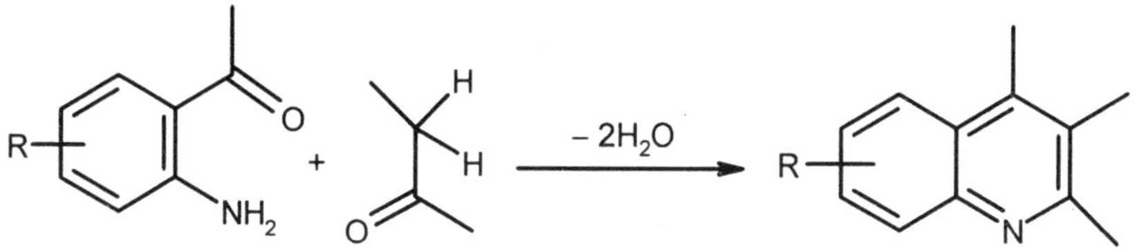

Mechanism

The orientation of condensation depends upon the conditions employed:

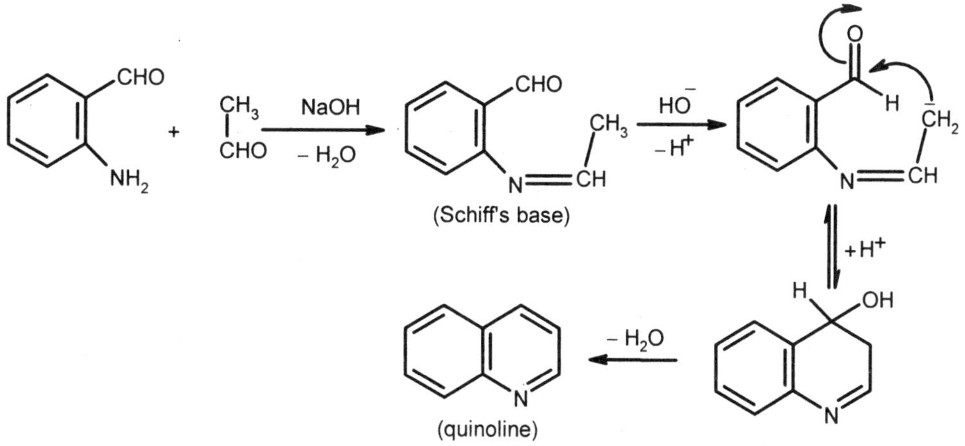

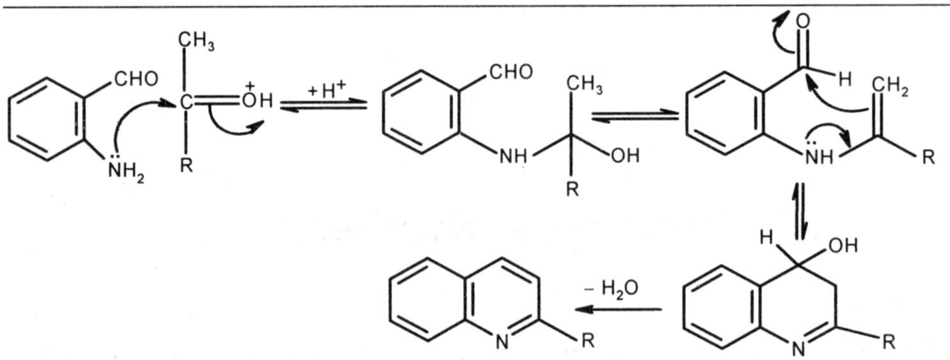

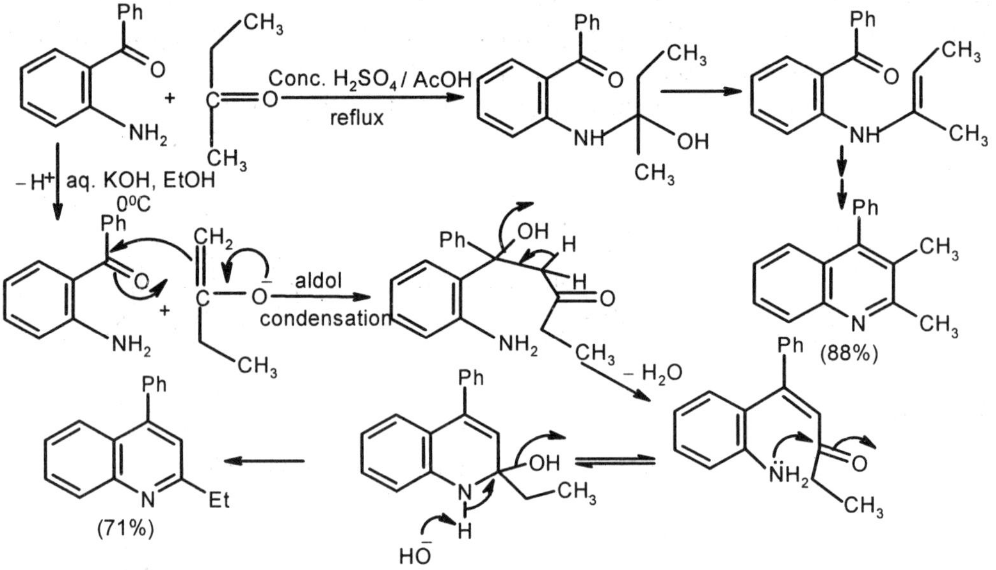

References

- Friedlander, P. and Ostermaier, H. (1881), *Ber. Dtsch. Chem. Ges.*, **14**, 1916.
- Friedlander, P. (1882), *Ber. Dtsch. Chem. Ges.*, **15**, 2572.
- Pfitzinger, W. (1886), *J. Prakt. Chem.*, **33**, 100.
- Manske, R. H. (1942), *Chem. Rev.*, **30**, 113. [Review]
- Fehnel, E. A. (1966), *J. Org. Chem.*, **31**, 2899.
- Rao, K. V. and Kuo, H.–S. (1979), *J. Heterocyclic Chem.*, **16**, 1241.
- Cheng, C.–C. and Yan, S.–J. (1982), *Org. React.*, **28**, 37. [Review]
- Katritzky, A. R., Ostercamp, D. L. and Yousaf, T. I. (1987), *Tetrahedron*, **43**, 5171. [Review]

- Okabe, M. and Sun, R.–C. (1995), *Tetrahedron*, **51**, 1861.
- Borger, D. L. and Chen, J.–H. (1995), *J. Org. Chem.*, **60**, 7369.
- Riesgo, E. C., Jin, X. and Thummel, R. P. (1996), *J. Org. Chem.*, **61**, 3017.
- Ubeda, J. I., Villacampa, M. and Avendano, C. (1998), *Synthesis*, 1176.
- Bu, X. and Deady, L. W. (1999), *Synth. Commun.*, **29**, 4223.
- Gladiali, S., Chelecci, G., Mudadu, M. S., Gastaut, M.–A. and Thummel, R. P. (2001), *J. Org. Chem.*, **66**, 400.
- Dormer, P. G., Eng, K. K., Farr, R. N., Humphrey, G. R., McWilliams, J. C., Reider, P. J., Sager, J. W. and Volante, R. P. (2003), *J. Org. Chem.*, **68**, 467.

Hoch-Campbell Ethyleneimine Synthesis

Reaction of ketoximes with Grignard reagents, followed by controlled hydrolysis of the organometallic complex formed, yields ethyleneimines (aziridines); this reaction is known as the *Hoch-Campbell ethyleneimine(or aziridine) synthesis*. The mechanistic scheme of the reaction is given as:

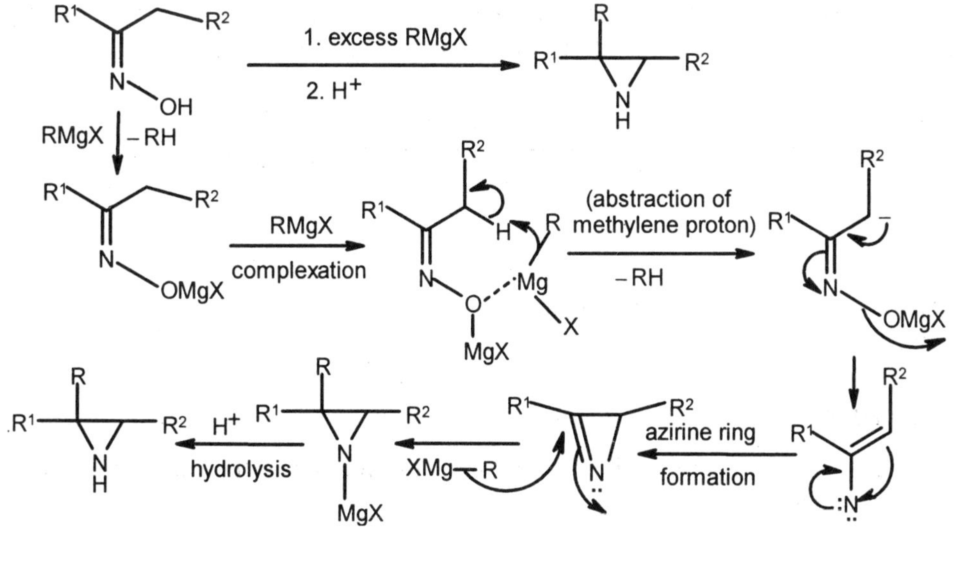

References

- Hoch, J. (1934), *Compt. Rend. Acad. Sci.*, **198**, 1865.
- Campbell, K. N. and McKenna, J. F. (1939), *J. Org. Chem.*, **4**, 198.
- Campbell, K. N., Campbell, B. K., Mckenna, J. F. and Chaput, E. P. (1943), *J. Org. Chem.*, **8**, 103.
- Campbell, K. N., Campbell, B. K., Hess, L. G. and Schaffner, I. J. (1944), *J. Org. Chem.*, **9**, 184.
- Dermer, O. C. and Ham, G. E., *Ethyleneimine and Other Aziridines*, Academic Press: New York, 1969, pp. 65-68. [Review]
- Freeman, J. P. (1973), *Chem. Rev.*, **73**, 283. [Review]
- Alvernhe, G. and Laurent, A. (1978), *J. Chem. Res. (S)*, 28.
- Laurent, A., Marsura, A. and Pierre, J.–L. (1980), *J. Heterocyclic Chem.*, **17**, 1009.

Hofmann-Loffler-Freytag Reaction

N-Haloamines, in which one alkyl at 4 or 5-position bears a hydrogen, on heating with sulphuric acid followed by treatment with a base produce pyrrolidines or piperidines; this reaction is called the *Hofmann-Loffler-Freytag reaction*.

$$\underset{R}{\overset{Cl}{N}}-(CH_2)_4-R' \xrightarrow[\text{2. HO}^-]{\text{1. Conc. H}_2\text{SO}_4/\text{heat}} \underset{\underset{R}{|}}{N}-R' \quad \text{(N-alkyl pyrrolidine derivative)}$$

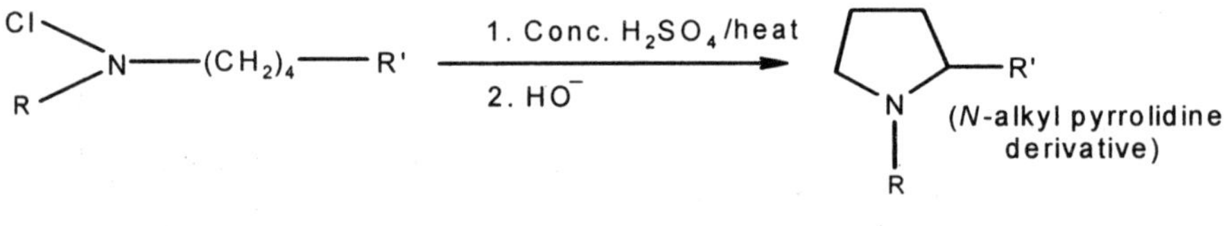

Photo-irradiation or use of chemical initiators (*e.g.* peroxides) can effect the transformation.

Mechanism

The mechanism is a free-radical process, with the key step involving internal hydrogen abstraction.

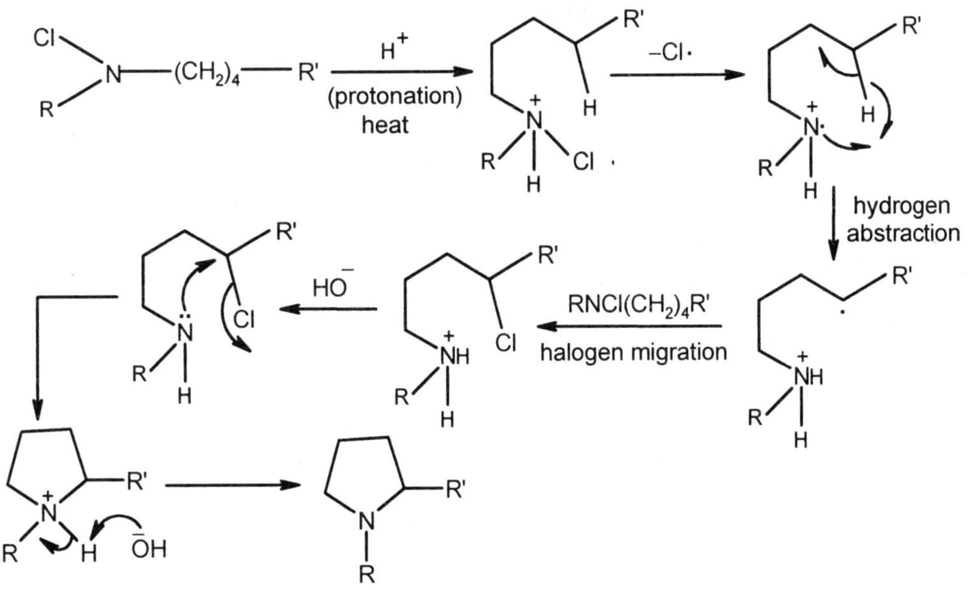

References

- Hofmann, A. (1879), *Ber. Dtsch. Chem. Ges.*, **12**, 984.
- Loffler, K. and Freytag, C. (1909), *Ber. Dtsch. Chem. Ges.*, **42**, 3727.
- Wawzonek, S. and Thelan, P. J. (1950), *J. Am. Chem. Soc.*, **72**, 2118.
- Corey, E. J. and Hertler, W. R. (1960), *J. Am. Chem. Soc.*, **82**, 1657.
- Wolff, M. E. (1963), *Chem. Rev.*, **63**, 55. [Review]
- Dupeyre, R. and Raṣsat, A. (1973), *Tetrahedron Lett.*, 2699.
- Deshpande, R. P. and Nayak, U. R. (1979), *Indian J. Chem.*, **17B**, 310.
- Stella, L. (1983), *Angew. Chem. Int. Ed. Engl.*, **22**, 337. [Review]
- Majetich, G. and Wheless, K. (1995), *Tetrahedron*, **51**, 7095.
- Togo, H. and Katohgi, M. (2001), *Synlett*, 565.

Jones Oxidation

Jones oxidation permits the oxidation of secondary alcohols to ketones rapidly and in high yield at room temperature or slightly above without appreciable oxidation or rearrangement of double or triple bonds that may already present and also without epimerizing adjacent chiral centre. The *Jones reagent* is usually prepared from CrO_3 and sulphuric acid, and oxidations are carried out in aqueous acetone. Primary alcohols on *Jones oxidation* would give aldehydes that become further oxidized to the corresponding carboxylic acids.

$$R-\underset{R'(H)}{\underset{|}{\overset{H}{\overset{|}{C}}}}-OH \xrightarrow[(acetone)]{CrO_3/H_2SO_4} R-\overset{O}{\overset{\|}{C}}-R'(H)$$

Mechanism

The mechanism involves the initial formation of a chromate ester that subsequently loses a proton and $HCrO_3^-$ ion to yield the carbonyl derivative.

$$R-\underset{R'(H)}{\underset{|}{\overset{H}{\overset{|}{C}}}}-OH \xrightarrow{\underset{"H_2CrO_4"}{CrO_3/H_2SO_4}} \underset{\text{(chromate ester)}}{R-\underset{R'(H)}{\underset{|}{\overset{H}{\overset{|}{C}}}}-O-\underset{O}{\underset{\|}{\overset{O}{\overset{\|}{Cr}}}}-OH} \xrightarrow[-H_3O^+]{-HCrO_3^-} R-\overset{O}{\overset{\|}{C}}-R'(H)$$

Aldehydes get further oxidized *via* their hydrates:

$$R-\overset{O}{\overset{\|}{C}}-H \xrightarrow{H_2O} \underset{\text{(hydrate)}}{R-\underset{OH}{\underset{|}{\overset{OH}{\overset{|}{C}}}}-H} \xrightarrow{\underset{"H_2CrO_4"}{CrO_3/H_2SO_4}} R-\underset{OH}{\underset{|}{\overset{H}{\overset{|}{C}}}}-O-\underset{O}{\underset{\|}{\overset{O}{\overset{\|}{Cr}}}}-OH \xrightarrow[-H_3O^+]{-HCrO_3^-} RCOOH$$

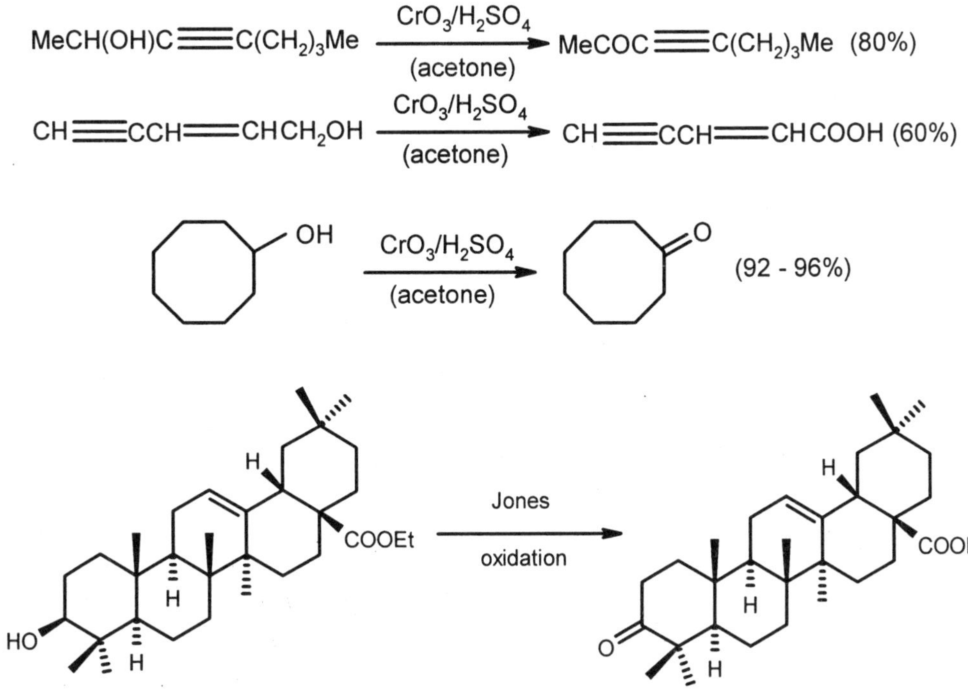

Jones Oxidation

MeCH(OH)C≡C(CH$_2$)$_3$Me →[CrO$_3$/H$_2$SO$_4$ (acetone)] MeCOC≡C(CH$_2$)$_3$Me (80%)

CH≡CCH=CHCH$_2$OH →[CrO$_3$/H$_2$SO$_4$ (acetone)] CH≡CCH=CHCOOH (60%)

cyclooctanol →[CrO$_3$/H$_2$SO$_4$ (acetone)] cyclooctanone (92 - 96%)

References

- Bowden, K., Heilbron, I. M., Jones, E. R. H. and Weedon, B. C. L. (1946), *J. Chem. Soc.*, 39.
- Bowers, A., Halsall, T. G., Jones, E. R. H. and Lemin, A. J. (1953), *J. Chem. Soc.*, 2548.
- Djerrasi, C., Hart, P. A. and Warawa, E. J. (1964), *J. Am. Chem. Soc.*, **86**, 78.
- Djerrasi, C., Engle, R. R. and Bowers, A. (1956), *J. Org. Chem.*, **21**, 1547.
- Eisenbraun, E. J. (1965), *Org. Synth.*, **45**, 28.
- Harding, K. E., May, L. M. and Dick, K. F. (1975), *J. Org. Chem.*, **40**, 1664.
- Grieco, P. A., Nishizawa, M., Oguri, T., Burke, S. D. and Marinovich, N. (1977), *J. Am. Chem. Soc.*, **99**, 5573
- Cacchi, S., La Torre, F. and Misiti, D. (1979), *Synthesis*, 356.
- Katoh, T. and Terashima, S. (1996), *Pure & Appl. Chem.*, **68**(3), 703.
- Gurjar, M. K. and Lalitha, S. V. S. (1998), *Pure & Appl. Chem.*, **70**(2), 303.
- Finlay, H. J., Honda, T. and Gribble, G. W. (2002), *ARKIVOC*, 38.

Koch Reaction

Koch reaction offers a method for hydrocarboxylating alkenes under acid-catalyzed condition. The reaction can be carried out in a number of ways —— an alkene is treated with carbon monoxide and water at elevated temperature (100-350^{0}C) under high pressure (500-1000 atm) or can also be carried out under milder condition (0-50^{0}C, 1-100 atm) if the alkene is first treated with carbon monoxide and catalyst, and then water is added.

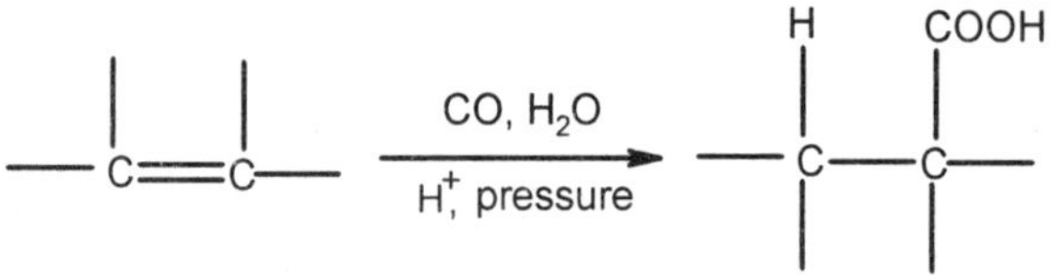

$$\underset{}{-\overset{|}{C}=\overset{|}{C}-} \xrightarrow[\text{H}^+,\text{ pressure}]{\text{CO, H}_2\text{O}} -\overset{H}{\underset{|}{C}}-\overset{COOH}{\underset{|}{C}}-$$

Mechanism

The reaction proceeds through carbocation intermediate, and hence the initially formed carbocation undergoes rearrangement to the most stable one during the reaction course. When the source of both the CO and water is formic acid, the reaction can be accomplished at room temperature and atmospheric pressure, and this procedure is called the *Koch-Haff reaction*. The reaction can also be applied to alcohols yielding trisubstituted acetic acids.

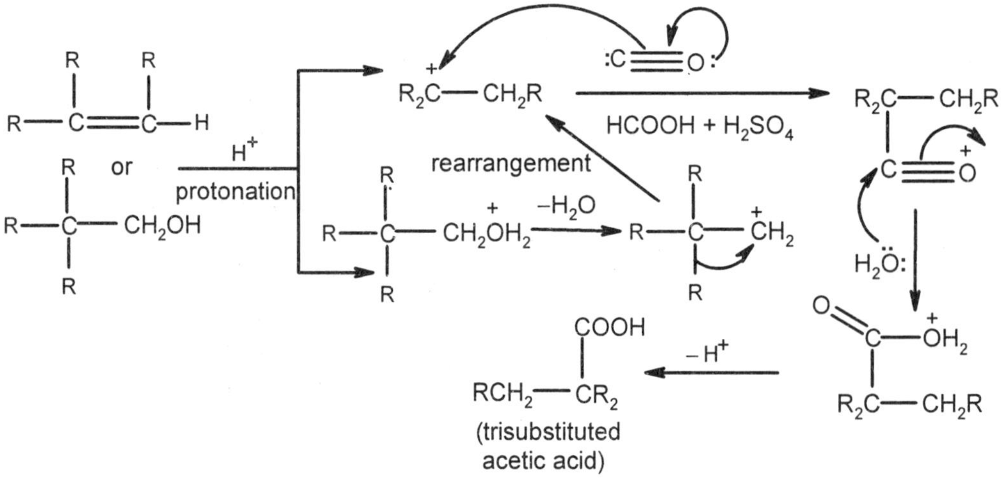

$$\text{Me}_2\text{C(OH)CH}_2\text{CH}_2\text{CH}_2\text{OH} \xrightarrow{\text{HCOOH + H}_2\text{SO}_4} \text{[lactone with gem-dimethyl]} \quad (94\%)$$

References

- Koch, H. and Haff, W. (1958), *Justus Liebigs Ann. Chem.*, **618**, 251.
- Haff, W. (1966), *Chem. Ber.*, **99**, 1149.
- Christol, H. and Solladie, G. (1966), *Bull. Soc. Chim. Fr.*, 1307.
- Norell, J. R. (1972), *J. Org. Chem.*, **37**, 1971.
- McQuillin, F. J. and Parker, D. G. (1974), *J. Chem. Soc., Perkin Trans. 1*, 809.
- Booth, B. and El-Fekky, T. A. (1979), *J. Chem. Soc., Perkin Trans. 1*, 2441.
- Bahrmann, H. in Falbe's *New Syntheses with Carbon Monoxide*, Springer : NY, 1980, p. 372.
- Takahashi, Y., Yoneda, N. and Nagai, H. (1982), *Chem. Lett.*, 1187.
- Farooq, O., Marcelli, M., Prakash, G. K. S. and Olah, G. A. (1988), *J. Am. Chem. Soc.*, **110**, 864.
- Lapidus, A. L. and Pirozhkov, S. D. (1989), *Russ. Chem. Rev.*, **58**, 117. [Review]
- Olah, G. A., Prakasg, G. K. S., Methew, T. and Marinez, E. R. (2000), *Angew. Chem. Int. Ed. Engl.*, **39**, 2547.
- Tsumori, N., Xu, Q., Souma, Y. and Mori, H. (2002), *J. Mol. Catalysis*, **179**, 271.

Kolbe Electrolysis Reaction

Kolbe electrolysis reaction relates to the synthesis of hydrocarbons by means of electrolysis of aqueous solutions of alkali salts of aliphatic carboxylic acids; the reaction involves decarboxylation of the substrate followed by subsequent combination of the resulting radicals. The reaction is also known as *Kolbe electrosynthesis*. The reaction is not successful with aromatic acids. Symmetrical and unsymmetrical alkanes, both can be synthesized by this method, but is usually preferred for symmetrical alkanes to avoide unnecessary side reactions.

$$2\text{RCOO}^- \xrightarrow[-\text{CO}_2]{\text{electrolysis}} \text{R}-\text{R}$$

Mechanism

The reaction proceeds through free-radical mechanism.

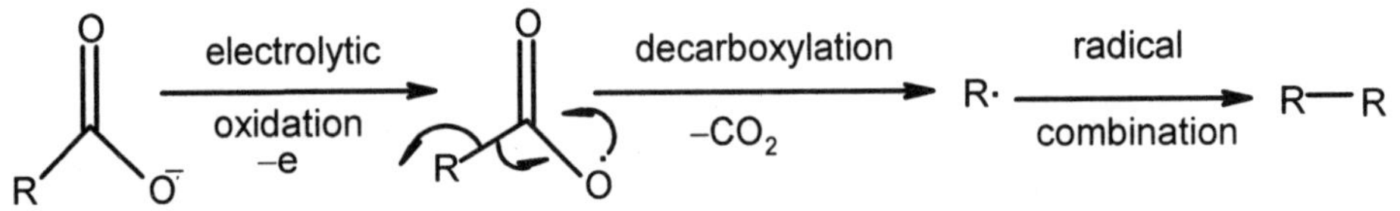

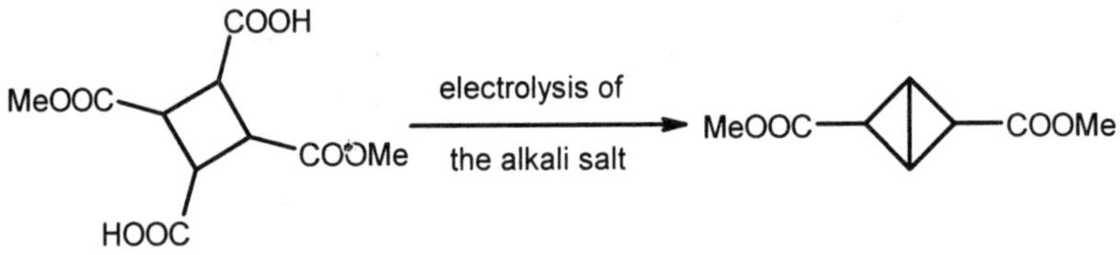

[Scheme: RCOO⁻ → (electrolytic oxidation, −e) → RCOO• → (decarboxylation, −CO_2) → R• → (radical combination) → R—R]

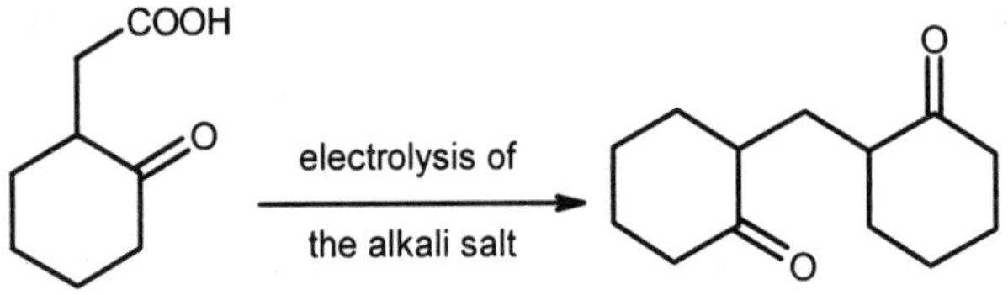

[Scheme: 2-(carboxymethyl)cyclohexanone → (electrolysis of the alkali salt) → 1,1'-methylenebis(cyclohexan-2-one) type dimer]

[Scheme: cyclobutane-1,2,3,4-tetracarboxylic acid dimethyl diester (two COOH, two COOMe) → (electrolysis of the alkali salt) → bicyclo[1.1.0]butane dimethyl dicarboxylate (MeOOC—◇—COOMe)]

References

- Kolbe, H. (1849), *Justus Liebigs Ann. Chem.*, **69**, 257.
- Corey, E. J., Bauld, N. L., La Londe, R. T., Casanova Jr., J. and Kaiser, E. T. (1960), *J. Am. Chem. Soc.*, **82**, 2645.
- Stork, G., Meiseis, A. and Davies, J. E. (1963), *J. Am. Chem. Soc.*, **85**, 3419.
- Vellturo, A. F. and Griffin, G. W. (1965), *J. Am. Chem. Soc.*, **87**, 3021.
- Vijh, A. K. and Conway, B. E. (1967), *Chem. Rev.*, **67**, 623. [Review]
- Kraeutler, B., Jaeger, C. D. and Bard, A. J. (1978), *J. Am. Chem. Soc.*, **100**, 4903.
- Raabjohn, N. and Flasch, G. W., Jr.(1981), *J. Org. Chem.*, **46**, 4082.
- Becking, L. and Schafer, H. J. (1988), *Tetrahedron Lett.*, **29**, 2797.
- Nuding, G., Vogtle, F., Danielmeier, K. and Steckman, E. (1996), *Synthesis*, 71. [Review]
- Hiebl, J., Blanka, M., Guttman, A., Kollmann, H., Leitner, K., Mayhofer, G., Rovenszky, F. and Winkler, K. (1998), *Tetrahedron*, **54**, 2059.

Kolbe-Schmitt Reaction

The method of carboxylation of sodium phenoxides by the action of carbon dioxide under pressure, mostly at the *ortho*-position is called the *Kolbe-Schmitt reaction*.

[Reaction scheme: Sodium phenoxide + CO_2 → (pressure) → sodium salicylate (o-COONa) → (H_3O^+ workup) → salicylic acid (o-COOH)]

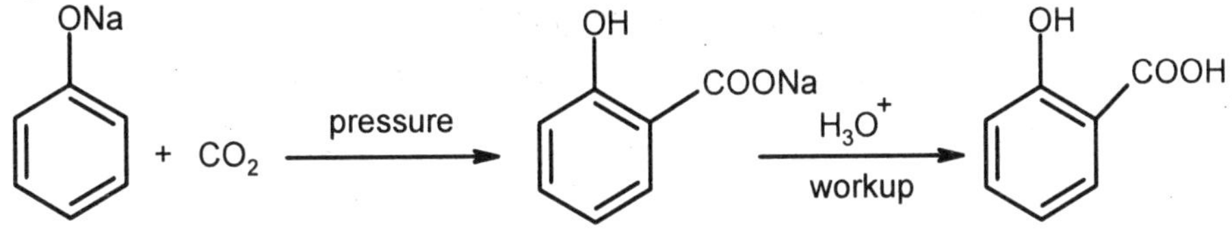

Mechanism

The mechanism is not clearly understood, but is supposed to proceed through the initial formation of a complex (**A**) between the two reactants so that the carbon atom of CO_2 becomes electron-deficient and takes up position to be attacked by the ring cloud at the *ortho*-ring carbon.

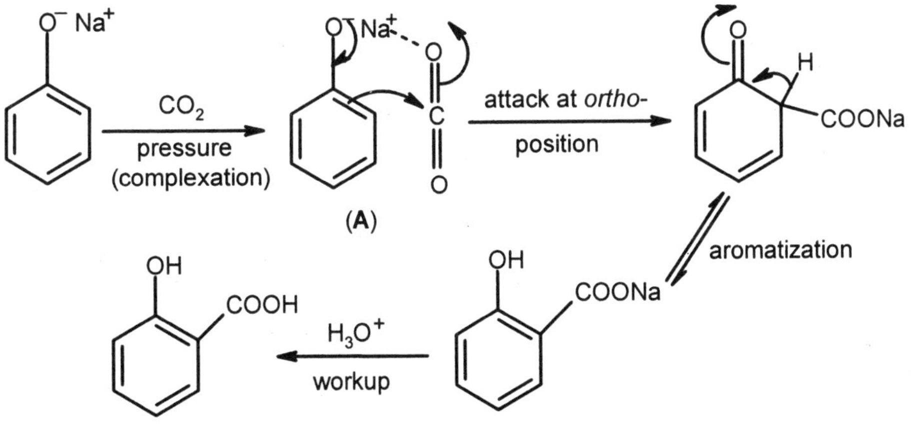

[Mechanism scheme showing: sodium phenoxide + CO_2 under pressure (complexation) forms complex (**A**); attack at *ortho*-position gives cyclohexadienone intermediate with COONa; aromatization gives sodium salicylate; H_3O^+ workup yields salicylic acid]

In case of potassium phenoxides such complex formation (**A**) is less likely to occur, and that's why carboxylation in this case takes place mostly at the *para*-position. Besides, initially formed amount of potassium salicylate rearranges to the *para*-isomer. On the contrary, sodium salicylate does not undergo such rearrngement.

References

- Kolbe, H. (1860), *Justus Liebigs Ann. Chem.*, **113**, 1125.
- Schmitt, R. (1885), *J. Prakt. Chem.*, **31**, 397.

- Cameron, D., Jeskey, H. and Baine, O. (1950), *J. Org. Chem.*, **15**, 233.
- Gal, E. M. (1950), *J. Am. Chem. Soc.*, **72**, 5315.
- Hales, J. L., Jonesa, J. I. And Lindsy, A. S. (1954), *J. Chem. Soc.*, 3145.
- Lindsy, A. S. and Jeskey, H. (1957), *Chem. Rev.*, **57**, 583. [Review].
- Shine, H. J., *Aromatic Rearrangements*, Elsevier : NY, 1967, p. 344.
- Hirao, I. And Kito, T. (1973), *Bull. Chem. Soc. Jpn.*, **46**, 3470.
- Ota, K. (1974), *Bull. Chem. Soc. Jpn.*, **47**, 2343.
- Rahim, M. A., Matsui, Y. and Kosugi, Y. (2002), *Bull. Chem. Soc. Jpn.*, **75**, 619.

Leuckart - Wallach Reaction

In *Leuckart-Wallach reaction* reductive alkylation of an amine occurs when it is treated with a cabonyl compound in the presence of excess of formic acid, which serves as the reducing agent by giving up a hydride ion.

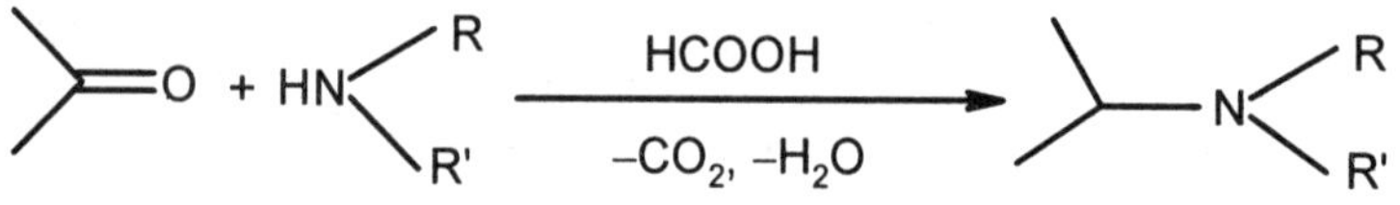

Mechanism

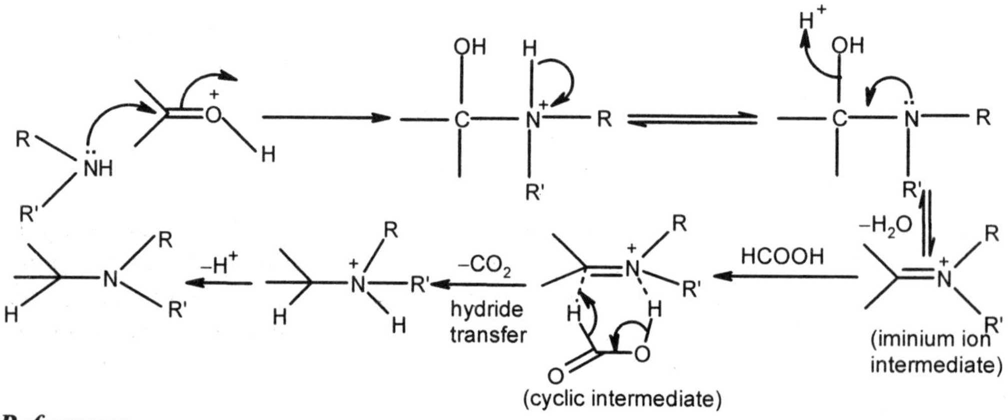

References

- Leuckart, R. (1885), *Ber. Dtsch. Chem. Ges.*, **18**, 2341.

- Wallach, O. (1892), *Justus Liebigs Ann. Chem.*, **272**, 99.
- Lukasiewiez, A. (1963), *Tetrahedron*, **19**, 1789.
- Bach, R. D. (1968), *J. Org. Chem.*, **33**, 1647.
- Klyuev, M. V. and Khidekelm M. L. (1980), *Russ. Chem. Rev.*, **49**, 14. [Review]
- Rylander, P. N., *Hydrogenation Methods*, Academic Press: NY, 1985, p. 82. [Review]
- Musumarra, G. and Sergi, C. (1994), *Heterocycles*, 37, 1033.
- Kitamura, M., Lee, D., Hayashi, S., Tanaka, S. and Yoshimura, M. (2002), *J. Org. Chem.*, **67**, 8685.

Lieben Haloform Reaction

Haloforms are obtained by the action of alkali hypohalites on acetaldehyde or methyl ketones, or their halogenated derivatives or on groups capable of being converted into these under the experimental conditions. The reaction is often called as the *Lieben haloform reaction.* Iodoform, a yellow precipitate in water, is often used for the detection of methyl ketones.

$$RCOMe \xrightarrow[NaOH]{X_2} CHX_3 + RCOONa + NaX \quad (X = I, Br, Cl)$$

$$RCOMe + 3\ NaOI \longrightarrow RCOCl_3 + 3\ NaOH$$

$$RCOCl_3 + NaOH \longrightarrow CHI_3\downarrow + RCOONa$$

Mechanism

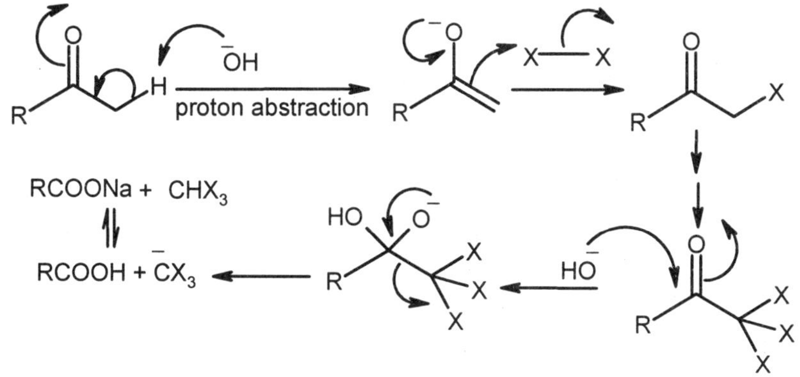

References

- Lieben, A. (1870), *Justus Liebigs Ann. Chem.*, **Suppl. 7**, 218.
- Levine, R. and Stephens, J. R. (1950), *J. Am. Chem. Soc.*, **72**, 1642.
- Pocker, Y. (1959), *Chem. Ind.* (London), 1383.
- Chakrabartty, S. K. in Trahanovsky's *Oxidation in Organic Chemistry*, pt. C, Academic Press : NY, 1978, p. 343.
- Zucco, C., Lima, C. F., Rezende, M. C., Vianna, J. F. and Nome, F. (1987), *J. Org. Chem.*, **52**, 5336.
- Tietze, L. F., Voss, E. and Hartfiel, U. (1990), *Org. Synth.*, **69**, 238.
- Rothenberg, G. and Sasson, Y. (1996), *Tetrahedron*, **52**, 13641.
- Jablonski, L., Billard, T. and Langlois, B. R. (2003), *Tetrahedron Lett.*, **44**, 1055.

Lossen Rearrangement

Hydroxamic acids or their *O*-acyl derivatives undergo rearrangement to form isocyanates either on treatment with bases or sometimes thermally, in a reaction known as *Lossen rearrangement*. The isocyanates may react further with the components of the system.

$$RCONHOCOR' \xrightarrow{\overline{O}H} R-N=C=O \xrightarrow{H_2O} RNH_2$$

Mechanism

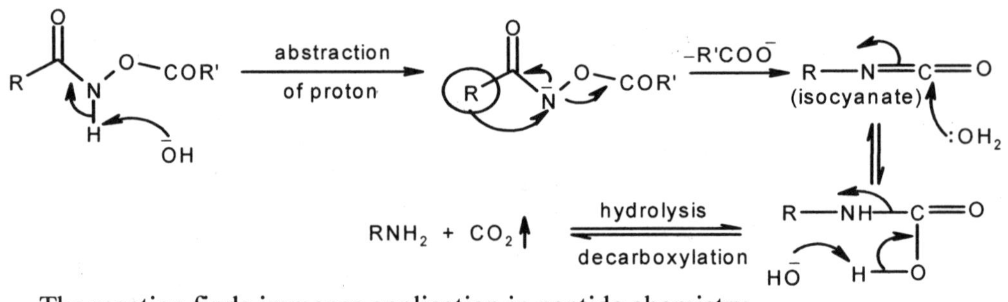

The reaction finds immense application in peptide chemistry.

References

- Lossen, W. (1872), *Justus Liebigs Ann. Chem.*, **161**, 347.
- Yale, H. L. (1943), *Chem. Rev.*, **33**, 209. [Review]
- Snyder, H. R., Elston, C. T. and Kellom, D. B. (1953), *J. Am. Chem. Soc.*, **75**, 2014.
- Bittner, S., Grinberg, S. and Karton, I. (1974), *Tetrahedron Lett.*, 1965.
- Bachmann, G. B. and Goldmacher, J. E. (1964), *J. Org. Chem.*, **29**, 2576.
- Bauer, L. and Exner, O. (1974), *Angew. Chem. Int. Ed. Engl.*, **13**, 376. [Review]
- Ulrich, H., Tucer, B. and Richter, R. (1978), *J. Org. Chem.*, **43**, 1544.
- Salomon, C. J. and Breur, E. (1997), *J. Org. Chem.*, **62**, 3858.
- Zalipsky, S. (1998), *J. Chem. Soc., Chem. Commun.*, 69.
- Anilkumar, R., Chandrasekar, S. and Sridhar, M. (2000), *Tetrahedron Lett.*, **41**, 5291.

McMurry Reaction

Reductive dimerization of carbonyl compounds on treatment with low-valent titanium (acts as reducing agent) leading to the formation of olefins is called the *McMurry reaction*. Low-valent titanium such as Ti(0) is derived from $TiCl_4$ or $TiCl_3$ by reaction with $LiAlH_4$, an alkali metal (Li, Na, K) and also with Zn-Cu couple.

$$Ti(III)Cl_4 + LiAlH_4 \longrightarrow Ti(0)$$

$$\diagdown_{\diagup}C=O \xrightarrow{Ti(0)} \diagdown_{\diagup}C=C_{\diagdown}^{\diagup}$$

Mechanism

Firstly, the carbonyl substrate binds to the metallic surface, and transfer of one electron takes place from the metal to the carbonyl group as a result of which the carbonyl group is reduced to a radical species —— two such species then dimerizes followed by cleavage of the C – O bonds to yield alkenes.

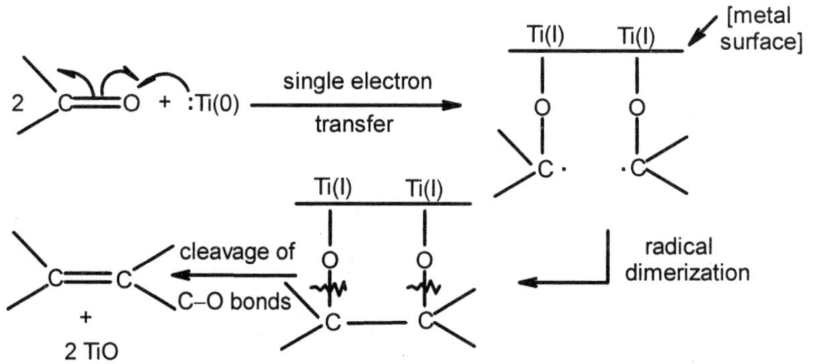

Unsymmetrical ketones on *McMurry reaction* form a mixture of two stereoisomeric alkenes — the ratio of which depends on the steric demand of the substituents.

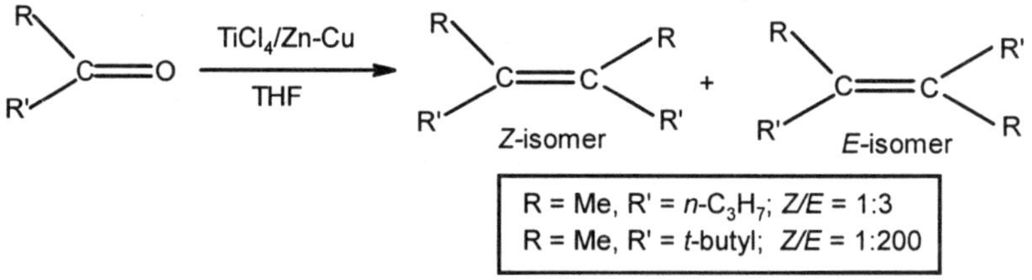

R = Me, R' = n-C$_3$H$_7$; Z/E = 1:3
R = Me, R' = t-butyl; Z/E = 1:200

An interesting example of *intramolecular McMurry coupling* is cited:

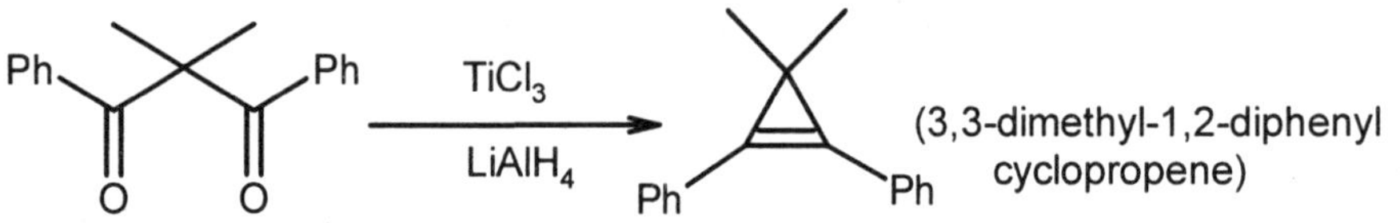

(3,3-dimethyl-1,2-diphenyl cyclopropene)

References

- McMurry, J. E. and Fleming, M. P. (1974), *J. Am. Chem. Soc.*, **96**, 4708.
- Corey, E. J., Danheiser, R. L. and Chandrasekaran, S. (1976), *J. Org. Chem.*, **41**, 260.
- McMurry, J. E. and Krepski, L. R. (1976), *J. Org. Chem.*, **41**, 3929.
- Baumstark, A. L., McCloskey, C. J. and Witt, K. E. (1978), *J. Org. Chem.*, **43**, 3609.
- McMurry, J. E., Lectka, T. and Rico, J. G. (1989), *J. Org. Chem.*, **54**, 3748.
- McMurry, J. E. (1989), *Chem. Rev.*, **89**, 1513. [Review]
- Lenoir, D. (1989), *Synthesis*, 883.
- Furstner, A. (1993), *Angew. Chem. Int. Ed. Engl.*, **32**, 164.
- Hirao, T. (1999), *Synlett*, 175.

- Yamato, T., Fujita, K. and Tsuzuki, H. (2001), *J. Chem. Soc., Perkin Trans. 1*, 2089.
- Sabelle, S., Hydrio, J., Leclerc, E., Mioskowski, C. and Renardm P.-Y. (2002), *Tetrahedron Lett.*, **43**, 3645.
- Williams, D. R. and Heidebrecht Jr., R. W. (2003), *J. Am. Chem. Soc.*, **125**, 1843.
- Honda, T., Namiki, H., Nagase, H. and Mizutani, H. (2003), *Tetrahedron Lett.*, **44**, 3035.

Mitsunobu Reaction

Mitsunobu reaction permits displacement of the –OH group of an alcohol by an incoming nucleophile in the presence of dialkyl azodicarboxylate and triphenylphosphine.

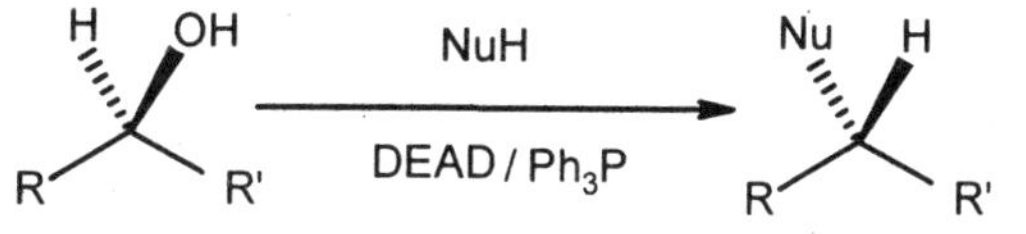

(Nu: nucleophile; DEAD: diethyl azodicarboxylate)

Mechanism

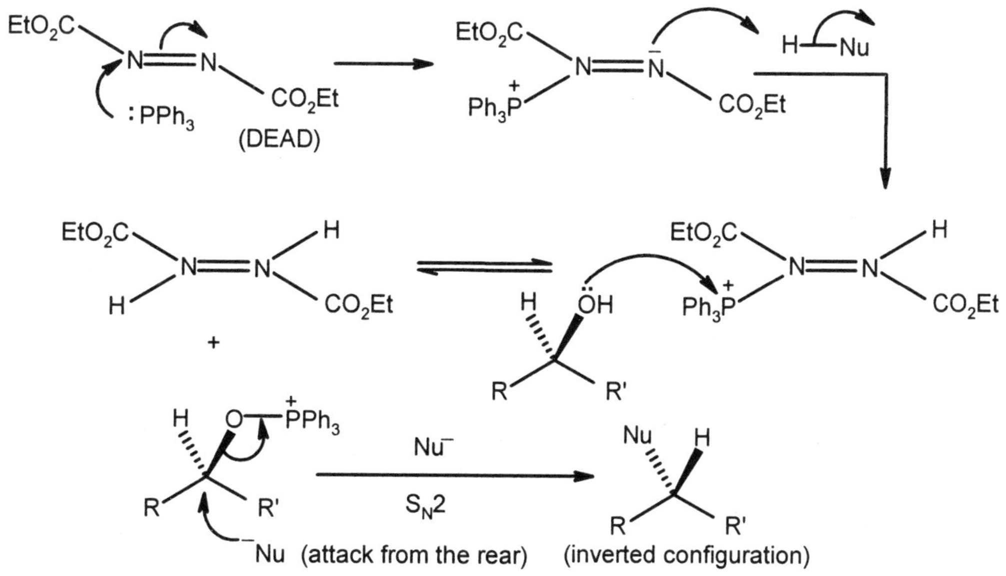

The major application of this reaction is the conversion of a chiral secondary alcohol into an ester having inverted configuration; the ester can be hydrolyzed to yield the inverted alcohol.

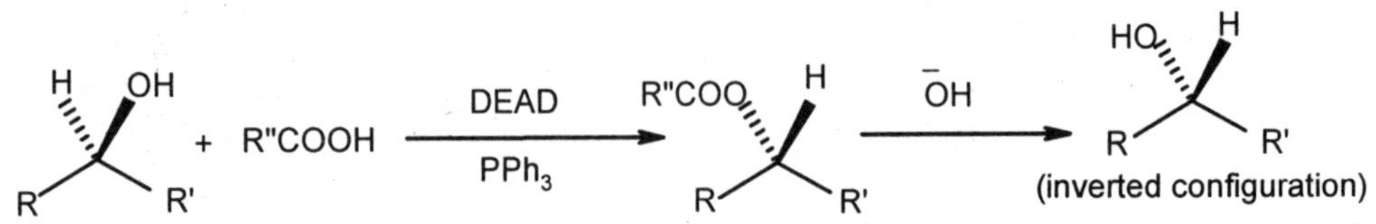

References

- Mitsunobu, O. and Yamada, M. (1967), *Bull. Chem. Soc. Jpn.*, **40**, 2380.
- Mitsunobu, O. (1967), *Bull. Chem. Soc. Jpn.*, **40**, 4235.
- Mitsunobu, O. (1981), *Synthesis*, 1. [Review]
- Hughes, D. L., Reamer, R. A., Bergan, J. J. and Grabowski, E. J. J. (1988), *J. Am. Chem. Soc.*, **110**, 6487.
- Crich, D., Dyker, H. and Harris, R. J. (1989), *J. Org. Chem.*, **54**, 257.
- Camp, D. J. and Jenkins, I. D. (1989), *J. Org. Chem.*, **54**, 3045, 3049.
- Hughes, D. L. (1982), *Org. React.*, **42**, 335. [Review]
- Hughes, D. L. (1996), *Org. Prep. Proc. Int.*, **28**, 127. [Review]
- Langlois, N. and Calvez, O. (2000), *Tetrahedron Lett.*, **41**, 8285.
- Ahn, C., Correia, R. and DeShong, P. (2002), *J. Org. Chem.*, **67**, 1751.
- Bitter, I. and Csokai, V. (2003), *Tetrahedron Lett.*, **44**, 2261.

Neber Rearrangement

Neber rearrangement permits the conversion of ketoxime tosylates into α-amino ketones by the action of a base such as ethoxide or pyridine.

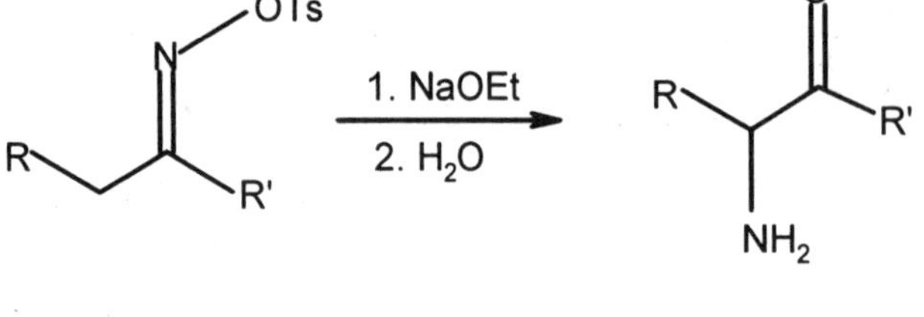

Substituent 'R' is usually aryl, but may also be alkyl or hydrogen. Again, substituent 'R'' may be aryl, alkyl, but not hydrogen.

Mechanism

The mechanistic pathway of the reaction involves azirine intermediate.

The α-amino ketone as obtained can be converted into an oxime tosylate, and after then may further be subjected to *Neber rearrangement* to form α,α'-diamino ketone derivative. Unlike *Beckmann rearrangement*, both *syn* and *anti* ketoximes give the same product. The reaction finds application in the synthesis of natural products.

References

- Neber, P. W. and Friedolsheim, A. (1926), *Justus Liebigs Ann. Chem.*, **449**, 109.
- Neber, P. W. and Burgard, A. (1932), *Justus Liebigs Ann. Chem.*, **493**, 281.
- O'Brien, C. (1964), *Chem. Rev.*, **64**, 81. [Review]
- Cram, D. J. and Hatch, M. J. (1953), *J. Am. Chem. Soc.*, **75**, 33.
- Hatch, M. J. and Cram, D. J. (1953), *J. Am. Chem. Soc.*, **75**, 38.
- House, O. and Berkowitz, W. F. (1963), *J. Org. Chem.*, **28**, 2271.
- Friis, P., Larsen, P. O. and Olsen, C. E.(1977), *J. Chem. Soc.,Perkin Trans 1*, 661.
- Parcell, R. F. and Sanchez, J. P. (1981), *J. Org. Chem.*, **45**, 3156.
- Ueda, S., Naruto, S., Yoshida, T., Sawayama, T. and Uno, H. (1985), *Chem. Commun.*, 218.
- Verstappen, M. M. H., Ariaans, G. J. A. and Zwanenburg, B. (1996), *J. Am. Chem. Soc.*, **118**, 8491.
- Banert, K., Hagedorn, M., Liedtke, C., Melzer, C. and Schoffler, C. (2000), *Eur. J. Org. Chem.*, 257.
- Ooi, T., Takahashi, M., Doda, K. and Maruoka, K. (2002), *J. Am. Chem. Soc.*, **124**, 7640.

Nef Reaction

The conversion of primary or secondary aliphatic and alicyclic nitro-compounds into the corresponding carbonyl compounds by means of hydrolysis of their conjugate bases (nitronate intermediates formed when treated with a base) with sulphuric acid is called the *Nef reaction*. Tertiary nitroalkanes do not respond the reaction because they cannot be converted into the corresponding conjugate bases.

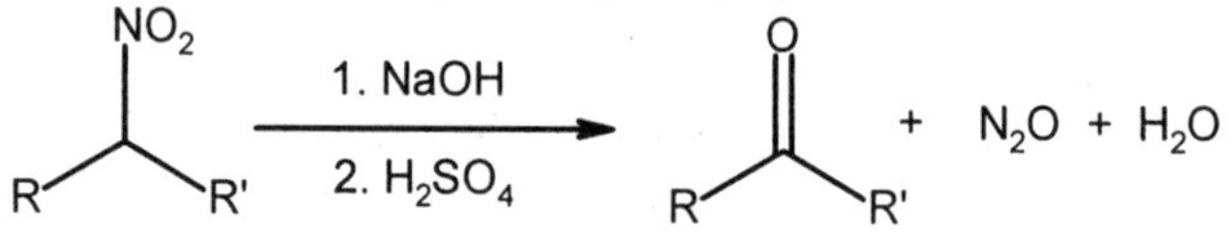

$$RR'CH\text{-}NO_2 \xrightarrow[2.\ H_2SO_4]{1.\ NaOH} RR'C{=}O + N_2O + H_2O$$

Mechanism

[Mechanism showing abstraction of proton by HO⁻ to form nitronate intermediate, followed by protonation, addition of water, loss of H⁺, protonation, and loss of water to give the carbonyl compound + HNO; then 2HNO ⇌ N₂O + H₂O]

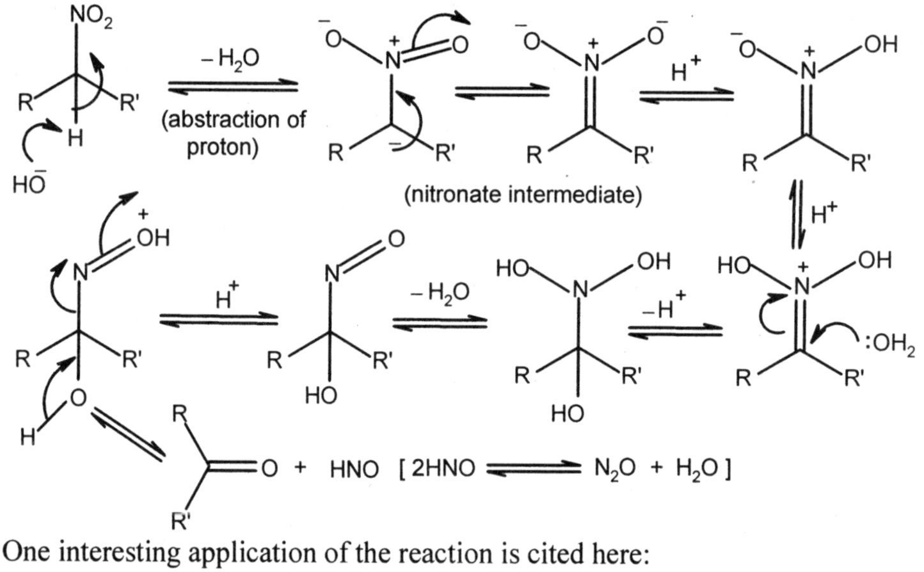

One interesting application of the reaction is cited here:

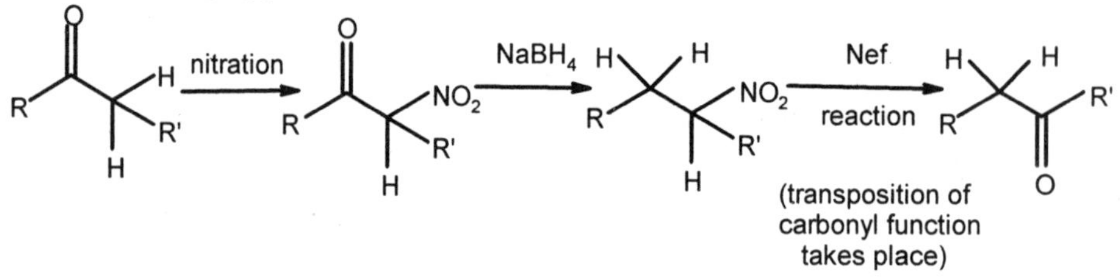

$$RCH(H)C(=O)R' \xrightarrow{\text{nitration}} RC(NO_2)(H)C(=O)R' \xrightarrow{NaBH_4} RCH(NO_2)CH(OH)R' \xrightarrow[\text{reaction}]{Nef} RCH(=O)CH(H)R'$$

(transposition of carbonyl function takes place)

References

- Nef, J. U. (1894), *Justus Liebigs Ann. Chem.*, **280**, 263.
- Hawthrone, M. F. (1957), *J. Am. Chem. Soc.*, **79**, 2510.
- Hassner, A., Larkin, J. M. and Dowd, J. E. (1968), *J. Org. Chem.*, **33**, 1733.
- Sun, S. F. and Folliard, J. T. (1971), *Tetrahedron*, **27**, 323.
- Lever Jr., O. W. (1976), *Tetrahedron*, **32**, 1943.
- Pinnick, H. W. (1990), *Org. React.*, **38**, 655. [Review]
- Hwu, J. R. and Gilbert, B. A. (1991), *J. Am. Chem. Soc.*, **113**, 5917.
- Adam, W., Makosza, M., Saha-Moeller, C. R. and Zhao, C.–G. (1998), *Synlett*, 1335.
- Shahi, S. P. and vankar, Y. D. (1999), *Synth. Commun.*, **29**, 4321.
- Capecchi, T., de Koning, C. B. and Michael, J. P. (2000), *J. Chem. Soc., Perkin Trans 1*, 2681.
- Ballini, R., Bosica, G., Fiorini, D. and Petrini, M. (2002), *Tetrahedron Lett.*, **43**, 5233.

Pechmann Reaction

The acid-catalyzed condensation of phenols with β-ketoesters to produce coumarins is called the *Pechmann reaction*. A variery of condensing reagents such as concentrated sulphuric acid, hydrogen fluoride, polyposphoric acid, Lewis-acids, *etc.* are used.

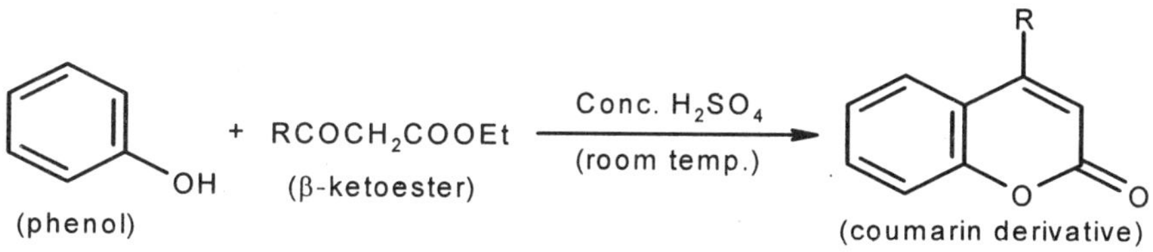

$$\text{(phenol)} + RCOCH_2COOEt \text{ (β-ketoester)} \xrightarrow[\text{(room temp.)}]{\text{Conc. } H_2SO_4} \text{(coumarin derivative)}$$

Mechanism
The reaction follows the mechanism as depicted below.

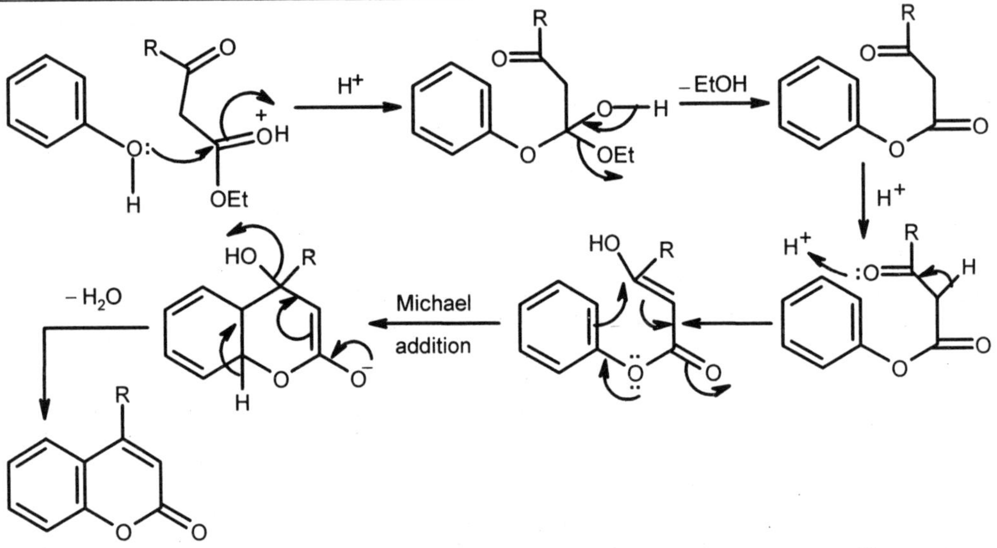

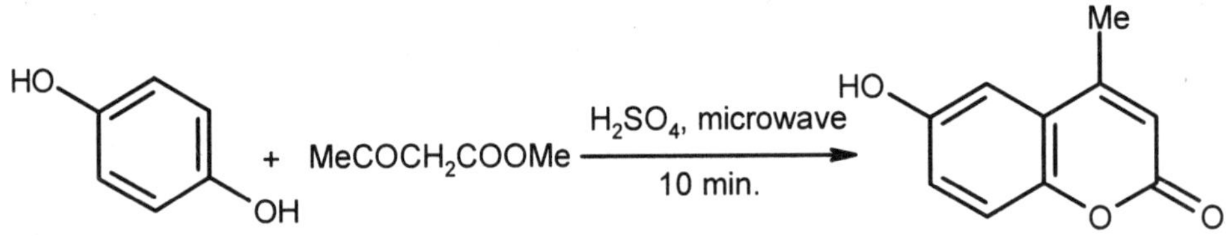

References

- Pechmann, H. V. and Duisberg, C. (1883), *Ber. Dstch. Chem. Ges.*, **16**, 2119.
- Sen, H. K. and Basu, U. (1928), *J. Indian Chem. Soc.*, **5**, 467.
- Sethna, S. and Phadke, R. (1953), *Org. React.*, 7, 1. [Review]
- Dann, O. and Mylius, G. (1954), *Justus Liebigs Ann. Chem.*, **587**, 1.
- John, E. V. O. and Israelstam, S. S. (1961), *J. Org. Chem.*, **26**, 240.
- Kappe, T. and Zeigler, E. (1969), *Org. Prep. Proceed. Int.*, **1**, 61.
- Kappe, T. and Mayer, C. (1981), *Synthesis*, 524.
- Chaudhury, D. D. (1983), *Chem. Ind.*, 568.
- Corrie, J. E. T. (1990), *J. Chem. Soc., Perkin Trans 1*, 2151.
- Biswas, G. K., Basu, K., Barua, A. K. and Bhattacharya, P. (1992), *Indian J. Chem., Sect. B.*, **31B**, 628.
- Singh, V., Singh, J., Kaur, K. P. and Kad, G. L. (1997), *J. Chem. Res. (S)*, 58.
- Holden, M. S. and Crouch, R. D. (1998), *J. Chemical Education*, **75**, 1631.
- Sugino, T. and Tanaka, K. (2001), *Chem. Lett.*, 110.
- Khandekar, A. C. and Khandekar, B. M. (2002), *Synlett*, 152.

Prins Reaction

Prins reaction is an acid-catalyzed addition of formaldehyde to an alkene leading to formation of a variety of products of which the three main components are 1,3-diol, 1,3-dioxane and allylic alcohol. The predominancy of the product depends on the alkene and the conditions applied.

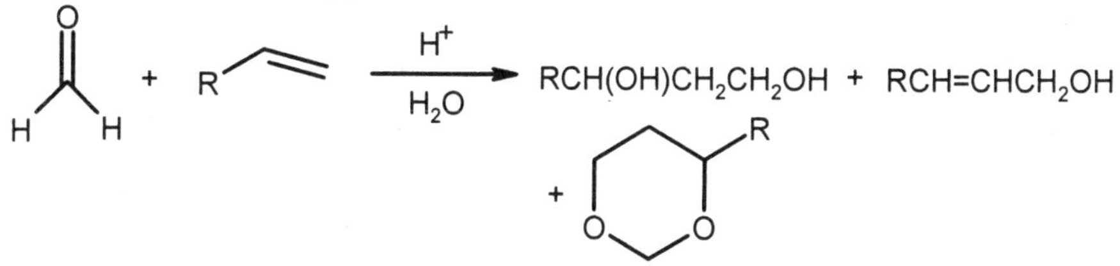

Mechanism

The reaction follows the following mechanistic pathway.

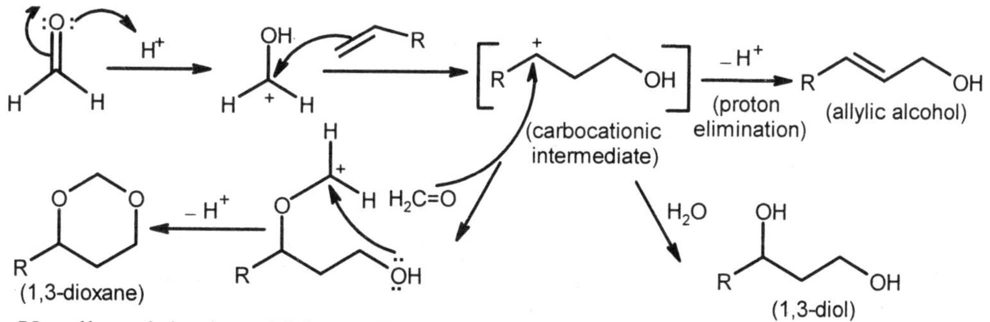

Usually sulphuric acid is used as the catalyst; however, phosphoric acid, BF_3 or acidic ion exchange resins are also in use. The main difficulty of the reaction is the formation of complex mixture of products. To get one of them as the major product, appropriate conditions would have to be applied. As such, below 70^0C the acid-catalyzed condensation of alkenes with aldehydes furnishes 1,3-dioxanes as the major products, while at higher temperatures the corresponding diols are the major ones.

References

- Prins, H. J. (1919), *Chem. Weekblad,* **16**, 64, 1072.
- Hellin, M., Davidson, M. and Coussemant, F. (1966), *Bull. Soc, Chim. Fr.,* 1890, 3217.

- Schowen, K. B., Smissman, E. E. and Schowen, R. L. (1968), *J. Org. Chem.*, **33**, 1873.
- Griegel, H. and Sieber, W. (1973), *Monatsh. Chem.*, **104**, 1008, 1027.
- Adam, D. R. and Bhatnagar, S. P. (1977), *Synthesis*, 661. [Review]
- El Gharbi, R. (1981), *Synthesis*, 361.
- Hanaki, N., Link, J. T., MacMillan, D. W. C., Overman, L. E., Trankle, W. G. and Wurster, J. A. (2000), *Org. Lett.*, **2**, 223.
- Yadav, J. S., Reddy, B. V. S., Kumar, G. M. and Murthy, C. V. S. R. (2001), *Tetrahedron Lett.*, **42**, 89.
- Davis, C. E. and Coates, R. M. (2002), *Angew. Chem. Int. Ed. Engl.*, **41**, 472.
- Braddock, D. C., Badine, D. M., Gottschalk, T., Matsuno, A. and Rodrihuez-Lens, M. (2003), *Synlett*, 345.

Ritter Reaction

Nitriles react with secondary or tertiary alcohols or with a variety of alkenes in strong acids to yield *N*-alkyl amide in a reaction known as *Ritter reaction*. The alcohol or an alkene is converted into a stable carbocation by a strong acid; the intermediate then adds to the nitrogen of a nitrile followed by water addition — the immediate product tautomerizes to the *N*-alkyl amide.

$$ROH + R'CN \xrightarrow[H_2O]{\text{concentrated } H_2SO_4} R'-\underset{OH}{C}=N-R \rightleftharpoons R'-\underset{O}{\overset{\|}{C}}-NHR$$

$$\begin{array}{c} Me_2C(OH)CH_3 \\ \text{or} \\ Me_2C=CH_2 \end{array} + MeCN \xrightarrow[H_2O]{\text{concentrated } H_2SO_4} MeCONHCMe_3$$

Mechanism

The reaction proceeds through the following mechanistic pathway.

$$\text{Me}_2\text{C(OH)CH}_3 \text{ or } \text{Me}_2\text{C}=\text{CH}_2 \xrightarrow{\text{H}^+} \underset{\text{(stable carbocation)}}{\text{Me}-\overset{+}{\text{CMe}_2}} \xrightarrow{:\text{N}\equiv\text{C}-\text{Me}} \text{Me}_3\text{C}-\underset{}{\text{N}}=\overset{+}{\text{C}}-\text{Me}$$

$$\text{Me}_3\text{C}-\text{N}=\underset{\overset{|}{\text{OH}}}{\text{C}}-\text{Me} \xleftarrow{-\text{H}^+} \text{Me}_3\text{C}-\text{N}=\underset{\overset{|}{\overset{+}{\text{OH}_2}}}{\text{C}}-\text{Me} \xleftarrow{\text{H}_2\ddot{\text{O}}:}$$

$$\Updownarrow \text{tautomerism}$$

$$\text{Me}_3\text{CNHCOMe} \quad (N\text{-alkyl amide})$$

Since amides can easily be converted into amines, this method provides a way for producing amines from alcohols or alkenes.

References

- Ritter, J. J. and Minieri, P. P. (1948), *J. Am. Chem. Soc.*, **70**, 4045.
- Ritter, J. J. and Kalish, J. (1948), *J. Am. Chem. Soc.*, **70**, 4048.
- Johnson, F. and Madronero, R. (1966), *Adv. Heterocycl. Chem.*, **6**, 95. [Review]
- Krimen, L. I. And Cota, D. J. (1969), *Org. React.*, **17**, 2123. [Review]
- Martinez, A. G., Alvarez, R. M., Vilar, E. T., Fraile, A. G., hanack, M. and Subramanian, L. R. (1989), *Tetrahedron Lett.*, **30**, 581.
- Chen, H. G., Goel, O. P. Kesten, S. and Knobelsdorf, J. (1996), *Tetrahedron Lett.*, **37**, 8129.
- Tongco, E. C., Prakash, G. K. S. and Olah, G. A. (1997), *Synlett*, 1193. [Review]
- Jirgensons, A., Kauss, V., Kalvinsh, I. And Gold, M. R. (2001), *Synthesis*, 1709.
- Nair, V., Rajan, R. and Rath, N. P. (2002), *Org. Lett.*, **4**, 1575.
- Reddy, K. L. (2003), *Tetrahedron Lett.*, **44**, 1453.

Rosenmund Reduction

The selective hydrogenation of an acid chloride leading to its reduction to aldehyde by using palladium catalyst on barium sulphate support is called the *Rosenmund reduction*. A suitable catalyst poisoner or regulator, such as quinoline and sulphur or thiourea, is to be added to inactivate the the catalyst so that it becomes unable to reduce the aldehyde (as formed) into alcohol. Thus, *Rosenmund reduction* offers a method for conversion of a carboxylic acid into the corresponding aldehyde *via* acid chloride.

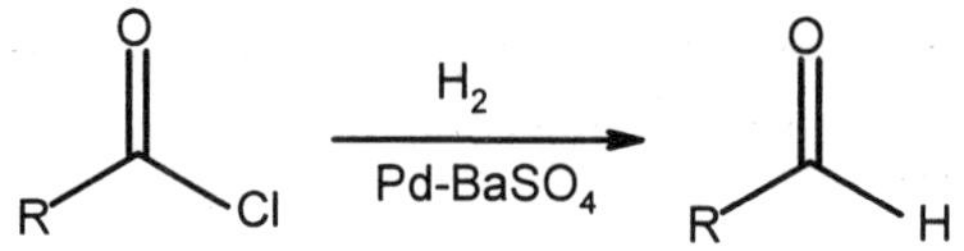

$$R-COCl \xrightarrow[\text{Pd-BaSO}_4]{H_2} R-CHO$$

Mechanism

The reaction proceeds presumably through the initial formation of an organopalladium species that then reacts with the hydrogen.

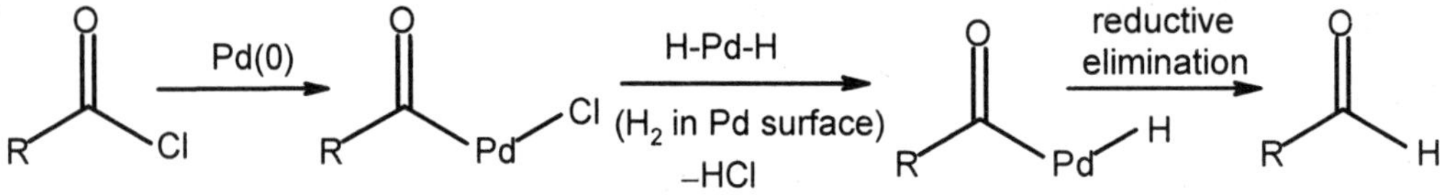

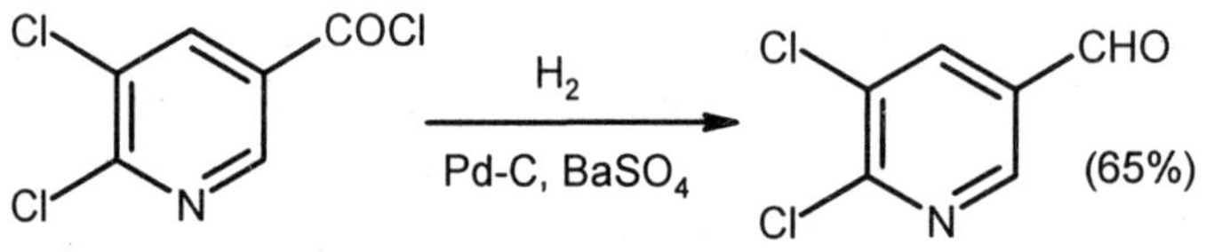

$$R-COCl \xrightarrow{Pd(0)} R-CO-Pd-Cl \xrightarrow[\text{(H}_2\text{ in Pd surface)}]{H-Pd-H} R-CO-Pd-H \xrightarrow{\text{reductive elimination}} R-CHO$$
(–HCl)

[3,5-dichloropyridine-COCl] $\xrightarrow[\text{Pd-C, BaSO}_4]{H_2}$ [3,5-dichloropyridine-CHO] (65%)

References

- Rosenmund, K. W. (1918), *Ber. Dtsch. Chem. Ges.*, **51**, 585.
- Mosetting, E. and Mozingo, R. (1948), *Org. React.*, **4**, 362. [Review]
- Burgstahler, A. W., Weigel, L. O. and Schafer, C. G. (1976), *Synthesis*, 767.
- Danishefsky, S., Hinama, M., Gombatz, K., Harayama, T., Berman, E. and Schirda, P. (1979), *J. Am. Chem. Soc.*, **101**, 7020.
- McEwen, A. B., Guttieri, M. J., Maier, W. L., Laine, R. M. and Shvo, Y. (1983), *J. Org. Chem.*, **48**, 4436.
- Maier, W. F., Chettle, S. J., Rai, R. S. and Thomas, G. (1986), *J. Am. Chem. Soc.*, **108**, 2608.
- Bold, G., Steiner, H., Moesch, L. Walliser, B., St. Pjau, A. and Plamttner, P. A. (1990), *Helv. Chim. Acta*, **73**, 405.
- Chandnani, K. H. and Chandalia, S. B. (1999), *Org. Proc. Res. Dev.*, **3**, 416.
- Chimichi, S., Boccalini, M. and Cosimelli, B. (2002), *Tetrahedron*, **58**, 4851.

Schotten - Boumann Reaction

Schotten-Boumann reaction offers a method for acylating a hydroxyl or amino group using acid halides under basic conditions.

R'OH + RCOCl + NaOH ⟶ R'OCOR + NaCl + H$_2$O

R'NH$_2$ + RCOCl + NaOH ⟶ R'NHCOR + NaCl + H$_2$O

Mechanism

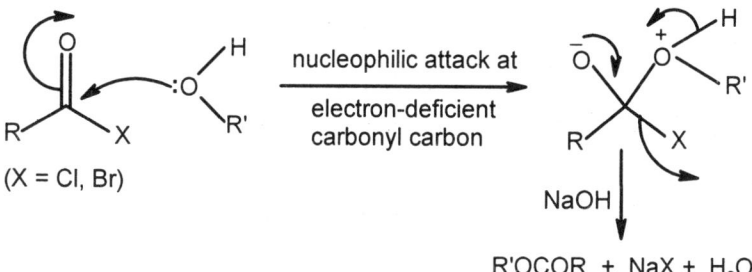

Both R and R' may be primary, secondary, or tertiary alkyl or aryl; in case of hindered acid halides or tertiary 'R' group, alkoxide of the alcohol is often used.

References

- Schotten, C. (1884), *Ber. Dstch. Chem. Ges.*, **17**, 2544.
- Sonntag, N. O. V. (1953), *Chem. Rev*, **52**, 237. [Review]
- Altman, J. and Ben-Ishai, D. (1968), *J. Hetercyclic Chem.*, **5**, 679.
- Tsuchiya, M., Yoshida, H., Ogata, T. and Inokawa, S. (1969), *Bull. Chem. Soc. Jpn.*, **42**, 1756.
- Kaiser, E. M. and Woodruff, R. A. (1970), *J. Org. Chem.*, **35**, 1198.
- Gutteridge, N. J. A. and Dales, J. R. N. (1971), *J. Chem. Soc.(C)*, 122.
- Kevill, D. N. and Knauss, D. C. (1993), *J. Chem. Soc., Perkin Trans 2*, 307.
- Fleming, I. and Winter, S. B. D. (1993), *Tetrahedron Lett.*, **34**, 7287.
- Bentley, T. W., Llewellyn, G. and McAliser, J. A. (1996), *J. Org. Chem.*, **61**, 7927.
- Sano, T., Sugaya, T., Inoue, K., Mizutaki, S., Ono, Y. and Kasai, M. (2000), *Org. Proc. Res. Dev.*, **4**, 147.

Simmons – Smith Reaction

Simmons-Smith reaction is an efficient way of cyclopropanation by transferring methylene from methylene iodide and zinc-copper couple to a C – C double bond. Free carbenes are not formed during the course of the reaction, and thus usual side-reactions arising out of free carbenes can thus be avoided.

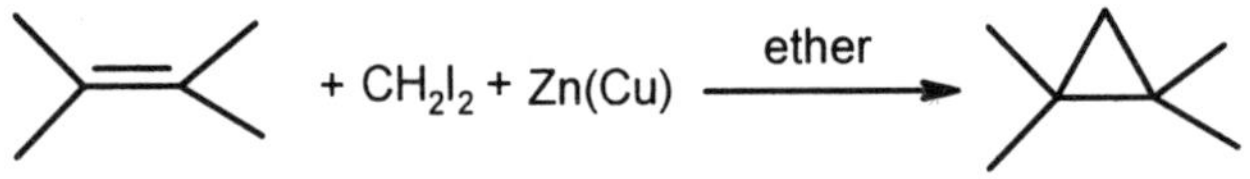

Mechanism

Initial formation of an intermediate, ICH_2ZnI is suggested; it behaves as an electrophilic reagent towards the double bond. The addition is stereospecifically *syn*, and follows a *concerted* mechanism.

$$2\ CH_2I_2 + 2\ Zn(Cu) \xrightarrow{ether} 2\ ICH_2ZnI \rightleftharpoons (ICH_2)_2Zn \cdot ZnI_2$$

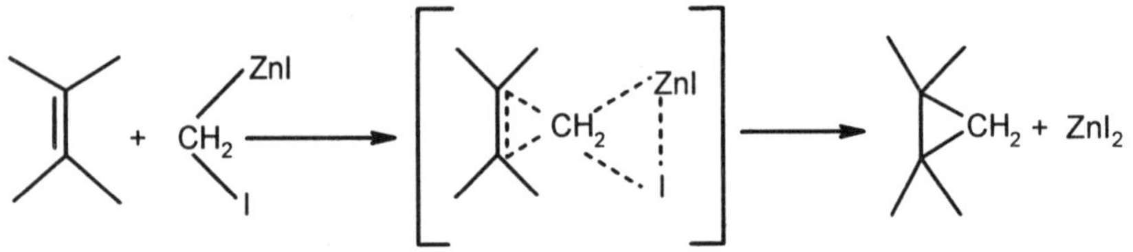

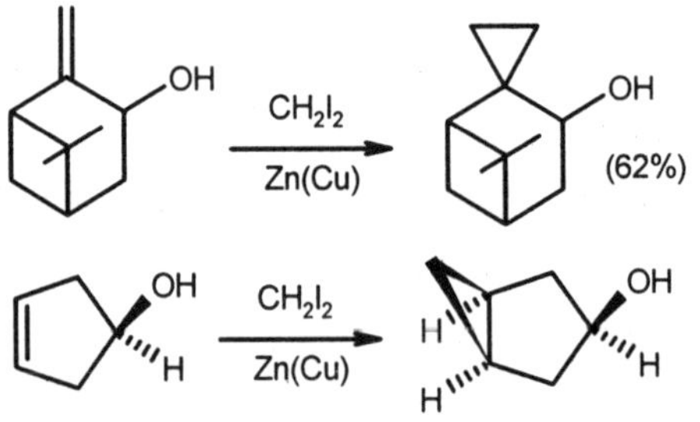

References

- Simmons, H. E. and Smith, R. D. (1958), *J. Am. Chem. Soc.*, **80**, 5323.
- Simmons, H. E. and Smith, R. D. (1959), *J. Am. Chem. Soc.*, **81**, 4256.
- Simmons, H. E., Blanchard, E. P. and Smith, R. D. (1964), *J. Am. Chem. Soc.*, **86**, 1347.

- Ripoll, J. E. and Conia, J. M. (1969), *Tetrahedron Lett.*, 979.
- Rawson, R. J. and Harrison, I. T. (1970), *J. Org. Chem.*, **35**, 2057.
- Simmons, H. E., Cairns, T. L., Vladuchick, S. A. and Hoiness, C. M. (1973), *Org. React.*, **20**, 1. [Review]
- Bee, L. K., Beeby, J., Everett, J. W. and Garratt, P. J. (1975), *J. Org. Chem.*, **40**, 2212.
- Takai, K., Kakiuchi, T. and Utimoto, K. (1994), *J. Org. Chem.*, **59**, 2671.
- Dargel, T. K. and Koch, W. (1996), *J. Chem. Soc., Perkin Trans 2*, 877.
- Nakamura, E., Hirai, A. and Nakamura, M. (1998), *J. Am. Chem. Soc.*, **120**, 5844.
- Kaye, P. T. and Molema, W. E. (1999), *Synth. Commun.*, **29**, 1889.
- Charette, A. B. and Beauchemin, A. (2001), *Org. React.*, **58**, 1. [Review]
- Nakamura, M., Hirai, A. and Nakamura, E. (2003), *J. Am. Chem. Soc.*, **125**, 2341.
- Mahata, P. K., Syam Kumar, U. K., Sriram, V., Iia, H. and Junjappa, H. (2003), *Tetrahedron*, **59**, 2631.

Simonini Reaction

Simonini reaction is an ester forming reaction by treating silver carboxylate with iodine in the ratio 2:1.

$$2\ RCOOAg + I_2 \longrightarrow 2RCOOR + CO_2 + 2AgI$$

Mechanism

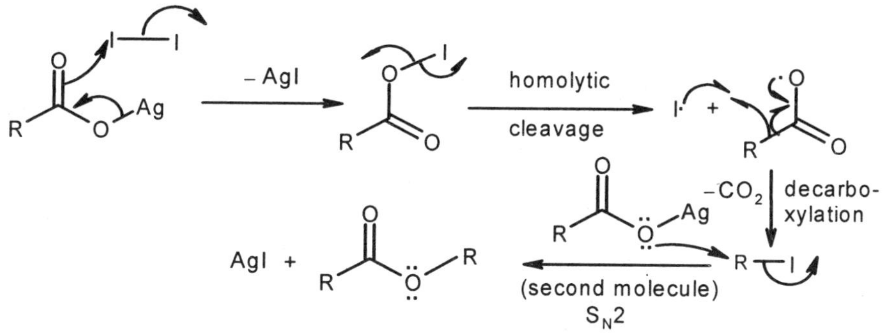

References

- Simonini, A. (1892), *Monatsh. Chem.*, **13**, 320; *ibid*, (1893), **14**, 81.
- Oldman, J. W. H. (1950), *J. Chem. Soc.*, 100; Wasserman, H. H. and Precopio, F. M. (1954), *J. Am. Chem. Soc.*, **76**, 1242.

- Chalmers, D. J. and Thomson, R. H. (1968), *J. Chem. Soc. (C)*, 848.
- Bunce, N. J. and Murray, N. G. (1971), *Tetrahedron*, **27**, 5323.

Sommelet – Hauser Rearrangement

Rearrangement of benzylic quaternary ammonium salts upon treatment of strong bases such as alkali metal amides or phenyllithium is called the *Sommelet-Hauser rearrangement*. The reaction is most often carried out with three methyl groups on the amine nitrogen. *Ortho*-position (s) of benzene ring of the benzyl group is involved in the rearragnement process.

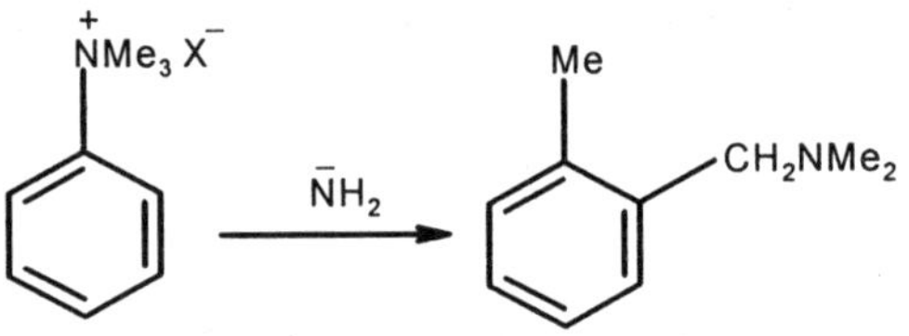

Mechanism

The benzylic hydrogen being more acidic is abstracted first producing species (**1**) that remains in equilibrium with (**2**); the latter is the actual attacking species and undergoes the rearragnement thereby shifting the equilibrium in its favour. The reaction involves a [2,3]-sigmatropic rearrangement.

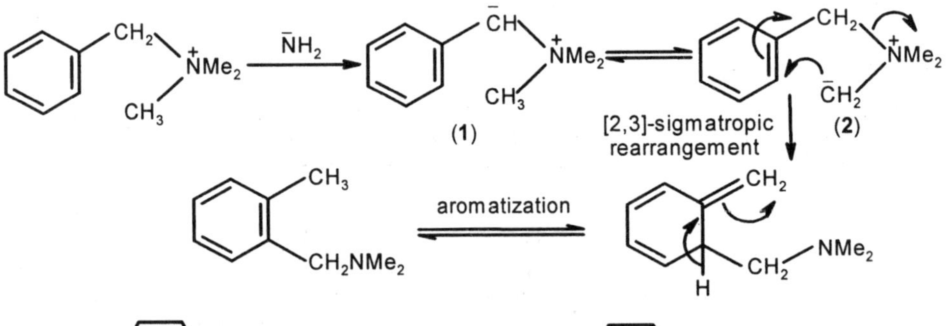

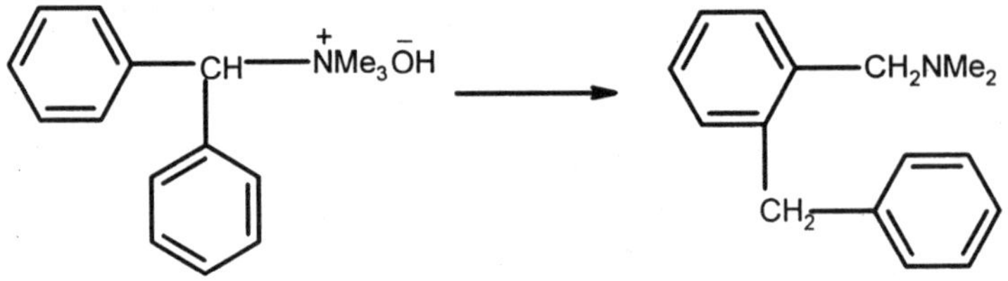

References

- Sommelet, M. (1937), *Compt. Rend.*, **205**, 56.
- Beard, W. Q. and Hauser, C. R. (1960), *J. Org. Chem.*, **25**, 334.
- Jones, F. N. and Hauser, C. R. (1961), *J. Org. Chem.*, **26**, 2979.
- Puterbaugh, W. H. and Hauser, C. R. (1964), *J. Am. Chem. Soc.*, **86**, 1105.
- Pine, S. H. (1967), *Tetrahedron Lett.*, 3393.
- Pine, S. H. (1970), *Org. React.*, **18**, 403. [Review]
- Robert, A. and Lucas-Thomas, M. T.(1980), *J. Chem. Soc., Chem. Commun.*, 629.
- Shirai, N., Watanabe, Y. and Sato, Y. (1990), *J. Org. Chem.*, **55**, 2767.
- Tanaka, T., Shirai, N., Sugimoro, J. and Sato, Y. (1992), *J. Org. Chem.*, **57**, 5034.
- Klunder, J. M. (1995), *J. Heterocyclic Chem.*, **32**, 1687. [Review]
- Maeda, Y. and Sato, Y. (1996), *J. Org. Chem.*, **61**, 5188.
- Endo, Y., Uchida, T. and Shudo, K. (1997), *Tetrahedron Lett.*, **38**, 2113.

Stephen Reaction

Reduction of nitriles by means of anhydrous stannous chloride dissolved in ether saturated with hydrogen chloride, followed by hydrolysis with water affords aldehydes in a reaction called as *Stephen reaction*.

$$R-C\equiv N \xrightarrow[2.\ H_2O]{1.\ SnCl_2/HCl\ \text{ether}} R-CHO$$

Mechanism

The reaction proceeds through the following mechanistic pathway.

$$R-C\equiv N + HCl \longrightarrow R-C=\overset{+}{N}H\ \overset{-}{Cl} \longrightarrow R-C=\overset{+}{N}H_2\ \overset{-}{Cl}$$

[R—CH=$\overset{+}{N}H_2$]$_2$SnCl$_6^{2-}$
(aldimine as a complex with SnCl$_4$, aldimine stannichloride)

$\xleftarrow{2\ H^+}$ SnCl$_4^{2-}$ with R—C=$\overset{+}{N}H_2$ Cl$^-$ ... SnCl$_2$/HCl ... SnCl$_2$ ... R—C=$\overset{+}{N}H_2$ Cl$^-$

hydrolysis $\Big\downarrow$ H$_3$O$^+$

R—CHO

The *Stephen reduction* is most useful when 'R' is aromatic; for aliphatic 'R' up to six carbon atoms the reaction is found to be smooth. Substrates of the type ArCONHPh can be made to undergo the reduction to the corresponding aldehydes by treating them firstly with PCl_5, and then usual *Stephen reagent*. The variation of the process is called *Sonn-Muller method*.

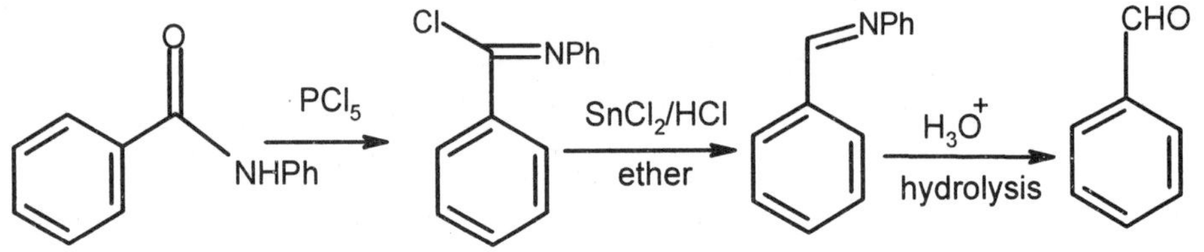

References

- Stephen, H. (1925), *J. Chem. Soc.*, 1874.
- Gardner, T. S., Smith, F. A., Wenis, E. and Lee, J. (1951), *J. Org. Chem.*, **16**, 1121.
- Knight, J. A. and Harry D. Zook, H. D. (1952), *J. Am. Chem. Soc.*, **74**, 4560.
- Williams, J., Witten, C. and Krynitsky, J. (1955), *Org. Synth.*, **Coll. Vol. 3**, 626.
- Zil'berman, E. N. and Pyryalova, P. S. (1963), *J. Gen. Chem., USSR*, **33**, 3348.
- Rabinovitz, M. in Rappoport, *The Chemistry of the Cyano Group*, Wiley: NY, 1970, p. 307.

Strecker Synthesis

Aldehydes or ketones on treatment with sodium cyanide and amine afford α-amino nitriles in a reaction known as *Strecker synthesis*; the products can easily be hydrolyzed to the corresponding α-amino acids.

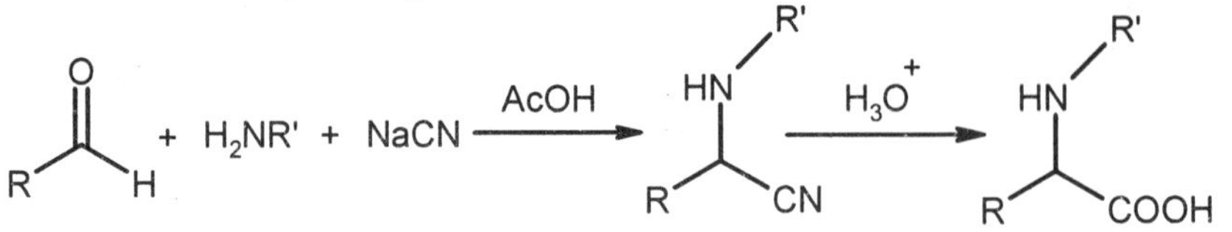

Mechanism

$$RCHO \xrightarrow{H^+} \xrightarrow{H_2NR'} \text{[tetrahedral intermediate]} \longrightarrow \underset{\text{(iminium intermediate)}}{R-CH=\overset{+}{N}HR'} \xrightarrow{NC^-} \underset{\text{(α-amino nitrile)}}{\underset{R}{HN(R')-CH-C\equiv N}} \xrightarrow[\text{hydrolysis}]{H_3O^+} \underset{\text{(α-amino acid)}}{HN(R')-CH(R)-COOH}$$

The reaction is a special case of *Mannich reaction*. It can also be carried out with ammonia and hydrogen cyanide, and with ammoniumcyanide. Use of primary and secondary amine salts instead of NH_4^+ affords *N*-substituted and *N,N*-disubstituted α-amino nitriles, respectively.

References

- Strecker, A. (1850), *Justus Liebigs Ann. Chem.*, **75**, 27.
- Williams, R. M., *Synthesis of Optically Active α-Amino Acids*, Pergamon: Elmsford, NY, 1989, p. 208. [Review]
- Shafran, Yu M., Bakulev, V. A. and Mokrushin, V. S. (1989), *Russ. Chem. Rev.*, **58**, 148. [Review]
- Altenbach, H.-J. in Mulzer, J., Altenbach, H. -J., Barun, M., Krohn, K. and Reissig, H.-U., *Organic Synthesis Highlights*, VCH, Weinheim, 1991, p. 300.
- Chakraborty, T. K., Hussain, K. and Reddy, G. V. (1995), *Tetrahedron*, **51**, 9179.
- Iyer, M. S., Gigstad, K. M., Namdev, N. D. and Lipton, M. J. (1996), *J. Am. Chem. Soc.*, **118**, 4910.
- Ishitani, H., Komiyama, S., Hasegawa, Y. and Kobayashi, S. (2000), *J. Am. Chem. Soc.*, **122**, 762.
- Davis, F. A., Lee, S., Zhang, H. and Fanelli, D. L. (2000), *J. Org. Chem.*, **65**, 8704.
- Ding, K. and Ma, D. (2001), *Tetrahedron*, **57**, 6361.
- Matrumoto, K., Kim, J. C., Hayashi, N. and Jenner, G. (2002), *Tetrahedron Lett.*, **43**, 9167.
- Volk, F. J., Wagner, M. and Frahm, A. W. (2003), *Tetrahedron: Asymmetry*, **14**, 497.

Suzuki Coupling

In recent years *Suzuki coupling reaction* has received increased importance in synthetic organic chemistry as a carbon – carbon bond forming reaction. It is a palladium-catalyzed cross-coupling reaction of organoborane compounds with an organic (aryl or alkenyl, or alkynyl) halide or triflate (OTf = trifluoromethane-sulphonate) in the presence of a base.

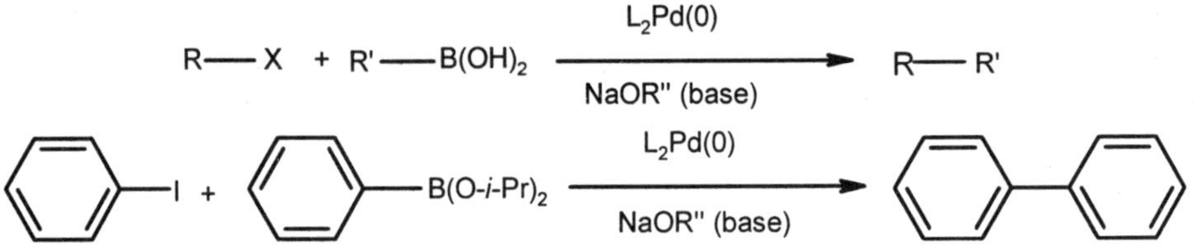

$$R-X + R'-B(OH)_2 \xrightarrow[\text{NaOR'' (base)}]{L_2Pd(0)} R-R'$$

$$Ph-I + Ph-B(O\text{-}i\text{-}Pr)_2 \xrightarrow[\text{NaOR'' (base)}]{L_2Pd(0)} Ph-Ph$$

Mechanism

The reaction occurs in the following mechanistic scheme.

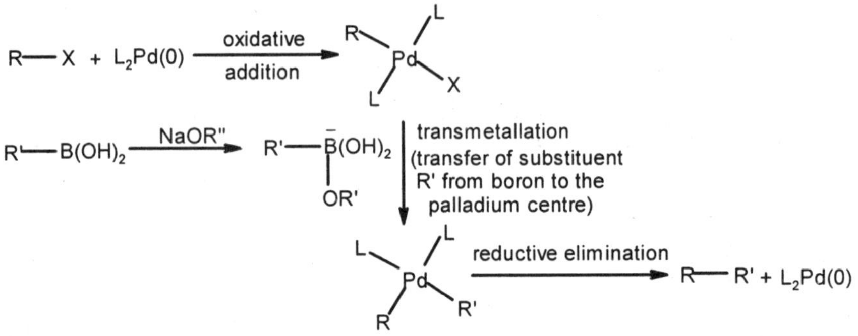

$$R-X + L_2Pd(0) \xrightarrow{\text{oxidative addition}} \underset{L}{\overset{R}{\underset{|}{Pd}}}\underset{X}{\overset{L}{}}$$

$$R'-B(OH)_2 \xrightarrow{NaOR''} R'-\bar{B}(OH)_2 \;|\; OR'$$

transmetallation (transfer of substituent R' from boron to the palladium centre)

reductive elimination → R—R' + $L_2Pd(0)$

As stated earlier, *Suzuki coupling* finds immense applications in carbon – carbon bond forming synthetic strategies.

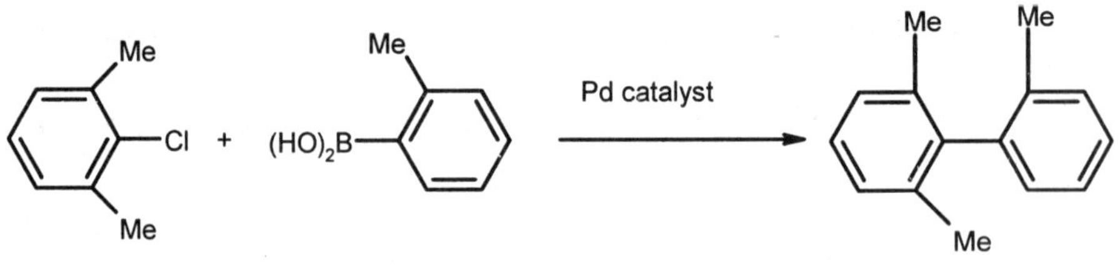

[2,6-dimethylphenyl chloride] + (HO)₂B—[2-methylphenyl] $\xrightarrow{\text{Pd catalyst}}$ [2,2',6-trimethyl biphenyl product]

References

- Suzuki, A. and Miyaura, N. (1979), *J. Chem. Soc., Chem. Commun.*, 866.

- Aliprantis, A. O. and Canary, J. W. (1994), *J. Am. Chem. Soc.*, **116**, 6985.
- Miyaura, N. and Suzuki, A. (1995), *Chem. Rev.*, **95**, 2457. [Review]
- Torrado, A., Iglesias, S., Lopez, S. and de Lera, A. R. (1995), *Tetrahedron*, **51**, 2435.
- Stanforth, S. P. (1998), *Tetrahedron*, **54**, 263. [Review]
- Groger, H. (2000), *J. Prakt. Chem.*, **342**, 334.
- Franzen, R. (2000), *Can. J. Chem.*, **78**, 957.
- Chemler, S. R., Trauner, D. and Danishefsky, S. J. (2001), *Angew. Chem. Int. Ed. Engl.*, **40**, 4544.
- LeBlond, C. R., Andrews, A. T., Sun, Y. and Sowa Jr., J. R. (2001), *Org. Lett.*, **3**, 1557.
- Collier, P. N., Campbell, A. D., Patel, I., Raynham, T. M. and Taylor, R. J. K. (2002), *J. Org. Chem.*, **67**, 1802.
- Agrofoglio, L. A., Gillaizeau, I. and Saito, Y. (2003), *Chem. Rev.*, **103**, 1875. [Review]
- Urawa, Y. and Ogura, K. (2003), *Tetrahedron Lett.*, **44**, 271.

Swern Oxidation

Oxidation of primary or secondary alcohols to the corresponding carbonyl compounds by activated dimethyl sulphoxide (as oxidant) in presence of oxalyl chloride followed by quenching with an amine is called the *Swern oxidation*.

$$\underset{R \quad R'}{OH} \xrightarrow[\text{2. Et}_3\text{N}]{\text{1. (COCl)}_2, \text{DMSO} \\ -78^\circ\text{C}} \underset{R \quad R'}{O}$$

Mechanism

DMSO reacts with oxalyl chloride to form activated species **1** or **2**; an alcohol is attacked by both of these species (called 'activated dimetyl sulphoxide') leading to a sulphonium salt (**3**), which then reacts with the base (triethylamine is most often used).

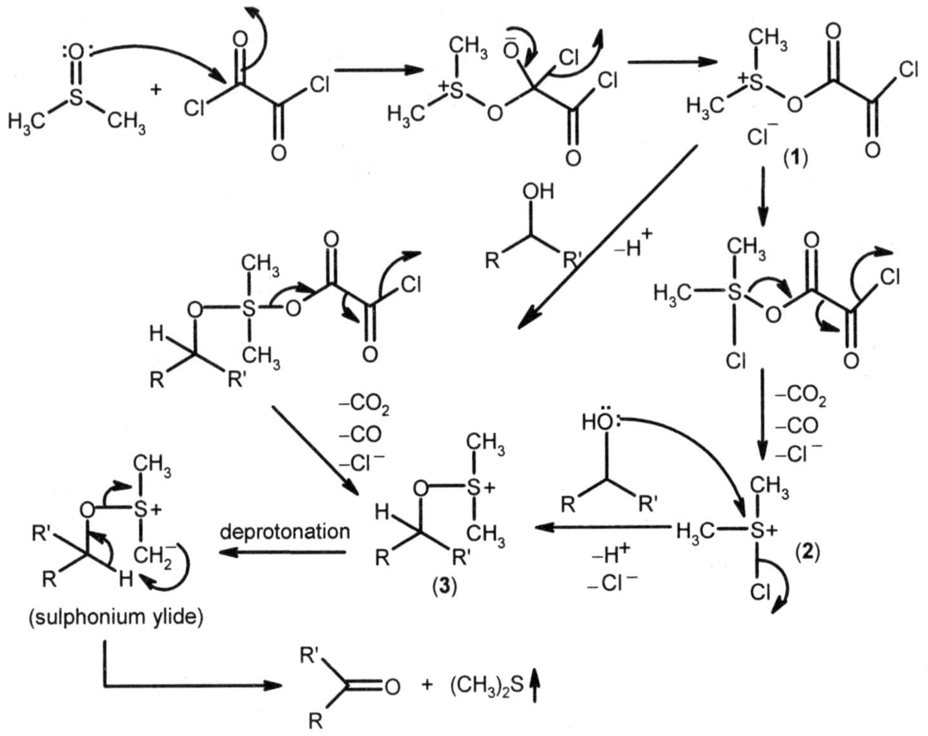

References

1. Huang, S. L., Omura, K. and Swern, D. (1976), *J. Org. Chem.*, **41**, 3329.
2. Omura, K. and Swern, D. (1978), *Tetrahedron*, **34**, 1651.
3. Huang, S. L., Omura, K. and Swern, D. (1978), *Synthesis*, **4**, 297.
4. Mancuso, A. J. and Swern, D. (1981), *Synthesis*, **7**, 165.
5. Marx, M. and Tidwell, T. T. (1984), *J. Org. Chem.*, **49**, 788.
6. Tidwell, T. T. (1990), *Org. React.*, **39**, 297. [Review]
7. Liu, Y. and Vederas, J. C. (1996), *J. Org. Chem.*, **61**, 7856.
8. Nakajima, N. and Ubukata, M. (1997), *Tetrahedron Lett.*, **38**, 2099.
9. Chrisman, W. and Singaram, B. (1997), *Tetrahedron Lett.*, **38**, 2053.
10. Rodriguez, A., Nomen, M., Spur, B. W. and Godfroid, J. J. (1999), *Tetrahedron Lett.*, **40**, 5161.
11. Dupont, J., Bemish, R. J., McCarthy, K. E., Payne, E. R., Pollard, E. B., Ripin, D. H. B., Vanderplas, B. C. and Watrous, R. M. (2001), *Tetrahedron Lett.*, **42**, 1453.
12. Nishide, K., Ohsugi, S.-I, Fudesaka, M., Kodama, S. and Node, M. (2002), *Tetrahedron Lett.*, **43**, 5177.

Thorpe (Ziegler) Reaction

Base-catalyzed condensation between two nitrile molecules, where α-carbon of one nitrile molecule is added to the cyano carbon of the other, is called the *Thorpe reaction*. The initially formed imino-product is tautomerized into enamine-form that is more stable. The enamine-form can easily be hydrolyzed to yield β-keto nitrile (or say α-cyano ketone), which can in turn be hydrolyzed and decarboxylated, if desired.

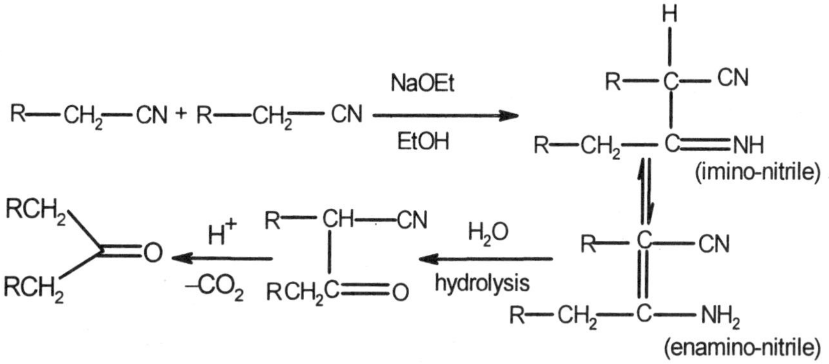

Thus, *Thorpe reaction* is run intermolecularly; when the reaction is carried out intramolecularly, the method is called *Thorpe-Ziegler reaction*. This is a useful method for closing large rings (5 to 8-membered).

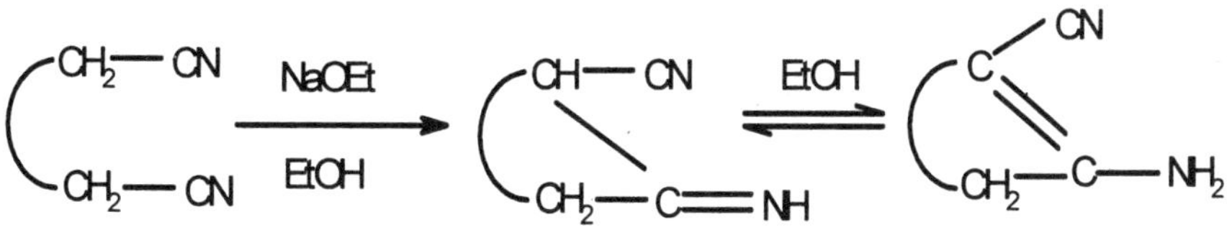

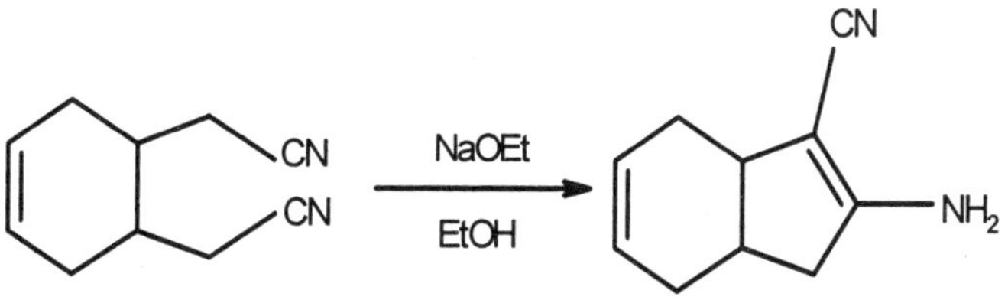

Mechanism

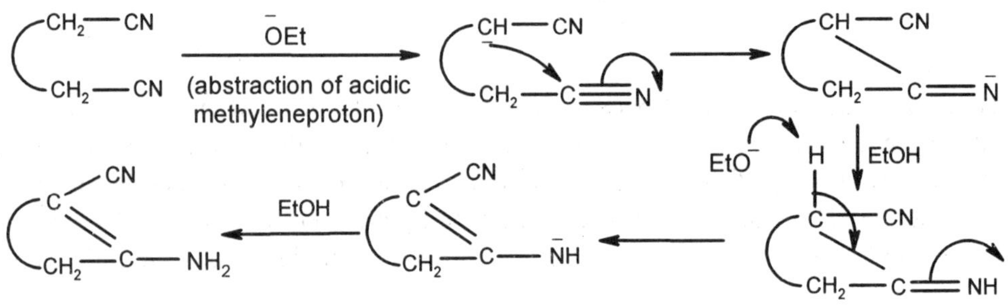

(mechanism scheme showing base-mediated cyclization of a dinitrile: abstraction of acidic methylene proton by OEt⁻, intramolecular attack on the nitrile, protonation by EtOH, tautomerization to give the cyclic enaminonitrile bearing CN and NH₂ groups)

References

- Baron, H., Remfry, F. G. P. and Thorpe, Y. F. (1904), *J. Chem. Soc.*, **85**, 1726.
- Bloomfield, J. J. and Fennessey, P. V. (1964), *Tetrahedron Lett.*, 2273.
- Schaefer, J. P. and Bloomfield, J. J. (1967), *Org. React.*, **15**, 1. [Review]
- Taylor, E. C. and McKillop, A., *The Chemistry of Cyclic Enaminonitriles and ortho-Amino Nitriles*, Wiley: NY, 1970. [Review]
- Doornbos, T. and Strating, J. (1971), *Synth. Commun.*, **1**, 193.
- Rodriguez-Hahn, L., Parra, M. M. and Martinez, M. (1984), *Synth. Commun.*, **14**, 967.
- Curran, D. P. and Liu, W.(1999), *Synlett*, 117.
- Kovacs, L. (2000), *Molecules*, **5**, 127.
- Gutschow, M. and Powers, J. C. (2001), *J. Heterocyclic Chem.*, **38**, 419.
- Malassene, R., Toupet, L., Hurvois, J.–P. and Moinet, C. (2002), *Synlett.*, 895.
- Malassene, R., Vanquelef, E., Toupet, L., Hurvois, J.–P. and Moinet, C. (2003), *Org. Biomol. Chem.*, **1**, 547.

Tiemann Rearrangement

Amidoximes, derived from amides and hydroxylamine, undergo rearrangement to ureas when treated first with benezenesulphonyl chloride and with water in a reaction called the *Tiemann rearrangement*.

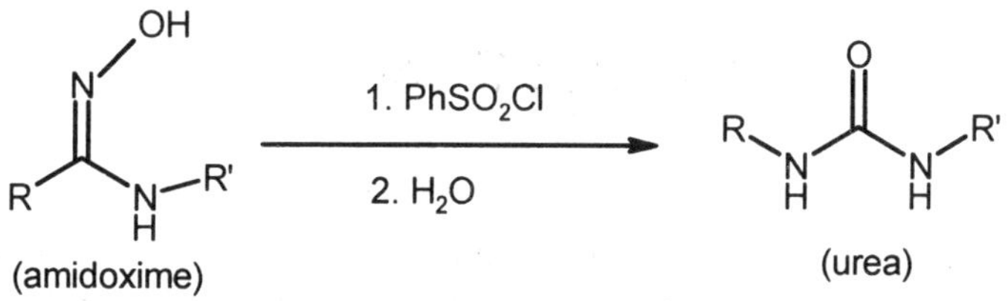

Tiemann Rearrangement

Mechanism

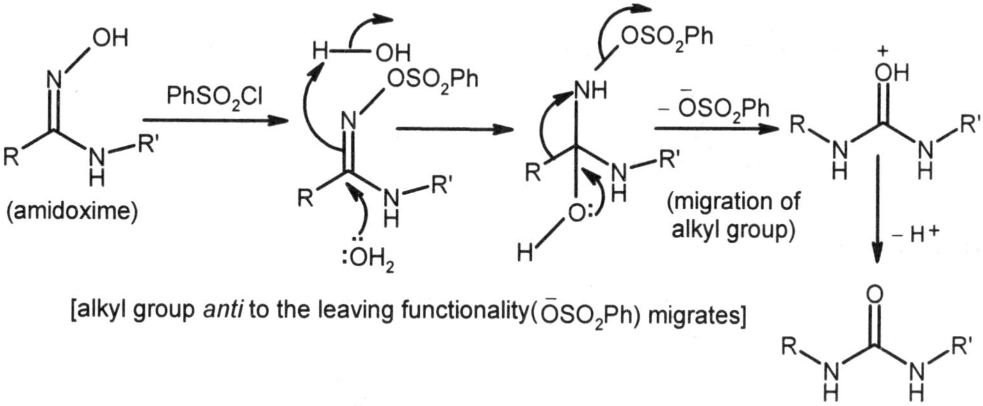

[alkyl group *anti* to the leaving functionality ($\bar{O}SO_2Ph$) migrates]

References

- Tiemann, F. (1891), *Ber. Dtsch. Chem. Ges.*, **24**, 4162.
- Garapon, J., Sillion, B. and Bonnier, J. M. (1970), *Tetrahedron Lett.*, 4905.
- Richter, R., Tucker, B. and Ulrich, H. (1983), *J. Org. Chem.*, **48**, 1694.
- Adams, G. W., Bowie, J. H., Hayes, R. N. and Gross, M. L. (1992), *J. Chem. Soc., Perkin Trans 2*, 897.
- Bakunov, S. A., Rukavishnikov, A. V. and Tkachev, A. V. (2000), *Synthesis*, 1148.

Tollens Reaction

Reductive condensation of carbonyl compounds possessing at least an α-hydrogen with formaldehyde in the presence of a base, most oftenly $Ca(OH)_2$, is called the *Tollens reaction*.

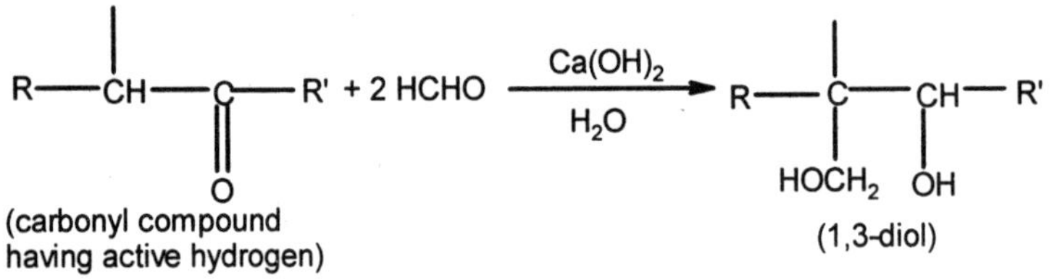

$$R-CH(R')-C(=O)- \; + \; 2\,HCHO \xrightarrow[H_2O]{Ca(OH)_2} R-C(CH_2OH)-CH(OH)-R'$$

(carbonyl compound having active hydrogen) → (1,3-diol)

Mechanism

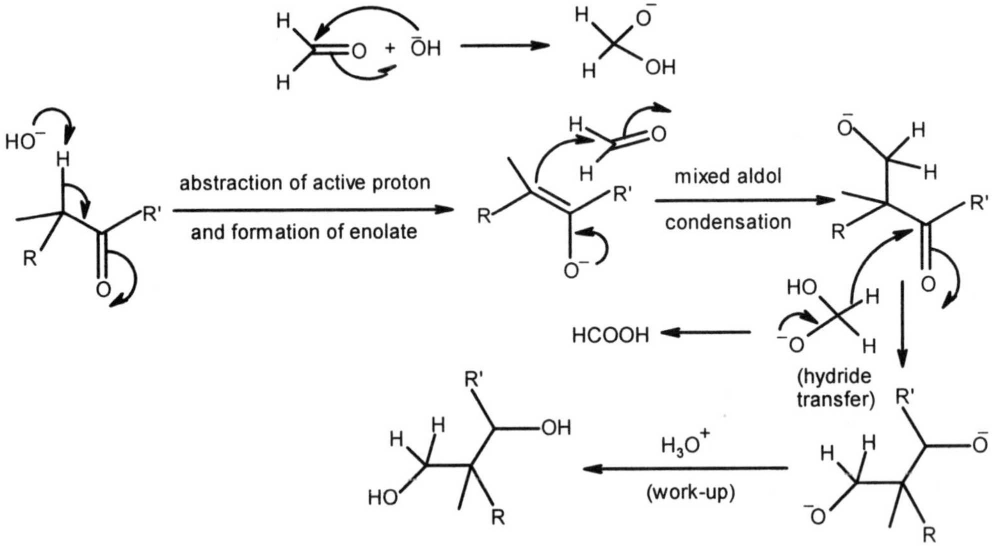

[Mechanism scheme: $HCHO + OH^- \rightarrow H_2C(O^-)(OH)$; abstraction of active proton from carbonyl compound and formation of enolate; mixed aldol condensation with formaldehyde; hydride transfer from formaldehyde giving HCOOH; work-up with H_3O^+ affords the 1,3-diol.]

In the case where the carbonyl compound bears several α-hydrogens, they all can be replaced. An interesting example of this kind is the condensation of acetaldehyde with formaldehyde (1:4 molar ratio) under this reaction condition to prepare pentaerythritol.

$$CH_3CHO + 4\,HCHO \xrightarrow{Ca(OH)_2} C(CH_2OH)_4 + HCOOH$$

References

- Stanley Wawzonek, S. and Donald A. Rees, D. A. (1948), *J. Am. Chem. Soc.*, **70**, 2433.
- Dermer, O. C. and Paul W. Solomon, P. W. (1954), *J. Am. Chem. Soc.*, **76**, 1697.
- Parry-Jones, R. and Kumar (1985), *J. Chem. Edu.*, **62**, 114.
- Jenkins, I. D. (1987), *J. Chem. Edu.*, **64**, 164.
- Munoz, S. and Gokel, G. W. (1993), *J. Am. Chem. Soc.*, **115**, 4899.
- Huang, S. and Mau, A. W. H. (2003), *J. Phys. Chem. B*, **107**, 3455.

Ullmann Diaryl Synthesis

The coupling of aryl halides with themselves or with other aryl halides in the presence of metals (*e.g.* copper) to afford diaryls is called the *Ullmann diaryl synthesis*. Most often aryl iodides are used, but bromides, chlorides, and even thiocyantaes are in use. The method is employed to synthesize both symmetrical and unsymmetrical diaryls.

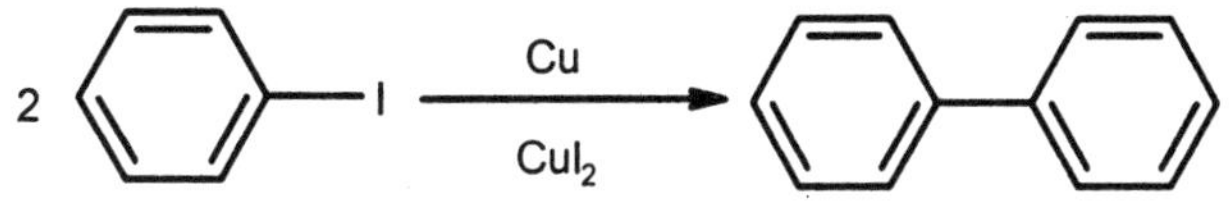

Mechanism

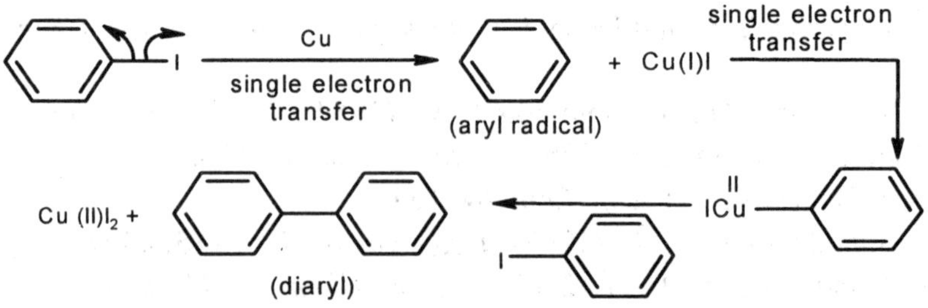

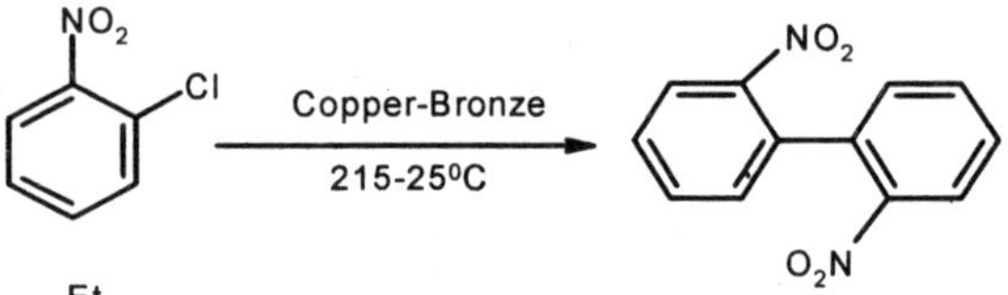

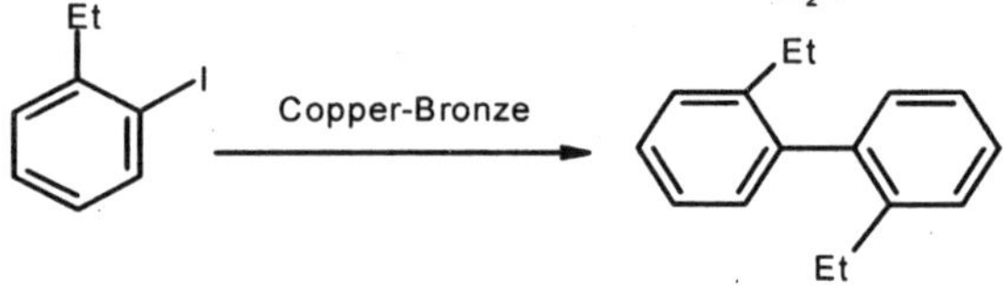

References

- Ullmann, F. (1904), *Justus Liebigs Ann. Chem.*, **332**, 38.
- Carlin, R. B. and Swakon, E. A. (1955), *J. Am. Chem. Soc.*, **77**, 966.
- Carlin, R. B. and George E. Foltz, G., E. (1956), *J. Am. Chem. Soc.*, **78**, 1997.
- Fanta, P. E. (1974), *Synthesis*, 9. [Review]
- Forrest, J. (1960), *J. Chem. Soc.*, 592.
- Lewin, A. H. and Cohen, T. (1965), *Tetrahedron Lett.*, 4531.
- Sainsbury, M. (1980), *Tetrahedron*, **36**, 3327. [Review]
- Semmelhack, M., Helqist, P., Jones, L., Keller, L., Mendelson, L., Ryono, L., Smith, J. and Stauffer, R. (1981), *J. Am. Chem. Soc.*, **103**, 6460.
- Bringmann, G., Walter, R. and Weirich, R. (1990), *Angew. Chem. Int. Ed. Engl.*, **29**, 977.
- Stark, L. M., Lin, X.–F. and Flippin, L. A. (2000), *J. Org. Chem.*, **65**, 3227.
- Venkataraman, S. and Li, C.–J. (2000), *Tetrahedron Lett.*, **41**, 4831.
- Ma, D. and Xia, C. (2001), *Org. Lett.*, **3**, 2583.
- Buck, E., Song, Z., J., Tschaen, D., Dormer, P. G. and Reider, P. J. (2002), *Org. Lett.*, **4**, 1623.
- Hameurlaine, A. and Dehaen, W. (2003), *Tetrahedron Lett.*, **44**, 957.

Vilsmeier – Hack Reaction

Formylation of electron-rich (*i.e.* activated) aromatic compounds with *N,N*-disubstituted formamides and phosphorus oxychloride is known as the *Vilsmeier-Hack reaction*. *N*-Phenyl-*N*-methylformamide is a common reageant, although other arylalkyl amides and dialkyl amides are also in use. Aromatic hydrocarbons and heterocycles, which are more active than simple benzene can also be formylated by this process.

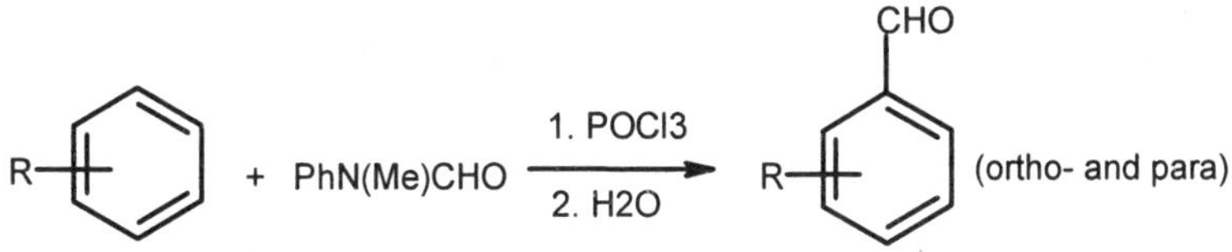

$$R\text{-}C_6H_4\text{-}H + PhN(Me)CHO \xrightarrow[\text{2. H}_2\text{O}]{\text{1. POCl}_3} R\text{-}C_6H_4\text{-}CHO \quad (\text{ortho- and para})$$

Mechanism

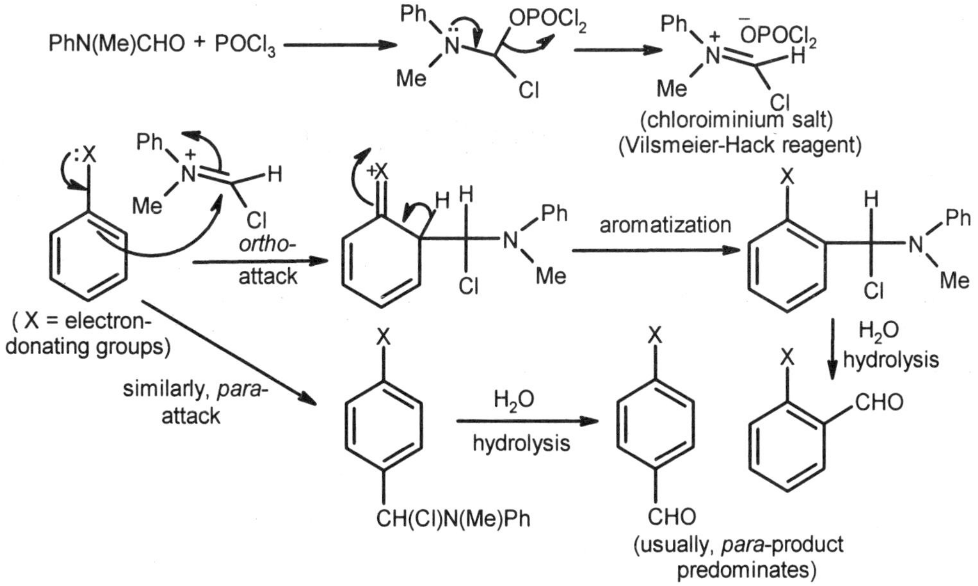

(chloroiminium salt)
(Vilsmeier-Hack reagent)

(X = electron-donating groups)

(usually, *para*-product predominates)

$$\text{indole} \xrightarrow[\text{POCl}_3]{\text{DMF}} \text{3-formylindole}$$

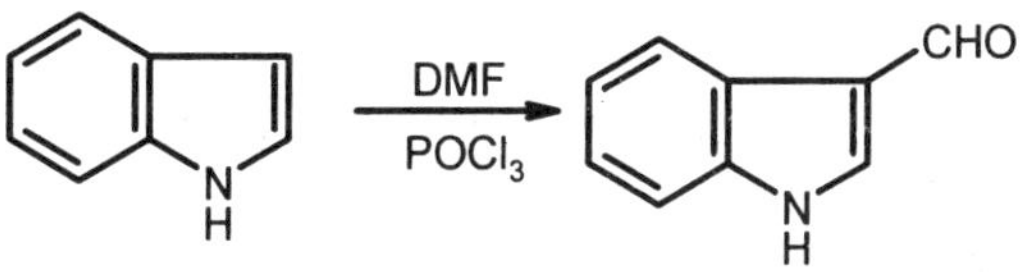

References

- Vilsmeier, A. and Hack, A. (1927), *Ber. Dtsch. Chem. Ges.*, **60**, 119.
- Smith, G. F. (1954), *J. Chem. Soc.*, 3842.
- Jugie, G., Smith, J. A. S. and Martin, G. J. (1975), *J. Chem. Soc., Perkin Trans. 2*, 925.
- Jutz, C. (1976), *Adv. Org. Chem.*, **9**, pt. 1, 225.
- Traas, P. C., Takken, H. J. and Boelens, H. (1977), *Tetrahedron Lett.*, 2129.
- Becalli, E. M., Marchesini, A. and Molinari, H. (1986), *Tetrahedron Lett.*, 627.
- Rao, M. S. C. and Rao, G. S. K. (1987), *Synthesis*, 231.
- Seybold, G. (1996), *J. Prakt. Chem.*, **338**, 392. [Review]
- Jones, G. and Stanforth, S. P. (1997), *Org. React.*, **49**, 1. [Review]
- Meth-Cohn, O. and Goon, S. (1997), *J. Chem. Soc., Perkin Trans. 1*, 85.
- Reichardt, C. (1999), *J. Prakt. Chem.*, **341**, 609.
- Ali, M. M., Tasneem, Rajanna, K. C. and Sai Prakash, P. K. (2001), *Synlett*, 251.
- Tasneem (2003), *Synlett*, 138.

von Braun Reaction

The *von Braun reaction* involves the cleavage of tertiary amines by cyanogen bromide to afford an alkyl bromide and a disubstituted cyanamide. Cyanogen bromide reacts with tertiary nitrogen compounds to break one carbon to nitrogen linkage.

$$R_3N + BrCN \longrightarrow R_2N\text{-}CN + R\text{-}Br$$

Usually, the smallest alkyl group or that would furnish the most reactive halide (*e.g.* benzyl or allyl) is cleaved. The reaction can also be run on secondary amines, but the yields have been found to be poor.

Mechanism

In this reaction cyanogen bromide, a single reagent, is responsible for two transformations in one reaction vessel; hence NC-Br is called a counterattack reagent.

References

- von Braun, J. (1907), *Ber. Dtsch. Chem. Ges.*, **40**, 3914.
- Von Braun, J. and Engelbertzm P. (1923), *Ber. Dtsch. Chem. Ges.*, **56**, 1573.
- Elderfield, R. C. and Hageman, H. A. (1949), *J. Org. Chem.*, **14**, 605.
- Hageman, H. A. (1953), *Org. React.*, **7**, 198. [Review]
- Rapoport, H., Lovell, C. H., Reist, H. R. and Warren Jr., E. (1967), *J. Am. Chem. Soc.*, **89**, 1942.
- Fodor, G. and Abidi, S. (1971), *Tetrahedron Lett.*, 1369.
- Paukstelis, J. V. and Kim, M. (1974), *J. Org. Chem.*, **39**, 1494.
- Nakahara, Y., Niwaguchi, T. and Ishii, H. (1977), *Tetrahedron*, **33**, 1591.
- Fodor, G. and Nagubandi, S. (1980), *Tetrahedron*, **36**, 1279.
- Cooley, J. H. and Evain, E. (1989), *Synthesis*, 1.
- Laabs, S., Scherrmann, A., Sudau, A., Diederich, M., Kierig, C. and Nubbemeyer, U. (1999), *Synlett*, 25.
- Chambert, S., Thamosson, F. and Decout, J.–L. (2002), *J. Org. Chem.*, **67**, 1898.

von Richter Rearrangement

Aromatic nitro compounds when treated with cyanide ion departs the nitro group resulting carboxylation exclusively at the *ortho*-position of the departing functionality in a reaction called as the *von Richter rearrangement*.

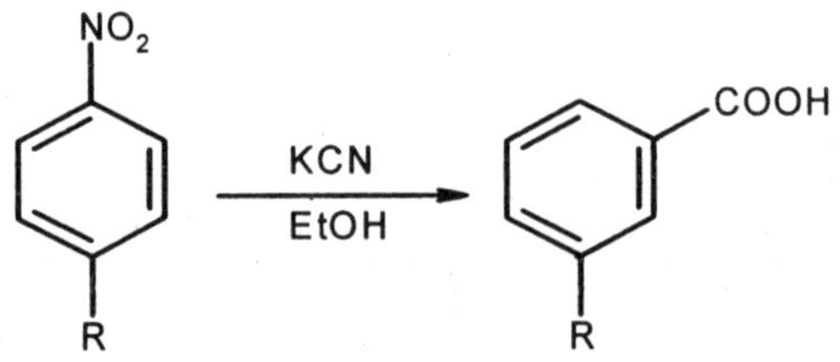

Mechanism

The reaction proceeds in the following mechanistic pathway:

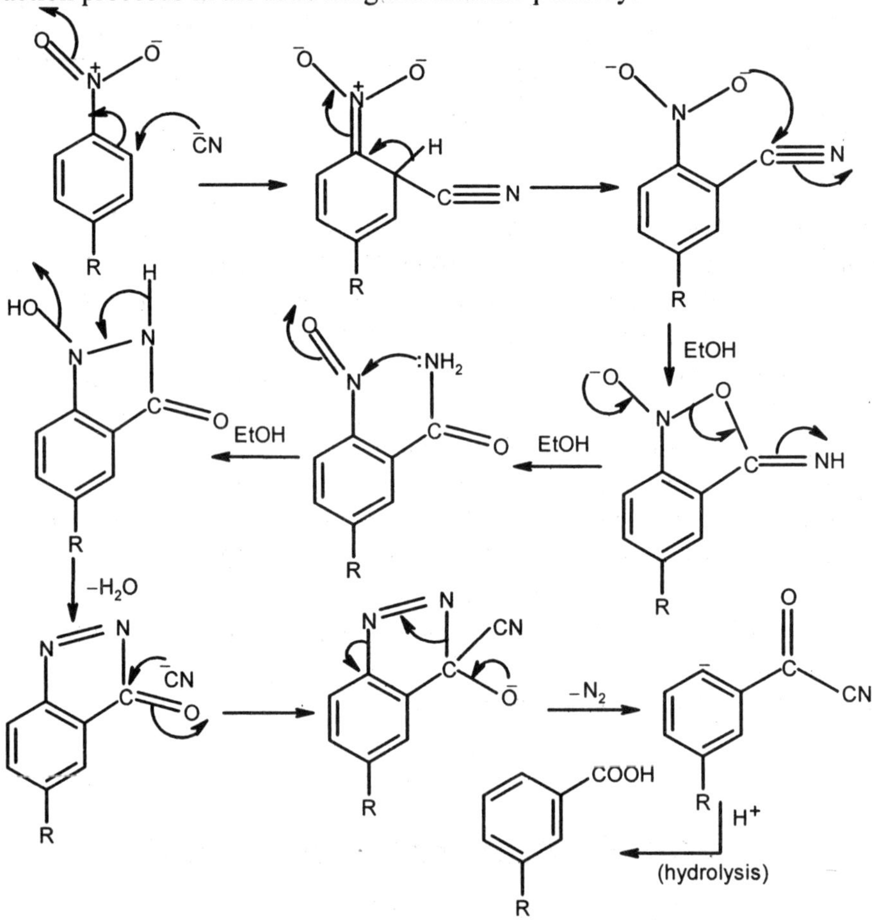

References

- von Richter, V. (1871), *Ber. Dtsch. Chem. Ges.*, **4**, 21, 459, 533.
- Rosenblum, M. (1960), *J. Am. Soc. Soc.*, **82**, 3796.

- Ibne-Rasa, K. M. and Koubek, E. (1963), *J. Org. Chem.*, **28**, 3240.
- Rogers, G. T. and Ulbricht, T. L. V. (1968), *Tetrahedron Lett.*, **23**, 1029.
- Ellis, A. C. and Rae, I. D. (1977), *J. Chem. Soc., Chem. Commun.*, 152.
- Tretyakov, E. V., Knight, D. W. and Vasilevsky, S. F. (1999), *J. Chem. Soc., Perkin. Trans. 1*, 3721.
- Brase, S., Dahmen, S. and Heuts, J. (1999), *Tetrahedron Lett.*, **40**, 6201.
- Shine, H. J., *Aromatic Rearrangements*, Elsevier : NY, 1967, p. 329. [Review]

Wagner – Meerwein Rearrangement

Skeletal rearrangement of a carbenium ion intermediate involving nucleophilic 1,2-migration of an alkyl group to yield a more stable carbenium ion, which in turn affords a more substituted olefin, is called *Wagner-Meerwein rearrangement*. A schematic representation of the rearrangement may be shown as:

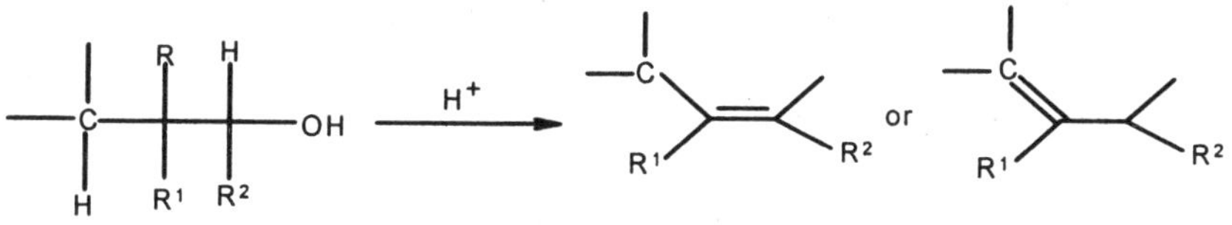

Mechanism

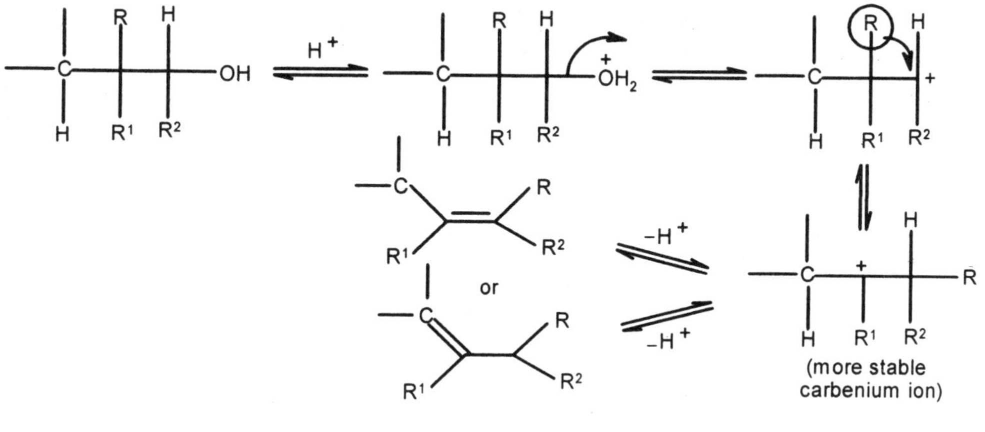

The aptitude of migration is found to be: phenyl > *t*-butyl > ethyl > methyl. The *Wagner-Meerwein rearrangment* finds immense application in the terpenoid chemistry.

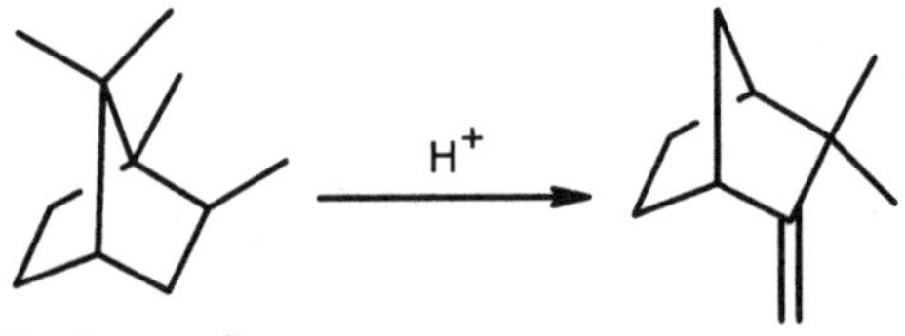

(isoborneol) →(H⁺) (camphene)

References

- Wagner, G. (1899), *J. Russ. Chem. Soc.*, **31**, 690.
- Meerwein, H. and Unkel, W. (1910), *Justus Liebigs Ann. Chem.*, **376**, 152.
- Streitwieser Jr., A. (1956), *Chem. Rev.*, **56**, 698. [Review]
- Sorensen, T. S. (1976), *Acc. Chem. Res.*, **9**, 257. [Review]
- Hogeveen, H. and van Kruchten, E. M. G. A. (1979), *Top. Curr. Chem.*, **80**, 89. [Review]
- Paquette, L. A., Waykole, L., Jendralla, H. and Cottrell, C. E. (1986), *J. Am. Chem. Soc.*, **108**, 3739.
- Kaupp, G. (1988), *Top. Curr. Chem.*, **146**, 57. [Review]
- Martinez, A. G., Vilar, E. T., Fraile, A. G., Fernandez, A. H., De La Moya Cerero, S. and Jimenez, F. M. (1998), *Tetrahedron*, **54**, 4607.
- Birladeanu, L. (2000), *J. Chem. Edu.*, **77**, 858.
- Trost, B. M. and Yasukata, T. (2001), *J. Am. Chem. Soc.*, **123**, 7162.
- Colombo, M. I., Bohn, M. L. and Ruveda, E. A. (2002), *J. Chem. Edu.*, **79**, 484.
- Roman, L. U., Cerda-Garcia-Rojas, C. M., Guzman, R., Armenta, C., Hernandez, J. D. and Joseph-Nathan, P. (2002), *J. Nat. Products*, **65**, 1540.
- Guizzardi, B., Mella, M., Fagnoni, M. and Albini, A. (2003), *J. Org. Chem.*, **68**, 1067.

Wohl – Ziegler Reaction

Bromination of alkenes at the allylic or benzylic position on treatment with *N*-bromosuccinimide (NBS) in the presence of an initiator (*e.g.* peroxides or ALBN) or under photolysis is known as *Wohl-Ziegler reaction*.

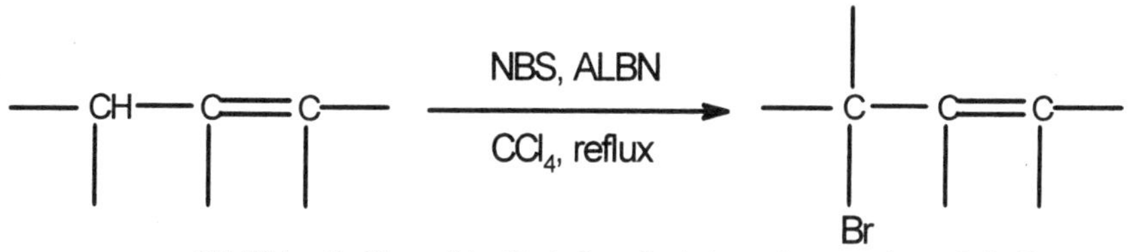

$$-CH-C=C- \quad \xrightarrow[CCl_4, \text{ reflux}]{NBS, \text{ ALBN}} \quad -\underset{Br}{C}-C=C-$$

[ALBN = 2, 2'-azobis-(2,4-dimethyl-4-methoxyvaleronitrile)]

Mechanism

The reaction proceeds through a free-radical mechanism as depicted below.

Initiation:

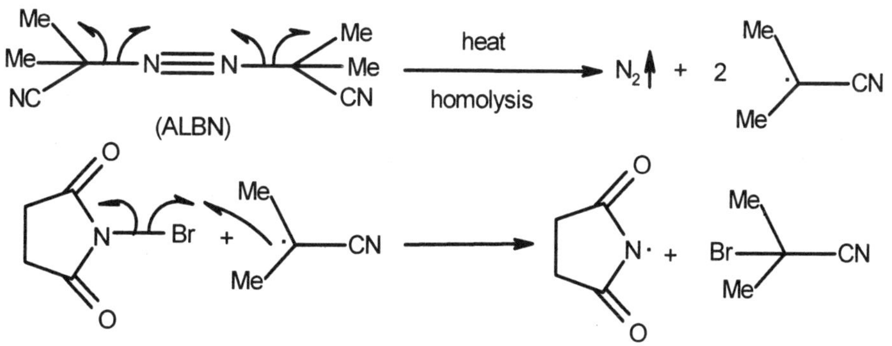

Propagation:

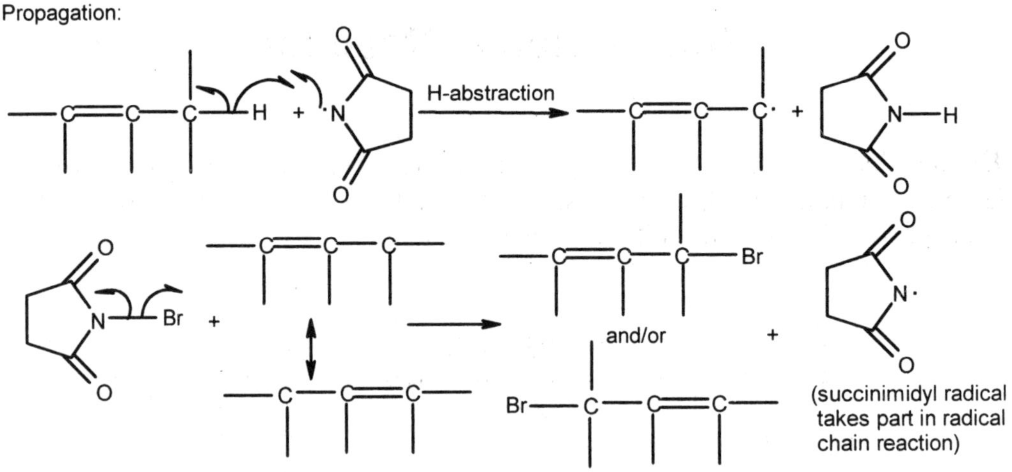

H-abstraction
H
N
O
Br
and/or
(succinimidyl radical takes part in radical chain reaction)

References

- Wohl, A. (1919), *Ber. Dtsch. Chem. Ges.*, **52**, 51.
- Ziegler, K., Spath, A., Schaaf, E., Schumann, W. and Winkelmann, E. H. (1942), *Justus Liebigs Ann. Chem.*, **551**, 80.
- Dauben Jr., H. J. and McCoy, L. L. (1959), *J. Am. Chem. Soc.*, **81**, 4863.
- Horner, L. and Winkelmann, E. H. (1959), *Angew. Chem.*, **71**, 349.
- Walling, C., Rieger, A. L. and Tanner, D. D. (1963), J. *Am. Chem. Soc.*, **85**, 3129.
- Hinz, J., Oberliner, A. and Ruchardt, C. (1973), *Tetrahedron Lett.*, 1975.
- Day, J. C., Lindstrom, M. J. and Skell, P. S. (1974), *J. Am. Chem. Soc.*, **96**, 5616.
- Ito, I. and Ueda, T. (1975), *Chem. Pharm. Bull.*, **23**, 1646.
- Pennanen, S. I. (1978), *Heterocycles*, **9**, 1047.
- Kita, Y., Sano, A., Yamaguchi, T., Oka, M., Gotanda, K. and Matsugi, M. (1997), *Tetrahedron Lett.*, **38**, 3549.
- Allen, J. G. and Danishefsky, S. J. (2001), *J. Am. Chem. Soc.*, **123**, 351.

Subject Index

A

Allan-Robinson reaction 364-365
  application 365
  mechanism 364
Amadori rearrangement 366-367
Angeli-Rimini hydroxamic acid synthesis 367-369
Arndt-Eistert homologation 166-171
  applications 169
  Barbier-Weiland degradation 168
  carbene-carbene rearrangement 168
  mechanism 167
  oxirene intermediate 168
  Wolff rearrangement 166

B

Baeyer-Villiger oxidation 1-23
  anti-fused ketone 20
  bicyclic ketones 15
  Criegee intermediate 5, 6, 7, 17, 18
  enzymatic *B-V* oxidation 14
  exo-attack 17
  insertion of oxygen 1
  mechanism 5
  migratory aptitude 9
  regioselectivity 9
  regiospecific *B-V* 20, 21
  retention of configuration 7
  stereochemistry 7
  syn-fused ketone 20
  transesterification 2
  UHP 5
  Zr (salen) complex 5
  α-diketones 1
  β-diketones 1
  π-participation 12
Baker–Venkataraman rearrangement 369-371
Bamberger rearrangement 371-372
Bamford-Stevens reaction 372-374
Bardhan-Sengupta synthesis 374-375
Bartoli indole synthesis 376-377
Barton reaction 24-38
  abnormal Barton reaction 34, 36
  applications 29
  Barton fragmentation 34, 35
  definition 25
  intramolecular hydrogen abstraction 24
  mechanism 25, 27
  remote functionalization and molecular modeling 28
Baylis-Hillman reaction 378-380
  mechanism 379
Beckmann rearrangement 39-55
  abnormal Beckmann rearrangement 47
  anti-group migration 45, 47
  anti-orientation 40
  applications 52
  Beckmann fragmentation 48
  examples 40

extensions 50
fragmentation-recombination mechanism 41, 43
intermolecular mechanism 42, 43
ionic intermediate 46
limitations 47
non-stereospecific Beckmann 49
second-order Beckmann 47
stereochemistry 47
synchronous Beckmann 42, 46
syn-migration 49
Benzilic acid rearrangement 172-176
   applications 175
   benzil-benzilic acid rearrangement 172, 174
   mechanism 172
Benzoin condensation 177-183
   mechanism 177
   N-alkylthiazolium salt 178
   crossed benzoin condensation 179-80
   thiamine hydrochloride 179
   applications 181
Birch reduction 184-190
   applications 188
   Birch conditions 187
   dissolving metal reduction 184
   Lithium (Li)-Birch reduction 187
   mechanism 184
Bischler-Napieralski reaction 191-197
   abnormal product formation 193
   applications 196
   Bischler-Napieralski synthesis 191
   isonitrolium salt 192
   mechanism 191
   Pictet-Gams modification 193
Blanc-Quelet chloromethylation reaction 380-381
Boord reaction 381-382
Bouvealt-Blanc reaction 198-202
  applications 201
Bouvealt-Blanc procedure 198, 200
buttressing effects 201
mechanism 198
SR_N1 pathway 201
Bouveault aldehyde synthesis 382-383
Brook rearrangement 383-384
Bucherer reaction 385-386

C

Cannizzaro reaction 203-208
   applications 206
   crossed Cannizzaro reaction 205-207
   hydride ion transfer 203
   intramolecular (internal) Cannizzaro reaction 205, 206
   mechanism 203
Carroll rearrangement 386-387
Chapman rearrangement 388-389
Chichibabin amination reaction 209-211
   applications 210
   intramolecular Chichibabin cyclization 210
   mechanism 209
   pyridyne type intermediate 210
   Suzuki reaction 210
Chugaev reaction 389-390
Claisen condensation 212-218
   applications 217
   crossed Claisen condensation 215
   intramolecular Claisen (Dickmann) 215
   mechanism 213
Claisen rearrangement 56-64
   abnormal Claisen rearrangement 59
   abnormal product 59
   amino-Claisen rearrangement 62
   applications 61
   aza-Claisen rearrangement 62
   chirality transfer 63
   Claisen-Ireland rearrangement 60, 61

mechanism 57
normal product 59
sigmatropic rearrangement 56, 57
[3,3]-sigmatropic rearrangement 59, 61, 62
stereoselective route 60
Clemmensen reduction 219-224
   applications 223
   Clemmensen conditions 222
   mechanism 219
   Zinc-amalgam 223
Cope rearrangement 225-232
   applications 229
   degenerate Cope rearrangement 229
   mechanism 225
   oxy-Cope rearrangement 227
   retro-ene reaction 227, 228
   stereospecificity 226
   zwitterionic intermediate 228, 229
Corey-Kim oxidation 390-391
Corey-Winter reaction 392-393

D

Dakin reaction 233-236
   applications 235
   Baeyer-Villiger reaction 234
   Dakin oxidation 233
   mechanism 233
de Mayo reaction 394-395
Demjanov rearrangement 395-396
Diels-Alder reaction 65-92
   concerted and non-concerted *D-A* reaction 73
   CTAB 84
   cycloaddition 65, 70
   diene 66, 74, 78, 84
   dienophile 65, 68, 71, 74, 78, 80, 81, 84
   endo and exo product 77
   Fisher carbene complex 86
   Homo Diels-Alder reaction 80
   homo-conjugated D-A 81
   homo-*D-A* adduct 82
   inverse electron demand 74
   kinetic effect 74
   Lewis-acid catalyst 74
   mechanism 72
   regioselectivity 76, 84
   retro-Diels-Alder fragmentation 77, 85
   stereochemistry 74
   stereoselectivity 73
   stereospecific 73, 74
   steric effect 68
   synthetic applications 85

Dienone-phenol rearrangement 397-398
Di-pi-methane rearrangement 237-241
   applications 240
   mechanism 237
   oxa-di-pi-methane rearrangement 239
   Zimmermann reaction 237

E

Eglinton reaction 399
Elbs persulphate oxidation 400
Etard reaction 242-244
   mechanism 242
   applications 243

F

Favorskii rearrangement 93-105
   applications 101
   cyclopropanone intermediate 95-98
   mechanism 94
   quasi Favorskii reaction 99-101
   semi-benzilic like mechanism 95, 100
   Wallach degradation 93
Fischer indole synthesis 245-250
   applications 249
   Fischer cyclization 246
   Fischer reaction 248

mechanism 245
Fischer-Hepp rearrangement 401-402
Friedel-Crafts reaction 251-259
   applications 255
   Friedel-Crafts acylation 251, 253, 255
   Friedel-Crafts alkylation 251, 252, 254
   Scholl reaction 255
Friedlander synthesis 403-405
Fries rearrangement 106-121
   abnormal Fries 110
   catalyst 106
   cross-over experiment 110
   mechanism 110
   meta-Fries rearrangement 109
   photo-Fries 115-117
   singlet state rection 117
   solvent 106
   thermodynamically controlled product 107
   π-complex 110-113

G

Gabriel Synthesis 260-264
   applications 261
   GABA 263
   Ing-Manske modification 261
   mechanism 261
Gattermann-Koch reaction 265-267
   applications 267
   Gattermann-Koch formylation 265
   mechanism 265

H

Haller-Bauer reaction 268-271
   applications 270
   H-B type cleavage 269, 270
   mechanism 268
Hell-Volhard-Zelinsky reaction 272-276
   applications 274
   mechanism 273

Hoch-Campbell ethyleneimine synthesis 405-406
Hofmann- Loffler-Freytag reaction 406-407
Hofmann rearrangement 277-282
   applications 281
   Curtius rearrangement 279
   Hofmann degradation 277, 279
   intramolecular process 279
   Lossen rearrangement 279
   mechanism 277
   retention of configuration 279
   synchronous migration 279
   synchronous process 281
Houben-Hoesch reaction 283-286
   applications 284
Hoesch reaction 284
   mechanism 283
Hunsdiecker reaction 287-290
   applications 288
   Cristol-Firth modification 288
   mechanism 287
   Simonini reaction 288

J

Jones oxidation 408-409

K

Knoevenagel reaction 291-296
   applications 293
   Cope modification 293
   Cope-Knoevenagel reaction 293
   Doebner modification 293
   mechanism 292
   Michael addition 293
Koch reaction 410-411
Kolbe elctrolysis reaction 411-412
Kolbe-Schmitt reaction 413-414

L

Leuckart-Wallach reaction 414-415
Lieben haloform reaction 415-416
Lossen rearrangement 416-417

M

Mannich reaction 297-303
  applications 299
  iminium salt 297-299
  Mannich base 297, 299, 302
  mechanism 297
McMurry reaction 417-419
Meerwein-Ponndorf-Verley reduction 304-306
  applications 305
  mechanism 304
  Oppenauer oxidation 304
Michael reaction 307-311
  abnormal Michael 309
  aldol condensation 310
  applications 309
  intramolecular Michael 309
  mechanism 308
  Michael addition 310
  pseudo-acidic addendum 307
  Robinson annulation 310
Mitsunobu reaction 419-420

N

Neber rearrangement 420-421
Nef reaction 422-423
Norrish type I and II reaction 122-137
  eclipsing interaction 134, 136
  entropy effect 129
  Norrish type I reaction 122, 123, 130
  OTCM 129
  quantum yield 127
  stereochemistry of type II 128
  stereoselective 135
  synthetic applicability 130-136
  type II reaction 122, 123, 128

O

Oppenauer oxidation 312-314
  applications 313
  mechanism 312
  Meerwein-Ponndorf-Verley reduction 312

P

Paterno-Buchi reaction 138-151
  competing energy transfer 143
  exciplex 142,143
  intersystem crossing 139, 141, 143
  intramolecular fashion 147
  mechanism 139
  stereochemical integrity 141
  stereoselective fashion 142
  stereospecific 140
Pechmann reaction 423-424
Perkin reaction 315-318
  applications 317
  Erlenmeyer-Plochl azalactone synthesis 316
  mechanism 315
Pinacol rearrangement 319-324
  applications 323
  carbenium intermediate 319
  mechanism 319
  migratory aptitude 320
Prins reaction 425-426

R

Reformatsky reaction 325-327
  applications 326
  Blaise reaction 326
  mechanism 325
Reimer-Tiemann reaction 328-331
  abnormal products 330
  applications 331
  mechanism 329
Ritter reaction 426-427
Rosenmund reduction 427-428

S

Sandmeyer reaction 332-335
   applications 334
   mechanism 333
   Schiemann salt 333
Schotten-Boumann reaction 429
Sharpless asymmetric epoxidation 336-339
   applications 338
   DET 336
   stereospecific reaction 336
   enantioselective epoxidation 337
   mechanism 337
Simmons-Smith reaction 430-431
Simonini reaction 431-432
Sommelet-Hauser rearrangement 432-433
Stephen reaction 433-434
Stevens rearrangement 340-344
   1,2-shift 342
   1,4-rearranged product 342
   applications 342
   intramolecular rearrangement 340
   mechanism 340
   radical-pair intermediate 341
Stobbe condensation 345-347
   aldol condensation 345
   applications 346
   lactone intermediate 346
   mechanism 345
Stork enamine reaction 152-165
   aminomercuration-demercuration 156
   base catalyzed rearrangement 155
   mechanism 160
   Horner-Wittig reaction 154
   iminium salt 161, 162
   molecular sieves 154
   preparation of enamines 153
   Stork reaction 153
   structure and reactivity 157
Strecker synthesis 434-435
Suzuki coupling 436-437
Swern oxidation 437-438

T

Thorpe (Ziegler) reaction 439-440
Tiemann rearrangement 440-441
Tollens reaction 442-443

U

Ullmann diaryl synthesis 443-444

V

Vilsmeier-Hack reaction 445-446
von Braun reaction 446-447
von Richter rearrangement 447-449

W

Wagner-Meerwein rearrangement 449-450
Williamson ether synthesis 348-350
   applications 350
   C-alkylation 349
   mechanism 348
   SET-mechanism 349
Wittig reaction 351-356
   applications 354
   betaine intermediate 352, 353
   E-selectivity 353
   Horner-Wadsworth-Emmons modification 353
   mechanism 352
   oxaphosphetane intermediate 352, 353
   phosphorus ylide 351
   Wittig olefination reaction 351
   Z-selectivity 353
Wohl-Ziegler reaction 451-452
Wolff-Kishner reduction 357-360
   applications 359
   Huang-Minlon modification 357
   mechanism 358

Y

Yamaguchi esterification 361-363
   applications 362
   mechanism 361
   Yamaguchi macrolactonization 362, 363

About the Author

After graduating from Visva-Bharati University in 1990 Dr. Goutam Brahmachari completed his masters in 1992 and received his Ph.D. degree in 1997 from the same University. With about eight years teaching experience, he has published nearly thirty research papers in the field of organic chemistry (Natural Products) in national and international journals, and has authored two invited book chapters. Dr. Brahmachari presently holds the position of Reader in Chemistry, Visva-Bharati University, Santiniketan, West Bengal.